Medicinal and Aromatic Plants of the World

Volume 5

Series Editor
Ákos Máthé
University of West Hungary
Faculty of Agriculture and Food Science
Mosonmagyarovar, Hungary

Medicinal and Aromatic Plants (MAPs) have been utilized in various forms since the earliest days of mankind. They have maintained their traditional basic curative role even in our modern societies. Apart from their traditional culinary and food industry uses, MAPs are intensively consumed as food supplements (food additives) and in animal husbandry, where feed additives are used to replace synthetic chemicals and production-increasing hormones. Importantly medicinal plants and their chemical ingredients can serve as starting and/or model materials for pharmaceutical research and medicine production. Current areas of utilization constitute powerful drivers for the exploitation of these natural resources. Today's demands, coupled with the already rather limited availability and potential exhaustion of these natural resources, make it necessary to take stock of them and our knowledge regarding research and development, production, trade and utilization, and especially from the viewpoint of sustainability. The series Medicinal and Aromatic Plants of the World is aimed to look carefully at our present knowledge of this vast interdisciplinary domain on a global scale. In the era of global climatic change, the series is expected to make an important contribution to the better knowledge and understanding of MAPs. The Editor of the series is indebted for all of the support and encouragement received in the course of international collaborations started with his ISHS involvement, in 1977. Special thanks are due to Professor D. Fritz, Germany for making it possible. The encouragement and assistance of Springer Editor, Mrs. Melanie van Overbeek, has been essential in realizing this challenging book project. Thanks are due to the publisher - Springer Science+Business Media, The Netherlands - for supporting this global collaboration in the domain of medicinal and aromatic plants. We sincerely hope this book series can contribute and give further impetus to the exploration and utilization of our mutual global, natural treasure of medicinal and aromatic plants. Budapest, Prof. Dr. Ákos Máthé.

More information about this series at http://www.springer.com/series/11192

Ulysses Paulino Albuquerque
Umesh Patil • Ákos Máthé
Editors

Medicinal and Aromatic Plants of South America

Brazil

Springer

Editors
Ulysses Paulino Albuquerque
Departamento de Botânica,
Centro de Biociências
Universidade Federal de Pernambuco
Recife, Brazil

Umesh Patil
Natural Product Research Laboratory
Dr. Hari Singh Gour University
Sagar, India

Ákos Máthé
Department of Botany, Faculty of
Agriculture & Food Science
West Hungarian University
Mosonmagyarovar, Hungary

ISSN 2352-6831 ISSN 2352-684X (electronic)
Medicinal and Aromatic Plants of the World
ISBN 978-94-024-1550-6 ISBN 978-94-024-1552-0 (eBook)
https://doi.org/10.1007/978-94-024-1552-0

Library of Congress Control Number: 2018958340

This Springer imprint is published by the registered company Springer Nature B.V.
The registered company address is: Van Godewijckstraat 30, 3311 GX Dordrecht, The Netherlands

Preface

This book gathers information about a small variety of medicinal and aromatic plants that spontaneously grow or are cultivated in South America, and it is part of the series Medicinal and Aromatic Plants of the World, conceived by Prof. Dr. Ákos Máthé. The plants are described in the form of short monographs and were selected according to the following criteria: (1) plants that are widely used in South America, and preferentially but not exclusively included in official programs of primary health care or (2) plants that are being investigated in the laboratories of researchers who accepted our invitation to collaborate on the present volume.

We tried to present state-of-the-art information for each of the 43 species included in this book. The reader will realize that although some species were extensively studied, several popular claims about their therapeutic potential have not been scientifically determined. In South America, we only study a very small fraction of the available plants with alleged medicinal properties and do not even exhaust all of the research possibilities in these cases, which is likely true in other continents as well.

We believe that several actions are required to change this scenario, including performing ethnobotanical and ethnopharmacological studies that are more theoretically and methodologically rigorous and performing systematic long-term studies of the species that exhibit at least one interesting biological activity. In the meantime, the present book, together with the remaining volumes of this series, may constitute a reference guide for future research and public health professionals.

Some chapters of this book are a contribution of the INCT Ethnobiology, Bioprospecting and Nature Conservation, certified by CNPq, with financial support from FACEPE (Foundation for Support to Science and Technology of the State of Pernambuco - Grant number: APQ-0562-2.01/17).

Recife, Brazil — Ulysses Paulino Albuquerque
Sagar, India — Umesh Patil
Mosonmagyarovar, Hungary — Ákos Máthé

Contents

Part I
Medicinal and Aromatic Plants of South America

South American Biodiversity and Its Potential in Medicinal and Aromatic Plants

Alissandra Trajano Nunes and Ulysses Paulino Albuquerque

Abstract The Americas are characterized by an array of ecosystems and are home to one of the most biologically diverse areas in the world, in addition to a vast cultural diversity represented by different ethnic groups. Historically, South American peoples have shown a high degree of dependence on natural resources, especially on plants, which are used for a variety of purposes. This relationship has resulted in potential sources for new natural products, possibly including the extraction of plant-derived chemical compounds for medicinal and aromatic purposes. The global herbal market is worth billions of dollars, but in South American countries, incentives for research and the development of bioproducts by domestic companies are lacking. Moreover, a lack of scientific knowledge on these resources causes native plants to be undervalued, and the high degree of environmental degradation threatens the biological diversity and associated traditional knowledge.

Keywords Ethnobotany · Sociobiodiversity · Traditional ecological knowledge · Diversity of useful plants

1 South American Sociobiodiversity

South America, whose wealth in biological and cultural diversity is distributed across a large area of the Americas (40%), is considered to be the largest territory in the Southern Hemisphere (Gardi et al. 2014). Its geography features a variety of environments, ranging from mountainous areas with high elevations, such as the

A. T. Nunes (✉)
Universidade de Pernambuco, Licenciatura em Ciências Biológicas, Grupo de Pesquisa em Biotecnologia e Inovação Terapêutica, Garanhuns, Brasil

U. P. Albuquerque
Departamento de Botânica, Centro de Biociências, Universidade Federal de Pernambuco, Recife, Brazil

U. P. Albuquerque et al. (eds.), *Medicinal and Aromatic Plants of South America*, Medicinal and Aromatic Plants of the World 5,
https://doi.org/10.1007/978-94-024-1552-0_1

Andes (Aconcágua reaches 6961 m), to plains and the basins of the Amazon, Orinoco and Paraná Rivers (Gardi et al. 2014).

Due to the changes in terrain, the climate changes dramatically, ranging from tropical humid to dry and cold, resulting in irregular rainfall, which in very dry areas of Chile, Bolivia and Argentina reaches only 250 mm annually; a complete contrast to this climate type is the wettest region on the planet, located in Colombia, where the greatest annual rainfall is recorded, approximately 8000 mm (Gardi et al. 2014). South America's soils form a mosaic of more than 30 types, directly affecting biodiversity and ecosystem function (Gardi et al. 2014).

Such environmental variations generate diverse landscape units, with forest formations ranging from Araucaria forests in colder regions (Paraná, Brazil and southern Chile) to shrub formations and grasslands in dry and arid regions in northern Chile, such as in the Atacama Desert, which is considered the driest place in the world (Prado 2003; PRHS 2006; Echeverria et al. 2007; Rey-Benayas et al. 2007; Gardi et al. 2014). The other formations in South America include savannas, *cerrados*, the Pantanal, tropical rainforests (Amazon rainforest) and the pampas and steppes found in the highlands of Ecuador and Peru (Prado 2003; PRHS 2006; Salazar et al. 2007; Gardi et al. 2014).

South America comprises biomes that are home to a large diversity of plants, estimated at 81000 species (Mittermeier et al. 2003; Myers et al. 2000). Of these, approximately 50,000 angiosperms are found in Brazil, representing 22% of the global species richness (MMA 2002; Giulietti et al. 2005; FAO 2011). Colombia, Peru, Ecuador and Venezuela, and Brazil form one of the most megadiverse region in the world (Table 1) (MMA 2002; Fioravanti 2013). It is estimated that half of all plant species worldwide occur in the Amazon Basin, which spans 6.9 million square kilometers across nine countries (Brazil, Bolivia, Peru, Colombia, Ecuador, Venezuela, Guyana, Suriname and French Guiana).

Bolivia has 12,000 native plant species, and the total diversity of angiosperms is distributed among 286 families, comprising 16% of the endemic flora of the country (Meneses et al. 2015). Although the countries in northern South America, namely, Guyana, French Guiana and Suriname, are small in area, they are home to a high diversity of plants (Boggan et al. 1997; Jørgensen et al. 2014).

Uruguay is an outlier among South American countries in that it has the lowest diversity of angiosperms. Finally, Chile has a high degree of endemism despite the low number of species (Zuloaga and Belgrano 2015) and also contains very rich sites, such as the Juan Fernández Archipelago National Park, whose flora includes 137 endemic and 213 native species (CONAF 2016).

The biodiversity of a region extends far beyond the variability of living organisms (Brasil 2000); it also includes a set of social and cultural activities associated with the knowledge, use and management of natural resources (Diegues and Arruda 2001). Thus, the diversity of plants in South America is certainly part of the life history of the inhabitants of this continent. This strong relationship between local populations and the environment is manifested in changes in landscape units for animal husbandry and crop cultivation, such as those performed by indigenous peoples such as the Incas, the oldest civilization on this continent, who lived in the Andes

Table 1 Diversity and endemism of South American angiosperm plant species

Country	Diversity	Endemism	Source
Brazil	50,000–56,000	33%	Mittermeier et al. (1997)
			MMA (2002)
			Giulietti et al. (2005)
Colombia	45,000–51,000	33%	Mittermeier et al. (1997)
			Giam et al. (2010)
			Fonnegra and Jiménez (2007)
Peru	18,000–20,000	29.8%	Mittermeier et al. (1997)
Ecuador	17,600–21,100	22,7–23,7%	Mittermeier et al. (1997)
			Jørgensen and León-Yánez (1999)
Venezuela	15,000–21,070	33.3–38%	Mittermeier et al. (1997)
Bolivia	15,345	15.3%	Davis et al. (1997)
			Jørgensen et al. (2014)
			Meneses et al. (2015)
Argentina	10,944	17.5%	Zuloaga and Belgrano (2015)
French Guiana and Guyana	8507	23,5%	Boggan et al. (1997)
			Davis et al. (1997)
Suriname	5000	–	Boggan et al. (1997)
			Davis et al. (1997)
Paraguay	6500–7000	24.6%	Basualdo et al. (1991)
			Zuloaga and Belgrano (2015)
Chile	4672	16.6%	Davis et al. (1997)
			Massardo and Rozzi (1996) and Gardner et al. (2015)
Uruguay	313	–	Haretche et al. (2012)

(Peru, Bolivia, Chile and Ecuador) and essentially dominated South America for centuries (Beyhaut 1994).

After the arrival of European colonists in the sixteenth century, the native people lost their territory, and the exploitation of natural resources expanded (Todorov 1993; Bueno and Dias 2015). Immense areas were devastated across the continent, and many biomes were degraded, with only small forest remnants remaining as environmental protection units, such as, for example, in Brazil (MMA 2002). Despite this destruction and the continuing deforestation, South America remains one of the most biologically diverse places on the planet.

Along with the loss of biological diversity, many ethnic groups have vanished, but there are still some ethnic remnants, such as in Colombia (120 indigenous groups), Peru (55 indigenous groups), Bolivia (35 indigenous groups), Venezuela (28 recognized ethnic groups), Ecuador (22 indigenous groups, Afro-Ecuadorians, Mestizos and Whites) and Paraguay (19 indigenous groups) (Vilca 2008; ACNUR 2009; DGEEC 2013; Zarur 2000; MIDIC 2016). In Brazil, there are more than 200 indigenous groups and many riverside, hinterland and *quilombola* (Maroon) communities, among others, bringing together an invaluable wealth of traditional knowledge of biodiversity (Diegues and Arruda 2001; Bosi 2000).

Colonization had a strong negative impact on native populations, consequently representing a strong threat to local knowledge. However, it resulted in a complex multicultural mosaic, in which different cultures and knowledge are interconnected. This knowledge may be a valuable tool in the struggle for biodiversity conservation (Diegues 2000), as many different groups of people depend on these resources (Posey 1984; Diegues and Arruda 2001; MMA 2002; Nogueira et al. 2010). The process of cultural exchange is dynamic and active in South America due to the contact among different groups of people through various types of migratory events (Neves et al. 2007). These processes enrich both the local biodiversity and the knowledge associated with it.

The importance of local knowledge of South American biological diversity is also evident in its contribution to advancing the field of bioprospecting. In this context, there is growing interest on ethnopharmacological research, as most manufactured drugs have a natural origin that often relies on information corresponding to the traditional uses of plants (Patwardhan 2005; Moore et al. 2017).

2 The Medicinal and Aromatic Plants of South America

Considering global biodiversity, it is estimated that there are between 50,000 and 70,000 medicinal and aromatic plants (MAPs) used worldwide by a majority of the planet's population. For example, in some South American countries, approximately 80% of the population uses medicinal plants (Firmo et al. 2011). Based on this estimate, it is clearly necessary to better understand the diversity of MAPs, especially given the lamentable destruction of ecosystems worldwide, which has resulted in approximately 15,000 species being threatened with extinction, according to the International Union for Conservation of Nature (IUCN 2000).

According to the World Health Organization (WHO 2007), more than 21,000 species are used worldwide for medicinal purposes, but there is no systematic data for South America (IUCN 2000). Another important aspect is that the uncontrolled exploitation of these countries has reduced biodiversity every year and many plant species are disappearing and with them, their associated traditional knowledge.

The study of MAPs allows the improved understanding of the local medical systems and thus the elucidation of gaps in the development of herbal medicines, contributing to the search for active compounds to develop drugs and increase therapeutic options for healthcare professionals (Elisabetsky and Moraes 1990; Klein et al. 2009; Tavares et al. 2013).

Despite the importance of the study of MAPs, the data in the literature are scattered and limited to a specific sector of the public. Further, even when information is gathered, as in one of the largest databases available on the Internet, "Plants for the future", with approximately 7000 useful species, the available information covers a limited number of plants (PFAF 2016). This database provides the scientific name and common name and information on the geographic distribution and uses of plants (PFAF 2016).

Table 2 Articles published on plants in the indicated Latin American countries during the period 1984–2004

South American countries	Year of publication							
	1984–1986	1987–1989	1990–1992	1993–1995	1996–1998	1999–2001	2002–2004	Total
Brazil	24	34	252	378	622	981	1431	3722
Argentina	15	15	98	176	339	495	603	1741
Chile	4	5	51	75	100	144	194	573
Venezuela	0	10	31	64	89	99	101	394
Colombia	1	2	28	39	48	72	75	265
Peru	5	5	27	24	39	43	71	214
Uruguay	1	1	1	5	13	22	26	69

Source: Adapted from Calixto (2005)

Most of the information available is primarily concentrated in books. Small percentages of the medicinal flora of some South American countries, such as Colombia, Venezuela, Chile, Ecuador, Bolivia and Peru, are described in books on medicinal plants of South America (see, for example, Roth and Lindorf 2002), which provide an overview of the phytochemistry of the plants common to these countries (Roth and Lindorf 2002). In Brazil, a considerable number of articles and books provide information on the use of and specific properties for very few species, usually the most common species or those with the most widespread use.

Calixto (2005) analyzed 25 years of research on the medicinal plants of Latin America (Table 2), finding records for seven of the 13 South American countries. In the last decade, the number of studies in Brazil has increased, and in the Scopus database alone, more than 1967 publications are found for this country when conducting a search using a combination of the keywords "medicinal plants" and Brazil. Research involving the MAPs in South American countries is of interest for pharmaceutical companies that seek to find active ingredients with the potential for the production of phytomedicines (Calixto 2005).

Table 3 presents records of native and exotic plants per country. Colombia, Brazil and Argentina are exceptional in that more than 1000 plant species in each country are recorded as being used for medicinal purposes (Table 3). In Peru, 4000 plant species used for medicinal and aromatic purposes have been recorded (Sanz-Biset et al. 2009; Gupta et al. 2014). A critical feature of these listings is that the records are incomplete regarding the origin of the species; therefore, the estimates are inaccurate regarding the diagnostic of the potential of the continent's native flora.

Despite South America's rich biodiversity and its pharmacological potential, there is a clear need to invest in research on plant species (Heinzmann and Barros 2007; Simões and Schenkel 2002). According to Calixto (2005), natural products originating from the continent's flora have been rapidly developed as a result of combined efforts between universities and the pharmaceutical industry to produce new effective and safe drugs. However, great effort is needed to establish the rational and sustainable exploitation of South American biodiversity in order to sustain-

Table 3 Estimated numbers of plant species used for medicinal purposes in South America

Country	Year/Period	No. of maps	Source
Brazil	2016	3000	MIDIC (2016)
Argentina	2009	1529	Barboza et al. (2009)
Chile	1996	469	Massardo and Rozzi (1996)
Peru	2009–2010	1500–4000	Sanz-Biset et al. (2009), Bussmann and Glenn (2010), and Gupta et al. (2014)
Colombia	2013–2015	5000	Fonnegra and Jiménez (2007), Cadena-González et al. (2013), and Jiménez et al. 2015
Uruguay	1993	22	González et al. (1993)
Venezuela	2002–2009	700	Giraldo et al. (2009)
Paraguay	1991	1500–3500	Basualdo et al. (1991)
Guyana			DeFilipps et al. (2004)
French Guiana	2004	1000–1200	DeFilipps et al. (2004)
Suriname	1982–2007	138	Verpoort and Dihal (1987), Hasrat et al. (1997), and Andel et al. (2007)
Ecuador	2006–2016	275	Torre et al. (2006) and Tinitana et al. (2016)

ably meet the needs of pharmaceutical companies and local people in these countries while also respecting the intellectual property rights that include the traditional knowledge associated with these plants.

The use of medicinal plants by people from different parts of South America is not random. The variety of medicinal plants reported is related to the richness within each botanical family, with different evidence from Brazil (Medeiros et al. 2014), Bolivia (Thomas et al. 2009) and Ecuador (Bennett and Husby 2008). These data reinforce the fact that the biodiversity in South America may mask the real abundance of MAPs. Despite these findings, there is also evidence for some plants, such as ferns and lycophytes, that, although used in accordance with their existing availability, are used less and less in local communities because they are perceived as inferior therapeutic resources (Reinaldo et al. 2015). This phenomenon suggests the need for detailed ethnobiological and ethnopharmacological studies to understand the roles of plants in different local medical systems in South America.

Notably, despite the high biodiversity in South America, few phytomedicines have been developed from the flora. This anomaly may be explained by the following criticisms of several researchers: a lack of systematic and continued studies with promising plants; a lack of collaboration among researchers; limitations related to research methods and misinterpretations of pharmacological tests; and confusing, misleading and limited procedures for collecting ethnobotanical data, which are often the basis for other research fields (Houghton et al. 2007; Gertsch 2009; Albuquerque et al. 2014). For example, Medeiros et al. (2014) found problems in several published studies on medicinal plants that were based on surveys of information from the local populations, which compromises the quality, reliability and clarity of the findings.

In the case of phytomedicines, one-quarter of products sold in pharmacies are manufactured from materials extracted from tropical plants (Abranches 2015). Thus, some researchers consider the value of natural products for society and the economy incalculable (Abranches 2015) and the losses of genetic resources through biopiracy also incalculable. To minimize these risks, the Interministerial Group on Industrial Property (Grupo Interministerial de Propriedade Industrial, or GIPI, appointed by the Brazilian Ministry of Development, Industry and Foreign Trade/2006), produced a "Non-Exhaustive List of Customary Names Used in Brazil Associated with Biodiversity" to track native species patented by other countries (GIPI 2016).

3 The Treasures of South America

South America's biodiversity is a valuable source of active ingredients that can be used as medicines, with only a few products that are currently commercially available, such as pilocarpine, which is extracted from the leaves of *Pilocarpus microphyllus* Stapf. (jaborandi), a native plant from Brazil (Valdez et al. 1993; Wynn 1996; Pinheiro 2002). Pilocarpine has been used for decades in the preparation of medication indicated for glaucoma (Merck 1998) and is also used to relieve some side effects of radiotherapy, such as dry mouth (xerostomia), by stimulating the secretion of saliva (Valdez et al. 1993; Wynn 1996).

An important contribution of medicinal flora is d-tubocurarine, a substance known as "curare", which is a preparation made with the species *Chondrodendron tomentosum* Ruiz and Pavon (Menispermaceae). Curare is used as poison by indigenous peoples and was introduced into the market for anesthesiology in 1940 due to its relaxant effect on skeletal muscles (Nogueira et al. 2010). Another phytomedicine recently introduced to the market is derived from the medicinal plant known as cordia, *Cordia verbenacea* DC. (Boraginaceae), which has anti-inflammatory activity with indications for tendonitis and muscle pain and is produced by a major Brazilian pharmaceutical company (Calixto 2005).

Myracrodruon urundeuva Allemão is one of the primary plants used in traditional medicine in northeast Brazil and in other South American countries, including Bolivia (Deharo et al. 2004). It is indicated as antimicrobial, anti-inflammatory and healing in the treatment of wounds, gastritis, gastric ulcers, cervicitis, vaginitis and hemorrhoids (Lorenzi and Matos 2002; Botelho et al. 2007; Bianco 2004). With properties similar to the Brazilian peppertree (*Schinus terebinthifolius* Raddi), it has antimicrobial, healing and anti-inflammatory indications. *M. urundeuva* is used as a drug in the treatment of cervicitis, vaginitis and cervical vaginitis in the form of gynecological gel and vaginal ovules (Brasil 2016).

As in the previous examples, in recent decades, research on medicinal plants has confirmed some traditional indications, but there is an urgent need to determine the actual diversity of medicinal plants and to protect and regulate access to the biological resources of South America (Marques 2000). Aiming to regulate the use of some species, the Brazilian National Health Surveillance Agency (Agencia Nacional de Vigilância Sanitária do Brasil – ANVISA 2011) published a list of the phytomedicines of the Brazilian pharmacopoeia, containing information on 47 plant species and their derivatives as phytomedicines for infusions and decoctions, tinctures, syrup, gels, ointment, soap and creams. Despite the important ANVISA initiative, this list is only a sample of all medicinal plants, many of which are exotic (Brasil 2011).

Unfortunately, the technological state of the products marketed by the pharmaceutical industry in Brazil, which may be one of the few South American countries with major advances in this area, is based on the popular use of plants rather than with pre-clinical proof of biological activities (Yunes et al. 2001; Firmo et al. 2011). To improve this situation, a policy committed to the development of scientific studies and incentives for the pharmaceutical industry is needed (Rates 2001; Yunes et al. 2001; Calixto 2005). In Brazil, for example, 74 native species are used by the industry in 300 diverse types of products, but "the lack of quantitative data indicating where these plants are harvested, the quantities involved, and their harvesting capacity will limit any attempts at establishing conservation strategies at a national level" (Melo et al. 2009).

The generation of patents requires additional attention, considering that in South America, there seems to be no effective culture or stimulus toward generating patents arising from scientific studies, which is also a concern with regard to biopiracy, as it endangers the genetic heritage of the continent (Marques 1999; Moreira et al. 2004).

Finally, South America has a valuable assortment of plant resources with the potential for bioprospecting and conservation (see Gonzales and Valerio 2006; Sülsen et al. 2011; Cruz et al. 2013) despite high levels of degradation and the improper exploitation of MAPs. In addition, it is necessary to gather information in a systematic way to advance analysis and propose strategic actions for development and research.

References

Abranches MV (2015) Plantas Medicinais e Fitoterápicos: abordagem teórica com ênfase em nutrição. Ed. A. S. Sistemas

ACNUR (2009) Indígenas en las Américas. El Trabajo del ACNUR con Pueblos Indígenas. Available at: http://www.acnur.org/t3/pueblosindigenas/pueblos-indigenas-en-colombia. Accessed 16 Aug 2016

Albuquerque UP, Medeiros PM, Ramos MA, Júnior WSF, Nascimento ALB, Avilez WMT, Melo JG (2014) Are ethnopharmacological surveys useful for the discovery and development of drugs from medicinal plants? Braz J Pharmacogn 24:110–1S15

Andel Van T, Behari-Ramdas J, Havinga R, Groenendijk S (2007) The medicinal plant trade in Suriname. Ethnobot Res Appl 5:351–372

Barboza GE, Cantero JJ, Núñez C, Pacciaroni A, Espinar LA (2009) Medicinal plants: a general review and a phytochemical and ethnopharmacological screening of the native argentine Flora. Kurtziana 34(1–2):7–365

Basualdo I, Zardini E, Ortiz M (1991) Medicinal plants of Paraguay: underground organs. Econ Bot 45(1):86–96

Bennett BC, Husby CE (2008) Patterns of medicinal plant use: an examination of the Ecuadorian Shuar medicinal flora using contingency table and binomial analyses. J Ethnopharmacol 116(3):422–300

Beyhaut G (1994) Dimensão cultural da integração na América Latina. Estudos Avançados 8(20):183–198

Boggan J, Funk V, Kelloff C, Hoff M, Cremers G, Feuillet C (1997) Checklist of the plants of the guianas (Guyana, Surinam, French Guiana), 2nd edn. University of Guyana, Georgetown

Bosi A (2000) História, etnias, culturas: 500 anos construindo o Brasil: Subsidio apresentado à 38ª Assembleia Geral da CNBB, Ed. Loyola

Botelho MA, Bastos GM, Fonseca SGC, Matos FJA, Montenegro D, Rao VS, Brito GAC (2007) Antimicrobial activity of the essential oil from *Lippia sidoides*, cavacrol and thymol against oral pathogens. Braz J Med Biol Res 40:349–356

BRASIL, Ministério do Meio Ambiente (2000) Política Nacional de Biodiversidade: roteiro de consulta para elaboração de uma proposta. Biodiversidade, 1: 48, Brasília

Brasil (2016) Agência Nacional de Vigilância Sanitária Memento Fitoterápico da Farmacopeia Brasileira. ANVISA, Brasília Available at: http://portal.anvisa.gov.br/documents/33832/2909630/Memento+Fitoterapico/a80ec477-bb36-4ae0-b1d2-e2461217e06b. Accessed 15 July 2016

Brasil (2011) Agência Nacional de Vigilância Sanitária. Formulário de Fitoterápicos da Farmacopeia Brasileira/Agência Nacional de Vigilância Sanitária. ANVISA, Brasília Available at: http://www.anvisa.gov.br/hotsite/farmacopeiabrasileira/conteudo/Formulario_de_Fitoterapicos_da_Farmacopeia_Brasileira.pdf. Accessed 15 July 2016

Bueno L, Dias A (2015) Povoamento inicial da América do Sul: contribuições do contexto brasileiro. Estudos Avançados 29(83):119–147

Bussmann W, Glenn A (2010) Medicinal plants used in Northern Peru for reproductive problems and female health Rainer. J Ethnobiol Ethnomed 6(30):1–12

Cadena-González AL, Sørensen M, Theilade I (2013) Use and valuation of native and introduced medicinal plant species in Campo Hermoso and Zetaquira, Boyacá, Colombia. J Ethnobiol Ethnomed 9(23):1–34

Calixto JB (2005) Twenty-five years of research on medicinal plants in Latin America: a personal view. J Ethnopharmacol 22,100(1–2):131–134

CONAF. Ministério da Agricultura. Parque Nacional Arquipélago de Juan Fernández (2016) Available at: http://www.conaf.cl/parques/parque-nacional-archipielago-de-juan-fernandez. Accessed 03.09.2016

Cruz LR, Spangenberg T, Lacerda MVG, Wells TNC (2013) Malaria in South America: a drug discovery perspective. Malar J 12(1):168

Davis SD, Heywood VH, Herrera O, Bryde M, Villalobos J, Hamilton AC (1997) Centres of plant diversity. A guide and strategy for their conservation. The World Wild Fund for Nature & The World Conservation Union, Oxford, p 596

DeFilipps RA, Maina SL, Crepin J (2004) Medicinal plants of the Guianas (Guyana, Surinam, French Guiana). Department of Botany National Museum of Natural History Smithsonian Institution, Washington, DC, p 15

Deharo E, Baelmans R, Gimenez A, Quenevo C, Bourdy G (2004) In vitro immunomodulatory activity of plants used by the Tacana ethnic group in Bolívia. Phytomedicine 11(6):516–522

DGEEC. Población originaria e indígena del Paraguay (primera parte) (2013) Available at: http://www.tierraviva.org.py/?pueblo=poblacion-originaria-e-indigena-del-paraguay-primera-parte. Accessed 23.08.2016

Diegues AC (2000) Os saberes tradicionais e a biodiversidade no Brasil. MMA/COBIO/NUPAUB/USP, São Paulo, p 211

Diegues AC, Arruda RSV (2001) Saberes tradicionais e biodiversidade no Brasil. Ministério do Meio Ambiente (Biodiversidade, 4). São Paulo: USP, Brasília, pp 176–188

Echeverria C, Cayuela L, Manson RH, Coomes DA, Lara A, Reys-Benayas JM, Newton AC (2007) Spatial and temportal patterns of forest loss and fragmentation in Mexico and Chile. In: Newton AC, Cabi H (eds) Biodiversity loss and conservation in fragmented forest: the forest of montane. Mexico and temperate south American. CAB International, Cambridge, p 370

Elisabetsky E, Moraes JAR (1990) Ethnopharmacology: a technological development strategy. In: Posey AD, Overal WL, Clement, CR, Plotkin, MJ, Elisabetsky E, Mota CN, Barros, JFP (eds) Ethnobiology: implacations and applications. Proceedings of the first international congress of ethnobiology. Belém: Museu Paraense Emílio Goeldi, p. 11–8

FAO State of the World's Forests (2011) pp 164. Available at: Available at: 12 Apr 2016

Fioravanti C (2013) Os primeiros passos de novas espécies. Plantas e animais se diferenciam por meio de mecanismos surpreendentes. Available at: http://revistapesquisa.fapesp.br/wp-content/uploads/2013/10/18-23-especiacao-212.pdf?9f3c9b Available at 12.09.2016

Firmo WCA, Menezes VJM, Passos CEC, Dias CN, Alves LPL, Dias ICL, Neto MS, Olea RSG (2011) Contexto Histórico, Uso Popular e Concepção Científica Sobre Plantas Medicinais. Cad Pesqui 18:90–95

Fonnegra RG, Jiménez SLR (2007) Plantas medicinales aprobadas en Colombia, 2ª edición. Editorial Universidad de Antioquia, Medellín

Gardi C, Angelini M, Barceló S, Comerma J, Cruz GC, Encina RA, Jones AKP, Mendonça SBML, Montanarella, L, Muniz UO, Schad P, Vara RMI, Vargas R (2014) Atlas de suelos de América Latina y el Caribe, Comisión Europea – Oficina de Publicaciones de la Unión, p. 176

Gardner MF, Hechenleitner PV, Hepp JC (2015) Plants from the woods and forests of Chile. Paintings the woods & forests of Chile. Royal Botanic Garden Edinburgh, Edinburgh

Gertsch J (2009) How scientific is the science in ethnopharmacology? Historical perspectives and epistemological problems. J Ethnopharmacol 122(2):177–183

Giam X, Bradshaw CJA, Tan TH, Sodhi NS (2010) Future habitat loss and the conservation of plant biodiversity. Biol Conserv 143(7):1594–1602

GIPI – Grupo Interministerial da Propriedade Industrial (n.d.) Lista Não-Exaustiva de Nomes Associados à Biodiversidade de Uso Costumeiro no Brasil. Available at: www.desenvolvimento.gov.br/arquivo/sti/publicacoes/lisBiodiversidade/ListaBiodivBrasilVer1.pdf. Accessed 20.08.2016.2016

Giraldo D, Baquero E, Bermúdez A, Oliveira-Miranda MA (2009) Medicinal plant trade characterization in popular markets of Caracas. Venezuela Acta Bot Venez 32(2):267–301

Giulietti AM, Harley RM, Queiroz LP, Wanderley MGL, Van Den CB (2005) Biodiversidade e conservação das plantas no Brasil. Megadiversidade 1(1):52–61

Gonzales GF, Valerio LG Jr (2006) Medicinal plants from Peru: a review of plants as potential agents against cancer. Anti Cancer Agents Med Chem 6(5):429–444

González A, Ferreira F, Vázquez A, Moyna P, Alonso Paz E (1993) Biological screening of Uruguayan medicinal plants. J Ethnopharmacol 39(21):217–220

Gupta MP, Handa SS, Longo G, Rakesh DD (2014) Compendium of medicinal and aromatic plants. In: Gupta MP, Sukhdev SH, Genaro L, Dev DR (eds) The Americas, 1st edn. Panama University, Peru, pp 151–169

Haretche F, Mai P, Brazeiro A (2012) Woody flora of Uruguay: inventory and implication within the Pampean region. Acta Bot Bras 26(3):537–552

Hasrat JA, DE Backer JP, Vauquelln G, Vlletinck AJ (1997) Medicinal plants in Suriname: screening of plant extracts for receptorbinding activity. Phytomedicine 4(1):59–65

Heinzmann BM, Barros MCB (2007) Potencial das plantas nativas brasileiras para o desenvolvimento de fitomedicamentos tendo como exemplo Lippia alba (Mill.) N. E. Brown (Verbenaceae). Saúde 33(1):43–48

Houghton PJ, Howes MJ, Lee CC, Steventon G (2007) Uses and abuses of in vitro tests in ethnopharmacology: visualizing an elephant. J Ethnopharmacol 110(3):391–400

IUCN (2000) Red list of threatened species. Guiding conservations for 50 years. Available at: http://www.iucnredlist.org/photos. Accessed 17.8.2016

Jiménez N, Carrillo-Hormaza L, Pujol A, Álzate F, Osorio E, Lara-Guzman O (2015) Antioxidant capacity and phenolic content of commonly used anti-inflammatory medicinal plants in Colombia. Ind Crop Prod 70:272

Jørgensen PM, León-Yánez S (1999) Catalogue of the vascular plants of Ecuador. Monogr Syst Bot Missouri Bot Gard 75:1–1182

Jørgensen PM, Harley MN, Beck SG (2014) In: Arrázola S, Saldias M, Hirth S, Swift V, Penagos JC, Romero C (eds) Catálogo de las plantas vasculares de Bolívia. Missouri Botanical Garden Press, St. Louis, pp 1–1744

Klein T, Longhini R, Bruschi ML, Mello JCP (2009) Fitoterápicos: um mercado promissor. Rev Ciênc Farm Básica Apl 30(3):241–248

Lorenzi H, Matos FJA (2002) Plantas Medicinais do Brasil: Nativas e Exóticas. Instituto Plantarum, Nova Odessa, p 512

Marques MB (1999) Planejamento e gestão da política de ciência e tecnologia: hora de rever? Revista Ciência & Saúde Coletiva 4(2):383–392

Marques MB (2000) Patentes farmacêuticas e acessibilidade aos medicamentos no Brasil. Hist Cienc Saúde Manguinhos 7(1):7–21

Massardo F, Rozzi R (1996) Valoración de la Biodiversidad: Usos medicinales dela flora nativa chilena. Ambiente y Desarrollo 7(3):76–81

Medeiros PM, Ladio AH, Albuquerque UP (2014) Sampling problems in Brazilian research: a critical evaluation of studies on medicinal plants. Rev Bras Farmacog 24(2):103–109

Melo JG, Amorim ELC, Albuquerque UP (2009) Native medicinal plants commercialized in Brazil – priorities for conservation. Environ Monit Assess 156(1–4):567–580

Meneses RI, Beck S, Garcia E, Mercado M, Araújo A, Serrano M (2015) Flora of Bolivia – where do we stand? Rodriguésia 66(4):1025–1031

Merck (1998) The Merck index: an encyclopedia of chemicals, drugs, and biologicals. Budavari S, O'Neil MJ (editors) 12th ed.

Michelin DC, Moreschi PE, Lima AC, Nascimento GGF, Paganelli MO, Chaud MV (2005) Avaliação da atividade antimicrobiana de extratos vegetais. Rev Bras Farmacog 15(4):316–320

MIDIC Ministério da Cultura, PERU (2016) Base de Datos Pueblos indígenas del Perú. Available at: http://bdpi.cultura.gob.pe/lista-de-pueblosindigenas. Accessed 24 Aug 2016

Ministério do Meio Ambiente (MMA) (2002) Avaliação e identificação de áreas e ações prioritárias para a conservação, utilização sustentável e repartição dos benefícios da biodiversidade nos biomas brasileiros. MMA/SBF, Brasília, p 404

Mittermeier RA, Mittermeier CG, Brooks TM, Pilgrim JD, Konstant WR, da Fonseca GAB, Kormos C (2003) Wilderness and biodiversity conservation. Proc Natl Acad Sci U S A 100(18):10309–10313

Mittermeier RA, Robles Gil P, Mittermeier CG (1997) Megadiversity: earth's biologically wealthiest nations. Mexico City, CEMEX and Agrupación Sierra Madre

Moore N, Hamza N, Berke B, Umar A (2017) News from Tartary: an ethnopharmacological approach to drug and therapeutic discovery. Br J Clin Pharmacol 83(1):33–37

Moreira AC, Antunes MAS, Pereira NJ (2004) Patentes extratos de plantas e derivados. Verdades e mentiras sobre as patentealidades do Brasil. Rev Biotecnol Ciênc Desenvlvimento 33:62–71

Myers N, Mittermeier RA, Mittermeier CG, Da Fonseca GA, Kent J (2000) Biodiversity hotspots for conservation priorities. Nature 403(6772):853

Neves WA, Bernardo DV, Okumura MMM (2007) A origem do homem americano vista a partir da América do Sul: uma ou duas migrações. Rev Antropol 50(1):10–44

Nogueira RC, Cerqueira HF, Soares MBP (2010) Patenting bioactive molecules from biodiversity: the Brazilian experience. Expert Opin Ther Pat 20(2):145–157

Patwardhan B (2005) Ethnopharmacology and drug discovery. J Ethnopharmacol 100(1):50–52

PFAF (2016) Plants for a future: earth, plants and people. Available at: http://www.pfaf.org/user/Default.aspx. Accessed 14.08.2016

Pinheiro CUB (2002) Extrativismo, Cultivo e Privatização do Jaborandi (*Pilocarpus microphyllus* Stapf Ex Holm. Rutaceae) no Maranhão, Brasil. Acta Bot Bras 16(2):141–150

Posey DA (1984) Os Kayapos e a natureza. Ciência Hoje 2(12):35–41

Prado DE (2003) As Caatingas da América do Sul. In: Leal IR, Tabarelli M, Cardoso JMS (eds) Ecologia e Conservação da Caatinga. Editora Universitária: UFPE, Recife, p 822

PRHS, Plano Nacional de Recursos Hídricos (2006) Síntese Executiva – português / Ministério do Meio Ambiente, Secretaria de Recursos Hídricos. MMA, Brasília

Rates SMK (2001) Promoção do uso racional de fitoterápicos: uma abordagem no ensino de Farmacognosia. Rev Bras Farmacog 11(2):57–69

Reinaldo RCPDS, Santiago ACP, Medeiros PM, Albuquerque UP (2015) Do ferns and lycophytes function as medicinal plants? A study of their low representation in traditional pharmacopoeias. J Ethnopharmacol 175:39–47

Rey-Benayas JM, Cayuela L, González-Espinosa M, Echeverria C, Manson RH, Williams-Linera G, Castillo DELRF, Ramiréz-Marciel N, Muniz-Castro MS, Blanco-Macías A, Lara A, Newton AC (2007) Plant diversity in highly frangmented forest landscapes in Mexico and Chile: implications for conservation. "Biodiversity loss and conservation in fragmented forest landscapes". The forests of montane Mexico and temperate South America. CABI, Wallingford, Oxfordshire, pp 43–68

Roth I, Lindorf H (2002) South American medicinal plants. Botany, remedial properties and general use. Springer, Heidelberg, p 492

Salazar LF, Nobre CA, Oyama MD (2007) Climate change consequences on the biome distribution in tropical South America. Geophys Res Lett 34(9)

Sanz-Biset J, Campos-de-la-Cruz J, Epiquién-Rivera MA, Cañigueral S (2009) A first survey on the medicinal plants of the Chazuta valley (Peruvian Amazon). J Ethnopharmacol 122(2):333–362

Simões CMO, Schenkel EP (2002) A pesquisa e a produção brasileira de medicamentos a partir de plantas medicinais: a necessária interação da indústria com a academia. Rev Bras Farmacogn 12(1):35–40

Sülsen VP, Cazorla SI, Frank FM, Anesini C, Muschietti LV, Martino VS (2011) South American medicinal flora: a promising source of novel compounds with antiprotozoal activity. Lat Am J Pharm 30(1):202

Tavares WS, Freitasb SS, Grazziottib GH, Parentec LML, Lião LM, Zanuncioe JC (2013) Ar-turmerone from *Curcuma longa* (Zingiberaceae) rhizomes and effects on Sitophilus zeamais (Coleoptera: Curculionidae) and Spodoptera frugiperda (Lepidoptera: Noctuidae). Ind Crop Prod 46:158–164

Thomas E, Vandebroek I, Sanca S, Van Damme P (2009) Cultural significance of medicinal plant families and species among Quechua farmers in Apillapampa, Bolivia. J Ethnopharmacol 122(1):60–67

Tinitana F, Montserrat R, Romero JC, Benavides DELA, Rot MC, Santayana MP (2016) Medicinal plants sold at traditional markets in Southern. J Ethnobiol Ethnomed 12(29):1–18

Todorov TA (1993) Conquista da América do Sul. A questão do outro. São Paulo, 2ª edn. Martins Fontes, São Paulo

Torre L, Muriel P, Balslev H (2006) Etnobotánica en los Andes del Ecuador. In: Moraes M, Øllgaard B, Kvist LP, Borchsenius F, Balslev H (eds) Botánica Económica de los Andes Centrales. Universidad Mayor de San Andrés, La Paz, pp 246–267

Valdez IH, Wolff A, Atkinson JC, Macynski AA, Fox PC (1993) Use of pilocarpine during head and neck radiation therapy to reduce xerostomia and salivary dysfunction. Cancer 71(5):1848–1851

Verpoorte R, Dihal PP (1987) Medicinal plants of Suriname IV: antimicrobial activity of some medicinal plants. J Ethnopharmacol 21(3):315–318

Vilca JCM (2008) Las formas de propiedad y su registro: las tierras indígenas y recursos naturales. AECID/Bolivia. Available at: http://www.territorioindigenaygobernanza.com/bov_10.html. Accessed 17.08.2016

World Health Organization (WHO) (2007) Monographs on selected medicinal plants, vol 3. World Health Organization, Geneva, pp 349–358
Wynn RL (1996) Oral pilocarpine (Salagen): a recently approved salivary stimulant. Gen Dent 44(1):29–30
Yunes RA, Pedrosa RC, Cechinel FV (2001) Fármacos e fitoterápicos: a necessidade do desenvolvimento da indústria de fitoterápicos e fitofármacos no Brasil. Quím Nova 24(1):147–152
Zarur GCL(2000) Raízes Étnicas do Brasil: Modelo de Integração. In: História, Etnias, Culturas: 500 Anos Construindo o Brasil. Ed Loyola, São Paulo
Zuloaga FO, Belgrano MJ (2015) The catalogue of vascular plants of the southern cone and the flora of Argentina: their contribution to the world Flora. Rodriguésia 66(4):989–1024

Chemical Diversity and Ethnopharmacological Survey of South American Medicinal and Aromatic Plant Species

Rodney Alexandre Ferreira Rodrigues, Glyn Mara Figueira, Adilson Sartoratto, Lais Thiemi Yamane, and Verônica Santana de Freitas-Blanco

Abstract The present chapter is a short review providing information about the chemical constituents of some South American plant species used by local communities in countries of this continent except the Falkland Islands and Surinam. Many plants found in the countries of Argentina, Bolivia, Brazil, Chile, Colombia, Ecuador, Guyana, Paraguay, Peru, Uruguay, and Venezuela have valuable phytotherapeutic applications in alternative medicine. This chapter presents information reported in the scientific literature concerning the most significant plant families used in folk medicine, considering their chemical compositions and highlighting the following categories: alkaloids, an important class of biologically active compounds; phenolics, especially flavonoids; and essential oils.

Keywords South American countries · Chemical composition · Ethnopharmacology · Traditional medicine · Phytotherapy · Medicinal and aromatic species · Biodiversity · Alkaloids · Phenolic compounds · Essential oils · Flavonoids

1 Introduction

The use of herbs as medicinal plants by humanity, as an alternative therapy for the treatment of diseases, has been commonplace for thousands of years. More recently, herbs have been used as models for novel therapeutic agents. Medicinal plants provided the basis for modern traditional medicine, with the earliest records, dating from 2600 BC, documenting the use of almost 1000 plant-derived substances in Mesopotamia and ancient Egypt, the region now known as the Middle East.

R. A. F. Rodrigues · G. M. Figueira (✉) · A. Sartoratto · L. T. Yamane
V. S. de Freitas-Blanco
CPQBA/UNICAMP, Chemical, Biological and Agricultural Research Center, University of Campinas, Paulinia, Brazil
e-mail: rodney@cpqba.unicamp.br; glyn@cpqba.unicamp.br; adilson@cpqba.unicamp.br

U. P. Albuquerque et al. (eds.), *Medicinal and Aromatic Plants of South America*, Medicinal and Aromatic Plants of the World 5,
https://doi.org/10.1007/978-94-024-1552-0_2

The secondary metabolites of plants are an economically important source of pharmaceuticals and can serve as models for synthetic drugs. They play vital roles in the physiology of plants, helping to protect against unexpected environmental hazards. Studies in this area have increased over the last decades, with many compounds being isolated and their chemical structures discovered.

The biodiversity of the terrestrial ecosystems of South America constitutes one of its essential features, and most important is the fact that this region still contains vast intact wild areas, where new chemical molecules can be discovered. The South American biosphere therefore has an enormous potential to provide phytochemicals with active components that can be used in industrial products.

2 Ethnopharmacological Overview and Chemical Compositions

Below we have compiled some of the scientific research from South American countries related to species of recognized importance in folk medicine, describing a great diversity of chemical compounds and their known and potential uses.

Among many applications of plants, Gonzales and Valerio Junior (2006) specifically considered the anti-cancer properties of species used in folk medicine by Peruvian populations from the Andean and Amazonian regions. The authors found evidence of the beneficial use of *cat's claw*, also known as *uña de gato* (*Uncaria tomentosa* (Willd.) DC.), *maca* (*Lepidium meyenii* Walp.), and *dragon's blood* (*Croton lechleri* Müll. Arg.). Major constituents identified in *cat's claw* include alkaloids, organic acids, anthocyanins, sterols, and triterpenes. The major constituents reported in *maca* include tannins, saponins, sterols, polyunsaturated fatty acids, β-carbolines, uridine, malic acid, prostaglandins, flavonoids, and anthocyanins. *Dragon's blood* contains alkaloids, phenolic compounds such as proanthocyanidins and flavonoids, and tannins such as catechin-(4α→8)-epigallocatechin, gallocatechin-(4α→8)-epicatechin, gallocatechin-(4α→6)-epigallocatechin, catechin-(4α→8)-gallocatechin-(4α→8)-gallocatechin, and gallocatechin-(4α→8)-gallocatechin-(4α→8)-epigallocatechin.

Bixa orellana L., known by its folk name *urucum*, has been used by native people in Brazil because of its food and biological uses. Its widespread dissemination, evidenced by crops grown in other South American countries including Colombia, Paraguay, Venezuela, Bolivia, Argentina, Peru, Guyana, and Ecuador, is due to the demand for its natural dye (bixin) by the food and pharmaceutical industries (Vilar et al. 2014).

Considering species found in Colombia, studies have listed 254 plants, including 127 wild species, used in the northwest Antioquia region for various medicinal purposes. The species in this list belong to 193 genera of 79 families, notably the Asteraceae, Lamiaceae, Poaceae, Apiaceae, and Solanaceae, and their uses have been divided into 131 categories (Fonnegra-Gómez and Villa-Londoño 2011). The Asteraceae family was also studied by Ribeiro et al. (2010), who investigated the uses of 102 species, as well as their chemical constitutions, in a phytochemical screening approach using crude extracts of the plants.

Chemical evaluation was made of species of the *Eremanthus* genus collected in an ecological reserve in Brazil. The chemical screening of *E. erythropappus* (DC.) Macleish, *E. incanus* (Less.) Less., and *E. glomerulatus* Less. revealed the presence of reducing sugars, carbohydrates, amino acids, tannins, flavonoids, glycosides, cardiotonics, carotenoids, steroids, triterpenoids, coumarin and its derivatives, saponins, alkaloids, purines, polysaccharides, and anthraquinones. The *Eremanthus* genus contains several species that are known by their folk name *candeia* and are mostly exploited for the production of an essential oil, whose main component, α-bisabolol, has antiphlogistic, antibacterial, antimycotic, dermatological, and spasmodic properties.

Alviz et al. (2013) studied *Ceratopteris pteridoides* (Hook) Hieron. and found evidence of its diuretic activity, corroborating the popular use of this plant in the northern districts of Colombia. Major components found in *C. pteridoides* were aromatic amines and tryptamines, esters, aldehydes, and ketones, with smaller amounts of tannins and cardiotonics. Lagos-López (2007) studied ethnobotanical aspects of species with medicinal properties in six municipalities of the Department of Boyacá, in a survey of 600 people who claimed to have knowledge of the use of these plants. The species most commonly used for stomach ache (employed by 80% of the population) was *Cape gooseberry* (*Physalis peruviana* L.), a member of the Solanaceae family. Franco et al. (2007) also studied *Cape gooseberry*, due to its high commercial value and medicinal properties including anticancer, antimycobacterial, antipyretic, diuretic, immunomodulatory, and anti-inflammatory activities. Its anti-inflammatory activity was confirmed and validated, and the compound 12-*O*-tetradecanoylphorbol-13-acetate was isolated and tested, showing statistically significant activity.

Quintero et al. (2015) studied herbs collected from eight different local markets in the Colombian capital and conducted semi-structured interviews with 16 sellers of medicinal plants. In these interviews, the herb vendors mentioned species such as *chitato* (*Muntingia calabura* L.), *alfalfa* (*Medicago sativa* L.), *laurel* (*Morella pubescens* Willd), *suelda consuelda* (*Symphytum officinale* L.), and *paico* (*Chenopodium ambrosioides* L.), which were not found in the National Colombian Formulary. *Alfalfa* is rich in nutrients such as provitamin A and vitamins B, C, D, and K, and is used to combat scurvy and rickets. The herbs mentioned were found in folk medicine, and their efficacy and safety of use have not been established scientifically (Lorenzi and Matos 2008).

Folk knowledge is a consistent theme in this review, and according to Quintero et al. (2015) the vendors showed little knowledge about possible side effects of the medicinal plants, which could be indicative of unsatisfactory practices in the community. Ignorance of the differences between the decoction and infusion forms of preparation was also evident. Plants that could be promising for new therapeutic uses were identified, including *albahaca* (*Ocimum basilicum* L., also *Ocimum campechianum* Mill.), *calendula* (*Calendula officinalis* L.), *cidrón* (*Aloysia triphylla* Royle), *cola de caballo* (*Lasiacis sorghoidea* (Desv. ex Ham.) Hitchc &

Chase, also *Equisetum arvense* L. or *Equisetum bogotense* Kunth.), and *manzanilla* (*Matricaria chamomilla* L.) (Quintero et al. 2015). Eleven herbs with essential oils in their composition were collected and investigated by Bueno-Sánchez et al. (2009) for their anti-tubercular activity. The authors concluded that the essential oils from *Achyrocline alata* (Kunth) DC., which contains 24.0% thymol, and *Swinglea glutinosa* (Blanco) Merr., which contains 49.6% α-pinene as well as other identified compounds, are candidates as potential phytotherapeutic agents against tuberculosis in humans. *Macela* is one of several popular names of *A. satureioides* (Lam.) DC., and this name is also used to describe *A. alata*, a typical species from southern Brazil, which also occurs in Uruguay, Paraguay, and Argentina. *A. satureioides* is used as an anti-inflammatory, antispasmodic, digestive, sedative, and carminative (Lorenzi and Matos 2008). Chemical investigations of *A. alata* and other species of *Achyrocline* collected in Argentina and Uruguay showed similar profiles in terms of their phenolic constituents, flavonoids, and quinic acid derivatives, compounds that justify the folk uses of these plants. The main compounds found were chlorogenic acid, isoquercitrin, 3,4-dicaffeoyl quinic acid, 3,5-dicaffeoyl quinic acid, 4,5-dicaffeoyl quinic acid, quercetin, 3-*O*-methylquercetin, 4,2′,4′-trihydroxy-6′-methoxychalcone, and gnaphalium (Grassi-Zampieron et al. 2010).

Arias (2012) investigated herbs used to treat common diseases in the vicinity of the Colombian city of Leticia, in the Amazon region, during the years 2008 and 2009. A total of 115 herbs with medicinal uses were reported, comprising 109 genera and 99 species. It was concluded that the families Arecaceae, Bignoniaceae, and Rubiaceae, and species such as *yarumo* (*Cecropia sciadophylla* Mart.), *carambolo* (*Averrhoa carambola* L.), *cat's claw* (*Uncaria tomentosa* Willd. DC.), *acapu* (*Minquartia guianensis* Aubl.), *lancetilla* (*Alternanthera brasiliana* (L.) Kuntze), and *amacizo* (*Erythrina fusca* Lour.) had considerable cultural value within this specific Amazon community. Carvajal-De Pabón et al. (2014) assessed different parts of *Passiflora ligularis* Juss., locally known as *granadilla*, including the pulp, flowers, leaves, flower cores, and stems. Substances detected in different proportions in the various plant tissues included phenolic compounds, coumarins, anthocyanins, saponins, tannins, flavonoids, triterpenes/steroids, quinones, alkaloids, and lactones. This information served as a starting point for a basic qualitative procedure to describe the biological activity of this species, including phytochemical, bromatological, and mineral analyses.

Lorenzi and Matos (2008) described the *Drimys* genus in Brazil, where *Drimys brasiliensis* Miers is used against dyspepsia, dysentery, nausea, intestinal pain and cramping, fever, and anemia. This plant, which is recognized worldwide as a carminative, stomachic, and tonic, contains tannins and sesquiterpenoids in its composition.

Hajdu and Hohmann (2012) described two species of the genus *Triplaris*, namely *T. peruviana* Fisch. & Meyer ex C.A. Meyer and *T. pavonii* Meisn., used for the treatment of dysentery and burns by the Bolivian Kallawaya ethnic group. A close relative, *Triplaris americana* L., whose local name is *palo santo*, was studied by Oliveira et al. (2008), who identified the chemical compounds present as triterpenes (friedeline and friedelinol), flavonoids (quercetin and quercetin-3-*O*-α-L-arabinofuranoside), a phenylpropanoid glycoside (vanicoside), an amide (moupamide), and gallic acid. Its application for the treatment of malaria in Peru is supported by the detected high in vivo activity of the ethanol extract of the bark against *Plasmodium vinckei petteri*, as well as its in vitro activity against *Plasmodium falciparum*.

Brazil has a broad and rich biodiversity, which is accompanied by a long-standing acceptance of medicinal plants and traditional knowledge by the population. Herbal medicines are regulated by the National Health Surveillance Agency (ANVISA) and by the Brazilian Agricultural Ministry. Since 2006, Brazil has two current public policies favoring the widespread use of herbal medicines, namely the National Policy on Integrative and Complementary Practices in the Public Health System, and the National Policy on Medicinal Plants and Herbal Medicines. Compounded herbal medicines are prepared in pharmacies according to good manufacturing practices, under authorization by the Health Surveillance secretariats (Carvalho et al. 2014). Despite the wide biodiversity of higher plants native to Brazil, with over 45,000 species, or 20–22% of the total global diversity, Brazil has hardly any medicines near the top of the list of commercially available herbal products. In fact, this market is still only worth about 260 million US dollars, which represents less than 5% of the medicines sold in this country (Dutra et al. 2016). Species such as *Cordia verbenacea* DC. (also named *Varronia verbenacea* (DC.) Borhidi), *Euphorbia tirucalli* L., *Mandevilla velutina* K. Schum., *Phyllanthus spp.*, *Euterpe oleracea* Mart., *Vitis labrusca* L., *Hypericum caprifoliatum* Cham. & Schltdl., *Hypericum polyanthemum* Klotzsch ex Reichardt, *Maytenus ilicifolia* Mart. ex Reissek, *Protium kleinii* Cuatrec., *Protium heptaphylium* (Aubl.) Marchand, *Myracrodruon urundeuva* Allemão, and *Trichilia catigua* A. Juss. were selected for evaluation by Dutra et al. (2016). It was concluded that very few studies have been dedicated to investigation of the mode of action of isolated compounds, with most studies being based on the in vitro and in vivo effects of crude extracts. The authors described the use of *Myracrodruon urundeuva* Allemão, popularly known as *aroeira*, which presents an anti-colitis effect and includes in its composition the compounds β-caryophyllene, euphol, and α,β-amyrin, responsible for this action in mice. Also reported was *Trichilia catigua* A. Juss., a native Brazilian plant commonly used as a neurostimulant and aphrodisiac, known by its folk name *catuaba*, whose chemical composition includes the presence of alkaloids, lactones, β-sitosterol, stigmasterol, and flavalignans.

The following section describes important classes of chemical compounds found in plant species from South America.

3 Important Chemical Groups Found in South American Plant Species

3.1 Alkaloids

The term alkaloid, meaning alkali-like substance, was introduced in 1819 by the pharmacist W. Meissner to describe nitrogenous compounds derived from plants. Alkaloids are a very large and heterogeneous group of compounds that are not only derived from plants, but also from microorganisms, insects, and animals. They are usually basic and often cause a physiological response (Ebadi 2006; Yang and Ren-Sheng 2011).

In plants, alkaloids generally act as a defense against predators, due to their toxicity, bitter flavor, and action on the central nervous system, resulting in improved species survival rates (Matsuura and Fett-Neto 2015). Interestingly, these toxic properties have been useful to indigenous South American populations, who employ a mixture of *Strychnos* species to make curare, a poison used in hunting and warfare (Silva et al. 2005).

Alkaloids have been used in medicine since ancient times to treat a variety of ailments, and remain the subject of research today. Some examples of alkaloids with medicinal properties are morphine (analgesic), derived from *Papaver somniferum* L., ephedrine (anti-asthma), from *Ephedra sinica* Stapf, and vincristine (antitumor), from *Catharanthus roseus* (L.) G. Don.

A variety of alkaloids with pharmacological and economic importance can be found in South America. One example is quinine, obtained from the dried bark of the *Cinchona* tree (Rubiaceae family), which has been used for centuries to treat malaria. In combination with other drugs, quinine is still used to treat uncomplicated malaria, and is also employed as a muscle relaxant and as a flavoring agent in foods and beverages (Achan et al. 2011; Schardein and Macina 2006).

Lycopodium clavatum (L.) and *Lycopodium thyoides* (Humb. & Bonpl. ex Willd) are species from the Lycopodiaceae family that are rich in alkaloids and are used popularly in South America to treat gastrointestinal disorders and to stimulate the central nervous system (Navarrete et al. 2006; Øllgaard and Windisch 2014). Konrath et al. (2012) isolated alkaloids from these two species and observed antioxidant effects and significant inhibition of acetylcholinesterase in *in vitro* and *ex vivo* experiments, making these species candidates for the treatment of neurodegenerative disorders such as Alzheimer's disease.

The roots from the species *Psychotria ipecacuanha* Standl., native to Brazil, mainly contain the alkaloids emetine, cephaeline, and psychotrine. This species, known as *ipecac*, is used in folk medicine as an emetic, amebicide, and expectorant (Daniel 2006). Studies also suggest anti-HIV activity (Valadão et al. 2015) and antitumor activity (Uzor 2016), among other biological effects (Akinboye and Bakare 2011).

Another important alkaloid is pilocarpine, isolated from the leaves of jaborandi (*Pilocarpus microphyllus* Stapf), native to the Amazon region of Brazil. This alkaloid is usually the first choice in cholinergic agents for the initial treatment of open-angle glaucoma, and is also used for the treatment of xerostomia in patients undergoing radiotherapy for cancer of the head and neck (Ebadi 2001; Yang et al. 2016).

Members of the genus *Cassia*, commonly found in the Atlantic forest of Brazil, are widely used as ornamental plants due to the beauty of the flowers. Some species of this genus are popularly used as sources of purgative and anti-inflammatory agents. Piperidine alkaloids, which are major components in the species *C. carnaval* Speg., *C. excelsa* Kunth, and *C. spectabilis* DC., showed *in vitro* inhibitory activity against mutant strains of *Saccharomyces cerevisiae* yeast, as well as analgesic activity in vivo, demonstrating the importance of this species in the search for new drugs (Viegas Junior et al. 2006).

Osorio et al. (2006) compiled a list of various alkaloids from South American species with antiprotozoal activity. Quinoline alkaloids isolated from *Galipea longiflora* K. Krause and *Dictyoloma peruvianum* Planch., species used in Bolivia for the treatment of leishmaniasis, demonstrated *in vitro* activity against *L. braziliensis* and *L. amazonensis*, respectively. Alkaloids from *Galipea officinalis* J. Hancock, a plant native to Venezuela, presented potent *in vitro* activity against *P. falciparum*, with IC_{50} between 0.24 and 6.12 μM. *Peschiera australis* (Müll. Arg.) Miers and the genus *Geissospermum,* both native to South America, are other examples of plants containing antiparasitic alkaloids.

There are more than 8000 natural and derivative alkaloids, and this number grows every year with the discovery of new molecules (Aniszewski 2007). It is clear that alkaloids are important as a source of medicines for the treatment of a variety of diseases, and that South American biodiversity plays an important role in this respect.

3.2 *Phenolic Compounds*

A great number of plants and their isolated compounds have medical applications and are beneficial for human health, helping in the prevention, treatment, and management of diseases such as cancer, diabetes, heart disease, and others. The pharmacological effects of the plants are related to the presence of various categories of chemical compounds, including phenolics, which are responsible for antioxidant activity, associated with the presence of phenols, aldehydes, vitamins, volatile compounds, fatty acids, and tocopherols (Ceylan and Alic 2015).

Phenolic compounds are secondary metabolites of plants that are widely distributed throughout the plant kingdom. They can be classified as simple phenols/benzoquinones (C_6 with 1 phenolic ring), phenolic acids (C_6-C_1, with 1 phenolic ring), condensed tannins, also known as flavolans ((C_6-C_3) $_n$, (C_6) $_n$, and (C_6-C_3-C_6) $_n$, with more than 12 phenolic rings), quinone pigments, flavonoids (C_6-C_3-C_6, with 2 phenolic rings), biflavonoids ((C_6-C_3-C_6)$_2$, with 4 phenolic rings), anthocyanins and anthocyanidines, xanthonoids (C_6-C_1-C_6, with 2 phenolic rings), and stilbene. Phenolic compounds are mostly found in vascular plants, including flowering plants and Gymnosperms, although they are also found in non-vascular land plants (Bryophytes) (Jain et al. 2013). In a recent review, Haminiuk et al. (2012) highlighted the health benefits of these phytochemicals when consumed on a regular daily basis.

Flavonoids are one of the major classes of phenolic compounds that occur naturally in higher plants. López et al. (2015) studied a rain forest fruit named *borojo* (*Borojoa patinoi* Cuatrecasas), native to Colombia, Brazil, and Ecuador, with potential antioxidant and antibacterial activity. HPLC/UV (high performance liquid chromatography with ultraviolet detection) was used to quantify different compounds with valuable biological activity, including flavonoids such as rutin, quercetin, luteolin, apigenin, and luteolin-7-*O*-glucoside, together with other components such as catechin, epi-catechin, and caffeic, ferulic, synapic, p-coumaric, gallic, and chlorogenic acids.

Other important natural products in the phenolic compounds category are anthocyanins, which usually possess antioxidant activity associated with different colors of fruits and vegetables, especially blue, violet, red, and purple. Anthocyanins present different structures, including the aglycone structure (a structure without a sugar ligand, characteristic of anthocyanidins), as well as substituted forms such as glycosides and acylglycosides (Ruiz et al. 2013). An example of a species with a high content of phenolic compounds including flavonoids and anthocyanins is *Arrabidaea chica* (Humb. & Bonpl.) B. Verlt. (Bignoniacea). This liana, found in the Brazilian Amazon rain forest, produces a red-colored dye used by indigenous communities in ritual body painting. The anthocyanins 6,7-dihydroxy-5,4-dimethoxy-flavone and 6,7,4-trihydroxy-5-methoxyflavone of *A. chica* were quantified by HPLC/DAD (diode array detection) by Jorge et al. (2008), who studied their wound healing properties, and more recently by Michel et al. (2015), who investigated their anti-inflammatory, anti-angiogenic, and anti-proliferative properties. Siraichi et al. (2013) determined the antioxidant activity in a hydro-alcoholic extract of plants cultivated in southern Brazil and used HPLC/DAD to detect the flavonoids isoscutellarein, 6-hydroxyluteolin, hispidulin, scutellarein, luteolin, and apigenin. Mafioleti et al. (2013) reported that this species was able to act as an antimicrobial due to its high content of phenolic compounds, concluding that *A. chica* could be used safely. Another species with antimicrobial activity attributed to phenolic compounds is *Ilex paraguariensis* A. St.-Hil., known locally as *yerba mate*, which is widely used in South America (Martin et al. 2013).

Berries and products derived from them usually have high phenolic compound contents and present a variety of biological activities. Berry-producing plants from South America such as *Aristotelia chilensis* (Molina) Stuntz, *Euterpe oleracea* Mart., *Malpighia emarginata* DC., *Ugni molinae* Turcz., *Fragaria chiloensis* (L.) Mill., *Rubus glaucus* Benth., *Rubus adenotrichus* Schltdl., and *Vaccinium floribundum* Kunth. are examples of berries that provide excellent health benefits and can be used as nutritional foods. The phytochemical compositions of these species were described by Schreckinger et al. (2010). In a similar approach, considering fruits with phenolic compounds that confer antioxidant capacity, Denardin et al. (2015) studied *araçá* (*Psidium cattleyanum* Sabine), *butiá* (*Butia eriospatha* (Mart. ex

Drude) Becc.), *pitanga* (*Eugenia uniflora* L.), and *blackberries* (*Rubus sp.*). Andrade et al. (2011) studied the antioxidant and anti-chemotactic potentials of *Myrcianthes pungens* Berg. Legr., known as *guabiyú*, *guabijú*, *guabirá*, *ibaviyú*, and *arrayán* (Myrtaceae), native to Brazil, Argentina, Uruguay, and Paraguay. Cecilia et al. (2015) investigated the phenolic content according to the stage of maturation of *Acca sellowiana* (Berg) Burret, a native Uruguayan species known by the name *guayabo*. Simirgiotis et al. (2013) investigated *Luma apiculata* DC. Burret and *L. chequen* A. Gray, native fruits from Chile and Argentina used to prepare *chicha*, a typical fermented beverage consumed by a group of indigenous inhabitants (Mapuche) of south-central Chile and southwest Argentina, identifying for the first time in these species the compound 3-*O*-(6″-*O*-galloyl)-hexose and derivatives of myricetin, quercetin, laricitrin, and isorhamnetin.

Echinochloa crus-galli (L.) P. Beauv. (Poaceae), *Casearia sylvestris* Swartz (Salicaceae), *Byrsonima verbascifolia* (L.) DC. (Malpighiaceae), *Haplopappus* spp. (Asteraceae), *Prosopis* spp. (Fabaceae), *Myracrodruon urundeuva* Fr. All. (Anacardiaceae), *Salvia officinalis* L. (Lamiaceae), and *Myrciaria dubia* (Kunth) McVaugh (Myrtaceae), known locally as *camu-camu*, are some examples from a long list of species reported in the literature in recent years. Studies concerning phenolic compounds and their composition include the works of Bueno et al. (2015), Castro et al. (2016), Fracassetti et al. (2013), Garcia et al. (2016), Molla et al. (2016), Schmeda-Hirschmann et al. (2015a, b), and Vieira et al. (2015).

3.2.1 Flavonoids

Flavonoids are a class of phenolic compounds synthesized in the phenylpropanoid and acetate pathway from precursors including aliphatic amino acids, terpenoids, and fatty acids. They consist of a skeleton of diphenyl propane (C6C3C6) with two benzene rings (A and B) bonded to a pyran ring (C) (Fig. 1). The flavonoid subclasses are chalcones, dihydrochalcones, aurones, flavones (apegenin, luteolin, diosmetin), flavonols (quercetin, myricetin, kaempferol), dihydroflavonol, flavanones

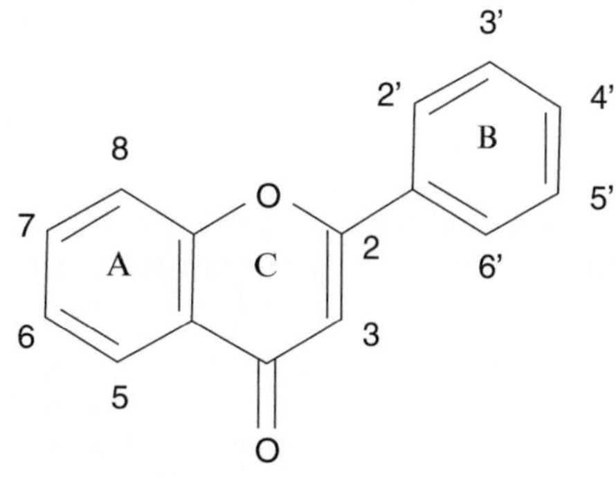

Fig. 1 Basic structure of flavonoids

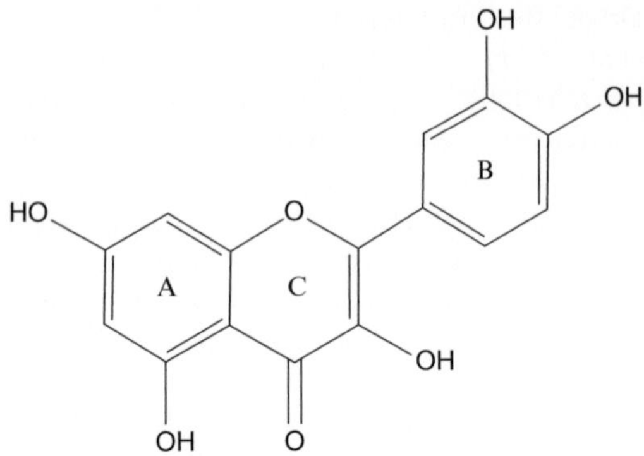

Fig. 2 Chemical structure of quercetin

(naringin, hesperidin), flavanol, flavandiol, isoflavones (genistein, daidzein), bioflavonoids, and proanthocyanins (Behling et al. 2004; Bravo 1998; Mann 1987).

Flavonoids are phytochemicals found in a variety of fruits, vegetables, grains, flowers, and medicinal teas, conferring color, flavor, and aroma, as well as nutritional and health benefits. Many flavonoids have been found to possess antioxidant, anti-inflammatory, anti-hepatotoxic, anti-ulcer, anti-mutagenic, and antidepressant activities in vivo (Behling et al. 2004; Guan and Liu. 2016; Nogueira et al. 2011). They act as protective scavengers against oxygen-derived free radicals by donating an electron to the free radical and converting it into an innocuous molecule. An increasing number of studies suggest that the consumption of fruits, vegetables, and beverages rich in phenolic antioxidants protects against cardiovascular disease and cancer (Haminiuk et al. 2012; Romanucci et al. 2016). Some of the flavonoids consumed are listed below.

Quercetin, the main flavonoid present in the human diet, is one of the most biologically active flavonoids, showing potent antioxidant and anti-inflammatory activities that provide beneficial health effects in cases of chronic illnesses such as cancer and cardiovascular disease (Behling et al. 2004; Park 2004). It is rarely found in plants in a free form, and is usually conjugated to sugar residues. The conjugation of quercetin and other flavonoids affects the mechanism by which the compound is absorbed by altering its basic physicochemical properties and hence its ability to enter cells and interact with transporters and cellular (lipo)proteins (Day and Williamson 2003). Quercetin belongs to the flavonol class, due to its hydroxylation at the 3-position of the C ring (Park 2004), as illustrated in Fig. 2.

Isoflavones, which are present in soybeans and soy foods, have potential health benefits including the prevention of heart disease and cancer, increase of bone mass density to prevent osteoporosis, and reduction of postmenopausal syndromes in women. The main difference between flavonoids and isoflavonoids lies in their basic skeleton structures. Flavonoids contain a 2-phenylchroman, whereas isoflavonoids contain a 3-phenylchroman (Chang 2002), as shown in Fig. 3.

Catechins, also known as tea polyphenols, are found in tea beverages. These compounds have been intensively investigated, with identification of many important biochemical and pharmacological activities. These include antioxidant

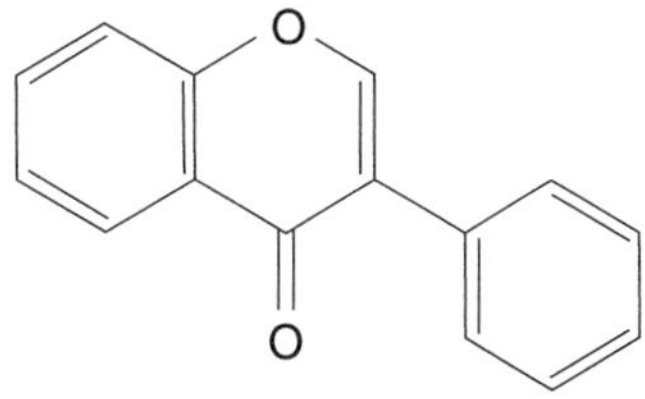

Fig. 3 Chemical structure of isoflavones

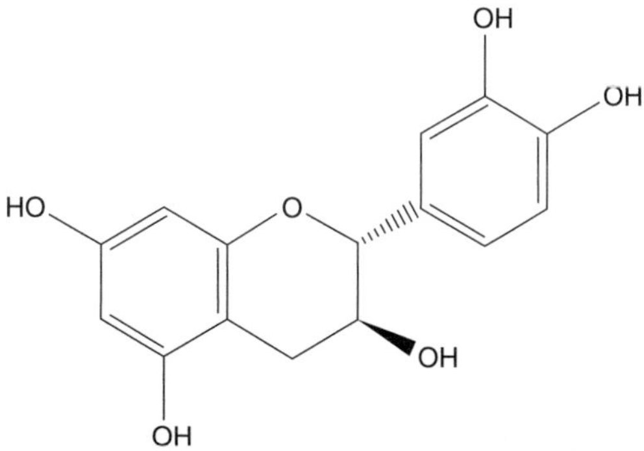

Fig. 4 Chemical structure of catechins

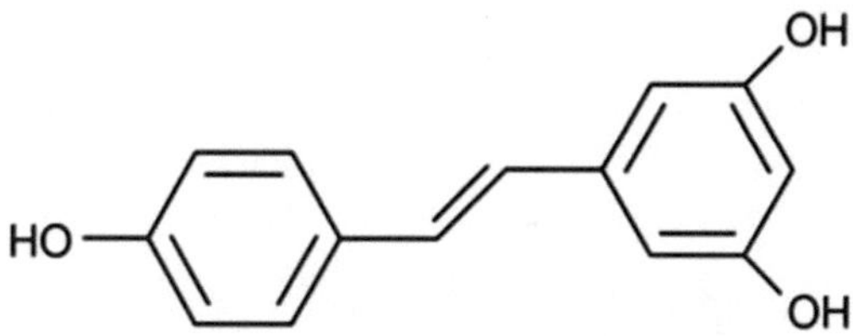

Fig. 5 Chemical structure of resveratrol

and pro-oxidant effects, induction of apoptosis and arrest of the cell cycle in cancer cells, and inhibition of cell proliferation and tumor progression by suppression of the epidermal growth factor receptor signaling pathway (Lin 2005). The catechin structure is shown in Fig. 4.

Resveratrol (Fig. 5), found in grapes and peanuts, has a wide range of beneficial medical activities in humans, including anti-inflammatory, cardiovascular protection, and anticancer effects. It has been shown to modulate the metabolism of lipids and to inhibit the oxidation of low-density lipoproteins and aggregation of platelets (Balanc et al. 2016; Liu et al. 2016).

Flavonoids are an important class of biologically active natural compounds found in the leaves and fruits of many species of plants used for human consumption. The impacts of these compounds on human health are of interest since they can act as chemoprotective adjuvants. One of the species that contains flavonoids is *Baccharis trimera* (Less.) DC., a plant that is widespread in South America and is popularly known in Brazil as *carqueja*. The anti-inflammatory action of *B. trimera* and its inhibitory effects on glutathione S-transferase have been at least partially attributed to the flavones genkwanin, cirsimaritin, hispidulin, and apigenin (de Souza et al. 2016; Nogueira et al. 2011).

Nutraceutical benefits of the extract of *Artemisia arborescens* L. (Asteraceae) have been attributed to the presence of flavonoids and phenolic compounds. The presence of flavonoids within the cell membrane and the resulting restriction on the fluidity of membrane components could hinder the diffusion of free radicals generated during estro-progestative oxidation. HPLC analysis has shown that *A. arborescens* is rich in phenolic acids (catechic, caffeic, epicatechic, vanillic, naringenic, coumaric, and cinamic) and flavonoids (quercetin, rutin, luteolin, kaempferol, and isorhamnetin). The beneficial effects of *A. arborescens* extract can be attributed to its free radical scavenging properties and the presence of polyphenols and flavonoids (Dhibi et al. 2016).

Another species that presents antioxidant activity is *Musa paradisiaca* L. (banana), which is common in most tropical and subtropical areas. Studies using rats fed on normal and high fat diets found that the flavonoids present in banana (catechin, gallocatechin, and epicatechin) acted as effective antioxidants (Singh et al. 2016).

Interest in the anticancer effects of flavonoids has been stimulated by in vitro and in vivo experimental evidence indicating they interfere in cancer processes such as proliferation, inflammation, angiogenesis, invasion, and metastasis. Use of the *Achyrocline* genus (*A. satureioides* and *A. lehmannii* Heiron) has been reported for anticancer therapy. *A. satureioides,* known locally as *macela*, is a medicinal plant grown in southern Brazil and elsewhere in South America. It is widely used in folk medicine as an anti-inflammatory, antibacterial, antispasmodic, digestive, and carminative agent. Most of the biological properties ascribed to *A. satureioides* extracts are related to the presence of flavonoids in its inflorescences. The main flavonoids found in extracts (normally hydroalcoholic preparations) are usually quercetin, luteolin, and 3-O-methylquercetin. The anticancer benefits of flavonoids from *A. satureioides* include effects on cell proliferation, cell cycle, apoptosis, angiogenesis, and migration/metastasis, as well as overcoming multidrug resistance. These effects were observed for flavonoids alone or in combination with commonly used chemotherapeutic drugs (Carini et al. 2014). Another plant native to Brazil, *Mimosa caesalpiniifolia* Benth, popularly known as *sabiá*, exhibits cytotoxic activity against human breast cancer and the ethanolic extract of its leaves is rich in catechins (Silva et al. 2014b).

Aristotelia chilensis (Molina) Stuntz (Eleocarpaceae), commonly known as *maqui berry* or *Chilean wineberry*, is native to Chile and is now distributed throughout tropical and temperate Asia, Australia, the Pacific, and South America. Its juice, which has important astringent, tonic, and antidiarrhoeal properties, is used in folk medicine for wound healing and as an analgesic. The berries are rich in anthocyanins (delphinidins and cyanidins), antioxidants responsible for their purple coloration and for many of the medicinal properties attributed to the plant. The fruits and products derived from them have shown positive effects in several chronic conditions, including obesity, cancer, and cardiovascular and neurodegenerative diseases. The biological properties have been mainly attributed to high levels of various phenolic compounds, as well as to interactive synergies between the natural phytochemical components (Romanucci et al. 2016).

Euterpe oleracea Mart. (Arecaceae) is a plant whose fruit, commonly known as *açaí*, is used in traditional Brazilian folk medicine to treat anemia, diarrhea, malaria, pain, inflammation, hepatitis, and kidney disease. Açaí fruit extracts have been found to induce a vasodilator effect in the rat mesenteric vascular bed, which suggests its possible use in the treatment of cardiovascular diseases. Chemical studies of açaí have shown that this fruit contains polyphenolic components with antioxidant properties, especially bioactive substances such as phenolics, flavonoids (quercetin and kaempferol), and anthocyanins (Marques et al. 2016).

Pyrostegia venusta (Ker Gawl.) Miers (Bignoniaceae), popularly known as *cipó-de-são-joão* is widely distributed in southern Brazil. The parts used in folk medicine include the stem, flowers, leaves, and roots. The aerial parts are used in infusions and decoctions, showing antioxidant, anti-inflammatory, antinociceptive, wound healing, antimicrobial, and melanogenic properties, and are used to treat diarrhea, uterine infections, and vitiligo. These therapeutic properties are associated with the presence of phenolic substances, mainly flavonoids, found in the leaves and stems (Braga et al. 2015; Moreira et al. 2015).

Maytenus ilicifolia Mart. ex Reissek and *M. aquifolia* Mart. (Celastraceae), popularly known as *espinheira-santa*, are widely used in Brazilian folk medicine in the form of aqueous infusions to combat ulcers and stomach diseases. Flavonoids identified in these species, including quercetin and catechins, have been found to be antiulcerogenic and to inhibit gastric acid secretion (Baggio et al. 2007; Marques and Mesia-Vela 2007; Leite et al. 2001).

3.3 Essential Oil Compounds

Terpenes are hydrocarbons present in plants and animals as multiples of a basic structural unit, isoprene (2-methylbuta-1,3-diene, Fig. 6), with the formula $(C_5H_8)_n$. Terpene biosynthesis occurs by the combination of two molecules of acetic acid to produce mevalonic acid, followed by the formation of pyrophosphate isopentenyl. Subsequent transformations of the isopentenyl compound produce terpenes and terpenoids.

The following terpenes have been identified, according to the number of isoprene units present in the molecule: monoterpenes ($C_{10}H_{16}$) such as limonene; sesquiterpenes ($C_{15}H_{24}$) such as bisabolene; diterpenes ($C_{20}H_{32}$) including vitamin A; sesterpenes ($C_{25}H_{40}$); triterpenes ($C_{30}H_{48}$); tetraterpenes ($C_{40}H_{64}$), among which

Fig. 6 Structure of isoprene

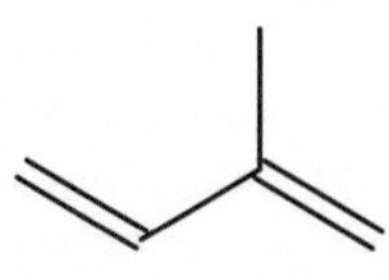

are the carotenoid pigments; and polyterpenes, such as natural rubber, which are composed of between 1000 and 5000 isoprene units.

These compounds can be acyclic, monocyclic, bicyclic, tricyclic, tetracyclic, and pentacyclic, as well as aromatic. Functionalization of the double bonds present in the chemical structures can lead to the formation of alcohols, ketones, aldehydes, esters, and carboxylic acids.

The chemical, physical, and biological properties of terpenes depend on the size of the molecules as well as the functional groups present. They are stored in the leaves, flowers, fruits, stems, and roots of many plants, and are also found in the odoriferous glands of animals. Terpenes are responsible for many of the odors found in nature. Below are some examples of terpenes from South American species and their properties.

Alpha-pinene, an alkene containing a reactive four-membered ring, is found in the oils of many plant species. At low levels of exposure, alpha-pinene is a bronchodilator in humans and is highly bioavailable, with pulmonary absorption of 60%, followed by rapid metabolization and redistribution. It is an anti-inflammatory agent, affecting prostaglandin E1 (PGE1), exhibits acetylcholinesterase inhibitory activity, and serves as auxiliary memory. Examples: *Baccharis dracunculifolia* DC. and Schinus *terebinthifolius* Raddi.

Beta-phellandrene is a cyclic monoterpene that is insoluble in water but soluble in ether. It is used in fragrances, due to its pleasant aroma, which has been described as peppermint. Its isomer can form dangerous and explosive peroxides in contact with air and high temperature. Examples: *Melaleuca alternifolia* Cheel and *Baccharis reticularia* DC.

Sabinene is a bicyclic monoterpene, present in the essential oils of a wide variety of plants. It is one of the substances that contribute to the flavor of black pepper, and is a major constituent of carrot seed oil. Examples: *Poiretia bahiana* Müll. Hal. and *Mikania smilacina* DC.

Beta-caryophyllene (trans-caryophyllene) and gamma-caryophyllene (cis-caryophyllene) are natural bicyclic sesquiterpenes that are present in many essential oils. Caryophyllene is notable for possessing a cyclobutane ring, which is rare in nature. Studies have reported that trans-caryophyllene is a selective agonist for cannabinoid receptor type 2 (CB2) and has significant pharmacological effects in rats, with anti-inflammatory activity. Examples: *Ageratum conyzoides* L., *E. uniflora*, and *C. verbenacea* (beta-caryophyllene); *Siparuna guianensis* Aubl. and *Baccharis crispa* Spreng (gamma-caryophyllene).

Germacrene B belongs to the class of volatile organic hydrocarbons, specifically sesquiterpenes. Germacrenes are produced in a large number of plant species that have antimicrobial and insecticidal properties, but also play a role as pheromones in insects (Matias et al. 2016). Examples: *E. uniflora* and *Myrcia multiflora* (Lam.) DC (Fig. 7).

Essential oils are produced in various genera distributed among 60 botanic families. They can be found in different parts of plants, including the leaves, flowers, fruits, and roots, and can vary in terms of both amount and composition. Essential oils are complex mixtures containing several tens or even hundreds of different

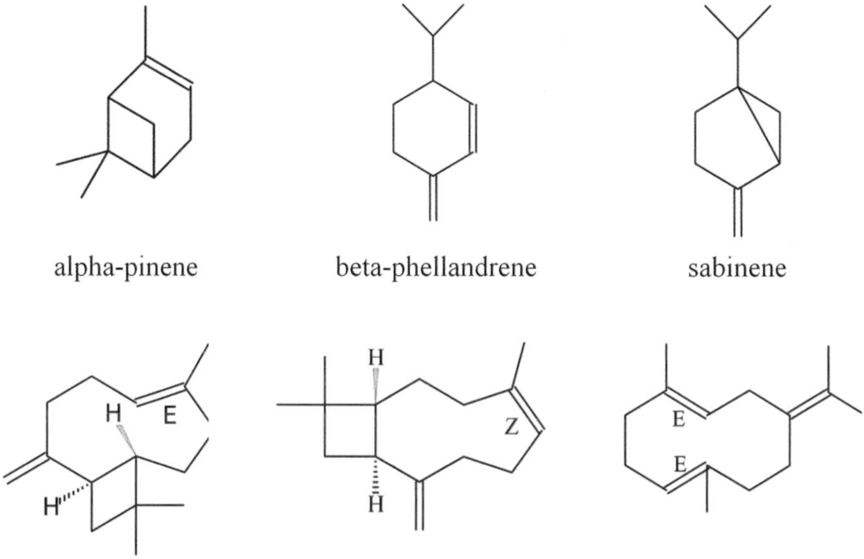

alpha-pinene beta-phellandrene sabinene

beta-caryophyllene gamma-caryophyllene germacrene B

Fig. 7 Terpenes of essential oils

compounds, of which terpenes are the main components. However, there is always a predominance of one to three substances that characterize a particular plant species, giving it a characteristic aroma. Terpenes are extensively used in the perfume, cosmetics, pharmaceuticals, and food industries, with useful compounds obtained from plant families including the Myrtaceae, Lauraceae, Lamiaceae, Asteraceae, and Piperaceae, amongst others (Kurdelas et al. 2012).

Eugenia and *Myrcia*, comprising about 550 and 250 species, respectively, are two of the main species of the genus Myrtaceae distributed in South and Central America. They play an important ecological role in tropical forests, where they provide edible fruits for many animals, and accumulate volatile compounds in their leaves and fruits.

Essential oils from the leaves of Eugenia acutata Miq. (araçá da serra, araçarana, laranjinha-do-cerrado), E. candolleana DC. (murtinha, murta, ameixa da mata, cereja roxa), E. copacabanensis Kiaersk (cambui de copacabana, goiabinha de copacabana, cambuijubá-guaçu), and Myrcia splendens (SW.) DC., present in the Atlantic forest of southeastern Brazil, contain mainly sesquiterpenes as major constituents. E. copacabanensis has the sesquiterpene oxygenates 1,10-di-epi-cubenol, caryophyllene oxide, and epi-α-cadinol as the main components, while E. candolleana contains muurola-4,10(14)-diene-1β-ol, 1-epi-cubenol, globulol, and α-cadinol. Trans-caryophyllene and germacrene D have been found to predominate in E. acutata, and cis-α-bisabolene in M. splendens (Nakamura et al. 2010) (Fig. 8).

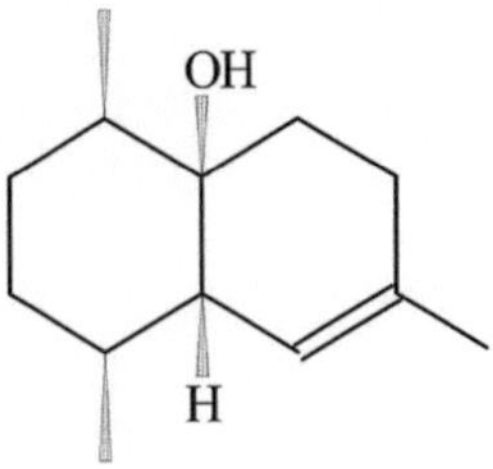

1,10-di-epi-cubenol
(*E. copacabanensis*)

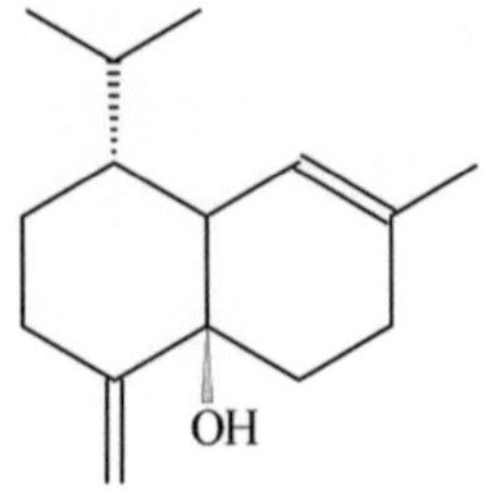

muurola-4,10(14)-diene-1β-ol
(*E. candolleana*)

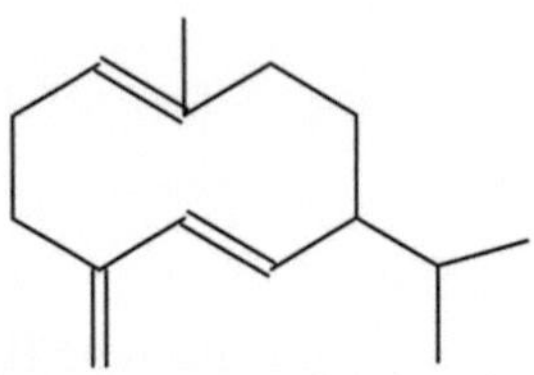

germacrene D
(*E. acutata*)

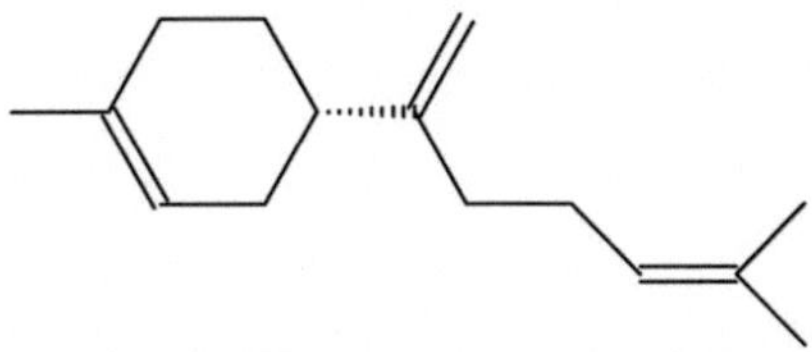

cis-α-bisabolene
(*M. splendens*)

Fig. 8 Terpenes of the genus Myrtaceae

The major compounds identified in the essential oil of *E. uniflora* were germacrene B (22%), selina-1,3,7-trien-8-one-oxide (19%), trans-caryophyllene (13%), germacrene A (11.6%), germacrene D (11.4%), selina-1,3,7-trien-8-one (9.5%), and curzerene (4%). This essential oil shows potent antioxidant activity and has therapeutic potential for the development of phytomedicines with antidepressant and antioxidant properties (Victoria et al. 2013).

Achyrocline alata (Asteraceae) is an aromatic plant of medium size with green leaves that produce small white flowers with a yellow center. It is widely used and found throughout Central and South America. The infusion of the flowers is used as an anti-inflammatory and the dried flowers are used for filling pillows and cushions, due to their calming effects. The great interest in the *Achyrocline* genus plants lies in their abundant biological activities, including analgesic, antimutagenic, anti-inflammatory, antiseptic, antitumor, antiviral, cytotoxic, digestive, hepatoprotective, hypoglycemic, insecticidal, muscle relaxant, sedative, and anthelmintic effects. The major constituents in both the leaves and the flowers are the sesquiterpenes trans-caryophyllene and α-humulene (Rodrigues et al. 2002).

Austroeupatorium inulifolium Kunth (Asteraceae), known as *salvia amarga*, is a plant native to South America and can be found from Panama to Argentina in savannas, swamps, and forests at altitudes of 100–2100 m, and is listed as an "agricultural

Fig. 9 Main compound of *A. inulifolium*

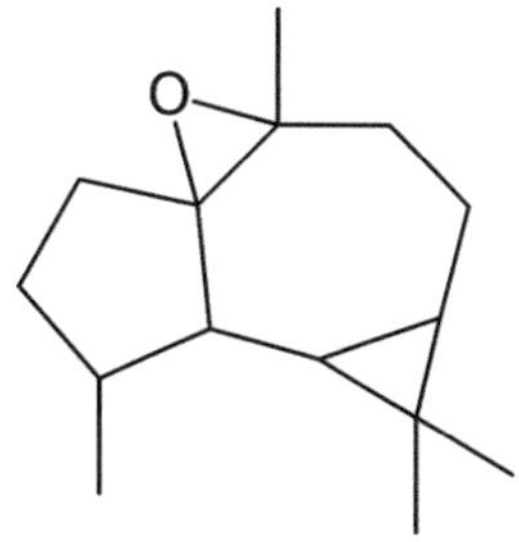

ledene oxide (II)

and environmental weed" in the Global Compendium of Weeds (Randall 2012). It is among the ten plants most widely used in folk medicine in rural areas of the Colombian Andes. It has been found that the essential oil and extracts obtained from this species show insecticidal, antibacterial, and anti-inflammatory activities. The main substances present in the essential oil are trans-caryophyllene and ledene oxide (II) (Tovar et al. 2016) (Fig. 9).

Baccharis, with over 500 species, is the largest genus in the Asteraceae and is mainly found in the warmer regions of Brazil, Argentina, Colombia, Chile, and Mexico. Essential oils from the *Baccharis* genus have been studied in several species from South America. The main compounds in *Baccharis salicifolia* Nutt. essential oil are cis-β-ocimene, germacrene D, muuroladiene, β-cubebene, α-thujene, α-phellandrene, and isoledene. The essential oil from *B. salicifolia* has shown post-ingestive toxicity towards *Spodoptera littoralis* larvae, without antifeedant effects (Sosa et al. 2012).

Carrillo-Hormaza et al. (2015) studied the essential oils from *Ageratina tinifolia*, *Baccharis antioquensis* Killip & Cuatrec., *B. brachylaenoides* DC., *B. tricuneata* (L.f.) Pers, *Diplostephium antioquense* Cuatrec., *D. rosmarinifolium* (Benth.) Wedd., *Pentacalia ledifolia* (Kunth) Cuatrec., and *P. trianae* (Klatt) Cuatrec. (Asteraceae). These species are native to the intertropical montane region of Páramos in Colombia. Eighty components were identified, with more than 45 constituents in each essential oil, including caryophyllene (trans-caryophyllene, α-caryophyllene, and caryophyllene oxide), α-copaene, (Z)-γ-bisabolene, δ-cadinene, and β-sesquiphellandrene. The major sesquiterpene metabolites accounted for percentages of 15.8–72.2%. The findings showed that under the altitude conditions of this eco-geographical area, metabolic diversity within the Asteraceae family was concentrated in this group of metabolites.

The biological properties of several *Lippia* species have been linked to the terpenes found in their essential oils. The viability of the mouse colon carcinoma CT26.WT cell line was significantly reduced following treatment with the essential oils of *L. sidoides* Cham., *L. salviifolia* Cham., and *L. rotundifolia* Cham., respectively. The viability of the human lung carcinoma A549 cell line was decreased by the action of the carvone chemotype essential oil of *L. alba* (Mill.) N.E. Br. ex Britton & P. Wilson. The essential oils did not compromise the viability of the normal CHO cell line (Gomide et al. 2013).

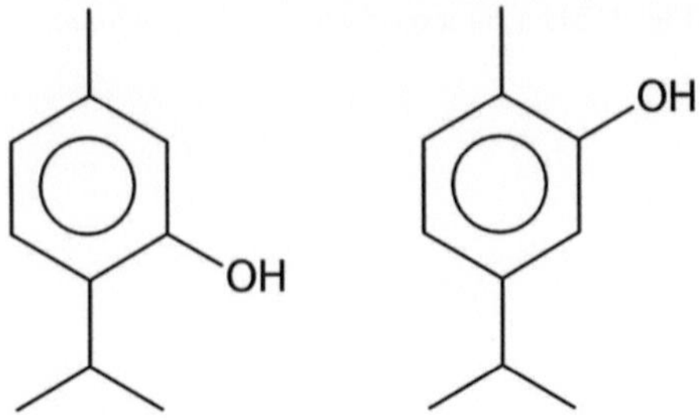

Fig. 10 Compounds present in the essential oil of *L. origanoides*

L. alba (Verbenaceae), known in Brazil as *erva cidreira*, is used in Central and South America as an eupeptic agent for indigestion. In Argentina, it is used by local populations in Chaco province. There are several chemotypes that differ in the chemical compositions of the essential oils. The plant is now cultivated experimentally in several countries of the region. The chemical composition and pharmacology of the essential oils reflect the medicinal usefulness of the *L. alba* chemotypes "citral" (CEO) and "linalool" (LEO), and evaluation of the potency and mechanism of action of these oils has validated their traditional use (Blanco et al. 2013).

L. sidoides (Verbenaceae), popularly known as *alecrim pimenta*, is a species native to northeastern Brazil. The essential oil is rich in thymol (50–70%), a phenolic compound with proven antifungal and antibacterial activity. It has antimicrobial activity against human pathogens that cause caries, as well as anti-inflammatory, leishmanicidal, anthelmintic, acaricidal, insecticidal, and antimalarial activities (Pinto et al. 2016).

Another *Lippia* species, *L. origanoides*, locally known as *orégano de monte* in Colombia, is an aromatic shrub native to northern South America. The major compounds identified in the essential oil are thymol and carvacrol (Vicuña et al. 2010) (Fig. 10).

The essential oils of Amazonian *Croton spp*. (Euphorbiaceae) such as *C. draco* Schltdl. & Cham., *C. zehntneri* Pax & K. Hoffm., *C. nepetifolius* Baill., *C. argyrophylloides* Müll. Arg., *C. urucurana* Baill., *C. cajucara* Benth., and *C. flavens* L. have been found to contain monoterpenes, sesquiterpenes, and diterpenes. The essential oil of *C. lechleri* Müll. Arg., containing sesquicineole, α-calacorene, 1,10-di-epi-cubenol, β-calacorene, and epicedrol, has been studied as a new flavoring ingredient for foods or dietary supplements, providing protection against potential mutagens formed during the cooking and/or processing of food (Rossi et al. 2011) (Fig. 11).

Cordia verbenacea DC (*Varronia verbenacea* (DC.) Borhidi.) (Boraginaceae), popularly known as *erva baleeira*, is a plant found on the Brazilian coast that has been studied in terms of its anti-inflammatory, anti-ulcer and analgesic properties. The essential oil has proven anti-inflammatory action related to the presence of α-humulene and trans-caryophyllene (Matias et al. 2016) (Fig. 12).

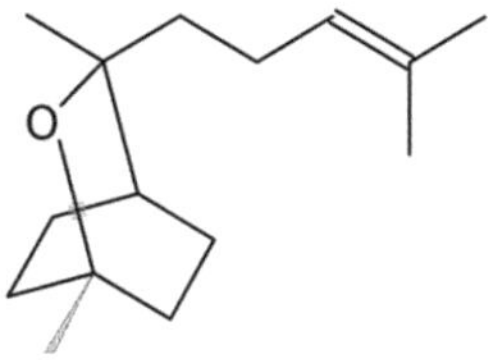

sesquicineole

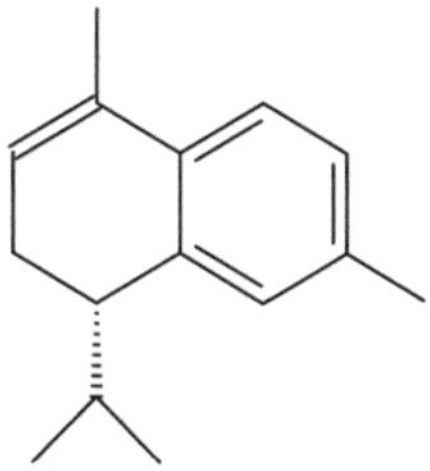

α-calacorene

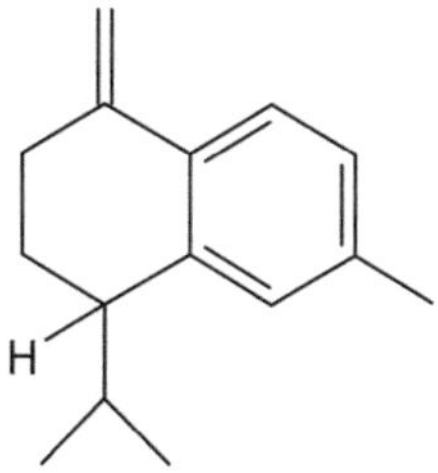

β-calacorene

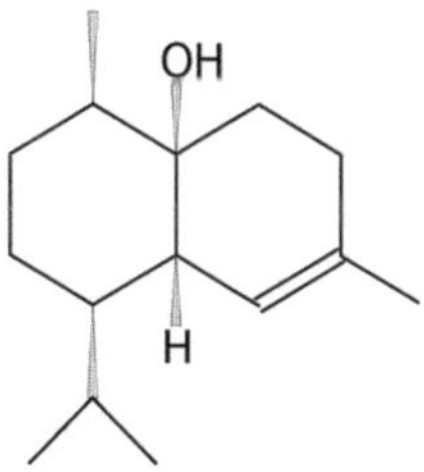

1,10-di-epi-cubenol

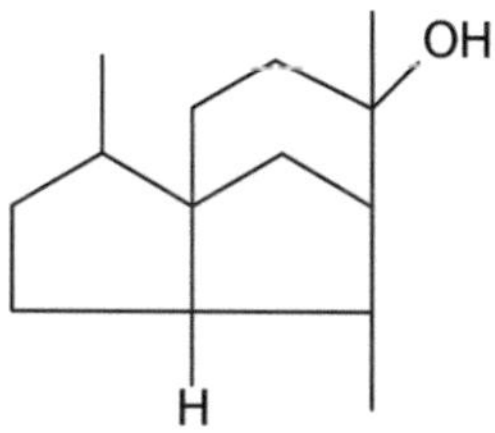

epicedrol

Fig. 11 Terpenes of *C. lechleri*

Fig. 12 Anti-inflammatory agent present in *C. verbenacea*

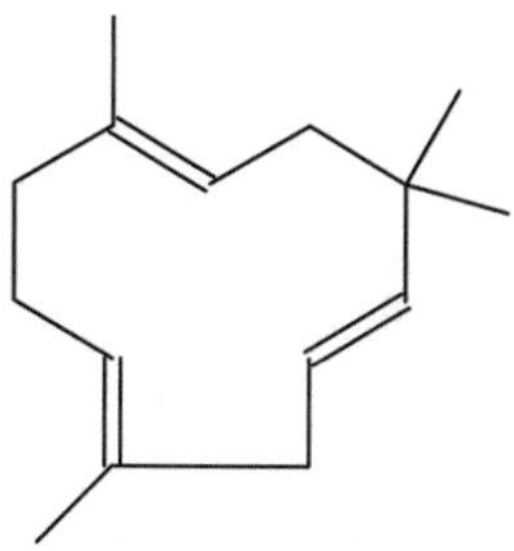

α-humulene

The genus *Cunila* (Lamiaceae) consists of 22 species, 10 native to Mexico and 12 to southern South America. In Brazil, *Cunila* species are found in the states of Rio Grande do Sul, Santa Catarina, and Paraná. The compositions of the essential oils of the South American species vary widely. The oils of *Cunila microcephala* Benth. and *C. fasciculata* Benth. contain high levels of menthofuran, while the main constituents of *C. menthoides* Benth. oil are isomenthone, menthone, and pulegone. The oil of *C. angustifolia* Benth. contains mainly sabinene, γ-terpinene, and limonene. The *C. galioides* Benth. species presents three distinct groups. The citral group, found on the Rio Grande do Sul plateau, contains high concentrations of neral and geranial. The ocimene group, present in high altitude pastures, has high concentrations of trans-β-ocimene, and the menthene group, which is present in the transition area between the two regions, contains 1,8-cineole, trans-2,8-p-menthadien-1-ol, 1,3,8-menthatriene, and 1,5,8-p-menthatriene as the main components (Echeverrigaray et al. 2003) (Fig. 13).

Minthostachys verticillata (Griseb.) Epling (Lamiaceae), known as *peperina*, is a South American aromatic and medicinal plant used to treat indigestion, vomiting, diarrhea, and abdominal pain. It is also known for its carminative and anti-rheumatic properties. It is used as an infusion or added to mate tea. The beneficial effects are attributed to its essential oil, whose main components are the monoterpenes pulegone, menthone, isomenthone, and limonene. The oil also contains smaller amounts

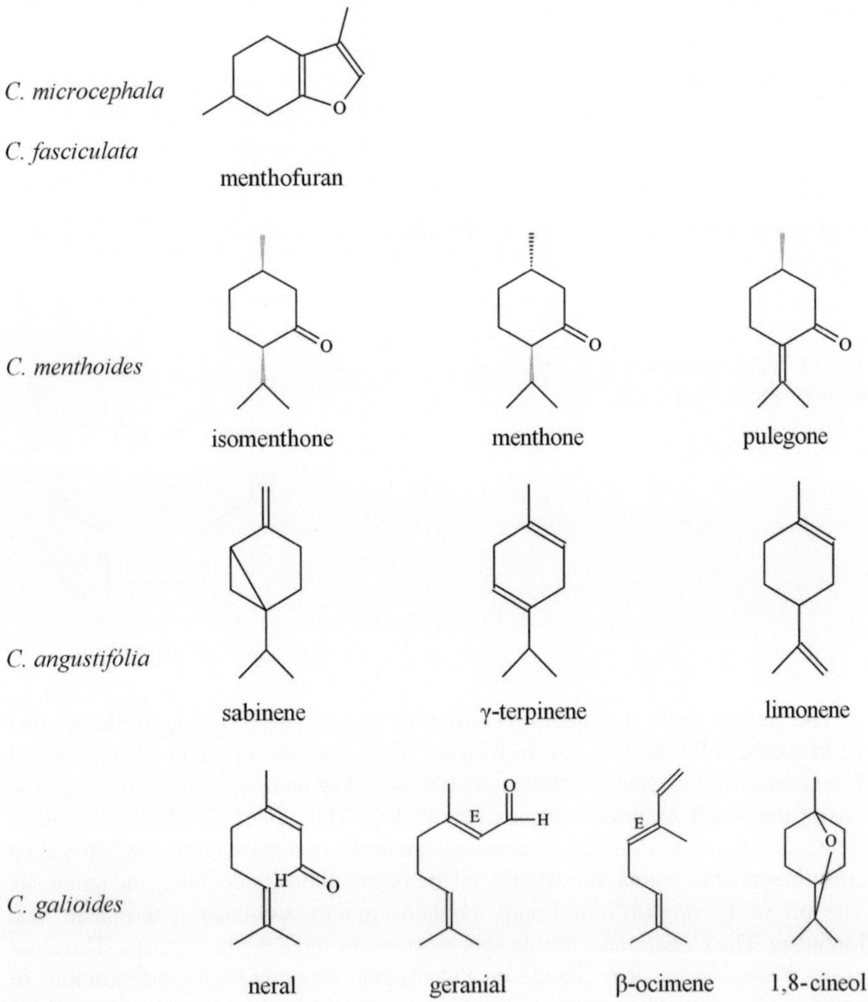

Fig. 13 Terpenes of the genus *Cunila*

of menthol, α-pinene, β-pinene, carvone, piperitenone, sabinene, myrcene, β-ocimene, thymol, and carvacrol (Escobar et al. 2015).

The Piperaceae family includes ten genera, of which five occur in shaded tropical regions of Brazil, and the *Piper* genus has 265 described species. Analysis of the essential oils of *Piper hispidum* Kunth, *P. aleyreanum* C. DC., and *P. anonifolium* (Kunth) C. DC., species commonly found in the Brazilian Amazon, revealed that the sesquiterpenes were the most highly represented classes and that the main compounds present were selin-11-en-4-α-ol, β-elemene, β-selinene, α-selinene, bicyclogermacrene, β-caryophyllene, α-humulene, and δ-elemene. The oils evaluated showed antifungal activity, in vitro cytotoxic activity against human melanoma cells, and antioxidant activity. The cell growth inhibition induced by the oil of *P. aleyreanum* is due to elemene (β-, δ-, and γ-elemene), which has previously been reported to inhibit proliferation, stimulate apoptosis, and interrupt the cell cycle in malignant cells (Silva et al. 2014a).

In *P. corcovadensis* (Miq.) C. DC., also known as *falso jaborandi*, the major constituents of the essential oil have been identified as 1-butyl-3,4-methylenedioxybenzene, terpinolene, trans-caryophyllene, α-pinene, δ-cadinene, and limonene. The leaf oil, terpinolene, and 1-butyl-3,4-methylenedioxybenzene have shown activity against larvae of the dengue fever mosquito (Aedes aegypti), interfering in the activity of proteases from the L4 gut enzymes. The essential oil also exhibited oviposition deterrent activity (Silva et al. 2016).

Another member of the *Piper* genus, *P. angustifolium* Lam., popularly known as *pimenta-de-macaco*, among other names in Brazil, has the compounds spathulenol and caryophyllene oxide as the main components of the essential oil. The oil has shown in vitro antileishmanial activity, suggesting its potential use as a drug to treat visceral leishmaniasis (Bosquiroli et al. 2015).

The essential oils of three species of *Piper* (*P. hispidum*, *P. anonifolium*, and *P. aleyreanum*), originated from the Carajás National Forest in Pará State, Brazil, show a predominance of sesquiterpenes in their compositions. In *P. hispidum*, the major analytes found were trans-caryophyllene and α-humulene. *P. aleyreanum* showed β-elemene, bicyclogermacrene, and δ-elemene. *P. anonifolium* contained selin-11-en-4-α-ol, β-selinene, and α-selinene. All the oils analyzed showed strong antifungal activity, with minimum inhibitory concentrations (MIC) of 0.1 to 1.0 μg against *Cladosporium cladosporioides* and *C. sphareospermum*. In an anticholinesterase evaluation, the oils of *P. anonifolium* (MIC = 0.01 ng) and *P. hispidum* (MIC = 0.01 ng) were 100-fold more potent than the standard physostigmine (MIC = 1.0 ng). The *P. aleyreanum* oil showed high in vitro cytotoxic activity against the human melanoma SKMEL-19 cell line (IC_{50} = 7.4 μg/mL) and significant antioxidant activity (DPPH = 412.2 mg TE/mL). The cell growth inhibition induced by *P. aleyreanum* oil is probably due to the presence of elemenes (β-, δ- and γ-elemene), which have been previously reported to inhibit the proliferation, stimulate the apoptosis, and interrupt the cell cycle in malignant cells (Silva et al. 2014a).

The *Annonaceae* family is represented by 29 genera and 386 species, with 27 genera and 280 species present in the Amazon region. Two genera and about 40 species are endemic to the Atlantic forest, and 10 genera and 47 species are found in the

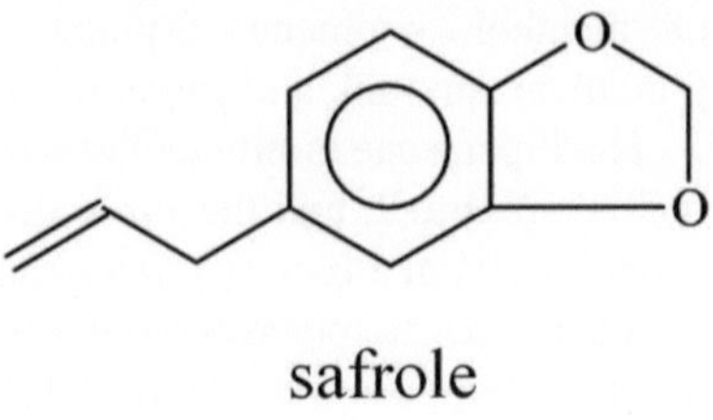

Fig. 14 Main compound present in *L. sempervirens* essential oil

Brazilian cerrado. In the essential oil of the medicinal plant *Xylopia frutescens* Aubl., popularly known as *embira*, the major compounds present are (E)-caryophyllene, bicyclogermacrene, germacrene D, δ-cadinene, viridiflorene, and α-copaene, which has interesting anticancer activity (Ferraz et al. 2013).

The essential oil of *Laurelia sempervirens* (Ruiz Pav.) Tul. (Atherospermataceae), a native Chilean tree, which contains safrole as the main compound, is used as a contact insecticide and a fumigant against stored grain pests and the pea aphid (Zapata et al. 2010) (Fig. 14).

Schinus molle L. (Anacardiaceae), commonly known as *pimenta rosa*, is native to subtropical regions of South America. The compounds present in leaf and fruit essential oil include α-phellandrene, β-phellandrene, β-myrcene, limonene, and α-pinene. This oil shows antioxidant and antimicrobial properties, and has potential for use in the food and pharmaceutical sectors (Martins et al. 2014).

4 Concluding Remarks

Medicinal plants can be used in alternative therapies for the treatment of various diseases, and the biodiversity present in South America offers a promising source of new drug models and new phytomedicines. The historical ethnobotanical knowledge of the population reveals the benefits of the use of herbs to treat a wide range of diseases. Herbs with stomachic, digestive, and tonic activities are used for gastrointestinal complaints such as dyspepsia and dysentery. Herbal preparations are employed as sedatives to treat central nervous system or general pains, to help against fever, to prevent problems in the genitourinary tract, and against dermatological disorders, respiratory system problems, nausea, and anemia. Their properties include carminative, anti-inflammatory, antispasmodic, and anti-tubercular activities. They can be used to assist the healing of broken bones, damaged tendons, wounds, and ulcerations, as well as to treat lung congestion and as anthelminthic agents.

Despite the knowledge of the population, there have been very few studies of the mechanisms of action of isolated compounds, with most investigations only considering the effects of crude extracts.

Popular knowledge sometimes neglects the toxicity of herbal medicines, and it is important to mention that although plants can cure, they can also be highly aggressive towards the human body. A lack of knowledge of vendors concerning potential side effects can result in unsatisfactory practices within the community.

Finally, it is clear that the biodiversity present in South America constitutes one of the subcontinent's most valuable features. Most importantly, this ecosystem still contains vast intact wild areas that should provide future opportunities for new developments in the application of natural substances derived from plants for medicinal purposes.

References

Achan J, Talisuna AO, Erhart A, Yeka A, Tibenderana JK, Baliraine FN, Rosenthal PJ, D'Alessandro U (2011) Quinine, an old anti-malarial drug in a modern world: role in the treatment of malaria. Malar J 10:144

Akinboye ES, Bakare O (2011) Biological activities of emetine. Open Nat Prod J 4(1):8–15

Alviz AA, Salas RD, Franco LA (2013) Efecto diurético agudo de los extractos etanólico y acuoso de *Ceratopteris pteridoides* (Hook) en ratas normales. Biomedica 33(1):115–121

Andrade JMM, Aboy A, Apel MA, Raseira MCB, Pereira JFM, Henriques AT (2011) Phenolic composition in different genotypes of Guabiju fruits (Myrcianthes pungens) and their potential as antioxidant and antichemotactic agents. J Food Sci 76(8):1181–1187

Aniszewski T (2007) Definition, typology and occurrence of alkaloids. In: Alkaloids – secrets of life: aklaloid chemistry, biological significance, applications and ecological role, 1st edn. Elsevier Science, Amsterdam, p 5

Arias RFQ (2012) Study of medicinal plants used by the Tikunas indigenous community of the upper Amazon (Macedonia). Ther Nova 10(18):135–250

Baggio CH, Freitas CS, Otofuji GM, Cipriani TR, Souza LM, Sassaki GL, Iacomini M, Marques MCA, Mesia-Vela S (2007) Flavonoid-rich fraction of Maytenus ilicifolia Mart. Ex. Reiss protects the gastric mucosa of rodents through inhibition of both H+, K+-ATPase activity and formation of nitric oxide. J Ethnopharmacol 181:433–440

Balanč B, Trifković K, Đorđević V, Marković S, Pjanović R, Nedović V, Bugarski B (2016) Novel resveratrol delivery systems based on alginatesucrose and alginate-chitosan microbeads containing liposomes. Food Hydrocoll 61:832–842

Behling EB, Sendão MC, Francescato HDC, Antunes LMG, Bianchi MLP (2004) Flavonóide quercetina: aspectos gerais e ações biológicas. Alim Nutr Araraquara 15(3):285–292

Blanco MA, Colareda GA, van Baren C, Bandoni AL, Ringuelet J, Consolini AE (2013) Antispasmodic effects and composition of the essential oils from two South American chemotypes of *Lippia alba*. J Ethnopharmacol 149(3):803–809

Bosquiroli LSS, Demarque DP, Rizk YS, Cunha MC, Marques MCS, Matos MFC, Kadri MCT, Carollo CA, Arruda CCP (2015) In vitro anti-Leishmania infantum activity of essential oil from *Piper angustifolium*. Rev Bras Farmacogn 25(2):124–128

Braga KQ, Coimbra MC, Castro AHF (2015) In vitro germination, callus induction and phenolic compounds contents from *Pyrostegia venusta* (Ker Gawl.). Miers Acta Sci Biol Sci 37(2):151–158

Bravo L (1998) Polyphenols: chemistry, dietary sources, metabolism and nutritional significance. Nutr Rev 56(11):317–333

Bueno PCP, Pereira FMV, Torres RB, Cavalheiro AJ (2015) Development of a comprehensive method for analyzing clerodane-type diterpenes and phenolic compounds from Casearia sylvestris Swartz (Salicaceae) based on ultra high performance liquid chromatography combined with chemometric tools. J Sep Sci 38(10):1649–1656

Bueno-Sánchez JG, Martínez-Morales JR, Stashenko EE, Ribón W (2009) Anti-tubercular activity of eleven aromatic and medicinal plants occurring in Colombia. Biomedica 29(1):51–60
Carini JP, Klamt F, Bassani VL (2014) Flavonoids from Achyrocline satureioides: promising biomolecules for anticancer therapy. RSC Adv 4:3131–3144
Carrillo-Hormaza L, Mora C, Alvarez R, Alzate F, Osorio E (2015) Chemical composition and antibacterial activity against Enterobactercloacae of essential oils from Asteraceae species growing in the Páramos of Colombia. Ind Crop Prod 77:108–115
Carvajal-De Pabón LM, Turbay S, Álvarez LM, Rodríguez A, Alvarez JM, Bonilla K, Restrepo S, Parra M (2014) Relationship between the folk uses of the granadilla plant (Passiflora ligularis Juss) and its phytochemical composition. Rev Bio Agro 12(2):185–196
Carvalho ACB, Ramalho LS, Marques RFO, Perfeito JPS (2014) Regulation of herbal medicines in Brazil. J Ethnopharmacol 158:503–506
Castro AHF, Braga KQ, Sousa FM, Coimbra MC, Chagas RCR (2016) Callus induction and bioactive phenolic compounds production from Byrsonima verbascifolia (L.) DC. (Malpighiaceae). Rev Ciênc Agron 47(1):143–151
Cecilia SA, Dennise O, Fernada Z, Mercedes R (2015) Determination of some quality attributes of Feijoa fruits [Acca sellowiana (Berg) Burret] at different ripening stages. Agrociencia Uruguay 19(1):24–30
Ceylan O, Alıc H (2015) Antibiofilm, antioxidant, antimutagenic activities and phenolic compounds of Allium orientale BOISS. Braz Arch Biol Technol 58(6):935–943
Chang SKC (2002) Isoflavones from soybeans and soy foods. In: Mazza G, Le Maguer M, Shi J (eds) Functional foods: biochemical and processing aspects. CRC Press, New York
Daniel M (2006) Alkaloids. In: Medicinal plants: chemistry and properties. Enfield, Science Publishers, pp 10–55
Day AJ, Williamson G (2003) Absorption of quercetin glycosides. In: Rice-Evans CA, Packer L (eds) Flavonoids in health and disease. CRC Press, New York
De Souza LM, Dartora N, Scoparo CT, Gorin PAJ, Iacomini M, Sassaki GL (2016) Differentiation of flavonol glucoside and galactoside isomers combining chemical isopropylidenation with liquid chromatography–mass spectrometry analysis. J Chromatogr A 1447:64–71
Denardin CC, Hirsch GE, Rocha RF, Vizzotto M, Henriques AT, Moreira JCF, Guma FTCR, Emanuelli T (2015) Antioxidant capacity and bioactive compounds of four Brazilian native fruits. J Food Drug Anal 23(3):387–398
Dhibi S, Bouzenna H, Samout N, Tlili Z, Elfeki A, Hfaiedh N (2016) Nephroprotective and antioxidant properties of Artemisia arborescens hydroalcoholic extract against oestroprogestative-induced kidney damages in rats. Biomed Pharmacother 82:520–527
Dutra RC, Campos MM, Santos ARS, Calixto JB (2016) Medicinal plants in Brazil: pharmacological studies, drug discovery, challenges and perspectives. Pharmacol Res 112:4–29
Ebadi M (2001) Pilocarpine and glaucoma. In: Pharmacodynamic basis of herbal medicine. CRC Press, Boca Raton
Ebadi M (2006) Alkaloids. In: Pharmacodynamic basis of herbal medicine. CRC Press, Boca Raton, pp143–49
Echeverrigaray S, Fracaro F, Santos ACA, Paroul N, Wasum R, Serafini LA (2003) Essential oil composition of south Brazilian populations of *Cunila galioides* and its relation with the geographic distribution. Biochem Syst Ecol 31(5):467–475
Escobar FM, Sabini MC, Cariddi LN, Sabini LI, Mañas F, Cristofolini A, Bagnis G, Gallucci MN, Cavaglieri LR (2015) Safety assessment of essential oil from *Minthostachys verticillata* (Griseb.) Epling (peperina): 90-days oral subchronic toxicity study in rats. Regul Toxicol Pharmacol 71(1):1–7
Ferraz RPC, Cardoso GMB, Silva TB, Fontes JEN, Prata APN, Carvalho AA, Moraes MO, Pessoa C, Costa EV, Bezerra DP (2013) Antitumor properties of the leaf essential oil of *Xylopia frutescens* Aubl. (Annonaceae). Food Chem 141(1):196–200
Fonnegra-Gómez R, Villa-Londoño J (2011) Medicinal plants used in some townships of municipalities in the high plains of eastern Antioquia, Colombia. Actual Biol 33(95):219–250

Fracassetti D, Costa C, Moulay L, Tomás-Barberán FA (2013) Ellagic acid derivatives, ellagitannins, proanthocyanidins and other phenolics, vitamin C and antioxidant capacity of two powder products from camu-camu fruit (*Myrciaria dubia*). Food Chem 139:578–588

Franco LA, Matiz GE, Calle J, Pinzón R, Ospina LF (2007) Actividad antinflamatoria de extractos y fracciones obtenidas de cálices de *Physalis peruviana* L. Biomedica 27(1):110–115

Garcia SCS, Menti C, Lambert APF, Barcellos T, Moura S, Calloni C, Branco CS, Salvador M, Roesch-Ely M, Henriques JAP (2016) Pharmacological perspectives from Brazilian *Salvia officinalis* (*Lamiaceae*): antioxidant, and antitumor in mammalian cells. An Acad Bras Cienc 88(1):281–292

Gomide MS, Lemos FO, Lopes MTP, Alves TMA, Viccini LF, Coelho CM (2013) The effect of the essential oils from five different Lippia species on the viability of tumor cell lines. Rev Bras Farmacogn 23(6):895–902

Gonzales GF, Valerio LG Jr (2006) Medicinal plants from Peru: a review of plants as potential agents against Cancer. Anti Cancer Agents Med Chem 6(5):429–444

Grassi-Zampieron R, França LV, Carollo CA, Vieira MC, Oliveros-Bastidas A, Siqueira JM (2010) Comparative profiles of *Achyrocline alata* (Kunth) DC. and *A. satureioides* (Lam.) DC., *Asteraceae*, applying HPLC-DAD-MS. Rev Bras Farmacogn 20(4):575–579

Guan LP, Liu BY (2016) Antidepressant-like effects and mechanisms of flavonoids and related analogues. Eur J Med Chem 121:47–57

Hajdu Z, Hohmann J (2012) An ethnopharmacological survey of the traditional medicine utilized in the community of Porvenir, Bajo Paragua Indian reservation, Bolivia. J Ethnopharmacol 139(3):838–857

Haminiuk CWI, Maciel GM, Plata-Oviedo MSV, Peralta RM (2012) Phenolic compounds in fruits – an overview. Int J Food Sci Technol 47(10):2023–2044

Jain P, Jain S, Pareek A, Sharma S (2013) A comprehensive study on the natural plant phenols: perception to current scenario. Bull Pharm Res 3(2):90–106

Jorge MP, Madjarof C, Ruiz ALTG, Fernandes AT, Rodrigues RAF, Sousa IMO, Foglio MA, Carvalho JE (2008) Evaluation of wound healing properties of *Arrabidaea chica* Verlot extract. J Ethnopharmacol 118(3):361–366

Konrath EL, Neves BM, Lunardi PS, Passos CS, Simoes-Pires A, Ortega MG, Gonçalves CA, Cabrera JL, Moreira JC, Henriques AT (2012) Investigation of the in vitro and ex vivo acetylcholinesterase and antioxidant activities of traditionally used Lycopodium species from South America on alkaloid extracts. J Ethnopharmacol 139(1):58–67

Kurdelas RR, López S, Lima B, Feresin GE, Zygadlo J, Zacchino S, López ML, Tapia A, Freile ML (2012) Chemical composition, anti-insect and antimicrobial activity of *Baccharis darwinii* essential oil from Argentina, Patagonia. Ind Crop Prod 40:261–267

Lagos-López MI (2007) Ethnobotany study of plant species with medicinal properties in six municipalities de Boyacá, Colombia. Actual Biol 29(86):87–96

Leite JPV, Rastrelli L, Romussi G, Oliveira AB, Vilegas JHY, Vilegas W, Pizza C (2001) Isolation and HPLC quantitative analysis of flavonoid glycosides from Brazilian beverages (*Maytenus ilicifolia* and *M. aquifolium*). J Agric Food Chem 49(8):3796–3801

Lin GK (2005) Catechins and inhibitory activity against carcinogenesis. In: Awad AB, Peter G (eds) Bradford nutrition and cancer prevention. CRC Press, New York

Liu Y, Nan L, Liu J, Yan H, Zhang D, Han X (2016) Isolation and identification of resveratrol-producing endophytes from wine grape cabernet sauvignon. Springerplus 5(1):1029

López CC, Mazzarrino G, Rodríguez A, Fernández-López J, Pérez-Álvarez JA, Viuda-Martos M (2015) Assessment of antioxidant and antibacterial potential of borojo fruit (*Borojoa patinoi* Cuatrecasas) from the rainforests of South America. Ind Crop Prod 63:79–86

Lorenzi H, Matos FJA (2008) Plantas medicinais no Brasil: nativas e exóticas. Plantarum, Nova Odessa, p 544

Mafioleti L, Silva Junior IF, Colodel EM, Flach A, Martins DTO (2013) Evaluation of the toxicity and antimicrobial activity of hydroethanolic extract of *Arrabidaea chica* (Humb. &Bonpl.) B.Verl. J Ethnopharmacol 150(2):576–582

Mann J (1987) Secondary metabolism. Clarendon Press, Oxford, p 374
Marques MCA, Mesia-Vela S (2007) Flavonoid-rich fraction of *Maytenus ilicifolia* Mart. ex. Reiss protects the gastric mucosa of rodents through inhibition of both H^+, K^+-ATPase activity and formation of nitric oxide. J Ethnopharmacol 113(3):433–440
Marques ES, Froder JG, Carvalho JCT, Rosa PCP, Perazzo FF, Maistro EL (2016) Evaluation of the genotoxicity of *Euterpe oleraceae* Mart. (Arecaceae) fruit oil (açaí), in mammalian cells in vivo. Food Chem Toxicol 93:13–19
Martin JGP, Porto E, Alencar SM, Glória EM, Corrêa CB, Cabral ISR (2013) Antimicrobial activity of yerba mate (*Ilex paraguariensis* St. Hil.) against food pathogens. Rev Argent Microbiol 45(2):93–98
Martins MR, Arantes S, Candeias F, Tinoco MT, Moraes JC (2014) Antioxidant, antimicrobial and toxicological properties of *Schinus molle* L. essential oils. J Ethnopharmacol 151(1):485–492
Matias EFF, Alves EF, Silva MKN, Carvalho VRA, Figueredo FG, Ferreira JVA, Coutinho HDM, Silva JMFL, Ribeiro-Filho F, Costa JGM (2016) Seasonal variation, chemical composition and biological activity of the essential oil of *Cordia verbenacea* DC (Boraginaceae) and the sabinene. Ind Crop Prod 87:45–53
Matsuura HN, Fett-Neto AG (2015) Plant alkaloids: main features, toxicity, and mechanisms of action. In: Gopalakrishnakone P, Carlini CR, Ligabue-Braun R (eds) Plant toxins. Springer, Netherlands
Michel AFRM, Melo MM, Campos PP, Oliveira MS, Oliveira FAZ, Cassali GD, Ferraz VP, Cota BB, Andrade SP, Souza-Fagundes EM (2015) Evaluation of anti-inflammatory, antiangiogenic and antiproliferative activities of *Arrabidaea chica* crude extracts. J Ethnopharmacol 165:29–38
Molla SGE, Motaal AA, Hefnawy HE, Fishawy AE (2016) Cytotoxic activity of phenolic constituents from Echinochloa crus-galli against four human cancer cell lines. Rev Bras 26(1):62–67
Moreira CG, Carrenho LZB, Pawloski PL, Soley BS, Cabrini DA, Otuki MF (2015) Pre-clinical evidences of *Pyrostegia venusta* in the treatment of vitiligo. J Ethnopharmacol 168:315–325
Nakamura MJ, Monteiro SS, Bizarri CHB, Siani AC, Ramos MFS (2010) Essential oils of four Myrtaceae species from the Brazilian southeast. Biochem Syst Ecol 38(6):1170–1175
Navarrete H, León B, Gonzales J, Aviles DK, Lecaro JS et al (2006) Helechos. In: Moraes R, Øllgaard B, Kvist LP, Borchsenius F, Balslev H (eds) Botánica Económica de los Andes Centrales. Universidad Mayor de San Andrés, La Paz, pp 385–411
Nogueira NPA, Reis PA, Laranja GAT, Pinto AC, Aiub CAF, Felzenszwalb I, Paes MC, Bastos FF, Bastos VLFC, Sabino KCC, Coelho MGP (2011) In vitro and in vivo toxicological evaluation of extract and fractions from *Baccharis trimera* with anti-inflammatory activity. J Ethnopharmacol 138(2):513–522
Oliveira PES, Conserva LM, Lemos RPL (2008) Chemical constituents from Triplaris americana L. (Polygonaceae). Biochem Syst Ecol 36(2):134–137
Øllgaard B, Windisch PG (2014) Lycopodiaceae in Brazil. Conspectus of the family I. The genera Lycopodium, Austrolycopodium, Diphasium, and Diphasiastrum. Rodriguésia 65(2):293–309
Osorio D, Edison J, Montoya P, Guillermo L, Arango A, Gabriel J (2006) Productos Naturales Alcaloidales com Actividad Antiprotozoaria. Vitae 13(1):61–84
Park JB (2004) Quercetin. In: Coates PM, Blackman MR, Cragg GM, Levine M, Moss J, White JD (eds) Encyclopedia of dietary supplements. CRC Press, New York
Pinto NOF, Rodrigues THS, Pereira RCA, Silva LMA, Cáceres CA, Azeredo HMC, Muniz CR, Brito ES, Canuto KM (2016) Production and physico-chemical characterization of nanocapsules of the essential oil from *Lippia sidoides* Cham. Ind Crop Prod 86:279–288
Quintero SEG, Lizarazú MCB, Robayo AM, Lobo AZP, Molano LG (2015) Traditional use of medicinal plants in markets from Bogotá, D.C. Nova 13(23):73–80
Randall RP (2012) A global compendium of weeds, 2nd edn. Department of Agriculture and Food, Western Australia

Ribeiro AO, Silva AF, Castro AHF (2010) Identificação de espécies da família *Asteraceae*, revisão sobre usos e triagem fitoquímica do gênero *Eremanthus* da Reserva Boqueirão, Ingaí-MG. Rev Bras PlMed 12(4):456–465

Rodrigues RAF, Queiroga CL, Rodrigues MVN, Foglio MA, Sartoratto A, Montanari I Jr (2002) Study of the variation of the composition of the essential oil of leaves and flowers of *Achyrocline alata* (D.C.) along a period of the day. J Essent Oil Res 14(4):280–281

Romanucci V, D'Alonzo D, Guaragna A, Marino C, Davinelli S, Scapagnini G, Di Fabio G, Zarrelli A (2016) Bioactive compounds of Aristotelia chilensis stuntz and their pharmacological effects. Curr Pharm Biotechnol 17(6):513–523

Rossi D, Guerrini A, Maietti S, Bruni R, Paganetto G, Poli F, Scalvenzi L, Radice M, Saro K, Sacchetti G (2011) Chemical fingerprinting and bioactivity of Amazonian Ecuador *Croton lechleri* Müll. Arg. (Euphorbiaceae) stem bark essential oil: a new functional food ingredient? Food Chem 126(3):837–848

Ruiz A, Hermosín-Gutiérrez I, Vergara C, von Baer D, Zapata M, Hitschfeld A, Obando L, Mardones C (2013) Anthocyanin profiles in south Patagonian wild berries by HPLC-DAD-ESI-MS/MS. Food Res Int 51(2):706–713

Schardein JL, Macina OT (2006) Quinine. In: Human developmental toxicants: aspects of toxicology and chemistry. CRC Press, Boca Raton

Schmeda-Hirschmann G, Quispe C, González B (2015a) Phenolic profiling of the South American "Baylahuen" tea (Haplopappus spp., Asteraceae) by HPLC-DAD-ESI-MS. Molecules 20(1):913–928

Schmeda-Hirschmann G, Quispe C, Soriano MPC, Theoduloz C, Jiménez-Aspée F, Pérez MJ, Cuello AS, Isla MI (2015b) Chilean Prosopis Mesocarp flour: phenolic profiling and antioxidant activity. Molecules 20(4):7017–7033

Schreckinger ME, Lotton J, Lila MA, Mejia EG (2010) Berries from South America: a comprehensive review on chemistry, health potential, and commercialization. J Med Food 13(2):233–246

Silva MA, Souza-Brito ARM, Hiruma-Lima CA, Santos LC, Sannomiya M, Vilegas W (2005) Strychnos L. da América do Sul e Central. Rev Bras 15(3):256–267

Silva JKR, Pinto LC, Burbano RMR, Montenegro RC, Guimarães EF, Andrade EHA, Maia JGS (2014a) Essential oils of Amazon Piper species and their cytotoxic, antifungal, antioxidant and anti-cholinesterase activities. Ind Crop Prod 58:55–60

Silva MJD, Carvalho AJS, Rocha CQ, Vilegas W, Silva MA, Gouvêa CMCP (2014b) Ethanolic extract of *Mimosa caesalpiniifolia* leaves: chemical characterization and cytotoxic effect on human breast cancer MCF-7 cell line. S Afr J Bot 93:64–69

Silva MFR, Silva PCB, Lira CS, Albuquerque BNL, Agra Neto AC, Pontual EV, Maciel JR, Paiva PMG, Navarro DMAF (2016) Composition and biological activities of the essential oil of *Piper corcovadensis* (Miq.) C. DC (Piperaceae). Exp Parasitol 165:64–70

Simirgiotis MJ, Bórquez J, Schmeda-Hirschmann G (2013) Antioxidant capacity, polyphenolic content and tandem HPLC–DAD–ESI/MS profiling of phenolic compounds from the South American berries Luma apiculata and L. chequén. Food Chem 139(1–4):289–299

Singh B, Singh JP, Kaur A, Singh N (2016) Bioactive compounds in banana and their associated health benefits – a review. Food Chem 206:1–11

Siraichi JTG, Felipe DF, Brambilla LZS, Gatto MJ, Terra VA, Cecchini AL, Cortez LER, Rodrigues-Filho E, Cortez DAG (2013) Antioxidant capacity of the leaf extract obtained from Arrabidaea chica cultivated in southern Brazil. PLoS One 8(8):e72733

Sosa ME, Lancelle HG, Tonn CE, Andres MF, Coloma AG (2012) Insecticidal and nematicidal essential oils from *Argentinean Eupatorium* and *Baccharis* spp. Biochem Syst Ecol 43:132–138

Tovar CDG, Lopez CC, Martos MV, Serio A, Ospina JD, Alvarez JAP, Ospina N, Tora S, Palmieri S, Paparella A (2016) Sub-lethal concentrations of Colombian *Austroeupatorium inulifolium* (H.B.K.) essential oil and its effect on fungal growth and the production of enzymes. Ind Crop Prod 87:315–323

Uzor PF (2016) Recent developments on potential new applications of emetine as anti-cancer agent. EXCLI J 15:323–328

Valadão ALC, Abreu CM, Dias JZ, Arantes P, Verli H, Tanuri A, de Aguiar RS (2015) Natural plant alkaloid (emetine) inhibits HIV-1 replication by interfering with reverse transcriptase activity. Molecules 20(6):11474–11489

Victoria FN, Brahm AS, Savegnago L, Lenardão EJ (2013) Involvement of serotoninergic and adrenergic systems on the antidepressant-like effect of *E. uniflora* L. leaves essential oil and further analysis of its antioxidant activity. Neurosci Lett 544:105–109

Vicuña GC, Stashenko EE, Fuentes JL (2010) Chemical composition of the *Lippia origanoides* essential oils and their antigenotoxicity against bleomycin-induced DNA damage. Fitoterapia 81(5):343–349

Viegas Junior C, Rezende A, Silva DHS, Castro-Gamboa I, Bolzani VS, Barreiro EJ et al (2006) Aspectos químicos, biológicos e etnofarmacológicos do gênero Cassia. Quim Nova 26(9):1279–1286

Vieira LM, Castro CFS, Dias ALB, Silva AR (2015) Fenóis totais, atividade antioxidante e inibição da enzima tirosinase de extratos de *Myracrodruon urundeuva* Fr. All. (*Anacardiaceae*). Rev Bras Plantas Med 17(4):521–527

Vilar DA, Vilar MSA, Moura TFAL, Raffin FN, Oliveira MR, Franco CFO, Athayde-Filho PF, Diniz MFFM, Barbosa-Filho JM (2014) Traditional uses, chemical constituents, and biological activities of *Bixa orellana* L.: a review. ScientificWorldJournal 2014:857292

Yang Y, Ren-Sheng X (2011) Alkaloids. In: Rensheng X, Yang Y, Weimin Z (eds) Introduction to natural products chemistry. CRC Press, Boca Raton, pp 55–79

Yang WF, Liao GQ, Hakim SG, Ouyang DQ, Ringash J, Su YX (2016) Is pilocarpine effective in preventing radiation-induced xerostomia? A systematic review and meta-analysis. Int J Radiat Oncol Biol Phys 94(3):503–511

Zapata N, Lognay G, Smagghe G (2010) Bioactivity of essential oils from leaves and bark of *Laurelia sempervirens* and *Drimys winteri* against *Acyrthosiphon pisum*. Pest Manag Sci 66:1324–1331

Part II
Medicinal and Aromatic Plants of Brazil

Introduction to Medicinal and Aromatic Plants in Brazil

Ákos Máthé and José Crisólogo de Sales Silva

Abstract MAPs have a long history in traditional medicine, and are still looked upon by certain Brazilian ethnic groups (e.g. Tupis and Guaranis), as "divine sources of healing". In relation to extreme diversity and long, as well as rich traditions, the public knowledge on this special group of economic plants, is still relatively scarce, although much has been done to explore and utilize MAPs. Two of the world's diversity hotspots (including the hottest of hotspots) can be found in the territory of Brazil (Mata Atlantica and Cerrado). These territories have been intensively studied to reveal the levels of habitat loss, rate of species extinction and to save their exceptional levels of plant endemism. In the past, there had been no reliable census of the plant species of Brazil flora. The first nationwide assessment of the naturalized flora of Brazil has revealed that as a result of human presence and actions, non-native species are widespread in all Brazilian biomes and regions. So called Mega-Developments taking place in certain domains of Brazil (e.g. the Amazon) already have major implications on the Global Climate Change. Traditional medicines, including herbal medicines, will continue to be used in Brazil to some capacity, similarly to several countries of the developing world, where 70–95% of the population rely on these traditional medicines for primary care. Brazil is one of the few countries in the world that provides public support for the payment for herbal medicines approved only on the basis of long-standing and widespread prior use. Brazil has a list of 12 herbal medicines funded by the government. The Ministry of Health of Brazil has presented a National Policy on Integrative and Complementary

Á. Máthé (✉)
Faculty of Agriculture and Food Science, University of West Hungary, Mosonmagyaróvár, Hungary
e-mail: akos.mathe@upcmail.hu

J. C. de Sales Silva
Course of Animal Science, State University of Alagoas – Uneal – Brazil, Santana do Ipanema, Brazil
e-mail: jose.crisologo@uneal.edu.br

U. P. Albuquerque et al. (eds.), *Medicinal and Aromatic Plants of South America*, Medicinal and Aromatic Plants of the World 5,
https://doi.org/10.1007/978-94-024-1552-0_3

Practices (PNPIC), in 2008, in order to coordinate a Unified Health System (SUS) in Brazil and to establish policies to ensure integrality of health care. This policy is expected to contribute to the farther exploration, safeguarding and sustainable/modern utilization of medicinal plant resources in Brazil.

Keywords Medicinal and aromatic plants · Flora of Brazil · Biomes · Biodiversity hot-spots · Endemic species · Conservation · Folk medicine · Integrative medicine

1 Introduction

The Federative Republic of Brazil is the largest country on the South American continent and this regards both its population size and geographic dimensions. Brazil is not only a large country but due to its most diverse topographic conditions and habitats it has also a diverse flora and fauna.

It is not frequently mentioned but according to literature resources, Brazil was given its name after the plant pau-brasil (*Caesalpinia echinata* L.), a member of the Fabaceae – Caesalpininioidae family which was used in the past as a source of a valuable dye-stuff known as "brasiline" (Goncalves De Lima et al. 1961 as cited by Mitra et al. 2007). Different parts of pau-brasil are commonly used in Brazil, as adstringent, healing agents, oral analgesics and tonics, with the bark of the trunk also being used to treat diarrhea and dysentery and to strengthen the gums (da Silva Gomes et al. 2014).

According to a still existing ancient tradition plants are looked upon as divine sources of healing, especially among the different ethnic groups like the Tupis in the north and the Guaranis in the south that inhabit the Amazon rain forests (Mitra et al. 2007). There were even times, when Bertoni, a nineteenth century botanist held strong convictions that the wild Guaranis had a better knowledge of plants compared to that of the Europeans of the sixteenth century (Marini-Bettòlo 1988 1977).

Public knowledge on the extreme and unique plant diversity, as well as rich traditions of their use by the native and later settler populations in Brazil is relatively scarce in relation to their values. The present chapter is to serve as a modest introduction to this wonderful world of natural wealth, with a special focus on medicinal plants. Due to the page limitations, this introduction cannot be complete, but can only aim at offering an insight into the recent information on the honorable amount of existing and ever enlarging knowledge.

2 Biodiversity Hotspots in Brazil

In a simplest way the expression biodiversity "hotspot" denotes a biogeographic region that is threatened by destruction. The concept takes its origin from the British ecologist Norman Myers, who in 1988, published a paper in which he identified 10 tropical forest so called "hotspots" (Myers 1988) with the aim to throw light on the mass extinction that is overtaking Earth's species.. These regions were characterized both by exceptional levels of plant endemism and serious levels of habitat loss.

In the years to come, the number of hotspots was expanded to 18 (Myers 1990). Conservation International, adopting Myers' hotspots as its institutional blueprint, in 1996, made the decision to undertake a reassessment of the hotspots concept. Three years later an extensive global review was undertaken, which introduced quantitative thresholds for the designation of biodiversity hotspots and resulted in the designation of 25 hotspots. Since then – with the recognition of the North American Coastal Plain, in 2016, the number of Earth's hotspots has arisen to 36.

Brazil is the home to the world's richest flora (40,989 species; 18,932 endemic) and includes two of the hottest hotspots (Mittermeier et al. 1997, 2004): Mata Atlântica (19,355 species) and Cerrado (12,669 species) (Forzza et al. 2012a, b). According to Begossi et al. (2000) hotspots in Brazil include a variety of ecosystems with mangroves, with savannah or cerrado or with forests.

Published estimates of described diversity were frequently divergent because the country lacked an authoritative inventory of plant, algal, and fungal species. In 2012, Rafaela C. Forzza et al. (2012a, b) published the results of their analyses with a focus on species endemism and the degree of threat. As a major and perhaps unexpectedly new conclusion they stated that **Brazil has fewer described species of plants**, algae, and fungi but higher levels of endemism than were previously reported. These analyses were assisted by the contributions of more than 100 scientists in the countries concerned and around 800 references in the professional literature. An area to qualify as a hotspot had to contain at least 0.5% or 1500 of the world's 300,000 plant species as endemics. It has been also concluded that 15 of the world's 25 hotspots contain at least 2500 endemic plant species, and 10 of them at least 5000.

3 Diversity of Plants in Brazil

Estimates of described diversity of Brazil are frequently widely divergent because of the lack of an authoritative inventory of plant, algal, and fungal species (Forzza et al. 2012a, b). According to Vieira (1999) with nearly 55,000 native species distributed over six major biomes, Brazil can be regarded as the country with the greatest biodiversity on our planet. The six major biomes as illustrated in (Fig. 1) are the following: Amazon (30,000); Cerrado (10,000); Caatinga (4000); Atlantic rainforest (10,000), Pantanal (10,000) and the subtropical forest (3000).

The Brazilian Amazon Forest (tropical rainforest) is a rather fragile ecosystem that covers nearly 40% of all national territory, with about 20% legally preserved. Its productivity and stability depend on the recycling of nutrients, and its efficiency is directly related to the biological diversity and the structural complexity of the forest Anon (1995) cited in (Vieira 1999). Giacometti (1990) estimated that there are about 800 plant species of economic or social value in the Amazon. Of these,

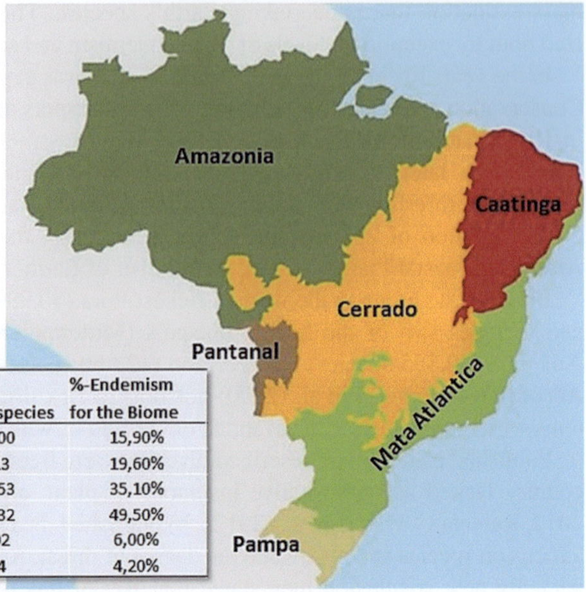

Biome	Number of		%-Endemism
	Seed plants	Endemic species	for the Biome
Amazonia	11912	1900	15,90%
Caatinga	4659	913	19,60%
Cerrado	12103	4253	35,10%
Atlantica	15004	7432	49,50%
Pampa	1687	102	6,00%
Pantanal	1277	54	4,20%

Fig. 1 Major biomes of Brazil. (After: R. C. Forzza et al. 2012a, b)

190 are fruit-bearing plants, 20 are oil plants, and there are hundreds of medicinal plants (van den Berg et al. 1988).

In a comprehensive 9-year market study on the impact of forest degradation on medicinal plant use and health care in Eastern Amazonia, Shanley and Luz (2003) stated that over the last three decades, forest degradation in the Brazilian Amazon has diminished the availability of some widely used medicinal plant species. One-third of the 300 species logged in eastern Amazonia are also valued for food, medicines, and gums and resins. Forests represent an important habitat for medicinal plants used in eastern Amazonia: 9 of the 12 top-selling medicinal plants are native species, and 8 are forest based. Five of the top-selling species have begun to be harvested for timber, decreasing their availability for medicinal purposes. Remarkably, several of these medicinal plants have no botanical substitutes, and frequently there are not pharmaceuticals that could substitute them in treating the diseases they are used. This verifies the following statement by Shanley and Luz (2003): "When rural communities sell timber, they often lose valuable fruit, medicinal, and game-attracting species".

In 2000, Laurance (2000) described mega-development trends in the Amazon and its implications for Global Climate Change. The study described four global-change phenomena that are having major impacts on Amazonian forests: (1) accelerating deforestation and logging that have increased from 1.1 million ha year^{-1} in the early 1990s, to nearly 1.5 million ha year^{-1} from 1992 to 1994, and to more than 1.9 million ha year^{-1} from 1995 to 1998. (2) patterns of forest loss and fragmentation are rapidly changing: The construction of major new highways is providing direct conduits into the heart of the Amazon and may largely bisect the forests of the

Amazon Basin. (3) climatic variability is interacting with human land uses, creating additional impacts on forest ecosystems. As an example, in 1997/1998 the El Niño drought led to a major increase in forest burning, with vast wildfires out of control (mainly in the northern Amazonian state of Roraima). (4) rapidly diminishing, intact Amazonian forests, which in turn, are a globally significant carbon sink. As a result of long term carbon-flux, as well as atmospheric CO_2 and isotope investigations it was established that not only is the destruction of these forests a major source of greenhouse gases, but it is reducing their intrinsic capacity to help buffer the rapid anthropogenic rise in CO_2 (Laurance 2000).

The "Cerrado" is the second largest ecological dominion of Brazil, where a continuous herbaceous stratum is joined to an arboreal stratum, with variable density of woody species. The cerrados cover a surface area of approximately 25% of Brazilian territory and around 220 species from cerrado are reported as used in the traditional medicine (Vieira 1999).

The "Caatinga" extends over areas of the states of the Brazilian Northeast and is characterized by a xerophitic vegetation that is typical of semi-arid climates. The soils that are fertile, due to the nature of their original materials and the low level of rainfall, experience minor runoff Anon (1995) cited by Vieira (1999). This northeastern region of Brazil comprises about one third of the country's territory. It is a semi-arid region with a flora rich in aromatic, toxic and medicinal plants. Various important medicinal plants (e.g. *Lippia* spp. and *Vanillosmopsis arborea*) have their centers of genetic diversity in this region, and the use of local folk medicines is common. Several important aromatic species are reported for this region (Craveiro et al. 2007).

The Atlantic Forest extends over nearly the entire Brazilian coastline. It is one of the most endangered ecosystems of the world, with less than 10% of the original vegetation remaining. The climate, here, is predominantly hot and tropical with a precipitation ranging between 1000 and 1750 mm. The landscape is composed of hills and coastal plains, accompanied by a mountain range (Vieira 1999). Several important medicinal species are found in this region, such as *Mikania glomerata, Bauhinia forficata, Psychotria ipecacuanha*, and *Ocotea odorifera*.

The Meridional Forests and Grasslands include the mesophytic tropical forests, the subtropical forests, and the meridional grasslands of the states of southern Brazil. The climate of this area is tropical and subtropical, humid, with some zones of temperate climate. Due to its naturally fertile soils and mild climate, this area had seen a rapid colonization mainly by European and, more recently, by Japanese immigrants, during the nineteenth century (Vieira 1999). As a consequence, several medicinal plants have been introduced, or naturalized, e.g.: chamomile (*Matricaria recutita*), calendula (*Calendula officinalis*), lemon balm (*Melissa officinalis*), rosemary (*Rosmarinus officinalis*), basil (*Ocimum basilicum*).

The Pantanal is a geologically lowered area filled with sediments which have settled in the basin of the Paraguay River. Pantanal flora is formed by species from both Cerrado and Amazon vegetation. More than 200 species useful for human and animal consumption as well as for industrial use have been recorded in this region (Vieira 1999).

4 The Flora of Brazil: Native vs. New Naturalized Species

The Brazilian flora, like many other floras in the world, are composed of both native and naturalized (introduced) species.

The need for a census of the Brazilian flora with sufficient scientific credibility to guide conservation planning has existed for a long time. According to Forzza et al. (2012a, b) the last complete inventory of Brazilian plants was the detailed and comprehensive Flora brasiliensis, published between 1833 and 1906, in which 19,958 species of plant, algae, and fungi were recorded for Brazil. Although, in the century to follow, virtually thousands of new species and their distributions were recorded, it was not followed by the sensus or comprehensive survey of the Brazilian flora for a long-long time. Existing knowledge was based mostly on estimates. According to this the number of described species of plants and fungi range between 60,700 and 70,210 (Lewinsohn and Prado 2005), while the most recent figures indicate 56,108 vascular species, with 12,400 (22%) species being endemic (Giam et al. 2010).

The largest plant families in Brazil, in terms of the number of species, are: Fabaceae (3200 spp. with 2144 endemics), Asteraceae (1900 spp.), Euphorbiaceae (1100), Myrtaceae (1038) and Rubiaceae (1000).

5 Naturalized Species in the Brazilian Flora

A recent study by Zenni (2015) has revealed that regarding the number of naturalized species, it was the Atlantic Forest had the largest number. In relation to the biome's total richness, it was the Pampa that had the highest proportion of naturalized species. The extent of naturalization expressed by the number of naturalized species seems to have been affected both by human population size and the proportion of remaining natural vegetation. Forty-six species were naturalized in five out of the six biomes and there were no records of species having naturalized in all six biomes. Remarkably, the Family Poaceae had the highest numbers of naturalized species in all biomes: nearly half of the recorded species belonged to this family, followed by the Asteraceae and Fabaceae. In fact, these species of these three Families were considered as top three families, in terms of the number of naturalized species in five out of the six biomes of Brazil.

In this context, it should be mentioned that the need to understand the patterns and drivers of species naturalizations and invasions has been expressed by many. Comprehensive reviews by Simberloff et al. (2013) and Zenni (2015) discuss the impacts of biological invasions that can be regarded as a pervasive component of global change. These studies have generated a remarkable understanding of the mechanisms and consequences of the spread of introduced populations and are useful in preventing and reducing the negative impacts caused by biological invasions.

Recognizing that human-mediated species introductions are important elements of the Anthropocene and that non-native species can form invasive populations that affect biodiversity, ecosystem services, or farming Zenni (2015) analyzed data on 32,634 identified vascular species in the Brazilian List of seed plants of which 525 were naturalized, non-native species. From this study the following important conclusions are worth highlighting: (1) the Atlantic Forest had the largest number of naturalized species, whereas the Pampa had the highest proportion of naturalized species in relation to the biome's total richness; (2) the number of naturalized species was affected both by human population size and proportion of remaining natural vegetation; (3) the plant Family Poaceae had the highest numbers of naturalized species in all biomes, and, together with Asteraceae and Fabaceae, forms the top three families in number of naturalized species in five of the biomes studied; (4) there were no records of species that have naturalized in all six biomes; (5) half of the 46 naturalized species, in five out of six biomes, belong to the Family Poaceae.

6 REFLORA Programme

The study of the Brazilian Flora, which is generally recognized as the richest in the world (Forzza et al. 2012a, b) has a long history. During the eighteenth and nineteenth centuries, European naturalists, who travelled to or resided in Brazil, and also a few Brazilian botanists, collected plant specimens and sent them to herbaria in Europe. Their main aim was to study and/or identify the plants found on that distant continent and explore their potential uses. Many of these plant collections served as the basis for the description of species or genera that were new to science (and these plants have become nomenclatural types), or simply formed part of the large collection of samples that were used to describe the over 22 thousand species of the Flora brasiliensis (Martius et al. 1840–1906).

Recognizing their scientific value, in 2010/2011, the Brazilian Government established the REFLORA (Brazilian Plants: Historic Rescue and Virtual Herbarium for Knowledge and Conservation of the Brazilian Flora) Program with the objective to rescue and make available images and information concerning Brazilian plants deposited mainly in overseas herbaria through an on-line facility, the REFLORA Virtual Herbarium.

To-date, the Rio de Janeiro Botanical Garden (JBRJ) hosts the physical structure of the Reflora Virtual Herbarium. It is responsible for receiving the repatriated images and transcribing the data associated with the samples. Thus, images and data derived from the repatriation process, together with images and data from the herbarium of the Jardin Botânico do Rio de Janeiro are made available to the scientific community and the general public (Anon n.d.).

Importantly, in addition to European and American herbaria, Brazilian herbaria have also begun the publication of their images and data in the REFLORA Virtual Herbarium, in 2014. The so called "Brazilian List", was concluded, in 2015 with the publication of five papers and their respective databases to open the way for a brand new system, the Brazilian Flora 2020 project, in 2016. The Brazilian Flora 2020 project is part of the Reflora program with some 700 scientists working in a network to prepare the monographs. The work platforms provided by the REFLORA Virtual Herbarium and the Brazilian Flora 2020 are meant to serve as fundamental tools that enable Brazil to meet the target No. 1 of the Global Strategy for Plant Conservation for 2020, i.e. the preparation of the Flora of Brazil online (Table 1).

The on-line plant identification tool of Reflora (Reflora Herbarium) and the English version of Flora do Brasil 2020 (Brazilian Flora 2020) are accessible at the following respective links: http://floradobrasil.jbrj.gov.br/reflora/herbarioVirtual/ConsultaPublicoHVUC/ConsultaPublicoHVUC.do, http://reflora.jbrj.gov.br/reflora/listaBrasil/PrincipalUC/PrincipalUC.do?lingua=en#CondicaoTaxonCP.

Table 1 Vegetation types and Phytogeographic domains, as recorded by Flora do Brazil 2020

Vegetation type	Phytogeographic domain
Antrhopic area	Amazon rainforest
Caatinga (stricto sensu)	Caatinga
Amazonian Campinarana	Central Brazilian savanna
High altitude grassland	Atlantic rainforest
Flooded field (Várzea)	Pampa
Grassland	Pantanal
Highland rocky field	Amazon rainforest
Carrasco vegetation	Caatinga
Cerrado (lato sensu)	Central Brazilian savanna
Riverine forest and/or gallery forest	Atlantic rainforest
Inundated forest (Igapo)	Pampa
Terra firme forest	Pantanal
Inundated forest (Várzea)	Amazon rainforest
Seasonal evergreen forest	Amazon rainforest
Seasonally semideciduous forest	Caatinga
Ombrophyllous forest (tropical rain forest)	Amazon rainforest
Mixed ombrophyllous forest	Central Brazilian savanna
Mangrove	Atlantic rainforest
Palm grove	Central Brazilian savanna
Coastal forest (Restinga)	Atlantic rainforest
Amazonian savanna	Amazon rainforest
Aquatic vegetation	Amazon rainforest
Rock outcrop vegetation	Central Brasilian savanna

7 Germplasm Conservation of MAPs in Brazil

The scientific literature seemingly does not abound in documents on the germplasm conservation of medicinal plants in Brazil. The 21 documents that our SCOPUS search has yielded for the period 1989–2017. seems to indicate that the first mention of this topic was in 2002, when Vieira (2002) illustrated some of the vast potentials of Brazilian flora and called attention to the enormous task to elaborate a program for genetic resource conservation of these species: a task that requires multi-institutional and multi-disciplinary collaboration. He also stated that plant collections will have an important role in the future, providing genetic material for chemical characterization, breeding of new crops, improving our understanding of secondary metabolism, and in preserving an important part of our cultural and national heritage pathways. The second publication by de Oliviera and Martins (2002) presented a methodology, on the example of ipecac (*Psychotria ipecacuanha*), by which the threat of genetic erosion to a wild plant species growing in a given geographic region can be assessed in a quantitative manner.

Remarkably, however, the conservation of medicinal plants in Brazil – independently of their germplasms – has been an increasingly frequent topic with 157 scientific publications (Fig. 2a), with two maxima (19 and 17 publications), in 2011. and 2017., respectively. As regards the sources of publications (Fig. 2b), the majority of papers were published in Brazilian scientific journals (in Revista Brasileira de Plantas Medicinals (23) and Acta Botanica Brasilica (9)).

An analysis of the frequency documents by subject areas reveals that the conservation of medicinal plants has been dealt with from various scientific approaches, quasi in the form of multi-institutional and multi-disciplinary collaboration, as foreseen by Vieira in 2002. The sciences involved and their share in the total number of 157 documents is as follows: Agricultural and Biological Sciences (42.4%), Medicine (27.8%), Pharmacology, Toxicology and Pharmaceutics (26.6%), Biochemistry, Genetics and Molecular Biology (24.1%), Environmental Science (15.8%), Social Sciences (11.4%), Multidisciplinary (2.5%), Veterinary (2.5%), Arts and Humanities (1.9%), Chemistry (1.9%), Other (7.6%).

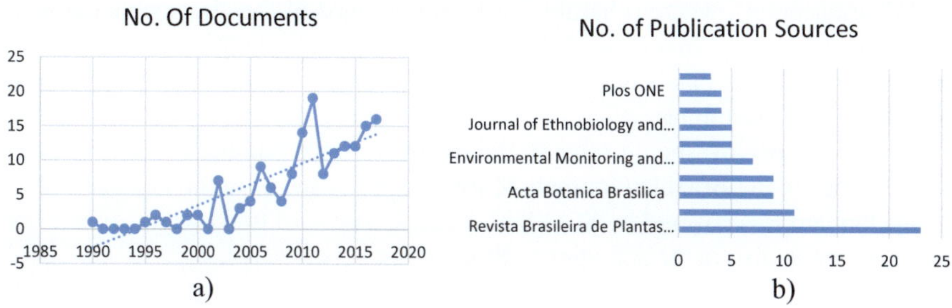

Fig. 2 Annual distribution (**a**) and Journal source (**b**) of documents on the conservation of medicinal plants, in Brazil (SCOPUS 1990–2017) n = 157

8 Effect of Chiang Mai Declaration

Since the early 1990s, roughly soon after the publication of the so called Chiang Mai Declaration, in 1988 (Akerele and Heywood 1991), serious efforts have been made to collect and preserve the genetic variability of medicinal plants also in Brazil. The National Center for Genetic Resources and Biotechnology (Cenargen), in collaboration with other research centers of the Brazilian Agricultural Research Corporation (Embrapa), and several universities, launched a program to establish germplasm banks for medicinal and aromatic species. An incomplete list of these, with a special focus on the endemic species of the Caatinga, is contained by Table 2.

In a historical perspective and also in order to properly appreciate the significance of above activities, we should reiterate that the need for the sustainable use of natural resources, including MAPs, was first duly recognized by the Chiang Mai Declaration (1988), when the international scientific community expressed alarm over the consequences in the loss of plant diversity (Máthé 2015). The Declaration highlighted the urgent need for international cooperation and coordination to establish programs for the conservation of medicinal plants with the ultimate aim to ensure that adequate quantities are available for future generations.

Remarkably, the subsequent decades were marked with an upsurge in activities, especially in the form of several declarations and sets of recommendations calling for the Conservation and Sustainable use of natural biodiversity, including medicinal plants.

As most of the crude drugs are sourced by wild-crafting (collection), the expectations vs. the "sustainable collection" of MAPs has gained on importance all over the world. This presumption seems to have been verified by the (slightly) increasing trend in the number of documents retrieved by the SCOPUS search (Fig. 3) that has yielded a total of 151 documents. Remarkably, this trend has become more expressed only as of the 2000s, i.e. following a lag-period of ca. 10 years. Focusing on Brazil, the SCOPUS search on sustainable collection of medicinal plants for the period 1995–2017. has yielded only a total of 14 documents (Fig. 3) from these 11 published in Brazil (11) and 1-1 in France, UK, US, respectively.

54.3% of the documents dealt with subjects belonging to agricultural and biological sciences, 30.5% sciences that can be ranked to Pharmacology, Toxicology and Pharmaceutics, whereas hardly less than one third of the documents (27.8%) with issues related to Medicine (Table 3). These data indicate that sustainable collection of medicinal plants is a multidisciplinary activity demanding the input of a broad range of disciplines.

In this regards we should refer to the relatively large number of surveys published either in the Brazilian journals of local significance or in the Portugese language that are less covered by SCOPUS. Thus, the upward trend seems to be even more apparent. also in the activities in Brazil (Assis et al. 2015).

Table 2 List of medicinal and aromatic species with high priority for germplasm collection and conservation in Brazil with a special focus on Caatinga

Species	Common name	Habit	Active substance/ pharmacological action	Region	Conservation form
Achyrocline satureioides L.	Macela	Herb	Hypotensive, spasmolytic	Cerrado	Field collection
Ageratum conyzoides L.	Mentrasto	Herb	Anti-inflammatory	Ruderal	Field collection
Amburana cearensis(Allemaõ) A.C. Sm	Cumaru	Tree	Sinusitis	Caatinga	In situ, field
Anadenanthera columbrina (Vell) Brenan	Sngico	Tree	Grippe	Atlantic Forest, Caatinga	Field collection
Aniba roseodora Ducke	Pau rosa	Tree	Linalool	Amazon forest	In situ
Apodanthera congestiflora Cogn	Cabeça-de-negro	climbing	Blood purifying,	Amazônia, Caatinga, Cerrado, Mata Atlântica, Pampa, Pantanal	Field
Astronium urundeuva (Fr. All.) Engl.	Aroeira	Tree	Anti-inflammatory, anti-ulceric	Cerrado	In situ, cold chamber
Baccharis trimera DC.	Carqueja	Herb	Hepatic disturbs	Ruderal	Field collection
Bauhinia forficata L.	Pata de Vaca	Tree	Diabetes	Atlantic forest	Cold chamber
Boerhavia diffusa L.	Pega-pinto	Herb	blood purifying, hepatitis and diarrhea	Atlantic forest	Cultivate field
Caryocar brasiliensis Camb.	Pequi	Tree	Anti-inflamatory	Cerrado	In situ
Chenopodium ambrosioides L.	Mastruz	Herb	fracture, gastritis, vermifuge	Atlantic Forest	Field
Copaifera langsdorffi Desf.	Copaiba	Tree	Oil, anti-inflamatory	Cerrado	In situ, cold chamber
Caesalpinia pyramidalis Tul.	Catingueira	Tree	Grippe	Caatinga	Field Caatinga
Cereus jamacaru	Mandacaru	Tree	Grippe, kidneys	Caatinga	In situ, field
Croton cajucara Benth.	Sacaca	Herb	Linalool	Amazon	Field collection
Croton zehntneri Pax et Hoff.	Cunha	Shrub	Anetol, eugenol	Caatinga	Field collection
Datura insignis B. Rodr.	Toe	Shrub	Escopolamina	Amazon forest	Cold chamber

(continued)

Table 2 (continued)

Species	Common name	Habit	Active substance/ pharmacological action	Region	Conservation form
Dimorphandra mollis Benth.	Faveiro	Tree	Rutin, anti-hemorragic	Cerrado	Cold chamber
Echinodorus macrophyllus(Kunth.) Mich	Chapeu de Couro	Herb	Diuretic	Cerrado	Field collection, cold chamber
Erythrina verna Vell. Conc.	Mulungu	Tree	anxiolytics and anticonvulsants	Atlantic Forest	Field collection
Harrisia adscendens	Rabo-de-raposa	Shrub	Kidneys, prostate, toothache	Caatinga	In situ, field
Jatropha elliptica (Pohl) Baill.	Batat de Tiu	Shrub	Jatrophone	Cerrado	In situ, field collection
Lippia spp.	Alecrim pimenta	Shrub	Source of volatile oils, anti-microbial	Caatinga	Field collection
Luffa operculata (L.) Cogn	Cabacinha, buchinha	Climbing	Sinusitis	All Brasil	Field and plantations
Lychnophora ericoides Mart.*; L. salicifolia* Mart.	Arnica do Cerrado	Shrub	Volatile oils	Cerrado	Field collection, in situ
Mandevilla vellutina Mart.	Serra dos Órgãos	Shrub	Anti-inflamatory, bradykynin antagonist	Cerrado	In situ, field collection
Maytenus ilicifolia Mart. ex. Reiss; *M. aquifolium* Mart.	Espinheira Santa	Tree	Anti-ulceric	Meridional forest	Cold chamber, *in situ*
Melocactus zehntneri	Cabeça-de-frade, coroa-de-frade	Cactus 20 cm	Gripe, mulligrubs	Caatinga	In situ, field
Mikania glomerata Spreng.	Guaco	Herb	Bronchitis, coughs	Atlantic forest	Field collection
Mimosa tenuiflora (Willd) Poir	Jurema Preta	Tree	Anti inflammatory	Caatinga	Field collection
Ocotea odorifera (Vell.) Rohwer	Canela Sassafraz	Tree	Safrol, metileugenol	Atlantic forest	In situ
Operculina macrocarpa (L.) Farwel	Batata de Purga	Herb	Purgative	Caatinga	Cold chamber
Opuntia palmadora	Palmatoria do sertão	Shrub	Urethra problem	Caatinga	In situ, field
Piper hispidinervum DC.	Pimenta longa	Herb	Safrol	Amazon	Cold chamber, field collection

(continued)

Table 2 (continued)

Species	Common name	Habit	Active substance/ pharmacological action	Region	Conservation form
Pfaffia paniculata (Martius) Kuntze	Ginseng brasileiro	Herb	Antitumor compounds	Margins of Parana river	Cold chamber, field collection
Phyllanthus niruri L.	Quebra pedra	Herb	Hepatitis B, renal calculus	Ruderal	Cold chamber
Phyllanthus niruri L.	Quebra pedra	Herb	Kidney desease	Atlantic Field	Field collection
Pilosocereus gounellei F.A.C. Weber ex K. Schum.	Xique-xique	Shrub	Rheumatism, crowfoot	Caatinga	In situ, field
Pilocarpus microphyllus Stapf.	Jaborandi	Shrub	Pilocarpine	Amazon forest	Cold chamber, in situ
Psychotria ipecacuanha (Brot.) Stokes	Ipecac	Herb	Emetin, cefaline	Amazon and Atlantic forest	Cold chamber, in situ
Pterodon emarginatus Vogel	Sucupira	Tree	Analgesic, antinoceptive, cercaricide	Cerrado	In situ, cold chamber
Senna ocidentallis (L.) Link	Manjerioba, Fedegoso	Shrub	Lowering the blood pressure as well as the blood Cholesterol	All Brasil	Field
Solanum mauritianum Scopoli	Cuvitinga	Shrub	Solasodine	Ruderal, southeast and southern Brazil	Cold chamber
Stryphnodendron adstringens	Barbatimao	Tree	Tannin, anti-inflamatory	Cerrado	In situ, cold
(Mart.) Coville					chamber
Tabebuia avellanedae (Lor.) ex. Griseb.	Ipe roxo	Tree	Lapachol	Cerrado	In situ
Vanillosmopsis arborea(Aguiar) Ducke	Candeia	Shrub	Bisabolol	Caatinga	In situ, field Collection
Vitex gardneriana Schauer	Jaramataia	Tree	Vermicide, soothing and anti-inflammatory	Caatinga	Field Collection
Ximenia Americana L.	Ameixa da caatinga	Shrub	Anti-inflammatory	Caatinga	Field Bahia

After: Vieira (1999), Roque et al. (2010), and Andrade et al. (2006)

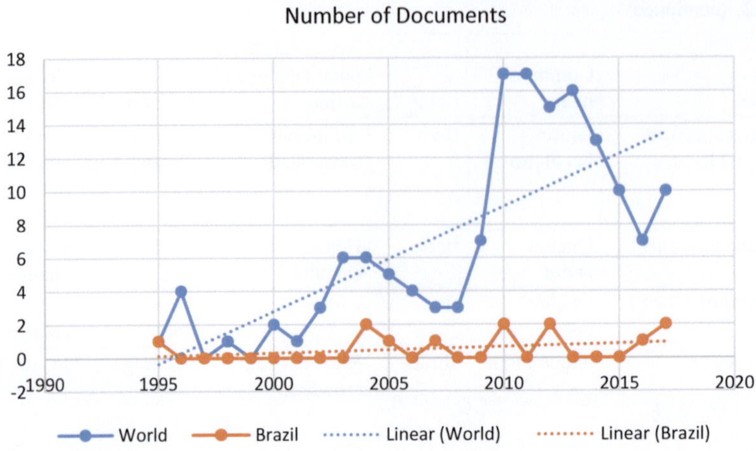

Fig. 3 Documents on the sustainable collection of MAPs in Brazil and the world (SCOPUS retrieved: 2018.02.24)

Table 3 Ten most frequently occurring type of documents by subject area in a SCOPUS search "sustainable collection of medicinal plants" (Retrieved: February 24, 2018)

No.	Subject area	Number of documents	
1.	Agricultural and Biological Sciences	82	54.3%
2.	Pharmacology, Toxicology and Pharmaceutics	46	30.5%
3.	Medicine	42	27.8%
4.	Social Sciences	30	19.9%
5.	Environmental Science	29	19.2%
6.	Biochemistry, Genetics and Molecular Biology	14	9.3%
7.	Engineering	6	4.0%
8.	Chemistry	3	2.0%
9.	Earth and Planetary Sciences	3	2.0%
10.	Energy	3	2.0%

9 Brazilian Medicinal Plants as Raw Materials for (Inter) National Trade

The world demand on medicinal plants has seen an exponential growth. There has been an increasing flow of medicinal plants from the southern hemisphere to developed countries, growing from 100 million dollars in 1979 to 35 billion dollars, in 2003. This increased interest in medicinal plants has put a dangerous pressure on the habitat of indigenous peoples (Caceres Guido et al. 2015).

Data on the international trade of medicinal and aromatic plants in Brazil are seemingly scarce. This fact is reflected by the SCOPUS search "medicinal plant trade in Brazil" that has yielded 19 documents for the period 1999–2016. As a *quasi* contrast, the diversity of MAPs traded in the local markets of different areas of Brazil is quite common subject of studies (de Albuquerque et al. 2007; Lós et al. 2012; Roque et al. 2010) In one of the most recent studies Carvalho et al. (2018) published a critical survey in the Brazilian Health Regulatory Agency (ANVISA) database to verify HMPs (Herbal Medicine Product) licensed in Brazil in September 2016. Their data were compared with previously published similar surveys. The survey has established that there are 332 single, 332 single, and 27 combined Herbal Medicines (HM), totaling 359 HM licensed in Brazil. There are no Traditional Herbal Products (THPs) notified in the Brazilian Health Regulatory Agency's (ANVISA) system, yet. Remarkably, however, there are 214 HMs classified as non-prescription (OTC) products, while 145 are sold under prescription (one of them with prescription retention). There are 101 plant species licensed as active in HM in Brazil, 39 of which are native, adapted or cultivated species. The most frequently licensed plant species is *Mikania glomerata* Spreng., with 25 HM licenses.

According to the somewhat too critical conclusion of Carvalho et al. (2018), there are few licensed HMs in Brazil, and this number has been decreasing in recent years. They expected that their survey, together with the changes promoted in sanitary and environmental rules, will help to develop as well as regulate HMP chain in Brazil.

In Brazil, several exotic plants are also used in formal commercial consumption, partly due to the fact that they are authorized by laboratories in other countries. As established by Albuquerque et al. (2007) markets conserve their basic repertoire while act as open and dynamic systems that is enriched by adding new plants and their respective use-indications.

Endemic native plants are most commonly used in popular markets, in small shops or in street markets, called "raizeiros", where also medicinal, aromatic, and spices are sold. Wilma et al. (2012), in a study on "raizers", in Arapiraca, state of Alagoas (northeastern Brazil), identified 103 main commercialized species belonging to 47 families. The most represented families were: Fabaceae (21 species), Lamiaceae (6 species) and Asteraceae, Cucurbitaceae, (5 -5 spécies), Apiaceae and Euphorbiaceae (4 species). Most of the species (66%) was used in the form of tea prepared from leaves and seeds (24–24%). According to Goncalces De Lima et al. (1961) ca. 80% of the identified plants were native and the predominantly of arboreous habitus. This study also shed light on the local pattern of MAP production and marketing and underpinned the need of minimum quality standards and the implementation of public policies.

In a recent study Alves et al. (2016) analyzed the marketing of medicinal plants and products by the healers of free fair in the city of Guarabira state of Paraiba. In evaluating their results they also applied an *Index of Relative Importance* (IR). The ethnobotanical survey of plants sold by sales-men public market Guarabira-PB, it identified 85 plants "in natura" Commercialization of medicinal plants: ethnobotanical study in the province of Guarabira, Paraíba, northeastern Brazil. Three hundred ninety-one sold dried, distributed in 44 families, totaling 246 citations of curative and preventive use of various diseases. Featured You are the plants used the bark, woody species such as Aroeira (*Myracrodruon urundeuva* Allemão) Barbatimão (*Stryphnodendron adstringens* (Mart) Coville.), purple Cashew (*Anacardium occidentale* L., Cumaru (*Amburana cearensis* (Allemão) AC Sm), Mulungu (*Erythrina verna*). Most represented botanical families – in terms of the number of species – were: Fabaceae (23%), Lamiaceae (19%), etc. The most frequently mentioned species were: chamomile (*Matricaria chamomilla* L.), Bilberry (*Plectranthus barbatus* Andrews), Rosemary (*Rosmarinus officinalis*).

Oliveira et al. (2013) – in a similar study – established that it was not the cultivation of medicinal plants but rather the purchase of their products that was characteristic of the markets of the city of Juazeiro do Norte and Fortaleza (Ceara). Medicinal plants marketed most, were: aroeira (*Myracrodruon urundeuva* Fr. All., Anacardiaceae), juazeiro (*Ziziphus joazeiro* Mart., Rhamnaceae and jatobá (*Hymenaea coubaril* L., Fabaceae).

10 Cultivation of Medicinal Plants in Brazil: Introduction and Domestication

In view of the complex and manifold possible implications (e.g.: biodiversity conservation, management and quality assurance), as well as sustainability issues, to date, MAP domestication and introduction into cultivation are increasing considered as methods that could secure the reliable raw material supplies (Máthé 2011). In the period 1989–2017. a total of 60 documents were published on "plant domestication in Brazil". These publications deal mostly with fruit, vegetable and ornamental species. As a contrast, the search "on medicinal plant domestication in Brazil" has yielded only 4 documents and a slightly altered search phrase (i.e. "medicinal plant introduction in Brazil") yielded 33 documents with an upward trend, to reach a maximum of 8 publications (in 2017.), during the last 10 years. Despite the encouraging trends it can be stated that in view of the vast MAP-diversity and -genetic potential of Brazil, these figures should denote only the beginnings of the huge tasks and opportunities ahead.

As the final aim of both introduction and domestication is to obtain a cultivated crop, we carried out a farther SCOPUS search on "medicinal plant production in Brazil". For the period 1982–2018., the search phrase has yielded 271 documents according to the following major groups of disciplines: 38.6% agriculture and biological sciences, 48.7% pharmacology, 32.1% medicine, 14.4% biochemistry, Veterinary implications: 2.6% (Fig. 4). These data verify the complex multi-disciplinary character of MAP domestication/introduction.

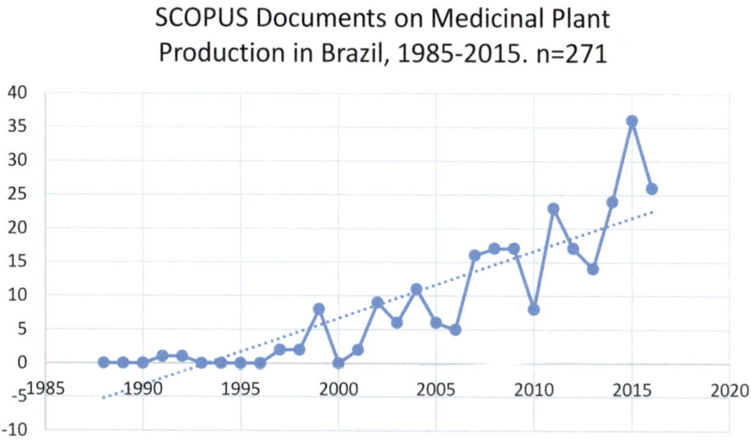

Fig. 4 SCOPUS documents on medicinal plant production in Brazil, 1985–2015

11 Medicinal Plants in the Brazilian Folk Medicine

Traditional medicines, including herbal medicines, have been, and continue to be, used in every country around the world in some capacity. In much of the developing world, 70–95% of the population rely on these traditional medicines for primary care. Developing countries, especially those in Asia, Africa, Latin America and the Middle East, use traditional medicine, including traditional and herbal medicines, for the management of health and as primary health care to address their health-care needs and concerns.

The use of Medicinal plants in folk (traditional) medicine has also long traditions in Brazil, where there is a still existing ancient tradition according to which plants are looked upon as divine sources of healing. According to Caceres Guido et al. (2015), it was marginalized for a long time. Due to the advancements in ethnobotanical and ethnomedical research, this situation has started to change at the end of the twentieth century. As such, especially, thanks to the important new contributions on traditional medicine, it is thus slowly being integrated into the clinical field.

Numerous interesting etnopharmacological studies have been published on various aspects of MAP usage. In 1994 (Elisabetsky and Shanley 1994) published a review of the ethnopharmacological and ethnobotanical studies that have been conducted in the Brazilian Amazon over the past 20 years. They discuss the role that ethnopharmacology can have in the discovery and development of new drugs from the Brazilian Amazon, a region hosting such enormous cultural and biological diversity.

A recent study by Mendes (2011) deals with species used as tonic, fortifier, aphrodisiac, anti-stress, among other uses that are similar to the indications of an adaptogen. Mendes provides a comparison of the main Brazilian plants used for such conditions, as follows: guarana (*Paullinia cupana* Kunth, family Sapindaceae), muirapuama (*Ptychopetalum olacoides* Benth., Olacaceae), catuaba (*Anemopaegma arvense* (Vell.) Stellfeld & J.F. Souza, Bignoniaceae, and *Trichilia catigua* A. Juss., Meliaceae), nó-decachorro (*Heteropterys aphrodisiaca* O. Mach, Malpighiaceae), damiana (*Turnera diffusa* Willd. ex Schult., Turneraceae) and pfaffia or Brazilian ginseng (*Pfaffia* sp., Amaranthaceae).

A similar ethnobotanical study was carried out into the antimalarial plants in the middle region of the Negro River, Amazonas by Tomchinsky et al. (2017) and as a result they state that the in the population of Barcelos there exists an extensive knowledge on the use of a diverse array of antimalarial plants, and may contribute to the development of novel antimalarial compounds.

According to Lopes et al. (2014a) a search in the database of the Brazilian Health Surveillance Agency (ANVISA) revealed that 15 species of herbal medicines are approved for treatment of acute cough from a URTI. Of these, Public Health System (SUS) funding is available for two. In view of the fact that there are no systematic reviews available that address the benefits and harms of the herbal medication approved by ANVISA for URTI, they implemented "the first" systematic review to assess Brazilian medicinal plants approved by the Brazilian Health Surveillance Agency (ANVISA) to treat upper respiratory tract and bronchial illness associated with cough and sputum. It is expected that the results of this systematic review will help clinicians in making decisions in clinical practice and also help patients with cough and sputum seeking **effective and safe treatment** options.

Antonio et al. (2014) analyzed 53 original studies on actions, programs, acceptance and use of phytotherapy and medicinal plants in the Brazilian Unified Health System. They state that over the past 25 years, there was a small increase in scientific production on actions/programs developed in primary care. Including phytotherapy in primary care services encourages interaction between health care users and professionals. It also contributes to the socialization of scientific research and the development of a critical vision about the use of phytotherapy and plant medicine, not only on the part of professionals but also of the population.

Finally it should be mentioned that a SCOPUS search on "traditional medicine plants Brazil" for the period 1988–2017. has yielded 789 documents (Fig. 5), out of which the following main subject areas (10%+) were represented: Pharmacology, Toxicology and Pharmaceutics (70.6%), Medicine (30.4%), Biochemistry, Genetics and Molecular biology (17.0%), Agricultural and Biological Sciences (12.8%), Chemistry (10.1%). This steady upward trend clearly underpins the increasing acceptance as well as popularity of the old still in the form of integrative medicine renewed science of healing with medicinal plants.

12 The Dawn of Use of Integrative Medicine

As in in high-income countries, there is increasing public interest in the use of therapies that lie outside the mainstream of traditional Western medical practice. Complementary and alternative medicine (CAM) has been growing rapidly over the

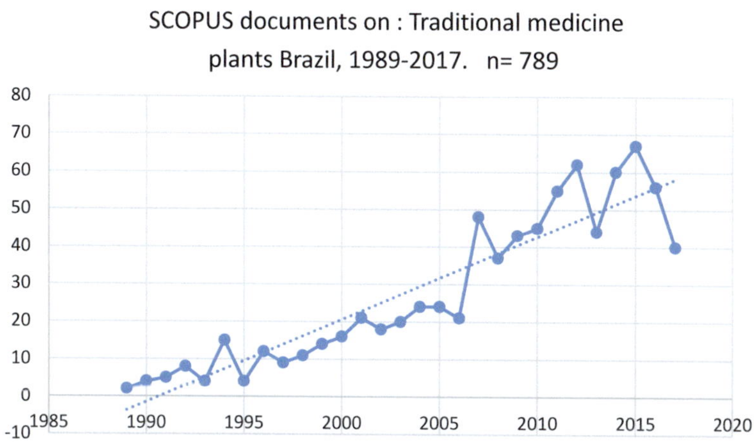

Fig. 5 SCOPUS documents on "traditional medicine plants Brazil" for the period 1989–2017

last decades (Lopes et al. 2014a). In the USA, an estimated 38% of adults and 12% of children are using some forms of CAM (Ekor 2014). Lopes et al. (2014b) cite that in Brazil, up to 25% of the total revenues of the pharmaceutical industry from sales of drugs, in the period from 1996 to 2014), came from preparations derived from plants. They also estimate that the government's decision to include herbal medicine in the list of publicly subsidized medicine in the Brazilian Health System (SUS) may have contributed to an increase in expenditures on herbal medicine in Brazil of 12% in 2012 over 2011, with a total of $1.147 billion.5.

In a recent review, entitled "The state of the integrative medicine in Latin America: The long road to include complementary, natural, and traditional practices in formal health systems" Caceres Guido et al. (2015) have estimated that more than 400 million people in Latin America use traditional/natural and/or complementary/alternative medicine (TN-CAM). The yearly expenditure on TN-CAM products of around 3 billion dollars illustrates that these practices have grown exponentially in this region as well. The quantity and quality of scientific studies on TN-CAM, although relatively scarce, has been steadily increasing. In Brazil, formal health systems - for different reasons - accept inclusion of TN-CAM. According to the authors, the immediate challenges are "how to improve multidisciplinary management, research, professional training, address legal/policy issues and a scientific approach to the extents and limitations of TN-CAM both in conventional health care and in the society as a whole."

13 Conclusions

Brazil, the largest country on the South American continent abounds in both diverse topographic conditions and diverse flora, including medicinal and aromatic plants.

MAPs have a long history in traditional medicine, and are still looked upon by certain ethnic groups (e.g. Tupis and Guaranis), as "divine sources of healing".

In relation to extreme diversity and long, as well as rich traditions, the public knowledge on this special group of economic plants, is still relatively scarce, although much has been done to explore and utilize MAPs.

Two of the world's diversity hotspots (the hottests of hotspots) can be found in the territory of Brazil. These are the Mata Atlantica (Atlantic Rainforest) and Cerrado (savannah). These territories have stood in the focus of especially intensive investigations with the aim to reveal the levels of habitat loss and the related rate of species extinction, ultimately to save their exceptional levels of plant endemism.

As for a long time in the past, there had been no reliable census of the plant species of Brazil flora, the estimates of diversity were frequently widely divergent. The first nationwide assessment of the naturalized flora of Brazil has, therefore, meant a scientific breakthrough, since it conveys important knowledge both for research and conservation prioritization. It was revealed by these studies that - also as a result of human presence and actions-, non-native species are widespread in all Brazilian biomes and regions.

It has been also recognized (Laurance 2000) that there are certain so called Mega-Developments taking place in certain domains of Brazil (e.g. the Amazon) that already have major implications on the Global Climate Change.

Traditional medicines, including herbal medicines, will continue to be used in Brazil to some capacity, similarly to several countries of the developing world, where 70–95% of the population rely on these traditional medicines for primary care.

It has been reported that Brazil is one of the few countries in the world that provides public support for the payment for herbal medicines approved only on the basis of long-standing and widespread prior use. Nowadays, Brazil has already a list of 12 such herbal medicines funded by the government (Lopes et al. 2014a).

The Ministry of Health of Brazil has presented a National Policy on Integrative and Complementary Practices (PNPIC), in 2008, in order to coordinate a Unified Health System (SUS) in Brazil and to establish policies that ensure integrality of health care, This policy, is based on public knowledge, support and incorporates as well as utilizes the rich experiences that had been developed in so far. It is to be hoped that it will also contribute to the farther exploration and modern utilization of medicinal plant resources of this vast country. As a consequence, it is expected to contribute to the safeguarding and/or sustainable use of natural resources in both Brazil and ultimately, in the world.

References

Akerele O, Heywood V (1991) Conservation of medicinal plants. Available at: https://books.google.com/books?hl=hu&lr=&id=mZZOAAAAIAAJ&oi=fnd&pg=PR11&dq=guidelines+on+the+conservation+of+medicinal+plants&ots=ouyFrvnIxh&sig=yxjNl3qyGK75bAgxlcK1hkWI9EE. Accessed 14 Feb 2016

Alves CAB et al (2016) Gaia Scientia. Gaia Scientia 10(4):390–407

Andrade CTS et al (2006) Utilização medicinal de cactáceas por sertanejos baianos. Rev Bras Pl Med 8(3):36–42. Available at: https://www.researchgate.net/profile/Daniela_Zappi/publication/283355113_The_use_of_medicinal_cacti_by_locals_at_the_semi-arid_in_Bahia_State_Brazil/links/5637425f08ae758841152285.pdf. Accessed 22 Feb 2018

Anon R (n.d.) Available at: http://floradobrasil.jbrj.gov.br/reflora/PrincipalUC/PrincipalUC.do?lingua=en. Accessed 5 Feb 2018

Antonio G, Tesser C, Moretti-Pires R (2014) Phytotherapy in primary health care. Rev Saude Publica 48(3):541–553. Available at: https://www.scopus.com/record/display.uri?eid=2-s2.0-84905911116&origin=resultslist&sort=plf-f&src=s&sid=422522a73e5092005f640fc52abe93ec&sot=b&sdt=cl&cluster=scosubtype%2C%22re%22%2Ct&sl=68&s=TITLE-ABS-KEY%28traditional+medicine+plants+Brazil%29+AND+PUBY

Assis MA, Morelli-Amaral VF, Pimenta FP (2015) Grupos de pesquisa e sua produção científica sobre plantas medicinais: um estudo exploratório no Estado do Rio de Janeiro. Research groups and their scientific literature on medicinal plants: an exploratory study in the state of Rio de Janeiro. Revista Fitos 9(1):1–72. Available at: https://www.arca.fiocruz.br/bitstream/icict/15933/2/19.pdf

Begossi A, Hanazaki N, Peroni N (2000) Knowledge and use of biodiversity in Brazilian hot spots. Environ Dev Sustain 2(3/4):177–193. Available at: http://link.springer.com/10.1023/A:1011409923520. Accessed 5 Feb 2018

Caceres Guido P et al (2015) The state of the integrative medicine in Latin America: the long road to include complementary, natural, and traditional practices in formal health systems. Eur J Integr Med 7(1):5–12. Available at: http://linkinghub.elsevier.com/retrieve/pii/S1876382014000961

Carvalho ACB et al (2018) The Brazilian market of herbal medicinal products and the impacts of the new legislation on traditional medicines. J Ethnopharmacol 212:29–35. Available at: http://linkinghub.elsevier.com/retrieve/pii/S037887411731797X. Accessed 6 Feb 2018

Craveiro AA et al (2007) Natural product chemistry in North-Eastern Brazil. Wiley, pp 95–105. Available at: http://doi.wiley.com/10.1002/9780470514634.ch7. Accessed 7 Feb 2018

da Silva Gomes ECB et al (2014) Evaluation of antioxidant and antiangiogenic properties of Caesalpinia echinata extracts. J Cancer 5(2):143–150. Available at: http://www.jcancer.org/v05p0143.htm

de Albuquerque UP et al (2007) Medicinal and magic plants from a public market in northeastern Brazil. J Ethnopharmacol 110(1):76–91. Available at: https://www.sciencedirect.com/science/article/pii/S0378874106004594. Accessed 26 Feb 2018

de Oliveira L, Martins ER (2002) A quantitative assessment of genetic erosion in ipecac (Psychotria ipecacuanha). Genet Resour Crop Evol 49(6):607–617

Ekor M (2014) The growing use of herbal medicines: issues relating to adverse reactions and challenges in monitoring safety. Front Pharmacol 4:177. Available at: http://journal.frontiersin.org/article/10.3389/fphar.2013.00177/abstract. Accessed 8 Feb 2018

Elisabetsky E, Shanley P (1994) Ethnopharmacology in the Brazilian Amazon. Pharmacol Ther 64(2):201–214. Available at: http://linkinghub.elsevier.com/retrieve/pii/0163725894900396. Accessed 6 Feb 2018

Forzza RC et al (2012a) New Brazilian floristic list highlights conservation challenges. Bioscience 62(1):39–45. Available at: http://reflora.jbrj.gov.br/downloads/FUNG1.pdf. Accessed 5 Feb 2018

Forzza RC et al (2012b) New Brazilian floristic list highlights conservation challenges. Bioscience 62(1):39–45. Available at: https://academic.oup.com/bioscience/article-lookup/doi/10.1525/bio.2012.62.1.8. Accessed 1 Mar 2018

Giacometti D (1990) Estrategias de coleta e conservacao de germoplasma horticola da America tropical. In: Proceedings of the Simposio Latinoamericano sobre recursos geneticos de especies horticolas, 1. Campinas/SP. Fundacao Cargill. pp 91–110

Giam X et al (2010) Future habitat loss and the conservation of plant biodiversity. Biol Conserv 143:1594–1602. Available at: https://pdfs.semanticscholar.org/650b/178fa829b06f163b6638bebe1c1a91d48f1b.pdf. Accessed 22 Feb 2018

Laurance WF (2000) Mega-development trends in the Amazon: implications for global change. Environ Monit Assess 61(1):113–122. Available at: http://link.springer.com/10.1023/A:1006374320085. Accessed 7 Feb 2018

Lewinsohn TL, Prado PI (2005) Quantas espécies há no Brasil? Megadiversidade 1:36–42

Lopes A et al (2014a) Brazilian medicinal plants to treat upper respiratory tract and bronchial illness: systematic review and meta-analyses—study protocol. BMJ 4(7):e005267. Available at: http://bmjopen.bmj.com/content/4/7/e005267

Lopes LC et al (2014b) Brazilian medicinal plants to treat upper respiratory tract and bronchial illness: systematic review and meta-analyses—study protocol. BMJ Open 4:e005267. Available at: http://dx.doi.org/ Accessed 8 Feb 2018

Lós DW d S, Barros RP, Neves JDS (2012) Comercialização de plantas medicinais: um estudo etnobotânico nas feiras livres do município de arapiraca–al. Revista de Biologia e Farmácia 7(2):38–51

Marini-Bettòlo GB (1977) Quimiotaxonomia e medicinia popular. Conferencia pronunciada na universidade federal de Alagôas, Maceio, Brazil. (Lecture delivered at the Federal University of Alagoas UFAL, Maceió, Brazil)

Máthé Á (2011) A new look at medicinal and aromatic plants. Acta Hortic 925:13–20.

Máthé A (2015) Medicinal and aromatic plants of the world: scientific, production, commercial and utilization aspects. Available at: https://books.google.com/books?hl=hu&lr=&id=LgqkCgAAQBAJ&oi=fnd&pg=PR5&dq=Medicinal+and+Aromatic+Plants+of+the+World:+Scientific,+Production,+Commercial+and+Utilization+Aspects&ots=V5eeol9FEj&sig=jW5qVfCCDHu6SQcSeQpT1noiz40. Accessed 8 Feb 2016

Mendes FR (2011) Tonic, fortifier and aphrodisiac: adaptogens in the Brazilian folk medicine. Rev Bras 21(4):754–763. Available at: http://www.scielo.br/scielo.php?script=sci_arttext&pid=S0102-695X2011000400028&lng=en&nrm=iso&tlng=en. Accessed 6 Feb 2018

Mitra R et al (2007) Medicinal plants of Brazil. Asia Pac Biotech News 11(11):689–743

Mittermeier RA, Robles Gil P, Hoffmann M, Pilgrim J, Brooks T, Mittermeier CG, Lamoreux J, da Fonseca GAB (2004) Hotspots revisited: Earth's biologically richest and most endangered ecoregions, 1st English edn. CEMEX, Mexico City, p 501

Mittermeier RA, Robles Gil P, Mittermeier CG (1997) Megadiversity: Earth's biologically wealthiest nations. CEMEX and Agrupación Sierra Madre, Mexico

Myers N (1988) Threatened biotas: "Hot spots" in tropical forests. Environmentalist 8(3):187–208. Available at: http://link.springer.com/10.1007/BF02240252 Accessed 6 Feb 2018

Myers N (1990) The biodiversity challenge: expanded hotspots analysis. Environmentalist 10(4):243–256. Available at: https://pdfs.semanticscholar.org/97ff/8edb8f2a51bd25e7113b7ed3b52a8f044ce6.pdf

Oliveira et al. (2013) Estudo etnobotânico sobre plantas medicinais nas feiras livres do município de limoeiro do norte. In: 64° Congresso Nacional de Botânica.2013. Belo Horizonte, 10–15 de Novembro de 2013. Available at: https://www.botanica.org.br/trabalhos-cientificos/64CNBot/resumo-ins18662-id5830.pdf

Roque AA, Rocha RM, Loiola B (2010) Uso e diversidade de plantas medicinais da Caatinga na comunidade rural de Laginhas, município de Caicó, Rio Grande do Norte (nordeste do Brasil). Rev Bras Pl Med 12(1):31–42. Available at: http://www.scielo.br/pdf/rbpm/v12n1/v12n1a06. Accessed 22 Feb 2018

Shanley P, Luz L (2003) The impacts of forest degradation on medicinal plant use and implications for health care in eastern Amazonia. Bioscience 53(6):573

Simberloff D et al (2013) Impacts of biological invasions: what's what and the way forward. Trends Ecol Evol 28(1):58–66 Available at: http://linkinghub.elsevier.com/retrieve/pii/S0169534712001747. Accessed 7 Feb 2018

Tomchinsky B et al (2017) Ethnobotanical study of antimalarial plants in the middle region of the Negro River, Amazonas, Brazil. Acta Amazon 47(3):203–212. Available at: http://www.scielo.br/scielo.php?script=sci_arttext&pid=S0044-59672017000300203&lng=en&tlng=en. Accessed 6 Feb 2018

van den Berg ME et al (1988) Contribuição à flora medicinal de Mato Grosso do Sul. Acta Amazon 18(suppl 1–2):9–22. Available at: http://www.scielo.br/scielo.php?script=sci_arttext&pid=S0044-59671988000500009&lng=pt&tlng=pt. Accessed 22 Feb 2018

Vieira R (1999) Conservation of medicinal and aromatic plants in Brazil. In: Janick J (ed) Perspectiv. ASHS Press, Alexandria. Available at: https://hort.purdue.edu/newcrop/proceedings1999/v4-152.html. Accessed 4 Feb 2018

Vieira RF (2002) Economic potential and conservation of medicinal and aromatic plants from Brazil | Conservamão de recursos genéticos de plantas medicinais e aromáticas brasileiras: Um desafio para o futuro

Wilma D et al (2012) Comercialização de plantas medicinais: um estudo etnobotânico nas feiras livres do município de arapiraca–al. BioFar 7(2):38–51. Available at: http://sites.uepb.edu.br/biofar/download/v7n2-2012/comercializacaodeplantasmedicinais.pdf. Accessed 26 Feb 2018

Zenni RD (2015) The naturalized flora of Brazil: a step towards identifying future invasive non-native species. Rodriguésia 66(4):1137–1144. Available at: https://books.google.hu/books?hl=hu&lr=&id=QzyBPA8SrN4C&oi=fnd&pg=PR5&dq=Invasive+species:+what+everyone+needs+to+know.+1ed&ots=88s8-8wWS-&sig=eKTGJ6lg1wbaxWnR6savhHDh0Dk&redir_esc=y#v=onepage&q&f=false. Accessed 6 Feb 2018

Medicinal Plants and State Policy in South America: The Case of Colonial Brazil

Maria Franco Trindade Medeiros

Abstract The premise that leads our reasoning rests on the view that modern Botany Science was constituted as an instrument of the modern State domination over colonial territories. We could expand this premise for the Portuguese and Spanish domains, especially in South America. However, to well illustrate the theme, this chapter will present a case study on the process of institutionalization of Botany in Portugal and its developments in Brazil, having the plant species and, especially medicinal plants, as the object of analysis. We will point the episteme of natural philosophy as guiding the formation of a new economic policy for the Lusitanian world, increasing the exploration of resources and natural products, including medicinal plants, greatly, from its colonies. Our intention is to address the Botany in a perspective of recognition of its practices and knowledge towards the service of political and economic interests of the State, which will bring implications to the domination process of the Crown in its colonial territories.

Keywords Flora · Natural history · Naturalists · Colonial project · Modern state

1 Science (Botany) and Colonial Project of Biodiversity Exploration (Medicinal Plants)

In the modern period, Europeans were approaching an "exotic" world full of different animals, plants and minerals from what they were used to. Many of these elements, taken as products, came from looting or colonial theft of various parts of the world and docked in Europe through the vessels of transoceanic traffic network (Dean 1991).

In this discovery process of the "new" in the eyes of Europe, Janeira (2005) states that *Natura*, earned a prominent spot in European collections as materialities of *Culture*. From the sixteenth century, the intense movement of natural products,

M. F. T. Medeiros (✉)
Museu Nacional da Universidade Federal do Rio de Janeiro, Rio de Janeiro, RJ, Brazil

U. P. Albuquerque et al. (eds.), *Medicinal and Aromatic Plants of South America*, Medicinal and Aromatic Plants of the World 5,
https://doi.org/10.1007/978-94-024-1552-0_4

trade and greed for States, nobles, collectors and scholars from different parts of Europe promoted the construction of botanical gardens which, at that time, would move from learning and rest spaces to configure as a representation of imperial power (Sanjad 2001).

Between the sixteenth and seventeenth centuries, Foucault (2007) identified work on the natural world that emphasized its symbolic dimension, addressing its admirable or historical aspects and combining the visible characters and the signs that have been deposited or discovered on natural resources. As a counterpoint to this procedure there is the work of Johnston (1657) that initiates a new episteme of Natural History, which focuses on the identification of the research object from observation and description of its own characteristics.

The study of how imperialism and new scientific ideas are imbricated and promoted the advancement, for example, of the British Empire (Drayton 2000), leads to the assertion that the relationship between science and modern empires gives us elements as there was, gradually, a shift of Renaissance conceptual paradigms to the construction of a Natural History based on these new epistemes and practices, in the eighteenth century. In the passage of the eighteenth to the nineteenth centuries, the botanical gardens had already been configured as a collection of plants from the most diverse locations of the empire and as a picture of the botanical discoveries period of naturalists who were operating a change in ideas about the world.

This is, therefore, a process of seeing science in its interface with politics, modern botany as a valued scientific space and assimilated by the economic policies of European preindustrial empires. In the specific case of the Portuguese empire, we see that they used the set of botanical knowledge and practices as a form of domination in their colonies, with a clear intention of exercising power to assimilate botany in its science policy. To Schiebinger and Swan (2005), the passage of the eighteenth to the nineteenth centuries witnessed the development of different scientific fields, including Botany and the dynamic relationship between plants, people, States and economies from that time should not be neglected. The same authors add that in this period the Natural History was strategically important on the global battle between States over territory and resources.

In Brazil, during the eighteenth century, as underlined by Lopes (2005), colonial reforms were linked to the establishment of Natural History consisting of knowledge areas, which are, botany, zoology and mineralogy, supported by local practices and global collections, exceeding only therapeutic interests. Thus, the Portuguese Crown assimilates this new scientific episteme and takes as input for the establishment of a set of reforms initiated in the Pombaline era, which provided for the organization of a reformist movement founded on the exploration of colony natural resources. In this same vein, the thought of Duarte (2004) is developed, by highlighting the role of science in Portugal in the late seventeenth century, due to the Portuguese Crown trying, at this time, to promote a relationship with Brazil through the acquisition of knowledge about its natural world as a maintenance strategy and increment of its colonial domain.

This way, Natural History reorganized relations between political power, knowledge about the natural environment and technical applications to serve the interests of the modern State. If such is this case, this science is manifested as an agent of the overseas empire, and also becomes an instrument of Colonies domination through observation and records in the form of inventory of their natural potential, including, certainly, medicinal and aromatic plants.

Highlighting this issue, Pratt (1999) points to the close connection there was in the second half of the eighteenth century between science, trade and colonial domination. The author shows that the exploration of the countryside and the systematic mapping of the world surface through Natural History would be related to the increasing search of resources that could be commercially exploited, and markets and lands to be colonized.

As Ladurie (1994) and Anderson (2004) enhance, an essential aspect in the State's experience over the modern era is the development and improvement of some technologies, such as the use of science. So, it's important for us to think about the role of scientific knowledge in the modern world, in the heart of the monarchical State, to discuss the relationship between science, trade and colonial domination.

The thought of this relationship between science and State leads to the analysis on the representation of Portuguese political and scientific center, facing the system peripheries, as the center-periphery model. In this case, from 1750, the center of the Empire that united the leading figures and institutions from the colonial policy of the Crown and, so to speak, from the scientific policy of the time, were Lisbon and Coimbra. The role of the center in the relation to the periphery was the accumulation of knowledge about the colonies, action for which the produced sources, whether travel, memories or crafts are seen as elements that can be the result of a previous cycle of accumulation, as can drive to a new knowledge, for example, on a plant species, a drug, a region, vegetation, etc. However, there is still the possibility of the periphery, at times, act as the center, the botanical center, which was independent from the accumulation cycles derived from Portugal to the knowledge and understanding of its nature, its plants and its therapeutic actions. Thus, different locations in Brazil may have been established, in its way, as centers within the Empire, forming their own accumulation cycles of knowledge about the environment, resources, medicinal and aromatic potential, and about many other approaches to the natural world.

The center of this established relationship between the Portuguese-Brazilian cycles have different status that allows it to decide or enforce policies and organize, in its own way, the colonial exploration with the science effort. In this context, although we may consider the existence of other centers in the context of the Brazilian colonial period, Portugal is regarded as the political center of a power network that formed its own Empire.

We must consider that the relationship between Brazil and Lisbon does not need, nor should be taken as linear, within a logic in which there is the possibility of several peripheries, that enclose at the term "Brazil", having negotiated this authority from the center, or having been seen as major actors in the production of knowledge about nature, about the medicinal plants. Thus, we must give recognition and space to the Indians, settlers, religious, and naturalists who resisted the center exploration project, questioning the policy methods established for this relationship between center and periphery (Portugal – Brazil). Russell-Wood (1998) suggests adopting a dynamic model with centers and peripheries alternating in the construction process of the relationship between Portugal and Brazil.

2 Natural History for the Kingdom Development

In the context of the Portuguese empire, the appreciation of knowledge and practices and the production of herbary knowledge were of interest to social actors, as travelers, military and staff. In the late eighteenth century, science or Natural History was consolidated as a compendiums development area that focused on the world's flora. In this period, studies on plants used in medicine, food, lighting and many other aspects of colonial social life were proliferated. It is interesting to note that this scientific output is presented today as a fertile field for the researcher. Contacting these sources, we notice clearly how in a pre-industrial society, elements as waxes, fibers, essential oils, pigments, fruits, seeds, roots, leaves, and other plant parts were essential to maintain the daily life of that society.

Deepening this issue of the scientific production in Europe, particularly in Portugal, it is said that it began with the patronage of King John V (1707–1750), which designed great efforts arising from the Brazilian mining (Schwarcz 2002) to advance and scientific renewal, creating a conducive environment for its consolidation. Between the seventeenth and eighteenth centuries, the presence of naturalists in Portugal was remarkable, particularly growing interest in Natural History in the latter period. With the increasingly frequent use of new plants from the Americas, with applications in medicine and food, and the creation of collectors' gardens, this field focused great scientific capital and, thus, became attractive to those who wanted to fight for its hegemony, establishing its gradual institutionalization throughout the eighteenth century (Bordieu 1983). In this scenario, the efforts of Dom João V, for example, through the establishment of a Royal library, the foundation of the Royal Academy of Portuguese History, the presence of foreign naturalists and their productions on national collections of plants and minerals, the presence of colonial nature in businessmen's and State administrators' minds were conceiving elements to this institutionalization process of Natural History as a scientific field in Portugal (Carvalho 1987; Schwarcz 2002; Furtado 2012).

It was at the following government, Dom José I's (1750–1777) that two important processes happened: First, there was the systematic positivation of sciences, related to teaching; and, secondly, natural philosophy was consolidated, taking precise contours and taking an active role in the Portuguese university framework. This consolidation and scientific progress of Portugal during its government was closely linked to the State reforms implemented by its Foreign Affairs Minister, Marquis of Pombal (1699–1782) (Cruz 2004). Pombal narrowed the relations between Natural History and the State, from the reforms operated in educational curriculum, in 1772, at the University of Coimbra (Gauer 1996). With this reform, scientific travel in the

Kingdom began to be used as practical activities in the course of natural philosophy, allowing a more accurate evaluation of the masters over their naturalist students, and enabling the discovery of new natural resources in the Empire. Contextualizing this historic moment, the picture was of the Portuguese economy crisis (between 1770 and 1780) and the travel and the concern about the dynamization of natural resources were included in this project on reduction of the trade deficit (Wehling 1976). In this Pombaline phase, which begins in 1764 and ends in 1779, year of foundation of the Lisbon Royal Academy of Sciences, is the crystallization of a process of collective intellectual sociability in the natural sciences, which was consolidated in the Marian (Dona Maria I, 1777–1792) and Joanine (Dom João VI, 1792–1808) period.

A second phase, between 1779 and 1808, is characterized by the appropriation of this Pombaline political reform, which allowed the naturalistic development, mainly through the philosophical travels over colonies. In this period of time, there was a fomentation and encouragement by the Crown, which was fully aware of the role of natural sciences to the development of the Kingdom and also an institutionalization process of Botany in the Brazilian colony, which had a spotlight among the others Sciences and was taken as an instrument in the overseas policy framework.

3 Imperial Network of (Medicinal) Plant Circulation

Immerging a little more on the issue of the role of plants in this greatest context recently described, the practices of plant species domestication have been associated with the consolidation of complex human societies and is important for the subsistence agricultural production (Diamond 2005). According to Crosby (1993), the plants also played a leading role in the European project of conquest of the Americas.

The back and forth of plants worldwide has been an issue addressed by different authors, for example, Brockway (1979), Osborne (1994), Drayton (2000), and Beinart and Midleton (2009), which present important issues for different knowledge areas for being interdisciplinary work, wide in their proposals.

The movement of organic materials was a constant among the achievements, between the domains of the Portuguese empire. The second half of the eighteenth century, as already noted, was extremely favorable to the exchange of plants and botanical knowledge. The Pombaline reforms promoted a favorable environment for the articulation of a large network of naturalists in the process of overseas conquests. This network of circulation of products and Natural History knowledge was formed by naturalist travelers and administrative agents represented by viceroys, Captaincy governors, ombudsmen and outside judges. Other social actors, such as scriveners, military and dealers, also participated in the network. In this set of social types, naturalist travelers were the ones who traveled the territories of Portuguese domain with the purpose of inventorying natural resources and send them to the scientific institutions of the Kingdom (Pataca 2006).

The naturalistic and participant in the reform of the Natural History teaching established at the University of Coimbra, the Italian Domenico Vandelli, points out in his writing entitled "Questions that must be answered about the productions of Brazil belonging to the three kingdoms of nature" (177-, pgs. 99–102) that the traveler naturalist's mission focused on registering "all the names of the useful herbs for drinks, and for application of the wounds. [...] Including [...] recipes that experience has made known useful for different diseases both external as internal, [...] [and] also insinuate the way to [...] [operate them]. " It is Important to think about the use of the term "herb" that, in the context of the documents produced by naturalists of the seventeenth, eighteenth and nineteenth centuries, take the meaning of plants used in medicine or food. The "herbs" were, then, those plant species used for the cure of diseases and symptoms such as, for example, wounds, fevers, pain in general; and practice of cooking, as a condiment and preservative; and also we can include in this term those vegetal resources with which there was a precaution in the use because these were considered toxic.

In order to guide future work in Botany that would take place from trips overseas, the first "Botanical Dictionary" (Munteal Son and Melo 2004) was organized by Julio Mattiazzi and Domenico Vandelli in 1780. This work aimed to systematize the knowledge about the New World plants having the primary reference in its uses and properties, identifying the causes and cures for diarrhea and constipation, impotence, generalized infections and therapeutic processes through healing systems adopted at the time, such as bloodletting, cupping, prepared and interventions from cuts (Munteal Son and Melo 2004).

These "herbs" were the plants that concentrate the greatest effort of studies by naturalists during this period. Perhaps the fact that the medicine be grounded in therapeutic practices that had the use of vegetal raw material as a fundamental basis since Antiquity, may justify this increased attention to the botanical production. But for Portugal in the last decades of the eighteenth century, there was also a political scenario that would influence this relevance attached to intellectual production aimed to "herbs". It is notable that from the 1780s, documentary sources can be found more often, including royal orders, addressing the collection, circulation (sending remittances of plants to the Kingdom) and production of knowledge about the "herbs". Elements common to these documents show: the importance the metropolitan authorities used to give to the recognition of Brazilian plants by its own locals, by recommending the registration of common names of species; the technique of circulating these plants between the colony and the metropolis (in the movement from the periphery to the center), which was always of using their matrix land to reach success in transposition and acclimatization of specimens; and the indication of their usefulness, especially medical, and economic application (Royal Order of Martin de Melo and Castro 1795).

At this historic moment there was a dynamization in the movement of plants, stimulated by the Department of Overseas Dominions, which encouraged the governors and officials of the Captaincy to raise useful plants and send remittances of these resources to Portugal. The conventional route of colonial products movement followed the inner pattern of the Captaincy, where the botanical collections and the gathering of information on the common names and usefulness took place, to the Kingdom, where remittances were destined to Help Botanical Gardens (Letter from Governor Don Fernando José de Portugal 1796).

It is interesting to say that the activities related to this process were placed in a rising mechanism in the bureaucracy of the Portuguese State, i.e., the natural history was a way to seek recognition of loyalty and obtaining "honor" in the old Luso-Brazilian regime society. Naturalist employees can be highlighted in history by recognizing indigenous skills, incorporating their knowledge and practices in the reports produced by these men who made science in a practical way, adapted to the colonial reality. In this sense, Marques (1999) says that the use of indigenous knowledge earned for naturalists, travelers and settlers from the sixteenth, seventeenth and eighteenth centuries to make their descriptions and identify medicinal and food plants.

Finally, we can say that this process of medicinal plants circulation considered as "exotic" and the knowledge about their therapeutic application took extensive features, directed to strengthening the Portuguese State. A central aspect of this policy was directly linked to the implementation of an imperial network of botanical gardens by the Portuguese possessions, the motivation of adaptive experiences among naturalists and colonial officials, and the domain of what could be achieved from these resources in curative and economic terms.

References

Anderson P (2004) Linhagens do estado absolutista. Brasiliense, São Paulo

Beinart W, Midleton K (2009) Transferências de plantas em uma perspectiva histórica: o estado da discussão. Topoi 10(19):160–180

Bordieu P (1983) O campo científico. In: Ortiz R (ed) Pierre Bordieu: sociologia. Ática, São Paulo, p 112

Brockway L (1979) Science and colonial expansion: the role of the British Royal botanic gardens. Academic, New York

Crosby AW (1993) Imperialismo ecológico: a expansão biológica da Europa (900–1900). Cia das Letras, São Paulo

da Cruz ALRB (2004) Verdades por mim vistas e observadas, oxalá foram fábulas sonhadas: cientistas brasileiros do setecentos, uma leitura auto-etnográfica [thesis]. Curitiba: Universidade Federal do Paraná

de Carvalho R (1987) A história natural em Portugal no século XVIII. Instituto de Cultura e Língua Portuguesa, Lisboa

Dean WA (1991) Botânica e Política Imperial: a introdução e a domesticação de plantas no Brasil. Rev Estud Históricos 4(8):216–228

Diamond J (2005) Armas, germes e aço. Os destinos das sociedades humanas, 7th edn. Record, Rio de Janeiro

Drayton R (2000) Nature's government: science, imperial Britain, and the improvement of the world. Yale University Press, New Haven

Duarte RH (2004) Facing the forest: European travelers crossing the Mucuri River Valley, Brazil in nineteenth century. Environment and History 10(1):31–58

Foucault M (2007) As palavras e as coisas, 9th edn. Martins Fontes, São Paulo

Furtado JF (2012) Oráculos da geografia iluminista. Dom Luis da Cunha e Jean-Baptiste Bourguignon D'na. Editora da UFMG, Belo Horizonte
Gauer RMC (1996) A modernidade portuguesa e a reforma pombalina de 1772. EDIPUCRS, Porto Alegre
Janeira AL (2005) Configurações epistémicas do colecionismo. Episteme 20(Special Suppl Jan./Jun.):231
Ladurie ELR (1994) O estado monárquico. França, 1460–1610. Companhia das Letras, São Paulo
Lopes MM (2005) Culturas das ciências naturais. Ciência & Educação 11(3):459
Marques VRB (1999) Natureza em boiões: medicinas e boticários no Brasil setecentista. Ed. Unicamp, Campinas
Munteal Filho O, de Melo MF (2004) Minas Gerais e a história natural das Colônias: política colonial e cultura científica no século XVIII. Fundação João Pinheiro, Belo Horizonte
Ofício do Governador Dom Fernando José de Portugal para Luiz Pinto de Sousa, no qual participa a remessa das plantas medicinais (1796) Manuscrito do Arquivo Histórico Ultramarino, Bahia, Caixa no. 85, Doc. 16665
Ordem Régia de Martinho de Melo e Castro para Dom Fernando José de Portugal para o envio de remessas de plantas (12 de setembro de 1795) Lisboa. Manuscrito do Arquivo Histórico Ultramarino, Bahia, Códice 606, fl. 64
Osborne M (1994) Nature, the exotic and the Science of french colonialismo. Indiana University Press, Indianapolis
Pataca EM (2006) Terra, água e ar nas viagens científicas portuguesas (1755–1808) [thesis]. Universidade Estadual de Campinas, Campinas
Pratt ML (1999) Os olhos do Império: relatos de viagem e transculturação. EDUSC, Bauru
Russel-Wood AJR (1998) Centros e periferias no mundo luso-brasileiro, 1500–1808. Rev Bras Hist 18(36):187–250
Sanjad NR (2001) Nos jardins de São José: uma história do Jardim Botânico do Grão-Pará, 1796–1873 [dissertation]. Universidade Estadual de Campinas, Campinas
Schiebinger L, Swan C (2005) Colonial botany: Science, commerce, and politics inthe early modern wolrd. University of Pennsulvania Press, Pennsylvania
Schwarcz LM (2002) A longa viagem da biblioteca dos reis: do terremoto de Lisboa à independência do Brasil. Companhia das Letras, São Paulo
Vandelli D. Perguntas a que se deve responder sobre as produções do Brasil pertencentes aos três reinos da natureza. Arquivo Nacional, Códice 807, v. 10, fls. 99–102.177
Wehling A (1976) O fomentismo português no final do século XVIII: doutrinas, mecanismos, exemplificações. Revista do IHGB 316:170–278

Part III
Selected Medicinal and Aromatic Plants of Brazil

Achyrocline satureioides (Lam.) DC.

Gabriela Granghelli Gonçalves, Maria Izabela Ferreira, and Lin Chau Ming

Achrocline satureiodes Lam. (DC.)
Photo: Fernando Alzate Guarín. Available in: http://www.tropicos.org/Image/100539810

G. G. Gonçalves · M. I. Ferreira · L. C. Ming (✉)
Horticulture Department, School of Agronomic Sciences, Universidade Estadual Paulista (UNESP), Botucatu, São Paulo, Brazil
e-mail: linming@fca.unesp.br

U. P. Albuquerque et al. (eds.), *Medicinal and Aromatic Plants of South America*, Medicinal and Aromatic Plants of the World 5,
https://doi.org/10.1007/978-94-024-1552-0_5

Abstract *Achyrocline satureioides* (Lam.) DC. belonging to the Asteraceae family, is a herb native to South America, heliophytic and ruderal, and grows wild in grasslands, wastelands and secondary forests. Naturally, its propagation structure constitutes the achenes, but propagation by apical cuttings is also feasible. The crude drug (Achyroclines flos) consists of its dried flowers, with golden yellow coloring and that presents aromatic and pleasant odor and slightly bitter taste, due to the presence of substances such as essential oils and flavonoids. Its flowers should be collected when fully developed, usually during autumn. In the Brazilian folk medicine, its tea is used as digestive, eupeptic, emmenagogue, antispasmodic, anti-inflammatory, expectorant and antidiarrheal. It is popular in treating disorders of the gastrointestinal tract and its anti-inflammatory action has been confirmed in pharmacological studies. Tests performed with isolated flavonoids (quercetin, quercetin 3-methyl ether and luteolin) of this species demonstrate that these compounds may be at least partially responsible for these activities.

Keywords *Achyrocline satureioides* · Asteraceae family · *Macela* · Aromatic plants · Flavonoids · Calming action

1 Taxonomic Characteristics

The Asteraceae family is one of the largest families, with about 1,600 genera and 23,000 species. It is distributed in tropical, subtropical and temperate regions and represents *ca.* 10% of the vascular flora of the world. It occurs throughout the Neotropics, but there are not many species in rainforest and aquatic habitats. Species of the family are common in mountain habitats, disturbed areas and semi-arid regions, and they can be found, also as common weeds, in the most cultivated regions. In Brazil, the family is represented by about 278 genera and 2,065 species in different biomes (Hind 2009; Nakajima et al. 2015). *Achyrocline satureioides* (Lam.) DC. was first described by French naturalist Lamarck with the name of *Gnaphalium satureioides* Lam., published in *Encyclopédie Methodique, Botanique* in 1788. Posteriorly, the Swiss botanist De Candolle listed the species as *A. satureioides*, currently valid name, which was published in Prodromus Systematis Naturalis Regni Vegetabilis 6: 220: 1837 [1838] (Tropicos.org. n.d.). Its crude drug name is *Achyroclines* flos. It belongs to the Equisetopsida class, Asterales order, Asteraceae family and *Achyrocline* genus. The basionym of *A. satureioides* is *Gnaphalium satureioides* Lam.

Synonyms *Achyrocline candicans* (Kunth) DC.; *Achyrocline satureioides* var. vargasiana (DC.) Baker; *Achyrocline vargasiana* DC.; *Gnaphalium candicas* Kunth; *Gnaphalium satureioides* Lam.; *Gnaphalium satureioides* var. candicans (Kunth) Kuntze)

2 Crude Drug Used

Its crude drug consists of the dried golden flowers, yellow in coloring but it can vary independently of the maturation stage. Of aromatic and pleasant odor, it has a slightly bitter flavor. It must contain a minimum of 1.7% total flavonoids (calculated as quercetin), 0.14% quercetin and 0.07% luteolin. The presence of peduncles and pedicels with a length up to 3 cm should not exceed 1% of the total dry weight (Brasil 2011).

3 Major Chemical Constituents and Bioactive Compounds

Essential oils which contain substances such as monoterpenes, sesquiterpenes, cadinene, caryophyllene, sesquiterpene, ocimene, pinene; alkaloids; flavonoids as flavonol, quercetin, and luteolin (Brasil 2011). Other substances have been isolated from aerial parts of *A. satureioides*, including galangin, chlorogenic acid, achifurano(5) galangin 3-methyl ether, quercetin 3-methyl ether, caffeic acid and two esters of calleryanin (3,4 dihydroxybenzylalcohol 4-glucoside), with caffeic acid and protocatechuic acid (Ferraro et al. 1981).

4 Morphological Description

The herb is 0.5–1 m tall; leafy, cylindrical, costate, woolly twigs. Its leaves are simple, alternate, sessile, 10–70 × 2–7 mm in limb, linear lanceolate; acuminated apex, entire margin, truncate base.; adaxial floribundum surface, abaxial canescent surface. Disciform, sessile capitula, in dense corymbs; cylindrical involucre, 5–6 mm long, 1–2 mm diam.; involucral hyaline bracts, 3-serial, 2.5–5 × 0.7–1 mm, ovate to lanceolate, glandular, external serial with acute apex, entire margin, woolly in the base; plane, foveolate, glabrous receptacle. Marginal, cream, corolla filiform flowers, tube 4.5 mm long, 0.1 mm diam., internally glabrous, 5-toothed; branches of the cylindrical style, truncated, glabrous apex. Cypsela is ellipsoid, 1 mm long, 0.5 mm diam., glabrous; pappus 5 mm long. Central, monoclinous, cream, corolla tubular flowers, tube 3.5 mm long, 0.6 mm diam., internally glabrous, lobes 0.5 × 0.1 mm, glandular; anthers with appendix of lanceolate connective, calcarate base; branches of cylindrical styles, truncated, penicillate, without hairy surface below the bifurcation point. Cypsela is cylindrical, 4–5-costate, 1 mm width, 0.4 mm diam.; pappus 5 mm long, 1-serial, setose, caduceus (Hattori and Nakajima 2008).

5 Geographical Distribution

It is native to South America, occurring in Argentina e Uruguay, in addition to Brazil, where it occurs in the Northeast (Bahia), the Southeast (Minas Gerais, São Paulo, Rio de Janeiro) and the South (Paraná, Santa Catarina, Rio Grande do Sul) (Nakajima et al. 2015).

6 Ecological Requirements

Heliophytic and ruderal species grows wild in grasslands, wastelands and secondary forests. It grows in sandy, clay, stony soil and even semi-halophytes areas, near the sea. However, it prefers fertile soils with good moisture content. It occurs in areas with different plant formations, Cerrado, Atlantic Forest and Pampa. It blooms in summer and in fall, bearing fruit in the same period (Flora do Brasil 2015; Flora SBS 2015). It grows from 0 to 2000 m elevation, adapting better at moderate climates (Martínez et al. 1999). Its natural propagation structure is constituted of achenes with an anatomical-morphological adaptation indicating anemochory, i.e.: the wind dispersion of seeds (Simões et al. 1988).The seeds are positive photoblastic with optimum temperatures for germination between 20 and 25 °C and can be stored for 10 months at room temperature (25 ± 5 °C). After this period there is a significant decrease in germination percentage (Ikuta and de Barros 1996).

7 Collection Practice

Flowers should be collected when fully developed, usually during autumn. In Rio Grande do Sul state, it is traditionally collected during the early hours of Good Friday, as it is believed that there is a potentialization of its medicinal properties (Mota 2011a, b). Dried flowers should be stored in tightly closed containers protected from light and heat, for a period not exceeding 1 year (Brasil 2011). Plant material for commercial purposes is mostly collected in the wild because it is not cultivated, but only in small plots in homegardens (Retta et al. 2012). There are some studies on domestication of this species, including the germination of seeds (Ikuta and Barros 1996; Marques and Inchausti 2000; Ajalla et al. 2009; Motta 2011; Vieira et al. 2015), and propagation by cuttings, that is also feasible. According to Ikuta (1998) apical cuttings are recommended, as these are more efficient than the side cuttings.

8 Traditional Use

A. *satureioides* is popularly known as macela or marcela, camomila-nacional, carrapichinho de agulha, chá de lagoa, losna do mato, macela amarela, macela da terra, macela do sertão, macelinha, macelinha do campo, marcela, marcela do campo, marcelinha, paina and Eloyatei-caá in the Guarani language (Lorenzi and de Matos 2002).

The use *A. satureioides* inflorescences is documented in the first edition of the Pharmacopoeia of Brazil (1928) and updated in the fourth edition (2001). Macela is also included in Phytotherapeutics Form from Brazilian Pharmacopoeia (Farmacopéia Brasileira 2001) indicated as anti-dyspeptic, anti-inflammatory and anti-spasmodic, and is recommended in the form of infusion of 1.5 g of flowers in 150 ml of water. It should be consumed immediately after preparation, two or three times a day. Children under 12 years of age, should not use it. In case of allergy the use should be discontinued (Brasil 2011).

Due to its gentle scent and calming action, dried inflorescences are used, in many parts of Brazil, for filling pillows and blankets (Lorenzi and de Matos 2002; Pio-Correa 1984).

In the state of Rio Grande do Sul, Brasil, *A. satureioides* is one of the most frequently used medicinal plants and due to its great importance for the population, it was considered by law (Lei n. 11,858) as the medicinal plant symbol of the state.

The tea of flowers is used in the Brazilian folk medicine as digestive, eupeptic, emmenagogue, antispasmodic, anti-inflammatory, expectorant and antidiarrheal (Pio-Correa 1984; Simões et al. 1988; Oliveira and Akisue 2009; Barata et al. 2009; Retta et al. 2012).

9 Modern Medicine Based on Its Traditional Use

A. satureoides is also included in the Medicinal Species List of ANVISA (2010), which formalizes and standardizes the use of these species as herbal medicines in Brazil. Due to its mild sedative and anti-inflammatory effect, its use is indicated for indigestion and intestinal colic.

Popular use of *A. satureioides* to treat disorders of the gastrointestinal tract and its anti-inflammatory action has been confirmed in pharmacological studies. The tests performed with isolated flavonoids (quercetin, quercetin 3-methyl ether and luteolin) of this plant demonstrate that these compounds may be at least partially responsible for these activities (Simões et al. 1988; De Souza et al. 2007).

The flavonoids of the extracts of the species are also responsible for its antioxidant action, that has been proved by chemical (Leal et al. 2006; Grassi-Zampieron et al. 2009) and biological assays (Desmarchelier et al. 1998; Polydoro et al. 2004; Arredondo et al. 2004).

Antimicrobial activity of its isolates (23-methyl-6-*O*-desmethylauricepyrone, quercetin and 3-*O*-methylquercetin) against *Staphylococcus aureus* and *Escherichia coli,* has been proven to have higher efficiency in the form of combined metabolites. These results indicate the synergism of the metabolites for the control of pathogenic bacteria (Joray et al. 2013).

The use of aerial parts of *A. satureoides* in the popular medicine and the presence of its derivate cafe oil showed hepatic protective activity in rats. Results obtained with aqueous extracts (5% (w/v) also support its use in popular medicine (Kadarian et al. 2002).

The extract also has neuroproctective effect, indicated for prevention and treatment of vascular isquemy, neurodegenerative diseases and brain lesions caused by aging (Heizen and Dajas 2003).

Retta et al. (2012) performed a bibliographic survey on the substantiated biological activities of extracts, infusions and decoctions of *A. satureioides*, and found the following activities: photo protection of ethanolic extracts (Morquio et al. 2005), antiviral- alcoholic extract (Zanon et al. 1999; Bettega et al. 2009), antiallergic – Leaves and flowers decoction (Maldonado et al. 2007), vein relaxant (Vecchio et al. 2002), protection of neuronal cells -infusion (Blasina et al. 2009), antitumoral-methanolic extract of aerial parts, flowers (Ruffa et al. 2002; Arisawa 1994), and antihyperglycemic – whole-plant extract (Carney et al. 2002).

10 Conclusions

Achyrocline satureioides Lam. (DC.) is widely used in folk medicine in several Brazilian regions, and there are pharmacological and clinical information that confirm its indications in popular uses. Due to its importance in Brazil, there are already studies on the propagation and genetic improvement of this species with the aim of obtaining plants with better chemical and agronomic quality (Ming et al. 2012).

References

Ajalla ACA, Vieira MC, Zarate NAH, Mota JH, Souza TM (2009) Produtividade da marcela [*Achyrocline satureioides* (Lam.) Dc.] em cultivo solteiro e consorciado com tanchagem (*Plantago major* L.). Ciênc Agrotec., Lavras 33(2):488–495

Arisawa M (1994) Cell growth inhibition of Kb cells by plant extracts. Nat Med 48:338–347

Arredondo MF, Blasina F, Echeverri C, Morquio A, Ferreira M, Abin-Carriquiry JA, Lafon L, Dajas F (2004) Cytoprotection by *Achyrocline satureioides* (Lam) D.C. and some of its main flavonoids against oxidative stress. J Ethnopharmacol 91:13–20

Barata LES, Alencar AAJ, Tascone M, Tamashiro J (2009) Plantas Medicinais Brasileiras. I. *Achyrocline satureioides* (Lam.) DC. (Macela). Rev Fitos 4(01):121–125

Blasina MF, Vaamonde L, Morquio A, Echeverry C, Arredondo F, Dajas F (2009) Differentiation induced by *Achyrocline satureioides* (Lam.) infusion in PC12 cells. Phytother Res 23:1263–1269

Bottega FC, Jeller AH, Cardoso CAL, Vieira M, Do Carmo L, Leite CMB, Zarate NAH, Oliveira EE (2009) Influencia de tratamentos agronomicos na produção de oleos essenciais em *Achyrocline satureoides*. In: Abstracts ofthe 32nd annual meeting of Brazilian Society of Chemistry, Fortaleza, Brazil

Brasil (2011) Formulário de Fitoterápicos Farmacopeia Brasileira

Carney JR, Krenisky JM, Williamson RT, Luo J (2002) Achyrofuran, a new antihyperglycemic dibenzofuran from the South American medicinal plant *Achyrocline satureioides*. J Nat Prod 65:203–205

De Souza KCB, Bassani VL, Schapoval EES (2007) Influence of excipients and technological process on anti-inflammatory activity of quercetin and *Achyrocline satureioides* (Lam.) D.C. extracts by oral route. Phytomedicine 14:102–108

Desmarchelier C, Coussio J, Cicci G (1998) Antioxidant and free radical scavenging effects in extracts of the medicinal herb *Achyrocline satureioides* (Lam.) DC. ("marcela"). Braz J Med Biol Res 31:1163–1170

Farmacopéia Brasileira (2001) Comissão Permanente de Revisão da Farmacopéia Brasileira, 2nd edn. Atheneu, São Paulo. 449 p

Ferraro GE, Norbedo C, Coussio JD (1981) Polyphenols from *Achyrocline satureioides*. Phytochemistry 20(8):2053–2054

Flora do Brasil: Lista de espécies do Brasil – *Achyrocline satureioides* (Lam.) DC (2015) [Internet]. [Cited 2015 Feb 11]. Available from: http://reflora.jbrj.gov.br/jabot/listaBrasil

Flora SBS – *Achyrocline satureioides* (Macela) (2015) [Internet]. [Cited 2015 Feb 11]. Available from. https://sites.google.com/site/florasbs/asteraceae/achyrocline-satureoides---macela

Grassi-Zampieron RF, Vieira MC, de Siqueira JM (2009) Atividade antioxidante e captora de radicais livres dos extratos de *Achyrocline alata* (Kunth.) DC. em comparação com extratos de *Achyrocline satureioides* (Lam.) DC. Rev Bras Farmacogn Braz J Pharmacogn 19(2B)

Hattori EKO, Nakajima JN (2008) A família Asteraceae na estação de Pesquisa e Desenvolvimento Ambiental Galheiros, Perdizes, Minas Gerais, Brasil. Rodriguesia 59:687–749

Heizen H, Dajas F (2003) Utilization of *Achyrocline satureioides* (marcela) extracts and liposomal preparations of natural and semi-synthetic flavonoids for the prevention and treatment of the consequences of stroke and neurodegenerative diseases. U.S. Patent application 10/190440. Protocolo n° 2003055103-A, Estados Unidos da América, 20/03/2003

Hind DJN (2009) Neotropical Asteraceae – Neotropikey from Kew [Internet]. In: Milliken W, Klitgård B, Baracat A (eds) Interactive key and information resources for flowering plants of the Neotropics. [Cited 2015 Mar 4]. Available from. http://www.kew.org/science/tropamerica/neotropikey/families/Asteraceae.htm

Ikuta ARY (1998) Estudos sobre propagação de Macela, *Achyrocline satureiodes* (Lam.) DC. (Asteraceae). In: Ming LC, Scheffer MC, Júnior CC, Barros IBI, Abreu Matos JK (eds) Plantas medicinais aromáticas e condimentares: avanços na pesquisa agronômica, vol 1. Unesp, Botucatu, p 238

Ikuta ARY, de Barros IBI (1996) Influência da temperatura e da luz sobre a germinação de marcela (*Achyrocline satureioides*). Pesqui Agrop Bras 31(12):859–862

Joray MB, Palacios SM, Carpinella MC (2013) Understanding the interactions between metabolites isolated from *Achyrocline satureioides* in relation to its antibacterial activity. Phytomedicine 20:258–261

Kadarian C, Broussalis AM, Miño J, López P, Gorzalczany S, Ferraro G, Acevedo C (2002) Hepatoprotective activity of Achyrocline satureioides (Lam.) DC. Pharmacol Res 45:57–61

Leal P, Queiroga C, Rodrigues M, Montanari I, Meireles MA (2006) Global yields, chemical compositions, and antioxidant activities of extracts from *Achyrocline alata* and *Achyrocline satureioides*. Pharmacogn Mag 2:153–159

Lorenzi H, de Matos FA (2002) Plantas medicinais no Brasil: nativas e exóticas. Plantarum, Nova Odessa. 512 p

Maldonado AM, Cariddi L, Alaniz F, Zigadlo J, Grosso M, Sabini L (2007) ArchAl. Inmun Clin 38:58–72

Marques FCB, Inchausti IB (2000) Qualidade de sementes de marcela (*Achyrocline satureioides*) provenientes de duas populações no Rio Grande do Sul. Ciênc Rural Santa Maria 30(2):241–247
Martínez JV, Yesid Bernal AH, Cáceres A (1999) Fundamentos de agrotecnología de cultivo de plantas medicinales iberoamericanas. Rev Cuba Plantas Med Editorial Cienc Méd 5(3):125
Ming LC, Ferreira MI, Gonçalves GG (2012) Pesquisas agronômicas das plantas medicinais da Mata Atlântica regulamentadas pela ANVISA. Rev Bras Plantas Med 14:131–137
Morquio A, Rivera-Megret F, Dajas F (2005) Photoprotection by topical application of *Achyrocline satureioides* ('Marcela'). Phytother Res 19:486–490
Mota F, Carvalho H, Wiest J (2011a) Atividade antibacteriana in vitro de inflorescências de *Achyrocline satureioides* (Lam.) DC.-Asteraceae ("macela","marcela") sobre agentes bacterianos de. Rev Bras Plantas Med 13(3):298–304
Mota JH, Vieira MC, Araújo C (2011b) Crescimento e produção de alface e marcela em cultivo solteiro e Consorciado, Acta Scientiarum. Agron Maringá 33(2):269–273
Nakajima JN, Loeuille B, Heiden G, Dematteis M, Hattori EKO, Magenta MAG (2015) Asteraceae in Lista de Espécies da Flora do Brasil. [Internet]. Jardim Botânico do Rio de Janeiro. [Cited 2015 Mar 4]. Available from: http://floradobrasil.jbrj.gov.br/jabot/listaBrasil
Oliveira FD, Akisue G (2009) Fundamentos de Farmacobotânica e Morfologia Vegetal, 3rd edn. Atheneu, São Paulo. 228 p
Pio-Correa M (1984) Dicionário de plantas úteis do Brasil e das exóticas cultivadas, 5th edn. Ministério da Agricultura, Instituto Brasileiro de Desenvolvimento Florestal, Rio de Janeiro. 687 p
Polydoro M, de Souza KCB, Andradesa ME, Da Silva EGB, Bonatto F, Heydrichb J, Dal-Pizzola F, Schapovalb EES, Bassanib VL, Moreira JCF (2004) Antioxidant, pro-oxidant and cytotoxic effects of *Achyrocline satureioides* extracts. Life Sci 74:2815–2826
Retta D, Dellacassa E, Villamil J, Suárez SA, Bandoni AL (2012) Marcela, a promising medicinal and aromatic plant from Latin America: a review. Ind Crop Prod 38:27–38
Ruffa MJ, Ferraro GE, Wagner ML, Calcagno ML, Campos RH, Cavallaro L (2002) Cytotoxic effect of Argentine medicinal plant extracts on human hepatocellular carcinoma cell line. J Ethnopharmacol 79:335–339
Simões CM, Schenkel EP, Bauer L, Langeloh A (1988) Pharmacological investigations on *Achyrocline satureioides* (LAM.) DC., Compositae. J Ethnopharmacol 22:281–293
Tropicos (n.d.) [Internet]. Missouri Botanical Garden. [Cited 2014 Dec 11]. Available from: http://www.tropicos.org/Name/2717100
Vecchio G, Moscatelli V, Castro J, Ferraro G, Acevedo C (2002) Efectos de "marcela" sobre la presión arterial y la frecuencia cardíaca de rata. In: I Congreso Latinoamericano de Fitoquímica; IV Reunión de la Sociedad Latinoamericana de Fitoquímica, Argentina, Buenos Aires
Vieira MC, Ramos MBM, Heredia Zárate NA, Luciano AT, Gonçalves WV, Rodrigues WB, Tabaldi LA, De Carvalho TM, Soares LF, De Siqueira JM (2015) Adubação fosfatada associada à cama de frango e sua influência na produtividade e no teor de flavonoides da Marcela (*Achyrocline satureioides* (Lam.) DC.) em duas épocas de colheita. Rev Bras Plant Med, Campinas 17(2):246–253
Zanon SM, Ceriatti FS, Rovera M, Sabini LJ, Ramos BA (1999) Search for antiviral activity of certain medicinal plants from Córdoba, Argentina. Rev Latinoam Microbiol 41:59–62

Adiantum raddianum C. Presl.

Rafael Corrêa Prota dos Santos Reinaldo, Ivanilda Soares Feitosa, Augusto César Pessôa Santiago, and Ulysses Paulino Albuquerque

Adiantum raddianum C. Presl.
Photo: Augusto Santiago.

R. C. P. d. S. Reinaldo (✉) · I. S. Feitosa
Laboratório de Ecologia e Evolução de Sistemas Socioecológicos, Departamento de Botânica, Universidade Federal de Pernambuco, Cidade Universitária, Recife, PE, Brazil

A. C. P. Santiago
Laboratory of Biodiversity (Laboratório de Biodiversidade), Biology Nucleus (Núcleo de Biologia), Federal University of Pernambuco (Universidade Federal de Pernambuco), Vitória Academic Center (Centro Acadêmico de Vitória),
Bela Vista, Vitória de Santo Antão, PE, Brazil

U. P. Albuquerque et al. (eds.), *Medicinal and Aromatic Plants of South America*, Medicinal and Aromatic Plants of the World 5,
https://doi.org/10.1007/978-94-024-1552-0_6

Abstract *Adiantum raddianum* C. Presl. is a fern that is used in popular medicine by several ethnic groups from different South American countries. Its most common traditional uses are as an analgesic, expectorant, and diuretic, as well as for the treatment of digestive problems. In addition to its medicinal uses, it is commonly used and cultivated for ornamental purposes. Similar to other species of the genus *Adiantum*, the species has confirmed pharmacological activities. Some of these activities are common to other species of this genus, which indicates the presence of an interesting chemical repertoire with therapeutic applications. Among its bioactive compounds, filicene is present in high quantities. It is indicated as one of the main compounds responsible for the strong analgesic activity observed in pharmacological studies.

Keywords *Adiantum cuneatum* Langsd. & Fisch · Pharmacological activity · Bioactive compounds · Medicinal tea

1 Taxonomic Characteristics

Adiantum raddianum C. Presl. is a fern that belongs to family Pteridaceae and order Polypodiales (Smith et al. 2006). The genus *Adiantum* is not monophyletic, specifically, because it includes the clade that contains *A. raddianum* (Prado et al. 2007), indicating that the species may be renamed in future studies. In Brazil, where the official language is Portuguese, this species is commonly known as "avenca" (Barros and Andrade 1997). In Ecuador, Argentina and other Spanish-speaking South American countries, it is known as "culantrillo" or "culantrillo del pozo" (Keller et al. 2011; Quattrocchi 2012). In English, it is known as "small cilantro," "Mexican maidenhair" or "maidenhair fern" (Quattrocchi 2012).

In addition to its most widely accepted scientific name, *Adiantum raddianum,* which was proposed by Carl Borivoj Presl. in 1836 (Prado 2015), several synonyms can be found in the special literature: *Adiantum amabile* Liebm.; *Adiantum amabile* Moore; *Adiantum boliviense* C. Chr. & Rosenst.; *Adiantum colpodes* T. Moore; *Adiantum cuneatum* G. Forst; *Adiantum cuneatum* Langsd. & Fisch. Nomileeg; *Adiantum decorum* Moore; *Adiantum decorum* var. *quadripinnatum* Rosenst; *Adiantum mexicanum* C. Presl.; *Adiantum moorei* Baker; *Adiantum remyanum* Esp., Bustus; *Adiantum rubellum* Moore; *Adiantum rufopunctatum* Mett. ex Kuhn; *Adiantum Tinctum* Moore; and *Adiantum werkleanum* H. Christ.

U. P. Albuquerque
Departamento de Botânica, Centro de Biociências, Universidade Federal de Pernambuco, Recife, Brazil

2 Crude Drug Used

Although some studies have reported the therapeutic potential of *A. raddianum,* the efficacy and safety of its medicinal use is not acknowledged in the current legislation from Brazil or any other South American country. In contrast, its medicinal uses are preserved in several local cultures. Although there are reports of the use of all parts of the plant in one single preparation (Basualdo et al. 2004), it is the fresh or dry fronds that are most commonly used for the preparation of medicinal teas (Keller et al. 2011; Santos et al. 2012).

3 Major Chemical Constituents and Bioactive Compounds

A. raddianum is increasingly being studied because it can produce large concentrations of filicene (Filic-3-ene), a steroidal triterpene that is one of the main constituents responsible for the analgesic activity attributed to *A. raddianum* by folk medicine (Souza et al. 2009). In addition, filicene has exhibited antihyperplasic and hypocholesterolemic activity in mice (Bresciani 2003). Filicene concentrations typically vary in different parts of the plant. Bresciani (2003) observed higher filicene concentrations in the fronds, although it also occurs in lower concentrations in the rhizomes. The filicene concentrations also exhibit high seasonal variations. *A. raddianum* tends to produce higher levels of filicene in winter. This variation may indicate that filicene is produced for plant growth, reproduction and defense and that its concentration increases under certain (more favorable) environmental conditions, or that filicene is a precursor of another compound (Bresciani 2003). Filicenal, another triterpene, has also been reported to be responsible for the analgesic activity of *A. raddianum*, although it is produced in lower quantities.

Filicene and filicenal are not the only triterpenes found in *A. raddianum.* Pan et al. (2011) performed a survey of the chemical constituents of genus *Adiantum*, and observed that the following triterpenes are abundant in *A. raddianum*: *Isohopane* and *neohopane [Neohop-12-eno; Neohop-18-en-12a-ol; 13-Epineohop-18-en-12a-ol; Neohop-13(18)-en-19a-ol; Neohopa-11,13(18)-diene], Norhopane (Trisnorhopane; Isoglaucanone; Glaucanol B acetate; 21-Hydroxy-30-norhopan-22-ona; Isoadiantol B), Fernano [Fern-9(11)-en-25-ol; Fern-9(11)-ene; Fern-7-en-25-ol; 7-fernene; 25-Norfern-7-en-10b-yl formate; 7α,8α-Epoxyfernan-25-ol; 7b,25-Epoxyfern-8-ene; 7β,25-Epoxyfern-9(11)-en-8 α-ol], Adiane,* and *Filicane [Adian-5-en-3a-ol; Adian-5-en-25-ol; Filicenal; 4,23-Bisnor-3,4-secofilic-5(24)-en-3-al; and 4,23-Bisnor-3,3-dimethoxy-3,4-secofilic-5(24)-ene].* Some flavonoids have also been isolated from *A. raddianum*, namely querciturone, kaempferol 3-glucuronide and astragalin.

4 Morphological Description

Adiantum is one of the most represented genera of medicinal ferns. Many of its species display delicate shapes with a highly decorative effect, conferring high ornamental value (Windisch 1990). A distinctive characteristic of this genus is that the sporangia are located above the indusium (the structure formed by the curving of the leaf blade) instead of under it, as it is observed in the remaining genera of this family. Due to the similarities between many species within this genus, there may be problems with the taxonomical delimitation of some species. Hybrids are also commonly found (Lellinger 1991).

A. raddianum is an herbaceous hemicryptophyte fern. It displays short rhizomes with acuminate blackened scales. It can be identified by the presence of tripinnate leaves and flabeliform segments. The fronds display shiny, fasciculate petioles that are approximately 10–20 cm in width and 30–40 cm in length and that have scales at the base. The leaflets possess a cuneate base, a rounded and wavy margin, and bifurcated veins. The plant exhibits numerous, very small sori that are surrounded by kidney-shaped subcircular indusia (Windisch 1990; Santos and Sylvestre 2006; Santos et al. 2012).

Some species may be mistaken for *A. raddianum*, such as *A. capillus-veneris*, *A. lorentzii* and *A. poiretii*. However, *A. raddianum* can be distinguished because it displays sterile pinnae veins ending in marginal sinuses (*versus* veins ending in marginal teeth, as in *A. capillus-veneris*), circular sori without yellow-colored powder between the sporangia (*versus* oblong sori with yellow-colored powder between the sporangia, as in *A. poiretii*) and incisions of the ultimate segments that reach up to half of the pinula, with rounded lobes (*versus* incision of the ultimate segments that reach up to 2/3 of the pinula, with linear lobes, as in *A. poiretii*) (Moran et al. 1995; Winter et al. 2011).

5 Geographical Distribution

A. raddianum displays a neotropical distribution, occurring from Southern Mexico to Argentina (Santos and Sylvestre 2006). It occurs in most countries of South America, including Argentina, Uruguay, Peru, Bolivia, Ecuador, Venezuela and Brazil (Winter et al. 2011). The primary area is in tropical climates and subtropical humid regions.

6 Ecological Requirements

A. raddianum grows in a wide variety of environmental conditions, and the light conditions are not important (Winter et al. 2011). It can be found in humid and shaded, partly shaded, and direct sunlight environments (Santos and Sylvestre 2006; Winter et al. 2011). It occurs in creek margins, swamp areas, humid ravines, cliffs next to waterfalls, roadsides, and on some palm trees (Sehnem 1972; Senna and Kasmirckaz 1997; Winter et al. 2011).

7 Collection Practice

Ideally, the plant should be collected during winter, when the production of filicene, the principle active component responsible for the pharmacological activities of *A. raddianum*, is higher (Bresciani 2003).

8 Traditional Use (Part(s) Used) and Common Knowledge

A. raddianum is included in the popular pharmacopoeia of different ethnic groups, in Latin America. The fronds (shoots) are the plant parts that are most commonly used for medicinal purposes in most of these cultures, and they are administered as infusions (Vendrusculo and Mentz 2006; Tribess et al. 2015). The medicinal properties attributed to this species are diverse. It is used in baths to treat colds, cough, food poisoning (vomiting and stomach pain), gynecological problems (irregular menstrual cycle), headaches, nausea, fever, nasal hemorrhage, diarrhea, and cancer and as a female contraceptive (Vendruscolo and Mentz 2006; Keller et al. 2011; de La Cruz et al. 2014; Tribess et al. 2015).

9 Modern Medicine Based on Its Traditional Medicine Uses

Sharma et al. (2013) evaluated the antimicrobial potential of *A. raddianum* and observed considerable antibacterial activity of its ethanol extracts against *Pseudomonas aeruginosa* and *Staphylococcus aureus*. This activity was comparable to the antibiotic netilmicin. The methanol extracts were also observed to inhibit *S. aureus* (Thomas 2014). Species of the genus *Adiantum* have been considered to be good sources of antimicrobial agents (Singh et al. 2008; Pan et al. 2011), and the methanol extracts of some species of *Adiantum* were found to exhibit higher antimicrobial activity than commercial antibiotics, such as gentamicin and ketoconazole (Singh et al. 2008).

A. raddianum also has antinociceptive activity. Sousa et al. (2009) observed that filicene was capable of inhibiting acetic acid-induced abdominal contractions in mice. Its analgesic effect was stronger than that of commercial analgesics, such as acetaminophen, diclofenac and acetylsalicylic acid. Although the mechanism of action of filicene is not completely clear, it is known to involve interactions with the cholinergic, dopaminergic, glutamatergic, GABAergic (gamma-aminobutyric acidergic) and tachykinergic systems (Sousa et al. 2009).

Another important medicinal property of *A. raddianum* is its antihyperplastic activity in mice. Crude extracts were shown to prevent prostate enlargement, as indicated by decreased acid phosphatase activity, which is a biochemical marker for prostate epithelial cell proliferation (Bresciani 2003). In the same study, the metabolic crude extract and ethyl acetate fraction were also shown to have diuretic activity using mice treated with water and hydrochlorothiazide as control.

A. raddianum also exhibits strong antioxidant activity (Lai and Lim 2011). The phenolic compounds of *A. raddianum* possess primary (the compounds react with peroxide radicals and convert them into stable substances) and secondary (oxygen scavengers suppress the formation of free radicals) antioxidant activity (Lai and Lim 2011). Other species of the genus *Adiantum*, such as *Adiantum caudatum* (Ahmed et al. 2015) and *Adiantum philippense L.*, have been reported to be good sources of antioxidants (Ali et al. 2013). Promising results have been demonstrated for the antioxidant activity of *Adiantum capillus-veneris* Linn. Kumar (2009) evaluated the antioxidant potential of *A. capillus-veneris* extracts in human lymphocytes under oxidative stress and observed that the extract was capable of inhibiting lipid peroxidation and improving the activity of the antioxidant enzymes in these cells.

10 Conclusions

A. raddianum may be an important source of pharmaceutical and phytotherapeutic drugs. The fact that it is used and has been validated as medicinal resource by several cultures in Latin America reinforces the importance of further studies for the evaluation of its pharmacological potential. The reported high medicinal potential of other *Adiantum* species is also indicative of the pharmacological potential of *A. raddianum*, as many of the pharmacological activities described for other *Adiantum* species are due to the classes of compounds isolated from *A. raddianum*.

Acknowledgements We are especially grateful to the National Institute of Science and Technology in Ethnobiology, Bioprospecting and Nature Conservation, certified by CNPq, with financial support from FACEPE (Foundation for the Support of Science and Technology of the State of Pernambuco).

References

Ahmed D, Khan MM, Saeed R (2015) Comparative analysis of phenolics, flavonoids, and antioxidant and antibacterial potential of methanolic, hexanic and aqueous extracts from *Adiantum caudatum* leaves. Antioxidants 4:394–409

Ali MS, Amin MR, Kamal CMI, Hossain MA (2013) In vitro antioxidant, cytotoxic, thrombolytic activities and phytochemical evaluation of methanol extract of the *A. philippense* L. leaves. Asian Pac J Trop Biomed 3(6):464–469

Barros ICL, Andrade LHC (1997) Pteridófitas Medicinais: samambaias, avencas e plantas afins. Ed. Universitária da Universidade Federal de Pernambuco, Recife

Basualdo IS, Ortiz N, Degen MR (2004) Plantas medicinales comercializadas en los mercados de Asunción y Gran Asunción. Rojasiana 6(1):95–114

Bresciani LF (2003) Análise qualitativa e quantitativa de metabólitos Secundários e propriedades farmacológicas de quatro Espécies de plantas medicinais da flora catarinense. Ed. da Universidade Federal de Santa Catarina, Florianópolis. https://repositorio.ufsc.br/bitstream/handle/123456789/84659/233234.pdf?sequence=1

De la Cruz MG, Malpartida SB, Santiago HB, Jullian V, Bourdy G (2014) Hot and cold: Medicinal plant uses in Quechua speaking communities in the high Andes (Callejón de Huaylas, Ancash, Perú). J Ethnopharmacol 155:1093–1117

Keller HA, Torres EIM, Prance GT (2011) Ethnopteridology of the Guaranís of Misiones Province. Am Fern J 101(3):193–204

Kumar A (2009) Antioxidant effect of *Adiantum capillus veneris* Linn. On human lymphocyte: an in vitro study. Cell Tissue Res 9(2):1899–1902

Lai HY, Lim YY (2011) Evaluation of antioxidant activities of the methanolic extracts of selected ferns in Malaysia. Int J Environ Sci Dev 2(6):442–447

Lellinger DB (1991) Common and confusing bipinnate-dimidiate *Adiantum* of tropical America. Am Fern J 81(3):99–102

Moran RC, Zimmer B, Jermy AC (1995) *Adiantum*. In: Moran RC, Riba R (eds) Psilotaceae a Salviniaceae. In: Davidse G, Sousa M, Knapp S (eds) Flora Mesoamericana. v.1. Ciudad de México: Universidad Nacional Autónoma de México, pp 106–108

Pan C, Chen YG, Ma XY, Jiang JH, He F, Zhang Y (2011) Phytochemical constituents and pharmacological activities of plants from the genus *Adiantum*: a review. Trop J Pharm Res 10(5):681–692

Prado J (2015) Pteridaceae in Lista de Espécies da Flora do Brasil. Jardim Botânico do Rio de Janeiro. Available at: http://floradobrasil.jbrj.gov.br/jabot/floradobrasil/FB91850. Accessed 30 Oct 2015

Prado J, Rodrigues CDN, Salatino A, Salatino MFL (2007) Phylogenetic relationships among Pteridaceae, including Brazilian species, inferred from rbcL sequences. Taxon 56:355–368

Quattrocchi UFLS (2012) World dictionary of plant names: common names, scientific names, eponyms, synonyms, and etymology. CRC Press Taylor & Francis Group, Boca Raton

Santos MG, Sylvestre LS (2006) Aspectos florísticos e econômicos das pteridófitas de um afloramento rochoso do estado do Rio de Janeiro, Brasil. Acta Bot Bras (São Paulo) 20(1):115–124

Santos RR, Rossato AE, Pirola EP, Borges MS, Cardoso PS, Pierini MM, Amaral PA, Citadini-Zanette V (2012) *Adiantum raddianum* C. Presl. avenca. In: Erna A, de Mattia PM, Aguiar AP, Santos RR, Citadini-Zanette V (eds) Organizadores. Fitoterapia racional: aspectos taxonômicos, agroecológicos, etnobotânicos e terapêuticos. DIOESC, Florianópolis, pp 40–55

Sehnem ASJ (1972) Pteridáceas. In: Reitz R (ed) Flora Ilustrada Catarinense. Tipografia e Livraria. Blumenauense, Itajaí

Senna RM, Kasmirckaz C (1997) Pteridófitas de um remanescente florestal no Morro da Extrema. Rev Fac Zoo Vet Agro (Porto Alegre). 1997 4(1):33–48

Sharma D, Bhatia VK, Patil S, Sharma PC (2013) Antimicrobial activity of selected cryptogams from Solan region. Int J Biol Pharm Res 4(6):448–454

Singh M, Singh N, Khare PB, Rawat AKS (2008) Antimicrobial activity of some important *Adiantum* species used traditionally in indigenous systems of medicine. J Ethnopharmacol 115:327–329

Smith R, Pryer KM, Schuettpelz E, Korall P, Schneider H, Wolf PG (2006) A classification for extant ferns. Taxon 55(3):705–731

Souza MM, Pereira MA, Ardenghi JV, Mora TC, Bresciani LF, Yunes RA, Delle Monache F, Cechinel-Filho V (2009) Filicene obtained from *Adiantum cuneatum* interacts with the cholinergic, dopaminergic, glutamatergic, GABAergic, and tachykinergic systems to exert antinociceptive effect in mice. Pharmacol Biochem Behav 93:40–46

Thomas T (2014) A study on antibacterial and phytochemical evaluation of fronds of *Aadiantum raddianum* c. Presl. J Biomol Screen 4(2):85–88

Tribess B, Pintarelli GM, Bini LA, Camargo A, Funez LA, Gasper AL, Zeni ALB (2015) Ethnobotanical study of plants used for therapeutic purposes in the Atlantic Forest region, Southern Brazil. J Ethnopharmacol. https://doi.org/10.1016/j.jep.2015.02.005i

Vendruscolo GS, Mentz LA (2006) Estudo da concordância das citações de uso e importância das espécies e famílias utilizadas como medicinais pela comunidade do bairro Ponta Grossa, Porto Alegre, RS, Brasil. Acta Bot Bras 20(2):367–382

Windisch PG (1990) Pteridófitas da região Norte-ocidental do Estado de São Paulo: guia para estudo e excursões. UNESP, São José do Rio Preto

Winter SLS, Sylvestre LS, Prado J (2011) O gênero *Adiantum* (Pteridaceae) no estado do Rio de Janeiro, Brasil. Rodriguésia 62(3):663–681

Aloysia citriodora Palau

Julio Alberto Hurrell

Aloysia citriodora Palau
Photo: Available in: http://www.dixpix.ca/sth_cordillera/flora/verbenas/002_lemonverbena.html

J. A. Hurrell (✉)
Laboratorio de Etnobotánica y Botánica Aplicada (LEBA), Facultad de Ciencias Naturales y Museo, Universidad Nacional de La Plata, La Plata, Argentina

Consejo Nacional de Investigaciones Científicas y Técnicas, Buenos Aires República, Argentina

U. P. Albuquerque et al. (eds.), *Medicinal and Aromatic Plants of South America*, Medicinal and Aromatic Plants of the World 5,
https://doi.org/10.1007/978-94-024-1552-0_7

Abstract *Aloysia citriodora* Palau, cedron or lemon verbena, is a South American aromatic species, widespread in North America, Eurasia, and Africa. It is appreciated because of its therapeutic and food (condiment, flavoring) uses. It is also valued for its ornamental, insect repellent properties and sometimes in perfumery. Its popular culinary and medicinal uses have been expanded from Latin America to the rest of the Western world. Its main active constituents are essential oils to which its lemon like aroma and flavor can be attributed. Farther constituents include flavonoids, verbascosides, iridoids heterosides. In folk medicine it is most frequently used to treat gastrointestinal disorders (digestive, antispasmodic, carminative, antidiarrheal), or used as a mild sedative, cardiotonic, febrifuge, analgesic, and antiseptic. Various experimental studies validate different effects, as eupeptic, spasmolytic, antimicrobial, anti-inflammatory, analgesic, hypotensive, among others. Its sedative/anxiolytic activity requires further studies. Of particular interest are its cancer-related effects (antimutagenic, antigenotoxic, and antiangiogenic), and its antioxidant activity linked in various ways to our health.

Keywords *Aloysia citriodora* · Verbenaceae · Cedron · Lemon verbena · Food and medicinal uses

1 Taxonomic Characteristics

Aloysia citriodora Palau is a widespread aromatic plant used for both medicinal and food purposes, due to its essential oils that confer to its leaves a fragrance and taste similar to lemon. The specific epithet *citriodora* refers to this characteristic (from Latin *citrus*, 'lemon', and *odoro*, 'perfuming'). Its best known vernacular names are: cedrón, cidrón, hierba Luisa, hierba de la princesa, María Luisa, verbena de Indias (Spanish), cidrão, cidrinha (Portuguese), verveine citronnelle, (French), cedron, lemon verbena (English). It is the type species of the genus *A. citriodora*, introduced to the Real Jardín Botánico de Madrid and described in 1784 by the Spanish physician and botanist Antonio Palau (1734–1793). The genus was named in honor of Maria Luisa of Parma (1751–1819), wife of King Carlos IV of Spain, to who, also some of its vernacular names refer (Dellacassa and Bandoni 2003; Hurrell et al. 2008, 2011).

A. citriodora comprises about 30 species of warm and temperate zones of America, and belongs to the Family Verbenaceae J. St.-Hil., Tribe Lantaneae (Schauer) Briq. This tribe includes plants with fleshy fruits with a 2-locular, 2-seeded pyrene (e.g. *Lantana* L.), or dry schizocarp fruits separating at maturity into two 1-seeded mericarps (e.g. *A. citriodora* and the related *Lippia* L.). *A. citriodora* has flowers in dense or lax spiciform racemes, calyx distinctly 4-dentated, corolla ± actinomorphic. *Lippia* has flowers in compact heads or spikes, calyx obscurely 2- or 4-lobed, corolla weakly zygomorphic, ± 2-lipped (Botta 1993; Múlgura et al. 2012; Atkins 2004; Siedo 2007).

Synonyms *Aloysia citriodora* Ortega ex Pers., hom. illeg., *A. sleumeri* Moldenke, *A. triphylla* (L'Hér.) Britton, *A. triphylla* (L'Hér.) Britton f. serrulata Moldenke, *Lippia citriodora* (Ortega ex Pers.) Kunth, nom. illeg., *L. triphylla* (L'Hér.) Kuntze, *Verbena citriodora* (Palau) Cav., *V. triphylla* L'Hér., *Zapania citriodora* Lam., nom. illeg.

2 Crude Drug Used

The drug consists of its whole or fragmented leaves (Folia Aloysiae citriodorae), sometimes with young stems and flowers. Both fresh and dried leaves are consumed as condiment and beverage flavoring, and used to make therapeutic preparations. The whole leaves must contain min. 0.20% essential oil while the fragmented leaves min. 0.15%. The drug should contain max. 2% strange matter. No common adulterants are known but may contain materials of inferior quality, easily recognizable by its less citric and fresh fragrance (Dellacassa and Bandoni 2003; Alonso and Desmarchelier 2005).

The crude drug of leaves was included in the pharmacopoeias of Mexico (1st edition), Argentina (6th edition), and France (10th edition). It is also included in the Argentine Food Code, the European Herbal Infusion Association, and different types of regulations of the United States, Colombia, Chile, and Uruguay, among other countries. The consume is considered safe, but it is not recommended in pregnancy, during lactation, children under 6 years of age, and adult patients with renal insufficiency (Muñoz et al. 2004; Alonso and Desmarchelier 2005; Fonnegra and Jiménez 2007; Hurrell et al. 2011).

The dry leaves are consumed mostly as infusions or decoctions (15 g per liter of water), 2–3 cups in daily intakes, also in tincture (20 g in 100 cc of 60° alcohol), 40 drops in water before meals, and extract fluid (1:1), 15–20 drops after meals (Burgstaller 1968; Alonso Paz et al. 1993; Alonso and Desmarchelier 2005). The most widespread commercial products are the dried leaves, whole or fragmented, sold in bulk or packaged, also as ingredient of herbal mixture, tea bags, mother tincture, and dietary supplements (Hernández Cano and Volpato 2004; Ragone et al. 2007, 2010; Hurrell et al. 2011).

3 Major Chemical Constituents and Bioactive Compounds

A. citriodora leaves contain essential oil which lends its lemon aroma and flavor, and its eupeptic and spasmolytic properties, by which the infusion is consumed to treat diverse gastrointestinal disorders. The chemical composition of the essential oil is variable and depends on the harvest periods and post-harvesting process, state and origin of the plant, cultivation conditions, among other factors (Díaz Fajardo

2007; Brant et al. 2009; Agah and Najafian 2012; Rojas et al. 2012; Moein et al. 2014; Nematian et al. 2014).

The main component citral is a pale yellow liquid with a strong lemon scent (Lewis 2007). The commercial product is a mixture of the isomers geranial (citral A) and neral (citral B). Other essential oil components mentioned are: limonene, citronellol, cymene, pinene, terpineol, borneol, linalool, verbenone, phellandrene, isosafrol, eucalyptol, thujone, caryophyllene. Other identified compounds include flavonoids, iridoids, heterosides, verbascosides, phytosterols, tannins, alkaloids (traces), and mucilage (Pascual et al. 2001; Dellacassa and Bandoni 2003; Alonso and Desmarchelier 2005; Fonnegra and Jiménez 2007; Di Leo Lira et al. 2008, 2013; Barboza et al. 2009; Rojas et al. 2010, 2012; Ganjewala et al. 2012).

The essential oil has been used in perfumery, but currently is not advised because of its possible skin irritant effect. The absolute is recommended only in a concentration not exceeding 1% (Dellacassa and Bandoni 2003; Alonso and Desmarchelier 2005).

4 Morphological Description

A. citriodora is an aromatic shrub, 1.5–4 (−7) m in height, with cylindrical, striated, glabrescent branches. Leaves ternate, deciduous; petiole 0.5–1.5 cm long; blade 2.5–8 (−10) cm long × 0.5–2.5 cm wide, narrowly elliptic, apex acute, margin entire or serrate, adaxially scabrous, abaxially with glandular-dotted and prominent veins. Flowers shortly pedicellate in lax spiciform racemes, 1.5–5 cm long, clustered in apical paniculiform inflorescence. Bracts 1–1.5 mm long, ovate, acute to acuminate, deciduous. Calyx 2–3 mm long, tubular, subactinomorphic, 4-dentate. Corolla subactinomorphic, hypocrateriform, white to pale lavender; tube 3.5–5 mm long, straight, upper half pubescent; limb 2.5–4.5 mm long, lobes 4, spreading, ovate, slightly equal. Stamens 4, inserted just above middle of tube, didynamous, the posterior pair slightly exserted. Ovary ca. 3 mm long, ovoid, glabrous or pubescent on the upper half, style short, stigma lateral, subcapitate. Fruit a dry schizocarp, 2–3 mm long × 1.0–1.5 mm wide, with persistent calyx, apicaly setose, separating at maturity into two 1-seeded mericarps, brown-reddish. $2n = 36$ (Atkins 2004; Siedo 2007; Múlgura et al. 2012).

5 Geographical Distribution

This species is native to warm-temperates and arid zones of the Northwest of Argentina (proposed as its origin area), in Jujuy, Salta, Tucumán, Catamarca, La Rioja, and San Juan provinces, also Bolivia and Uruguay (Botta 1993; Siedo 2007; Múlgura et al. 2012; Di Leo Lira et al. 2013).

It is cultivated in different countries from southern United States and Mexico to northern Chile, warm-temperate central Argentina, Paraguay and southern Brazil, also in central-southern Europe, northern Africa, and Asia. In many of these areas it is also found as cultivation escaped or naturalized (Muñoz et al. 2004; Hurrell et al. 2011; Randall 2005; Salimena and Múlgura 2014).

6 Ecological Requirements

A. citriodora prospers in warm temperate and temperate zones from sea-level up to about 2000 m altitude. It grows well in soils of medium consistency, loose, permeable, deep, pH between 6.5 and 7.2, rather cool but not wet, because excess water promotes root rot. In culture, the well-lighted environments have influence on the synthesis and accumulation of essential oils: shading produces larger leaves poor in bioactive compounds. Excessive wind is unfavorable because it increases the rate of evaporation of the essential oils and decreases the production per unit area. *A. citriodora* is propagated in spring by cuttings, layering or dividing clumps. In vitro micropropagation has also been tested in order to increase the biomass and quality of its essential oils. The seeds have limited or null germination power (Alonso and Desmarchelier 2005; Severin et al. 2005; Díaz Fajardo 2007; Berardi 2010).

7 Collection Practice

Aloysia citriodora is wild-crafted in areas where it grows spontaneously. According to Severin et al. (2005) it is already overexploited in Argentina. The leaves are harvested when they have reached its highest development, just before flowering (spring-early summer). The branches are cut and the leaves are removed at the same time to seize the cuttings or left to dry in the shade, protected from dust and moisture until it strip off the leaves. Shoots of the 2nd year are mainly used. The product quality is improved when leaves are dried in thin layers, in shaded and ventilated places, until desiccation is complete. The material retains its fragrance for many years in good storage conditions (Dellacassa and Bandoni 2003; Fonnegra and Jiménez 2007; Elechosa 2009).

8 Traditional Use and Common Knowledge

A. citriodora leaves have a long record of use in folk medicine in various parts of Latin America, from Mexico to Argentina. Frequently, it is forms also part of different local culinary traditions. This botanical knowledge persists in current different communities even in some countries of the Old World (e.g. is one of the main components of the digestive and sedative infusion called *zhourat* in the Middle East, Obon et al. 2014). In pluricultural contexts, cedron are commercialized both in traditional markets in urban areas (Macía et al. 2005; Pochettino et al. 2012; Parodi et al. 2013), as well as in herb shops and health food stores (Hurrell et al. 2008, 2011).

The most widespread popular medicinal uses of the infusion include: digestive, eupeptic, stomachic, antispasmodic, carminative, hypotensive, cardiotonic, against heart palpitations, nausea, vomiting, dizziness, fainting, vertigo, nervous disorders, hysteria, hypochondria, mild sedative, anxiolytic, antidepressant, hypnotic, anticonvulsant, diuretic, febrifuge, antimalarial, expectorant, anti-asthmatic, antiseptic, analgesic, insect repellent (Hieronymus 1882; Alonso Paz et al. 1993; Dellacassa and Bandoni 2003; Muñoz et al. 2004; González Torres 2005; Osuna Torres et al. 2005; Díaz Fajardo 2007; Rondina et al. 2008; Angulo et al. 2012). In external use, is applied in poultices for toothache, varicose veins and haemorrhoids (Alonso and Desmarchelier 2005; Fonnegra and Jiménez 2007).

In addition, it has been indicated as antidiarrheal, antidysenteric, and vermifuge in Mexico (Osuna Torres et al. 2005), anti-catarrhal in Cuba (Hernández Cano and Volpato 2004), for herpes zoster treatments in Colombia (Fonnegra and Jiménez 2007), as emmenagogue in Brazil (Mors et al. 2000) and Mexico (Ponce-Monter et al. 2010), to relieve headache in Ecuador (Tene et al. 2007) and Peru (Rodríguez Quezada 2011), for prevention of atherosclerosis in Peru (Ono et al. 2008), against bites poisonous animals in Bolivia (Dellacassa and Bandoni 2003), and diabetes in Argelia (Rachid et al. 2012), and Morocco (Bousta et al. 2014). In Ecuador it is also utilized as analgesic in cases of rheumatism, cramps and involuntary muscle contractions (Álvarez Sarmiento 2012). In the Andean region and in Mexico, cedron infusion is used for combating the *susto*. According to the Andean oral tradition, it is a condition expressed in several symptoms such as weakness, dejection, depression, headache, insomnia, chills, fever, lack of appetite, vomiting, among others, awarded to the loss of the soul because of a big impression or a deep fear (Dellacassa and Bandoni 2003; Koss-Chioino et al. 2003).

In urban contexts, the herb (and as an ingredient in herbal mixture) is marketed as slimming or for weight-losing (Turano and Cambi 2009; Madaleno and Montero 2012).

Cedron leaves have wide dissemination in various Latin American culinary traditions, and it is also used in cuisines of Western Europe. The fresh leaves are used to prepare marinated fish and poultry, fruit salads, jellies, jams, puddings, desserts, and to flavor the water of the *mate*. The dried and chopped leaves are used to make sauces and dressings, and to prepare aromatic and digestive infusions (nutraceuticals). Powdered dried leaves are used to flavor beverages and liqueurs. In Ecuador it is also used to flavor *chicha* (alcoholic beverage derived mainly from non-distilled fermentation of corn) and to prepare *colada morada* (traditional drink made of corn, fruits, and aromatic herbs). In Argentine puna is employed to make liquors based on a mixture of grape must and alcohol, sugar, dyes, and aromatic herbs, called *mistelas* (Dellacassa and Bandoni 2003; Hurrell et al. 2008; Álvarez Sarmiento 2012).

A. citriodora is also locally cultivated in homegardens as ornamental, aromatic and insect repellent (Pochettino et al. 2014).

9 Modern Medicine Based on Its Traditional Medicine Uses

Several uses of the popular medicine have been supported by different studies about the active principles and its mechanisms of action, mostly under in vivo or in vitro conditions, in animals. However, clinical trials in humans would be required.

Regarding gastrointestinal disorders, some effects were analyzed: antispasmodic, eupeptic, digestive (Pascual et al. 2001; Velázquez et al. 2006; Ragone et al. 2007; Berardi 2010; Mamadou et al. 2011; Lenoir et al. 2012), *Helicobacter pylori* inhibitor (Ohno et al. 2003), antidiarrheal (Calzada et al. 2010). The tannin content may have effect on the bioavailability of certain trace elements such as Fe, Cu, Zn (Pizarro et al. 1994).

Studies, in relation to its antibiotic activity include: antibacterial (Ohno et al. 2003; Duarte et al. 2007; Rodríguez Vaquero et al. 2010; Ali et al. 2011; Parodi et al. 2013), against bacteria responsible for caries (Pellecuer et al. 1980), and genito-urinary pathogenic bacteria (Rojas et al. 2010), antimycotic (Duarte et al. 2005; Oliva et al. 2011), anti-*Trypanosoma cruzi* (Chagas disease agent) (Rojas et al. 2012). Apparently, the crude drug has no antimalarial action (Muñoz et al. 2000). Its inhibitory effect on dengue virus was evaluated by Ocazionez et al. (2010).

The insect repellent and insecticidal activity were checked by Gillij et al. (2008), Palacios et al. (2009), and Toloza et al. (2010).

The analgesic action has been analyzed by Nakamura et al. (1997), Pascual et al. (2001), Qnais et al. (2009), and Isacchi et al. (2011). Anti-inflammatory activity was the subject of both in vivo and in vitro studies, in cases of dysmenorrhea (Ponce-Monter et al. 2010). The anesthetic action on crustaceans has been assessed by Parodi et al. (2012). The extracts of the fresh aerial parts showed analgesic, anti-inflammatory, antipyretic and antioxidant properties (El-Hawary et al. 2012).

The antioxidant activity of cedron is supported by diverse studies (Díaz Fajardo 2007; Funes et al. 2009; Rodríguez Vaquero et al. 2010; Abderrahim et al. 2011; Ali et al. 2011; Portmann et al. 2012; Lasagni et al. 2014).

With regard to its effects on the cardiovascular system, the hypotensive activity on mice and rats has been validated (Ragone et al. 2010). Its popular use as cardiotonic remedy has, however, not yet been experimentally demonstrated (Dellacassa and Bandoni 2003).

Despite its diffused popular use as a sedative/anxiolytic, this action has been asserted by some authors and denied by others (Wannmacher et al. 1990; Zeichen et al. 1997; Ragone et al. 2010). More recently its antidepressant effect has been reported (Eram et al. 2012). For other species of the genus, like *Aploysia polystachya* (Griseb.) Moldenke (commonly called 'burrito'), and *A. gratissima* (Gillies & Hook.) Tronc. (known as 'cedrón del monte'), its anxiolytic and antidepressant activities have been studied in mice (Hellión-Ibarrola et al. 2006, 2008; Zeni et al. 2011). Both species are used in folk medicine and are marketed in herb shops with similar purposes to those of *A. citriodora* (Hurrell et al. 2008, 2011).

Its possible application against cancer, has been studied with regards to its effects as antimutagenic (Natake et al. 1989), antigenotoxic (Zamorano-Ponce et al. 2006), and antiangiogenic (Zihlif et al. 2012).

10 Conclusions

A. citriodora Palau is an aromatic species known almost globally. However, its leaves are used in gastronomy (condiment, flavor) and phytotherapy mainly in Latin America, and also in the United States, Eurasia and Africa. Its popular therapeutic use for treating gastrointestinal disorders is supported by several experimental studies. The same is valid also for their uses as anti-inflammatory, analgesic, antipyretic, antibiotic (antiseptic), and hypotensive. Its insect repellent and insecticidal effects have also been investigated. There are no studies that support the traditional uses as cardiotonic, expectorant, anti-catarrhal, anti-asthmatic, antidote, for herpes zoster treatments, anti-atherosclerosis, and anti-diabetic.

Popular wide spread uses related to nervous disorders: sedative, anxiolytic, antidepressant, anticonvulsant, hypnotic, and others related: heart palpitations, nausea, vomiting, dizziness, vertigo, hysteria, and hypochondria, have few or lack supporting studies. The available scientific evidence is frequently controversial, like in the case of the assessments of its sedative/anxiolytic effects, which seem to indicate the need of further in-depth studies.

Regarding the results of studies into its antioxidant capacity and cancer-related effects, there seems to be an undoubtedly a promising future for new research.

References

Abderrahim F, Estrella S, Susín C, Arribas S, González MC, Condezo-Hoyos L (2011) The antioxidant activity and thermal stability of lemon verbena (*Aloysia triphylla*) infusion. J Med Food 14(5):517–527

Agah M, Najafian S (2012) Essential oil content and composition of *Lippa citriodora* as affected by drying method before flowering stages. Eur J Exp Biol 2(5):1771–1777

Ali HFM, El-Beltagi HS, Nasr NF (2011) Evaluation of antioxidant and antimicrobial activity of *Aloysia triphylla*. Elect J Environ Agric Food Chem 10(8):2689–2699

Alonso J, Desmarchelier C (2005) Plantas medicinales autóctonas de la Argentina. Editorial Lola, Buenos Aires

Alonso Paz E, Bassagoda M, Ferreira F (1993) Yuyos: Uso racional de las plantas medicinales. Editorial Fin de Siglo, Montevideo

Álvarez Sarmiento XP (2012) Identificación, historia, características y aplicaciones culinarias de cinco plantas aromáticas endémicas de América. Universidad de Cuenca, Cuenca

Angulo AF, Rosero RA, González MS (2012) Estudio etnobotánico de las plantas medicinales utilizadas por los habitantes del corregimiento de Genoy, Municipio de Pasto, Colombia. Rev Univ Salud 14(2):168–185

Atkins S (2004) Verbenaceae. In: Kubitzki K (ed) The families and genera of vascular plants VII. Springer, Berlin, pp 449–468

Barboza GE, Cantero JJ, Núñez C, Pacciaroni A, Ariza Espinar L (2009) Medicinal plants: a general review and a phytochemical and ethnopharmacological screening of the native Argentine Flora. Kurtziana 34(1–2):7–365

Berardi A (2010) Etnofarmacología gastrointestinal de plantas medicinales argentinas del género *Aloysia*, familia Verbenaceae: mecanismos de acción y relación con los principios activos. Facultad de Ciencias Exactas, Universidad Nacional de La Plata, La Plata

Botta SM (1993) *Aloysia*. In: Cabrera AL (ed) Flora del la Provincia de Jujuy. Colecc. Cient. Inst. Nac. Tecnol. Agropecu. vol 13, no 9, pp 36–46

Bousta D, Boukhira S, Aafi A, Ghanmi M, El-Mansouri L (2014) Ethnopharmacological Study of anti-diabetic medicinal plants used in the Middle-Atlas region of Morocco (Sefrou region). Int J Pharma Res Health Sci 2(1):75–79

Brant R, Pereira Pinto JE, Vilela Bertolucci S, da Silva A, Brant Albuquerque CJ (2009) Teores do óleo essencial de cidrão [*Aloysia triphylla* (L'Hérit) Britton (Verbenaceae)] em diferentes horários de colheita e processamentos pós-colheita. Ciênc Agrotec 33:2065–2068

Burgstaller CH (1968) La vuelta a los vegetales. Dinizo, Buenos Aires

Calzada F, Arista R, Pérez H (2010) Effect of plants used in Mexico to treat gastrointestinal disorders on charcoal-gum acacia-induced hyperperistalsis in rats. J Ethnopharmacol 128(1):49–51

Dellacassa E, Bandoni AL (2003) Hierbaluisa. *Aloysia citriodora* Palau. Rev Fitoterapia 3(1):19–25

Di Leo Lira P, van Baren CM, Retta D, Bandoni AL, Gil A, Gattuso M, Gattuso S (2008) Characterization of Lemon Verbena (*Aloysia citriodora* Palau) from Argentina by the Essential Oil. J Essent Oil Res 20(4):350–353

Di Leo Lira P, van Baren CM, López S, Molina A, Heit C, Viturro C, de Lampasona MP, Catalán CA, Bandoni A (2013) Northwestern Argentina: a center of genetic diversity of lemon verbena (*Aloysia citriodora*, Verbenaceae). Chem Biodivers 10(2):251–261

Díaz Fajardo OL (2007) Estudio comparativo de la composición química y evaluación de la actividad antioxidante del aceite esencial de *Aloysia triphylla* (L'Hér.) Britton, cultivada en tres regiones de Colombia. Universidad Industrial de Santander, Bucaramanga

Duarte MC, Figueira GM, Sartoratto A, Rehder VL, Delarmelina C (2005) Anti-*Candida* activity of Brazilian medicinal plants. J Ethnopharmacol 97(2):305–311

Duarte MC, Leme EE, Delarmelina C, Soares AA, Figueira GM, Sartoratto A (2007) Activity of essential oils from Brazilian medicinal plants on *Escherichia coli*. J Ethnopharmacol 111(2):197–201

Elechosa MA (ed) (2009) Manual de recolección sustentable de plantas aromáticas nativas de la región central y noroeste de la Argentina. Inst Nac Tecnol Agropecu, Buenos Aires

El-Hawary SS, Yousif MF, Abdel Motaal AA, Abd-Hameed LM (2012) Bioactivities, phenolic compounds and in-vitro propagation of *Lippia citriodora* Kunth cultivated in Egypt. Bull Fac Pharm Cairo Univ 50(1):1–6

Eram S, Abbasi Maleki S, Mohammadi Motamed S, Abbasi Maleki M, Hanare Kheliany H (2012) Antidepressant activity of ethanolic extract, chloroform extract and aqueous extract of *Aloysia triphylla* in the FST and in the TST in male mice. Res Pharm Sci 7(Suppl 5):S856

Fonnegra R, Jiménez SL (2007) Plantas medicinales aprobadas en Colombia, 2nd edn. Editorial Universidad de Antioquía, Medellín

Funes L, Férnandez-Arroyo S, Laporta O, Pons A, Roche E, Segura-Carretero A, Fernández-Gutiérrez A, Micol V (2009) Correlation between plasma antioxidant capacity and verbascoside levels in rats after oral administration of lemon verbena extract. Food Chem 117(4):589–598

Ganjewala D, Gupta AK, Muhury R (2012) An update on bioactive potential of a monoterpene aldehyde citral. J Biol Act Prod Nat 2(4):186–199

Gillij YG, Gleiser RM, Zygadlo JA (2008) Mosquito repellent activity of essential oils of aromatic plants growing in Argentina. Bioresour Technol 99(7):2507–2515

González Torres DM (2005) Catálogo de plantas medicinales (y alimenticias y útiles) usadas en Paraguay, 2nd edn. Servilibro, Asunción

Hellión-Ibarrola MC, Ibarrola DA, Montalbetti Y, Kennedy ML, Heinichen O, Campuzano M, Tortoriello J, Fernández S, Wasowski C, Marder M, De Lima TC, Mora S (2006) The anxiolytic-like effects of *Aloysia polystachya* (Griseb.) Moldenke (Verbenaceae) in mice. J Ethnopharmacol 105(3):400–408

Hellión-Ibarrola MC, Ibarrola DA, Montalbetti Y, Kennedy ML, Heinichen O, Campuzano M, Ferro EA, Alvarenga N, Tortoriello J, De Lima TC, Mora S (2008) The antidepressant-like effects of *Aloysia polystachya* (Griseb.) Moldenke (Verbenaceae) in mice. Phytomedicine 15(6–7):478–483

Hernández Cano J, Volpato G (2004) Herbal mixtures in the traditional medicine of Eastern Cuba. J Ethnopharmacol 90(2–3):293–316

Hieronymus J (1882) Plantas diafóricas. Flora Argentina. G. Kraft, Buenos Aires

Hurrell JA, Ulibarri EA, Delucchi G, Pochettino ML (2008) Plantas aromáticas condimenticias. In: Hurrell JA (ed) Biota rioplatense XIII. Editorial Lola, Buenos Aires

Hurrell JA, Ulibarri EA, Arenas PM, Pochettino ML (2011) Plantas de Herboristería. Editorial Lola, Buenos Aires

Isacchi B, Iacopi R, Bergonzi MC, Ghelardini C, Galeotti N, Norcini M, Vivoli E, Vincieri FF, Bilia AR (2011) Antihyperalgesic activity of verbascoside in two models of neuropathic pain. J Pharm Pharmacol 63(4):594–601

Koss-Chioino J, Leatherman TL, Greenway C (2003) Medical pluralism in the Andes. Routledge, London

Lasagni Vitar RM, Reides CG, Ferreira SM, Llesuy SF (2014) The protective effect of *Aloysia triphylla* aqueous extracts against brain lipid-peroxidation. Food Funct 5(3):557–563

Lenoir L, Joubert-Zakeyh J, Texier O, Lamaison JL, Vasson MP, Felgines C (2012) *Aloysia triphylla* infusion protects rats against dextran sulfate sodium-induced colonic damage. J Sci Food Agric 92(7):1570–1572

Lewis RJ (2007) Citral. In: Hawley's condensed chemical dictionary, 15th edn. New York: Wiley

Macía MJ, García E, Vidaurre P (2005) An ethnobotanical survey of medicinal plants commercialized in markets of La Paz and El Alto, Bolivia. J Ethnopharmacol 97(2):337–350

Madaleno IM, Montero MC (2012) El cultivo urbano de plantas medicinales, su comercialización y usos terapéuticos en la ciudad de Río Cuarto, provincia de Córdoba, Argentina. Cuad Geográficos (Granada) 50(1):63–85

Mamadou G, Meddah B, Limas-Nzouzi N, Ait El Haj A, Bipolo S, Mokondjimobé E, Mahraoui L, Faouzi MA, Ducroc R, Cherrah Y, Eto B (2011) Antispasmodic phytomedicine, from traditional utilization to rational formulation: functional approach. Phytopharmacology 1(3):20–35

Moein M, Zarshenas MM, Etemadfard H (2014) Essential oil composition and total flavonoid content of *Aloysia citriodora* Palau under different cultivation systems. Int J Plant Anim Environ Sci 4(1):353–358

Mors WB, Rizzini CT, Alvarez P (2000) Medicinal plants of Brazil. Reference, Michigan

Múlgura ME, O'Leary N, Rotman A (2012) Verbenaceae. In: Anton AM, Zuloaga FO (eds) Flora Argentina. Flora Vascular de la República Argentina, vol 14, pp 1–220

Muñoz V, Sauvain M, Bourdy G, Arrázola S, Callapa J, Ruiz G, Choque J, Deharo E (2000) A search for natural bioactive compounds in Bolivia through a multidisciplinary approach. Part III. Evaluation of the antimalarial activity of plants used by Alteños Indians. J Ethnopharmacol 71(1–2):123–131

Muñoz O, Montes M, Wilkomirsky T (2004) Plantas medicinales de uso en Chile, 2nd edn. Editorial Universitaria, Santiago de Chile

Nakamura T, Okuyama E, Tsukada A, Yamazaki M, Satake M, Nishibe S, Deyama T, Moriya A, Maruno M, Nishimura H (1997) Acteoside as the analgesic principle of cedron, *Lippia triphylla*, a Peruvian medicinal plant. Chem Pharm Bull 45(3):499–504

Natake M, Kanazawa K, Mizuno M, Veno N, Kobayashi T, Danno G, Minamoto S (1989) Herb water-extracts markedly suppress the mutagenicity of Trp-P-2. Agric Biol Chem 53(5):1423–1425

Nematian A, Dalvandi GR, Shariati MA (2014) Effect of planting density and sowing date on the essential oil content and composition of lemon verbena (*Lippia citriodora*). Int J Biosci 5(2):56–63

Obon C, Rivera D, Alcaraz F, Attieh L (2014) Beverage and culture. "Zhourat", a multivariate analysis of the globalization of a herbal tea from the Middle East. Appetite 79:1–10

Ocazionez RE, Meneses R, Torres FA, Stashenko E (2010) Virucidal activity of Colombian *Lippia* essential oils on dengue virus replication in vitro. Mem Inst Oswaldo Cruz 105(3):304–309

Ohno T, Kita M, Yamaoka Y, Imamura S, Yamamoto T, Mitsufuji S, Kodama T, Kashima K, Imanishi J (2003) Antimicrobial activity of essential oils against *Helicobacter pylori*. Helicobacter 8(3):207–215

Oliva ML, Carezzano ME, Gallucci MN, Demo MS (2011) Antimycotic effect of the essential oil of *Aloysia triphylla* against *Candida* species obtained from human pathologies. Nat Prod Commun 6(7):1039–1043

Ono M, Oda E, Tanaka T, Iida Y, Yamasaki T, Masuoka C, Ikeda T, Nohara T (2008) DPPH radical-scavenging effect on some constituents from the aerial parts of *Lippia triphylla*. J Nat Med 62(1):101–106

Osuna Torres L, Tapia Pérez ME, Aguilar Contreras A (2005) Plantas medicinales de la medicina tradicional mexicana para tratar afecciones gastrointestinales. Edicions de la Universitat de Barcelona, Barcelona

Palacios SM, Bertoni A, Rossi Y, Santander R, Urzúa A (2009) Insecticidal activity of essential oils from native medicinal plants of Central Argentina against the house fly, *Musca domestica*. Parasitol Res 106(1):207–212

Parodi TV, Cunha MA, Heldwein CG, de Souza DM, Martins ÁC, de Garcia LO, Wasielesky W Jr, Monserrat JM, Schmidt D, Caron BO, Heinzmann B, Baldisserotto B (2012) The anesthetic efficacy of eugenol and the essential oils of *Lippia alba* and *Aloysia triphylla* in post-larvae and sub-adults of *Litopenaeus vannamei* (Crustacea, Penaeidae). Comp Biochem Physiol C Toxicol Pharmacol 155(3):462–468

Parodi TV, Castagna Vargas AP, Krewer C, Moraes Flores EM, Baldisserotto B, Heinzmann VM, Oliveira J, Secco Popiolski A, Minozzo M (2013) Chemical composition and antibacterial activity of *Aloysia triphylla* extracts obtained by pressurized CO_2 extraction. Braz Arch Biol Technol 56(2):283–292

Pascual ME, Slowing K, Carretero E, Sánches MD, Villar A (2001) *Lippia*: traditional uses, chemistry and pharmacology: a review. J Ethnopharmacol 76:201–214

Pellecuer J, Jacob M, Simeon DM, Dusart G, Attisto M, Barthez M, Gourgas L, Pascal B, Tomei R (1980) Essais d'utilisations d'huiles essentielles de plantes aromatiques Méditerranéennes en odontologie conservatrice. Plant Med Phytother 14:83–98

Pizarro F, Olivares M, Hertramp E, Walter T (1994) Factores que modifican el estado de nutrición del hierro: contenido de taninos de infusiones de hierbas. Arch Latinoam Nutr 44(4):277–280

Pochettino ML, Puentes JP, Buet Costantino F, Arenas PM, Ulibarri EA, Hurrell JA (2012) Functional foods and nutraceuticals in a market of Bolivian immigrants in Buenos Aires (Argentina). Evid Based Complement Altern Med. https://doi.org/10.1155/2012/320193

Pochettino ML, Hurrell JA, Bonicatto MM (2014) Horticultura periurbana: estudios etnobotánicos en huertos familiares y comerciales de la Argentina. Ambienta 107:86–99

Ponce-Monter H, Fernández-Martínez E, Ortiz MI, Ramírez-Montiel ML, Cruz-Elizalde D, Pérez-Hernández N, Cariño-Cortés R (2010) Spasmolytic and anti-inflammatory effects of Aloysia triphylla and citral, in vitro and in vivo studies. J Smooth Muscle Res 46(6):309–319

Portmann E, Nigro MM, Reides CG, Llesuy S, Ricco RA, Wagner ML, Gurni AA, Carballo MA (2012) Aqueous extracts of *Lippia turbinata* and *Aloysia citriodora* (Verbenaceae): assessment of antioxidant capacity and DNA damage. Int J Toxicol 31(2):192–202

Qnais E, Abu-Safieh K, Abu-Dieyeh MH, Abdulla FA (2009) Antinociceptive effect of two flavonoids from *Aloysia triphylla*. Jordan J Biol Sci 2(4):167–170

Rachid A, Rabah D, Farid L, Zohra SF, Houcine B, Nacéra B (2012) Ethnopharmacological survey of medicinal plants used in the traditional treatment of diabetes mellitus in the North Western and South Western Algeria. J Med Plants Res 6(10):2041–2050

Ragone MI, Stella M, Conforti P, Volonté MG, Consolini AE (2007) The spasmolytic effect of *Aloysia citriodora* Palau (South American cedrón) is partially due to its vitexin but not isovitexin on rat duodenus. J Ethnopharmacol 113(2):258–266

Ragone MI, Stella M, Pastore A, Consolini AE (2010) Sedative and cardiovascular effects of *Aloysia citriodora* Palau on mice and rats. Lat Am J Pharm 29(1):79–86

Randall RP (2005) A global compendium of weeds, 2nd edn. Department of Agriculture and Food of Western Australia, Perth

Rodríguez Quezada MP (2011) Manejo de plantas medicinales en el nororiente amazónico peruano. ECIPerú 8(2):150–157

Rodríguez Vaquero MJ, Tomassini Serravalle LR, Manca de Nadra MC, Strasser de Saad AM (2010) Antioxidant capacity and antibacterial activity of phenolic compounds from Argentinean herbs infusions. Food Control 21(5):779–785

Rojas LB, Velasco J, Díaz T, Gil Otaiza R, Carmona J, Usubillaga A (2010) Composición química y efecto antibacteriano del aceite esencial de *Aloysia triphylla* contra patógenos genito-urinarios. Bol Latinoam Caribe Plant Med Aromat 9(1):56–62

Rojas J, Palacios O, Ronceros S (2012) Efecto del aceite esencial de *Aloysia triphylla* (cedrón) sobre el *Trypanosoma cruzi* en ratones. Rev Perú Med Exp Salud Públ 29(1):61–68

Rondina R, Bandoni AL, Coussio JD (2008) Especies medicinales argentinas con potencial actividad analgésica. Dominguezia 24(1):47–69

Salimena FRG, Múlgura ME (2014) *Aloysia*. In: Lista de Espécies da Flora do Brasil. Jardim Botânico do Rio de Janeiro. Available in: http://www.floradobrasil.jbrj.gov.b. Accessed 27 Dec 2014

Severin C, Bruzzese D, Di Sapio O, Goubileo MG, Gattuso S (2005) Regeneración in vitro de plantas de *Aloysia citriodora* Palau (Verbenaceae). Rev Invest Fac Ci Agrarias (UNR) 5(8):61–66

Siedo SJ (2007) Systematics of *Aloysia* (Verbenaceae). ProQuest, Ann Arbor

Tene V, Malagón O, Vita PV, Vidari G, Armijos C, Zaragoza T (2007) An ethnobotanical survey of medicinal plants used in Loja and Zamora-Chinchipe, Ecuador. J Ethnopharmacol 111(1):63–81

Toloza AC, Zygadlo J, Biurrun F, Rotman A, Picollo MI (2010) Bioactivity of Argentinean essential oils against permethrin-resistant head lice, *Pediculus humanus capitis*. J Insect Sci 10:185. https://doi.org/10.1673/031.010.14145

Turano FA, Cambi VN (2009) Control de calidad de mezclas de hierbas medicinales que se comercializan como adelgazantes y/o reductoras en Bahía Blanca, Argentina. Lat Am J Pharm 28(1):10–18

Velázquez C, Calzada F, Torres J, Gonzalez F, Ceballos G (2006) Antisecretory activity of plants used to treat gastrointestinal disorders in Mexico. J Ethnopharmacol 103(1):66–70

Wannmacher L, Fuchs FD, Paoli CL, Fillman HS, Gianlupi A, Lubianca Neto JF, Hassegawa CY, Guimarães FS (1990) Plants employed in the treatment of anxiety and insomnia: II. Effect of infusions of *Aloysia triphylla* on experimental anxiety in normal volunteers. Fitoperapia 61(5):449–453

Zamorano-Ponce E, Morales C, Ramos D, Sepúlveda C, Cares S, Rivera P, Fernández J, Carballo MA (2006) Anti-genotoxic effect of *Aloysia triphylla* infusion against acrylamide-induced DNA damage as shown by the comet assay technique. Mutat Res 603(2):145–150

Zeichen R, De Emilio S, Bindstein E, Chiale C (1997) Efectos sobre el sistema nervioso de *Aloysia triphylla* (cedrón, yerba Luisa) en ratones. WOCMAP II (Mendoza, Argentina). Abs. p 306

Zeni AL, Zomkowski AD, Dal-Cim T, Maraschin M, Rodrigues AL, Tasca CI (2011) Antidepressant-like and neuroprotective effects of *Aloysia gratissima*: investigation of involvement of L-arginine-nitric oxide-cyclic guanosine monophosphate pathway. J Ethnopharmacol 137(1):864–874

Zihlif M, Afifi F, Muhtaseb R, Al-Khatib S, Abaza I, Naffa R (2012) Screening the antiangiogenic activity of medicinal plants grown and sold in Jordan. Planta Med 78(3):297–301

Anemopaegma arvense (Vell.) Stellfeld ex De Souza

Fúlvio Rieli Mendes and Luis Carlos Marques

Anemopaegma arvense (Vell.) Stellfeld ex De Souza

F. R. Mendes (✉)
Centro de Ciências Naturais e Humanas, Universidade Federal do ABC,
São Bernardo do Campo, SP, Brazil

L. C. Marques
Fitoscience Consulting Ltd., São Paulo, SP, Brazil
e-mail: luis.marques@anhanguera.com

U. P. Albuquerque et al. (eds.), *Medicinal and Aromatic Plants of South America*, Medicinal and Aromatic Plants of the World 5,
https://doi.org/10.1007/978-94-024-1552-0_8

Abstract The *Anemopaegma arvense* (Vell.) Stellfeld ex De Souza, synonym *Anemopaegma mirandum* (Cham.) DC, is a small shrub widely distributed in the Brazilian Cerrado, but currently considered an endangered species. It is popularly known as catuaba, tatuaba, verga-tesa, among other names, and is used as an aphrodisiac and a tonic for nervous debility and memory loss. The usually employed part is the root, although the aerial parts are also used. The adulteration of *A. arvense* crude drug is frequent and this has led to the implementation of several quality-control studies. The species contains triterpenes, flavonoids, proanthocyanins, and phenylpropanoid-substituted epicatechins, for which antimicrobial, antioxidant, and cytoprotective effects have been reported. Pre-clinical toxicological studies were performed with an herbal medicine containing both *A. arvense* and other species and the formulation was considered safe. However, there are no studies validating its popular use as an aphrodisiac.

Keywords *Anemopaegma arvense* · *Anemopaegma mirandum* · *Anemopaegma* · Bignoniaceae · Catuaba · Aphrodisiac

1 Taxonomic Characteristics

One of the first mentions of the name "catuaba" is attributed to the Brazilian botanist Freire Alemão, who published in a local newspaper, in 1860, a work titled "The Catuaba" (Ducke 1966). The species originally cited was identified as *Erythroxylum vaccinifolium* Mart. At the beginning of the 20th century Silva (1906) published a work about the "catuaba-da-Bahia", identifying it as a new species: *Erythroxylum catuaba* AJ da Silva. This identification, however, was later considered as a *nomen nudum*, that is, the name does not match any existing species (Ducke 1966). The definitive clarification occurred only when Marques (1998) got the flowered material and identified the species in question as *Trichilia catigua* Adr. Juss. (Meliaceae).

At the same time, in southeastern Brazil the use of roots from a species of *Anemopaegma*, also referred to as catuaba or "caatuyba" (Hoehne 1920) became popular. This species – *Anemopaegma arvense* (Vell.) Stellfeld ex De Souza (synonym *Anemopaegma mirandum* [Cham.] DC) (Bignoniaceae) – was selected and made official in the monograph of catuaba in the first Brazilian Pharmacopoeia (Silva 1926), thus becoming regarded as "catuaba verdadeira" (the true or official catuaba). Accordingly, the name catuaba refers to several species of different botanical families, representing one of the most remarkable cases of botanical confusion in Brazilian phytomedicine (Marques 1998; Kletter et al. 2004; Tabanca et al. 2007; Mauro et al. 2007; Mendes 2011).

According to the Brazilian indigenous language, *catuaba, catuíba*, or *caatuyba* mean "good leaf" or "good plant" (Silva 1926, 1927). Charan (1987) referred to catuaba as meaning "true man", derived from the Tupi language. Other popular names of *A. arvense* are "cataíba, tatuaba, catuaba verdadeira, catuabinha, alecrim do campo, verga-tesa", among other less-frequently used names (Corrêa 1931; Lohmann 2015; Plantamed 2015).

The *A. arvense* is an Angiosperm belonging to the Equisetopsida class, subclass Magnoliidae, superorder Asteranae, order Lamianales, family Bignoniaceae, genus *Anemopaegma* Mart. ex Meisn (Lohmann 2015; Tropics 2015). It has as synonyms *Anemopaegma mirandum* (Cham.) DC, *A. mirandum* (Cham.) Mart. ex DC., *A. sessilifolium* Mart. ex DC, *A. sessilifolium* Mart., *A. subundulatum* Bureau & K. Schum., *Bignonia arvensis* Vell., *Bignonia miranda* Cham., and *Jacaranda arvensis* (Vell.) Steud. (Lohmann 2015; Tropicos 2015). Names assigned to varieties are also recognized as synonyms, as *A. mirandum* var. *angustifolium* DC, *A. mirandum* var. *glabrum* DC, *A. mirandum* var. *hirsuta* Hassl., *A. mirandum* var. *latifolium* DC, *A. mirandum* var. *petiolatum* Bureau, *A. mirandum* var. *puberum* Burcau, *A. mirandum* var. *pubescens* DC, *A. mirandum* var. *sessilifolium* (Mart. ex DC) Bureau, and *A. mirandum* var. *verticellatum* Bureau.

2 Crude Drug Used

Catuaba was officially recognized in the first edition of the Brazilian Pharmacopoeia, although the plant drug was referred to as a rhizome (Silva 1926). In fact, it has poorly branched taproots, irregularly cylindrical and twisted, 6–10 cm long and 8–15 cm wide (Fig. 1). When the root is dry, the external surface has a yellowish-tan with shallow longitudinal grooves, a few transversal slits and numerous verrucose projections. The dry roots are virtually odorless and have a lightly astringent and weakly bitter flavor (Silva 1926; Hyakutake and Grotta 1965).

Under the microscope it is possible to observe that mature catuaba roots have a well-developed cork composed of two to ten layers of rectangular cells. There are round or tangentially elongated cells in the cortical parenchyma; thickened cana-

Fig. 1 Dry sample of *Anemopaegma arvense* showing the aerial part and roots

liculate sclereids, isolated or grouped into small clusters; small groups of fibers and numerous cells containing simple round starch grains. The cambium functions irregularly, producing four xylem and four phloem wedges and imparting a cross-shaped xylem to the root; the phloem also has groups of fibers arranged concentrically and in parallel; the secondary phloem is traversed by vascular rays up to three cells wide, which contain calcium oxalate prismatic crystals. The xylem has wide vessels, either isolated or in small groups and they are enveloped by scanty parenchyma and abundant starch-containing fibers with angular outlines and thin walls (Silva 1926; Hyakutabe and Grotta 1965).

3 Major Chemical Constituents and Bioactive Compounds

Employing precipitation tests with Dragendorff and Meyer reagents, Rizzini (1956) detected alkaloids only in the fresh root's bark of *A. arvense* but not within the plant subjected to drying. Jorge et al. (1989) reported the presence of phenolic compounds, saponins, coumarins, quinones, steroidal nucleus, and pentagonal lactones in leaves and roots of *A. arvense*, while the tests were negative for alkaloids and inconclusive for flavonoids.

The flavonoids rutin and quercetin 3-*O*-α-L-rhamnopyranosyl-(1→6)-β-D-galactopyranoside were identified from a methanolic extract of *A. arvense* leaves and subjected to various biological tests in which they showed moderate antifungal activity (Costanzo et al. 2013). Pro-anthocyanins and the phenylpropanoid-substituted epicatechins cinchonain Ia, cinchonain Ib, cinchonain IIa, cinchonain IIb (Uchino et al. 2004), as well as the compounds kandelin A1 and a flavan-3-ol-type lignoid trivially named catuabin A (Tabanca et al. 2007) also were described in the species. The catuabin A present in *A. arvense* is a flavonoid, but an alkaloid also named catuabin A is described in *Erythroxilum vaccinifolium*, an another species known as catuaba (Silva et al. 2012), which contribute to the confusion among the species.

The triterpenes oleanoic acid, ursolic acid, and betuline were identified in methanolic extracts from the roots and aerial parts of catuaba and the content of these constituents was ten times larger in the aerial parts (Pereira et al. 2007). According to the authors, these results suggest that aerial parts can be used successfully instead of using the roots, thus contributing to the preservation of the species since the aerial parts are renewable.

4 Morphological Description

Anemopaegma arvense is a subshrub with woody, hard and light-colored roots, rarely subscandent; stems quadrangular or sub-cylindrical, pubescent, rough, rarely glabrous, up to 40 cm tall (Corrêa 1931; Ferri 1969; Mauro et al. 2007). The leaves

Fig. 2 Two varieties of *Anemopaegma arvense*. (**a**) Variety with linear-oblong leaflets; (**b**) flowering variety with lanceolate-obtuse leaflets

are compound, trifoliate, sessile; leaflets narrow, linear or oblong-linear, acute or obtuse, narrow at the base, margins revolute, glabrous and rough to the touch (Mauro et al. 2007). The flowers are axillary, large, solitary, and pedunculate; calyx 5-lobed, corolla infundibuliform, petals yellow with a white or sulfurous face, 4.5–5.0 cm long. The fruits are flattened capsules 7–8 cm long and 4–5 cm wide formed by thick woody valves; the seeds are elliptical with hyaline wings (Corrêa 1931; Ferri 1969).

Firetti-Leggieri et al. (2014) proposed the use of leaf anatomy as a key for the identification of *Anemopaegma* taxa. Figure 2 shows two varieties of *A. arvense*, one with linear-oblong leaflets and a flowering plant with lineolate-obtuse leaflets. Corrêa (1931) describes the occurrence of the following varieties: *angustifolia*, with glabrous branches and linear-oblong leaflets; *lanceaefolia*, with velvety-pubescent branches and leaflets and linear-oblong leaflets; *petiolata*, with long-petiolate leaves and narrow-lanceolate, obtuse leaflets; *puberula*, with pubescent stems and oblong, very obtuse leaves; *sessilifolia*; and *verticillata*, with sessile leaves and very narrow, reticulate and glabrous leaflets. Only four of these six varieties have been confirmed using modern techniques based on taxonomic keys and genetic analysis (Batistini 2006), but a considerable level of genetic diversity can be observed in natural populations of catuaba (Batistini et al. 2009).

5 Geographical Distribution

A. arvense is widely distributed in the cerrado biome, in the southeastern and midwest states of Brazil, such as Goiás, Mato Grosso, Minas Gerais, and São Paulo (Corrêa 1931; Batistini 2006; Lohmann 2015). Specimens of *A. arvense* have been collected in Bolivia and Paraguay (Tropicos 2015).

6 Ecological Requirements

The species *A. arvense* is endangered due to the heavy extractivism and the reduction of the cerrado due to the expansion of urban and agricultural areas in addition to wild crafting, which substantiates the studies about the domestication of the species and management of its native populations. Pereira et al. (2003) evaluated the reproduction of *A. arvense* by asexual propagation and were able to obtain satisfactory proliferation using nodal segments. They also confirmed the feasibility of establishing a plant germplasm bank. In another study the same authors evaluated the germination rate of three varieties of *A. arvense* collected in different Brazilian states using controlled conditions of substrate, temperature, and humidity over 3 months (Pereira et al. 2007). They also reviewed the storage conditions of the seeds and found that dehydration by dry air flow and storage at −20 or −196 °C for 6 months did not affect the viability of the seeds. Souza et al. (2013) reported the occurrence of mycorrhizal fungi in the roots of plants from both different populations and varieties of *A. arvense.* They suggested that the symbiosis between these species is beneficial to the development of the plant and could provide a strategy for the cultivation of seedlings in greenhouses.

7 Traditional Use (Part(s) Used) and Common Knowledge

According to the Brazilian Pharmacopoeia, the medicinal parts of *A. arvense* are the roots (Silva 1926). However, other parts of the plant are also used, such as the stem, stem bark, leaves, or aerial parts in general (Silva 1927; Mendes and Carlini 2007; Silva et al. 2012). The uses of *A. arvense* as an aphrodisiac or as a tonic against nervous debility and loss of memory, among other uses, are listed in several books on Brazilian folk medicine (Mendes and Carlini 2007). Hoehne (1920) reports the use of roots of *A. arvense* as a nervous stimulant, useful as an aphrodisiac without harming the human body. Silva (1927) refers to the use of this species as a general stimulant for nerve diseases, gastrointestinal and circulatory asthenia, dysentery, in locomotor ataxia, persistent neuralgia, chronic rheumatism, and partial paralysis.

The main popular use of *A. arvense* is as an aphrodisiac. *A. arvense* has acquired a fame as a powerful sexual stimulant that in Minas Gerais, the Brazilian state where the plant is most widely used, there is also a popular saying according to which "up to 60 years the children are father's babies, and after this age they are catuaba's baby" (Silva 1927). When used as a nerve tonic and sexual stimulant, the usually employed part is the root, prepared as a tea (decoction), tinctures, and especially in "garrafadas" (alcoholic-based preparations).

8 Modern Medicine Based on Its Traditional Medicine Uses

The first pharmacological studies were carried out by Hamet (1938). They found hypotensive and bradycardiac effects in guinea pigs and dogs after treatment with aqueous extracts of *A. arvense*. Markus et al. (1980) carried out extensive pharmacological research in rodents with a crude aqueous extract of *A. arvense* roots. These authors confirmed the hypotensive and bradycardiac effects previously reported, but also showed opposite effects (positive inotropic and chronotropic effects in vitro). The negative effects were blocked by atropine and the positive by propranolol, demonstrating a profile of muscarinic and stimulating adrenergic activities. Acute *i.p.* administration of the same extract (1–500 mg/kg) to rats and mice did not alter the motor activity or excitability, but higher doses induced writhing and hypertonicity in the tail (Straub effect) in some animals (Markus et al. 1980). The chronic administration of the extract (25 or 50 μg/kg) did not modify the pharmacological response of the seminal vesicles or alter the weight of the organs sensitive to hormonal change (seminal vesicles, prostate, testis, among others). The estrous cycle in female rats was not modified by such treatment and the mating of these female rats with untreated male rats generated normal litters in number, weight, and development (Markus et al. 1980).

In another study, Chieregatto (2005) evaluated the effects of *Heteropterys aphrodisiaca* extracts (nó-de-cachorro) and *A. arvense* on the testis and in the spermatogenic process of Wistar rats of reproductive age. Treatment with an infusion of *A. arvense* for 56 days induced an increase of the seminiferous tubule diameter and seminiferous epithelium thickness and induced a significant increase in body weight, in the weight of testis, testicular parenchyma, and vesicular glands, among other effects, depending on the dose used. The total sperm reserves and daily sperm production were lower in all treatments compared to the control group (Chieregatto 2005).

Specifically in relation to sexual behavior, Abreu et al. (1980) evaluated the acute effect of an infusion of 2.5 and 5% of *A. arvense* roots in rats exposed to receptive females, with respect to parameters as mount, intromission, and ejaculation latencies, number and frequency of intromissions, post-ejaculatory interval, and number of ejaculatory series. There was no statistical difference between the groups, so the popular reputation of this plant as an aphrodisiac drug could not be confirmed.

Uchino et al. (2004) evaluated the effect of eight fractions and sub-fractions extracted with ethyl-acetate from the methanol extract of *A. arvense* on the viability of cells incubated with squalene mono-hydro-peroxide, a lipid hydro-peroxide. The treatment with the fractions containing cinchonain Ia, Ib, IIa, and IIb prevented most of the changes induced by hydroperoxide on the tested cell lines (Uchino et al. 2004). In another study the pre-incubation of SH-SY5Y human neuroblastoma cells with *A. arvense* roots extracted in DMSO at concentrations of 0.312 and 1.250 mg/

ml showed a partially protective effect on cell viability after incubation with rotenone, a drug used in experimental parkinsonism (Andrade et al. 2008). Ultrastructural analysis by electronic microscopy showed that concomitant treatment with catuaba induced a protective effect on the damage caused by rotenone on the cell and mitochondrial membranes and reduced the occurrence of apoptotic signals in cells incubated with rotenone (Andrade et al. 2008). The authors suggest that the neuroprotective effect of catuaba can be due to the antioxidant activity of its active principles. A bioguided assay led to the isolation of four compounds with antioxidant activity: kandelin A1, cinchonain Ia, cinchonain IIa, and catuabin A, the last two being comparable to the positive controls (vitamin C and Trolox) in potencies (Tabanca et al. 2007).

Tabanca et al. (2007) evaluated various biological activities for the methanol, hexane, and ethyl acetate extracts of *A. arvense* stem bark. None of the extracts showed significant activity in in vitro assays against different bacteria and fungi, nor did they show cytotoxic activity against tumor and non-tumor cell lines investigated with the concentrations evaluated. In another study a weak antibacterial activity against *Pseudomonas aeruginosa* was found for the alcoholic extract 96% of the aerial parts of *A. arvense* (Marques et al. 2013). The same authors also found a mild antifungal activity against *Crytococcus neoformans* for the aerial part and a weak activity against *Candida albicans* using the root extracts of *A. arvense* (Marques et al. 2013). The antifungal activity against *Trichophyton rubrum* was found by Costanzo et al. (2013) for the flavonoid-rich fraction obtained from the methanol extract of the leaves of *A. arvense*, and for two isolated flavonoids. The same study also evaluated the effect of the methanol extract of *A. arvense* and its isolated flavonoids against several bacteria and found inhibition using concentrations up to 2.5 mg/ml (Costanzo et al. 2013). According to Bastitini et al. (2009) the *A. arvense* was intensively studied by Japanese groups as regard its antitumoral and cell rejuvenation activities, which led to several patents.

The quality control of the roots of *A. arvense* should initially follow the organoleptic, macroscopic, and microscopic descriptions cited in the first edition of the Brazilian Pharmacopoeia (Silva 1926), supplemented by descriptions by Hyakutake and Grotta (1965). Physical-chemical data published by Jorge et al. (1989) are available, although these data need to be confirmed.

Beltrame et al. (2004) evaluated the roots of *A. arvense* and the barks of *T. catigua* by high-performance liquid chromatography. Their aim was to develop methodologies and profiles to differentiate the two species. The evaluation of three commercial samples offered as "catuaba" showed that they were all from *T. catigua* barks, although sold as roots of *A. arvense* or barks of *Erythroxylum catuaba*, representing typical cases of adulteration. In a similar study Daolio et al. (2008) concluded that the herbal medicine industry in Brazil does not employ the roots of *A. arvense* to manufacture the phytomedicine catuaba, but instead they use the bark of *T. catigua*.

A similar situation relating to the mixture and tampering of commercial samples of catuaba sold in Brazil was verified by Kletter et al. (2004). These authors assessed

14 commercial samples of catuaba and none of them showed the presence of *A. arvense* roots, but did reveal botanical material from other species known as catuaba or from unknown sources. Beltrame et al. (2010) also performed a morpho-anatomic study using roots of *A. arvense* and barks of *T. catigua* and showed that the commercial samples of catuaba were similar to *T. catigua*.

In contrast to the previous cited studies, all six commercial samples of catuaba evaluated by Tabanca et al. (2007) matched with a sample certified as *A. arvense*, without showing similarities to *T. catigua*. However, the authors used barks for comparison, not roots. The discrepancies between these studies and recurrent cases of adulteration indicate that we still need more studies regarding the identification and quality control of *A. arvense*.

In Brazil the sale of energy drinks prepared with catuaba is common, although in many cases the species used is not described. A biscuit formulation containing catuaba (*A. arvense*) and guarana (*Paullinia cupana*) was developed as an energetic and functional food: source of fiber, copper, iron, and zinc (Oliveira et al. 2009). Since it does not contain gluten it can be used as an alternative to conventional crackers.

A preclinical toxicology study evaluated the effect of oral administration for 30 days of an herbal medicine containing *A. arvense*, *Cola nitida*, *Passiflora alata*, *Paullinia cupana*, *Ptychopetalum olacoides*, and thiamin in male and female rabbits (Mello et al. 2010). The study evaluated general signs of toxicity, rectal temperature, food and water consumption, body and organ weights, as well as biochemistry, hematology, pathology, and urinalysis. Oral administration for 30 days in a dose ten times as high as prescribed for human use was considered innocuous (Mello et al. 2010). There are also other phytotherapic preparations containing *A. arvense* sold in Brazil, but to-date we have not found relevant studies on these formulations.

9 Conclusions

Anemopaegma arvense is the species considered to be the official "catuaba", according to Brazilian Pharmacopoeia. There are only few studies evaluating its chemical composition and its biological effects. Some biological activities such as an antioxidant, antimicrobial, cytoprotective, etc. have been recorded, but the main popular uses attributed to catuaba, especially its aphrodisiac action, have not been proven in clinical studies. The evaluation of commercially available materials seems essential to avoid the use of an adulterated botanical drug. Furthermore, it is suggested that more phytochemical and quality-control studies should be performed. Similarly, farther management and cultivation studies should be conducted with the aim of economic exploitation of this valuable species.

Acknowledgments The authors thank Profa Ana Maria Soares Pereira (UNAERP) for providing some photos and Prof. Wayne Losano for the linguistic review.

References

Abreu BC, Sollero L, Pereira NA (1980) O comportamento sexual dos ratos machos como modelo experimental para o estudo das plantas afrodisíacas. A catuaba (*Anemopaegma arvense* [Vell.] Stellf.). Cien Cult 33(Suppl):39–40 Portuguese

Andrade DVG, Oliveria DM, Barreto G, Bertolino LA, Saraceno E, Capani F, Giraldez LD (2008) Effects of the extract of *Anemopaegma mirandum* (Catuaba) on rotenone-induced apoptosis in human neuroblastomas SH-SY5Y cells. Brain Res 1198:188–196

Batistini AP (2006) Diversidade morfológica, genética e química de populações naturais de *Anemopaegma arvense* (Vell.) Stellf. [PhD Thesis] Universidade Estadual Paulista, Jaboticabal

Batistini AP, Telles MPC, Bertoni BW, Coppede JS, Môro FV, França SC, Pereira MAS (2009) Genetic diversity of natural populations of *Anemopaegma arvense* (Bignoniaceae) in the Cerrado of São Paulo State, Brazil. Genet Mol Res 8(1):52–63

Beltrame FL, Cass QB, Rodrigues Filho E, Barros F, Cortez DAG (2004) Análisis de produtos fitoterapéuticos comerciales de "catuaba" por LC-UV-MS. Not Tec Lab 12(3):14–16 Spanish

Beltrame FL, Rocha DC, Albiero ALM, Carmo MRB, Cass QB (2010) Estudos morfo-anatômicos de amostras comerciais de catuaba. Publ UEPG Ci Biol Saúde 16(2):111–118 Portuguese

Charan I (1987) Há ações afrodisíacas nas plantas medicinais do Brasil? Folha Med 94(5):303–309 Portuguese

Chieregatto LC (2005) Efeito do tratamento crônico com extratos de Heteropterys aphrodisiaca O. Mach. e *Anemopaegma arvense* (Vell.) Stellf. no testículo de ratos wistar adultos. Dissertation, Universidade Federal de Viçosa, Viçosa

Corrêa MP (1931) Dicionário das plantas úteis do Brasil e das exóticas cultivadas, vol 2. Ministério da Agricultura, Rio de Janeiro, pp 150–152

Costanzo CDG, Fernandes VC, Zingaretti S, Beleboni RO, Pereira AMS, Marins M et al (2013) Isolation of flavonoids from *Anemopaegma arvense* (Vell) Stellf. ex de Souza and their antifungal activity against *Trichophyton rubrum*. Braz J Pharm Sci 49(3):559–565

Daolio C, Beltrame FL, Ferreira AG, Cass Q, Cortez DAG, Ferreira MMC (2008) Classification of commercial catuaba samples by NMR, HPLC and chemometrics. Phytochem Anal 19(3):218–228

Ducke A (1966) A catuaba na botânica sistemática, científica e pseudocientífica. Rev Bras Farm 47(5):267–272 Portuguese

Ferri MG (1969) Plantas do Brasil: espécies do cerrado. São Paulo, USP/Edgard Blüchen

Firetti-Leggieri F, Lohmann LG, Semir J, Demarco D, Castro MM (2014) Using leaf anatomy to solve taxonomic problems within the *Anemopaegma arvense* species complex (Bignonieae, Bignoniaceae). Nord J Bot 32:620–631

Hamet R (1938) Sobre alguns efeitos fisiológicos da droga brasileira conhecida pelo nome de 'folhas de catuaba. Rev Flora Med 4(4):235–242 Portuguese

Hoehne FC (1920) O que vendem os hervanários da cidade de São Paulo. Casa Duprat, São Paulo, pp 78–79

Hyakutake S, Grotta AS (1965) Contribuição ao estudo morfológico e anatômico de *Anemopaegma arvense* (Vell.) Stellfeld. var. petiolata Bur. – Bignoniaceae. Rev Fac Farm Bioquim 3(1):51–78 Portuguese

Jorge LIF, Ferro VO, Sakuma AM (1989) Determinação das principais características estruturais e químicas da droga *Anemopaegma arvense* (Vell.) Stellfeld (catuaba). Rev Inst Adolfo Lutz 49(2):183–191 Portuguese

Kletter C, Glasl S, Presser A, Werner I, Reznicek G, Narantuya S et al (2004) Morphological, chemical and functional analysis of catuaba preparations. Planta Med 70:993–1000

Lohmann LG (2015) [Internet]. *Bignoniaceae*. In: Lista de Espécies da Flora do Brasil. Jardim Botânico do Rio de Janeiro. Available from: http://reflora.jbrj.gov.br/jabot/floradobrasil/FB112500. Accessed on 27 Feb 2015

Markus RP, Gonçalo MC, Lapa AJ, de Souza LCB, do Valle JR (1980) Atividade farmacológica dos extratos da catuaba *Anemopaegma arvense* (Vell.) Stell. Cien Cult 33(supl):130–135 Portuguese

Marques LC (1998) Contribuição ao esclarecimento da identidade botânica da droga vegetal catuaba. Rev Racine 8(43):8–11 Portuguese

Marques MCS, Hamerski L, Garcez FR, Tieppo C, Vasconcelos M, Torres-Santos EC, Chang M, Garcez WS (2013) In vitro biological screening and evaluation of free radical scavenging activities of medicinal plants from the Brazilian Cerrado. J Med Plant Res 7(15):957–962

Mauro C, Pereira AMS, Silva CP, Missima J, Ohnuki T, Rinaldi RB (2007) Estudo anatômico das espécies de cerrado Anemopaegma arvense (Vell.) Stellf. ex de Souza (catuaba), Zeyheria montana Mart. (bolsa-de-pastor) e Jacaranda decurrens Chamisso (caroba) – Bignoniaceae. Rev Bras Farmacog 17(2):262–265

Mello JRB, Mello FB, Langeloh A (2010) Toxicity study of a phytotherapic with *Anemopaegma mirandum*, *Cola nitida*, *Passiflora alata*, *Paullinia cupana*, *Ptychopetalum olacoides* and thiamin in rabbits. Lat Am J Pharm 29(8):1431–1435

Mendes FR (2011) Tonic, fortifier and aphrodisiac: adaptogens in the Brazilian folk medicine. Rev Bras Farmacog 21(4):754–763

Mendes FR, Carlini EA (2007) Brazilian plants as possible adaptogens: an ethnopharmacological survey of books edited in Brazil. J Ethnopharmacol 109:493–500

Oliveira KEO, Takase I, Gonçalves ECBA (2009) Development of gluten-free cookie from medicinal plants (Guaraná – Paullinea cupana and Catuaba – *Anemopaegma mirandum*) aiming at copper, iron, and zinc supplementation. Cienc Tecnol Aliment 29(3):631–635

Pereira AMS, Amui SF, Bertoni BW, Moraes RM, França SC (2003) Micropropagation of *Anemopaegma arvense*: conservation of an endangered medicinal plant. Planta Med 69(6):571–573

Pereira AMS, Salomão AN, Januario AH, Bertoni BW, Amui SA, França SC et al (2007) Seed germination and triterpenoid content of *Anemopaegma arvense* (Vell.) Stellfeld varieties. Genet Resour Crop Evol 54(4):849–854

Plantamed (2015) [Internet]. *Anemopaegma arvense* (Vell.) Stellfeld ex de Souza – Catuaba. Available from: http://www.plantamed.com.br/plantaservas/especies/Anemopaegma_arvense.htm. Accessed on 02 Mar 2015

Rizzini CT (1956) Catuaba. Rodriguesia 18–19(30–31):5–6 Portuguese

Silva AJ (1906) Estudo botânico e chímico da catuaba (Erythroxylaceae catuaba do norte). Dissertation, Faculdade de Medicina da Bahia, Salvador

Silva RAD (1926) Catuaba. In: Pharmacopeia dos Estados Unidos do Brasil, 1st edn. Companhia Editora Nacional, São Paulo

Silva RAD (1927) Plantas medicinaes brasileiras. Estudo botanico e pharmacognostico. Catuaba. Rev Bras Med Pharm 3(7/8):55–62 Portuguese

Silva CV, Borges FM, Velozo ES (2012) Phytochemistry of some Brazilian plants with aphrodisiac activity. In: Rao V (ed) Phytochemicals – a global perspective of their role in nutritional and health. Intech, pp 307–326, https://doi.org/10.5772/26989

Souza AV, Oliveira FJV, Bertoni BW, França SC, AMS P (2013) Ocorrência de fungos micorrízicos em catuaba (*Anemopaegma arvense* (Vell.) Stell. ex de Souza-Bignoniaceae), uma planta medicinal do Cerrado em risco de extinção. Rev Bras Pl Med 15((4) Suppl. 1):646–654 Portuguese

Tabanca N, Pawar RS, Ferreira D, Morais JP, Khan SI, Joshi V et al (2007) Flavan 3-ol-phenylpropanoid conjugates from *Anemopaegma arvense* and their antioxidant activities. Planta Med 73:1107–1111

Tropicos.org (2015) [Internet]. Missouri Botanical Garden. Available from: http://www.tropicos.org/. Accessed on 12 Mar 2015

Uchino T, Kawahara N, Sekita S, Satake M, Saito Y, Tokunaga H et al (2004) Potent protecting effects of catuaba (*Anemopaegma mirandum*) extracts against hydroperoxide-induced cytotoxicity. Toxicol In Vitro 18(3):255–263

Aniba canellila (Kunth) Mez.

Lidiam Maia Leandro, Paula Cristina Souza Barbosa, Simone Braga Carneiro, Larissa Silveira Moreira Wiedemann, and Valdir Florêncio da Veiga-Junior

Aniba canellila (Kunth) Mez.
Photo: Denisa Sasaki
Available in: https://www.kew.org/science/tropamerica/neotropikey/families/Lauraceae.htm

L. M. Leandro · P. C. S. Barbosa · S. B. Carneiro · L. S. M. Wiedemann
V. F. da Veiga-Junior (✉)
Chemistry Department, Institute of Exact Sciences, Amazonas Federal University, Manaus, AM, Brazil
e-mail: larissasmw@ufam.edu.br

U. P. Albuquerque et al. (eds.), *Medicinal and Aromatic Plants of South America*, Medicinal and Aromatic Plants of the World 5,
https://doi.org/10.1007/978-94-024-1552-0_9

Abstract *Aniba canellila* (Kunth) Mez. known as "casca-preciosa" (precious wood) is an important and historical species in the Amazon region. It is a large tree endemic to South America. It is a medicinal plant used in the Amazon traditional folk medicine. Monoterpenes, sesquiterpenes and benzenoids are classes of compounds present in the essentials oils of *A. canellila*. Of special interest is 1-nitro-2-phenylethane, the major constituent, with cardiovascular, fungistatic, cytotoxicity and antileishmanial activities. Methyl-eugenol, another important constituent, presents antispasmodic, hypotensive, anesthetic, cytotoxic, and genotoxic activities. The information summarized in this chapter intends to serve as a reference tool to chemistry and biological activities of the essential oil obtained from *A. canellila*.

Keywords 1-nitro-2-phenylethane · Methyleugenol · Precious bark · Stick-precious · False cinnamon

1 Taxonomic Characteristics

In the history of South America, as well as in the history of chemistry of natural products, there is one botanical species of Angiosperm belonging to Lauraceae family that stands out for its aroma, chemical composition and economic use: it is *Aniba canellila* (Kunth) Mez. Popularly known as shell-precious, precious sheet, false cinnamon, bark of Maranhão, amapaiama, pereiorá and stick-precious (Maia et al. 2001).

Synonyms Aniba elliptica AC Sm; Cryptocarya canelilla Kunth

2 Major Chemical Constituents and Bioactive Compounds

Several studies have reported that the constituents detected in the essential oil of this species belong to three classes of substances: monoterpenes, sesquiterpenes and benzenoids. The percentages of classes of constituents for each part of the plant are shown in Table 1.

The main benzenoids are phenylacetaldehyde, (E) -methyl-cinnamate, benzyl benzoate and especially 1-nitro-2-phenylethane, together with their precursor molecules, such as benzonitrile, benzoacetaldehyde and benzoacetonitrile (Maia et al. 1996). The other benzenoids present in oils belong to the class of phenylpropanoids: safrole, eugenol and methyleugenol (Silva et al. 2007; Taveira et al. 2003).

Approximately 17 monoterpenes have been reported in *A. canellila* essential oils, namely: α-pinene, β-pinene, myrcene, δ-3-carene, p-cymene, limonene, β-phellandrene, 1,8-cineole, (*Z*)-β-ocimene, (*E*)-β-ocimene, linalool oxide, linalool, trans-p-menth-2-ene-1-ol, terpinen-4-ol isomentol, α-terpineol and geraniol.

Table 1 Percentages of the classes of constituents detected in oils from different parts of *A. canellila*

Plant parts	Monoterpenes (%)	Sesquiterpenes (%)	Benzenoids (%)	References
Leaves	4.6	13.7	78.7	Lima et al. (2004)
	2.4	6.4	89.3	Silva et al. (2009)
	2.7	8.4	88.8	Silva et al. (2009)
Barks	0.9	4.6	93.5	Taveira et al. (2003)
	0.4	1.0	96.6	Oger et al. (1994)
	1.6	3.5	94.0	Silva et al. (2007)
Trunk wood	1.1	0.4	98.0	Silva et al. (2007)
	0.8	1.2	97.3	Silva et al. (2007)
Stems	9.9	14.3	75.0	Lima et al. (2004)

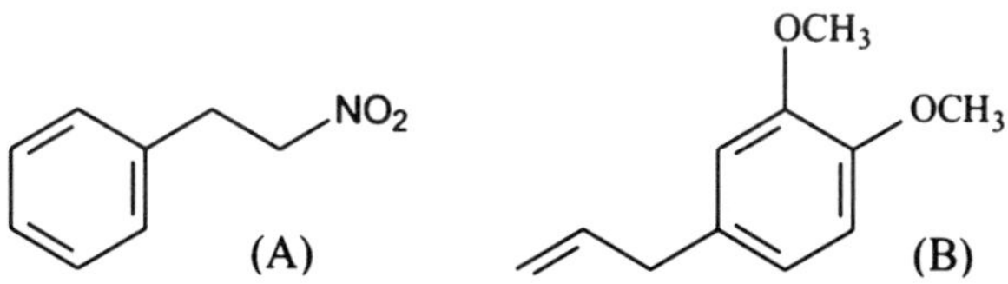

Fig. 1 Major constituents of *A. canellila* essential oils: (**a**) 1-nitro-phenylethane (**b**) methyleugenol

About 30 sesquiterpenes have been identified and described in this species. 18 of these are sesquiterpene hydrocarbons: α-cubebene, β-elemene, α-copaene, (*Z*)-caryophyllene, longifolene, α-gurjunene, (*E*)-caryophyllene, aromadendrene, α-humulene, β -chamigrene, β-selinene, α-selinene, β-bisabolene, delta-cadinene, cis-calamenene, β-sesquiphelandrene, trans-calamenene, cadin-1,4-diene. The 12 oxygenated sesquiterpenes already described are: elemol, (*E*)-nerodiol, spathulenol, caryophyllene oxide, globulol, guaiol, humulene epoxide, 1-epi-cubenol, cubenol, epi-α-muurulol, Selina-11-ene-4α-ol and bulnesol.

Vilegas et al. (1998) studied the bark essential oil extracted by supercritical fluid (CO_2), detecting 1-nitro-2-phenylethane, eugenol, methyleugenol, calamenene and cadinene, essential oils constituents commonly reported in *A. canellila*. In addition to these constituents, were detected for the first and only in this study the sesquiterpenes curcumene, γ-eudesmol and bisabolol, the latter with content of 7%.

The composition of the essential oil from *A. canellila* was first described by Gottlieb and Magalhães (1959). They reported, for the very first time, a molecule with nitro group in natural products: 1-nitro-2-phenylethane (Fig. 1). Similarly, Taveira et al. (2003) described the presence of methyl-eugenol in essential oils from leaves of *A. canellila*. The content of 1-nitro-2-phenylethane and methyl-eugenol in leaf samples, bark and wood of the trunk showed to be quantitatively different for each sampled area and often depend on seasonality. The highest levels of 1-nitro-2-phenylethane were observed during the rainy season whose values reached 95.3%. The largest methyl-eugenol contents were observed during the dry season reaching 45%.

3 Morphological Description

The botanical species *A. canellila* presents itself as an evergreen tree that can reach 25 m in height. Its stem has a diameter between 40 and 70 cm and is coated with highly aromatic reddish bark. Its leaves are simple, glabrous, and can reach a length of 20 cm. Flowers are small and yellowish. Fruits are ovoid berries of dark color (Manhães et al. (2012)).

4 Geographical Distribution

The main geographical region of its distribution is the Amazon-Region. The species is a native tree from Western Amazonia, at Peru, to Eastern Amazonia, at Macapá and Pará Brazilian States. Barks and leaves are commonly found at popular markets in the Amazon region and even at medicinal markets all over Brazil. (Maia et al. 2001).

5 Ecological Requirements

Environmental factors such as light and humidity have significant effects on A. canellila and its essential oil production (Sangwan et al. 2001). These can be ascribed to changes in seasonality (Duarte et al. 2009). An experiment carried out (Atroch 2008) with seedlings of A. canellila showed that moisture deficiency and light irradiation reduces the oil yield in roots and leaves, respectively.

6 Collection Practice

The commercial essential oil of *A. canellila* is extracted from the wood of the trunk. The high oil yields (Manhães et al. 2012) have led to the indiscriminate cutting of mature trees of reproductive age: a similar situation to *Aniba rosaeodora* Ducke (rosewood), which was already on the list of endangered species (IBAMA Ordinance No. 37-N, of April 3, 1992). *A. canellila* runs the same risk of extinction by predatory exploitation and extraction.

The essential oil extracted from branches and leaves of *A. canellila* has presented a viable alternative for sustainable use of the species. It facilitates field work and is economically more viable, which is due to the high essential oil yield (Silva et al. 2009; Manhães et al. 2012).

7 Traditional Use (Part(s) Used) and Common Knowledge

In popular usage, the use of seeds, bark and leaves was described mentioning the use of powdered seeds as antidiarrhoeal. The bark is used for treating problems such as poor digestion and aerophagia, arthritis, cough, chronic sputum, syphilis, leukorrhea, dropsy, heart ailments, memory loss, injuries, inflammation and stimulating the nervous system and also has carminative properties (Lorenzi and Matos 2008; Lima et al. 2004, 2009; Perazzo et al. 2009). They also recorded the use of barks for the treatment of malaria (Botsaris 2007) and Alzheimer's disease (Madaleno 2011).

Lorenzi and Matos (2008) described the use of essential oil from *A. canellila* to alleviate pain after tooth extraction, is indicated for use in acne, dermatitis and skin care, as well as cold, cough, fever, headaches, various infections, injuries, nervous tension and nausea.

In addition to medicinal properties, this species has a high value in the food market, as well as cosmetics and perfumes. Due to its strong aroma, the wood of the trunk, twigs and leaves, are used as seasonings and ingredients for local dishes, fragrances and flavoring sachets of clothes (Silva et al. 2007).

8 Modern Medicine Based on Its Traditional Medicine Uses

The essential oil of the bark of the tree of this species carries a relaxing effect on intestinal smooth muscle Lahlou et al. (2005) showed cardiovascular effects in normotensive rats induced by the essential oil, causing a decrease in the heart rate. In a subsequent study Siqueira et al. (2010) investigated the mechanisms underlying the cardiovascular responses to 1-nitro-2-phenylethane and in vitro data suggested that the phase 2 response to hypotensive *iv* 1-nitro-2-phenylethane resulted, at least in part, from a direct vasodilatory effect of 1-nitro-2-phenylethane in the peripheral smooth muscle.

According to the study by Silva et al. (2009), the essential oil of leaves presented leishmanicidal activity. The oil from the stem wood (Silva et al. 2007 has a cytotoxic effect against *Artemia salina* (Silva et al. 2009). Studies with 1-nitro-2-phenylethane showed anti-inflammatory activity (Vale et al. 2013); fungistatic activity against *Candida albicans*, as studied by Oger et al. (1994); high cytotoxicity study (Silva et al. 2007), in addition to an analgesic effect (Silva et al. 2009).

Other biological activities reported for methyleugenol include: antibacterial, antifungal, induce hypothermic, myorelaxant, antispasmodic, anticonvulsant, hypotensive, anesthetic, cytotoxicity, and genotoxicity anti-feedant activity (Sell and Carlini 1976; Dallmeier and Carlini 1981; Sousa et al. 1990; Sayyah et al. 2002; Burkey et al. 2000; Yano and Kamimura 1993; Fontenelle et al. 2011; Lahlou et al. 2004).

9 Conclusions

Despite the high risk of joining the list of endangered species, *A. canellila* is still one of the most important medicinal plant species in the Amazon region, since the essential oil it produces, as well as the, the major constituents of its oil (1-nitro-2-phenylethane and the methyl-eugenol) have wide ranging applications in the pharmaceutical and cosmetics industries. So, the sustainable management and sustainable extraction of leaves and branches of these trees to replace the extraction of wood from the trunk, accompanied by the overthrow of trees, have become an essential need for the commercialization of this raw material.

References

Atroch EMAC (2008) Efeitos de Fatores abióticos sobre o Crescimento, Características fotossintéticas e Síntese de Óleos voláteis em plantas Jovens de Espécies de lauraceae na Amazônia Central. Tese de Doutorado, Instituto Nacional de Pesquisas da Amazônia/Universidade Federal do Amazonas, Manaus, p 109

Botsaris AS (2007) Plants used traditionally to treat malaria in Brazil: the archives of Flora Medicinal. J Ethobio Ethnomed 3:1–18

Burkey JL, Sauer JM, McQueen CA, Sipes IG (2000) Cytotoxicity and genotoxicity of methyleugenol and related congeners – a mechanism of activation for methyleugenol. Mutat Res 453:25–33

Dallmeier K, Carlini EA (1981) Anesthetic, hypothermic, myorelaxant and anticonvulsant effects of synthetic eugenol derivatives and natural analogues. Pharmacol Ther 22(2):113–127

Duarte AR, Naves RR, Santos SC, Seraphin JC, Ferri PH (2009) Seasonal influence on the essential oil variability of *Eugenia dysenterica*. J Braz Chem Soc 20:967–974

Fontenelle ROS, Morais SM, Brito EHS, Brilhante RSN, Cordeiro RA, Lima YC, Brasil NVGPS, Monteiro AJ, Sidrim JJC, Rocha MFG (2011) Alkylphenol Activity against *Candida* spp. and *Microsporum canis*: a focus on the antifungal activity of thymol, eugenol and *O*-methyl derivatives. Molecules 16:6422–6431

Gottlieb OR, Magalhães MT (1959) Occurrence of 1-nitro-2-phenylethane in *Ocotea pretiosa* and *Aniba canellila*. J Organomet Chem 24:2070–2071

Lahlou S, Figueiredo AF, Magalhães PJC, Leal-Cardoso JH, Duarte GP (2004) Cardiovascular effects of methyleugenol, a natural constituent of many plant essential oils, in normotensive rats. Life Sci 74:2401–2412

Lahlou S, Magalhães PJC, Siqueira RJB, Figueiredo AF, Interaminense LFL, Maia JGS, Sousa PJC (2005) Cardiovascular effects of the essential oil of *Aniba canellila* bark in normotensive rats. J Cardiovasc Pharmacol 46(4):412–421

Lima MP, Silva TMD, Silva JD, Zoghbi MG, Andrade EH (2004) Essential oil composition of leaf and fine stem of Aniba canellila (Kunth) Mez from Manaus, Brazil. Acta Amaz 34(2):329–330

Lima AB, Santana MB, Cardoso AS, Silva JKR, Maia JGS, Carvalho JCT, Sousa PJC (2009) Antinociceptive activity of 1-nitro-2 phenylethane, the main component of *Aniba canellila* essential oil. Phytomedicine 16(6–7):555–559

Lorenzi H, Matos FJA (2008) Plantas medicinais no Brasil: nativas e exóticas. Ed. Instituto Plantarum, Nova Odessa, Brasil, p 337

Madaleno IM (2011) Plantas da medicina popular de São Luís, Brasil. Bol Mus Para Emílio Goeldi Cienc Hum 6(2):273–286

Maia JGS, Zoghbi MGB, Andrade EHA (2001) Plantas aromáticas na Amazônia e seus óleos essenciais. Museu Paraense Emílio Goeldi, Belém, p 200

Maia JGS, Taveira FSN, Zohbi MGB, Santos AS, Luz AIR (1996) Óleo essencial de casca-preciosa. Summaries of The XIV Simpósio de Plantas Medicinais do Brasil 1996 (Florianópolis, Brasil, 17–20 September), p 197

Manhães AP, Veiga Junior VF, Wiedermann LSM, Fernandes KS, Sampaio PT (2012) Biomass production and essential oil yield from leaves, fine stems and resprouts using pruning the crown of Aniba canelilla (H.B.K.) (Lauraceae) in the Central Amazon. Acta Amazon 42:355–362

Oger JM, Richomme P, Guinaudeau H, Bouchara JP, Fournet A (1994) *Aniba canellila* (H.B.K.) Mez essential oil: analysis of chemical constituents, fungistatic properties. J Essent Oil Res 6(5):493–497

Perazzo FF, Carvalho JCT, Sousa PJC, Araújo JS, Pereira LLS, Modro MNR, Maia JGS, Araújo MTF (2009) Phytochemical toxicological evaluations of the essential oil from the bark of Aniba canellila (H.B.K.) Mez. J Essent Oil Res 21(4):381–384

Sangwan NS, Farooqi AHA, Shabih F, Sangwan RS (2001) Regulation of essential oil production in plants. Plant Growth Regul 34:3–21

Sayyah M, Valizadeh J, Kamalinejad M (2002) Anticonvulsant activity of the leaf essential oil of *Laurus nobilis* against pentylenetetrazole- and maximal electroshock-induced seizures. Phytomedicine 9(3):212–216

Sell AB, Carlini EA (1976) Anesthetic action of methyleugenol and other eugenol derivatives. Pharmacol Ther 14(4):367–367

Silva JKR, Sousa PJC, Andrade EHA, Maia JGS (2007) Antioxidant capacity cytotoxicity of essential oil and methanol extract of *Aniba canellila* (H.B.K.) Mez. J Agric Food Chem 55(23):9422–9426

Silva JRA, Carmo DFM, Reis EM, Machado GMC, Leon LL, Silva BO, Ferreira JLP, Amaral ACF (2009) Chemical biological evaluation of essential oils with economic value from Lauraceae species. J Braz Chem Soc 20(6):1071–1076

Siqueira RJB, Macedo FIB, Interaminense LFL, Duarte GP, Magalhães PJC, Brito TS, Silva JKR, Maia JGS, Sousa PJC, Leal-Cardoso JH, Lahlou S (2010) 1-Nitro-2-phenylethane, the main constituent of the essential oil of *Aniba canellila*, elicits a vago-vagal bradycardiac and depressor reflex in normotensive rats. Eur J Pharmacol 638(1–3):90–98

Sousa MB, Ximenes MF, Mota MT, Moreira LF, Menezes AA (1990) Circadian variation of methyleugenol anesthesia in albino rats. Braz J Med Biol Res 23(5):423–425

Taveira FSN, Lima WN, Andrade EHA, Maia JGS (2003) Seasonal essential oil variation of *Aniba canellila*. Biochem Syst Ecol 31(1):69–75

Vale JKL, Lima AB, Pinheiro BG, Cardoso AS, Silva JKR, Maia JGS, De Souza GEP, Da Silva ABF, Souza PJC, Borges RS (2013) Evaluation and theoretical study on the anti-inflammatory mechanism of 1-nitro-2-phenylethane. Planta Med 79(8):628–633

Vilegas JHY, Lanças FM, Vilegas W (1998) Composition of the volatile compounds from *Aniba canellila* (H. B. K.) Mez. Extracted by CO_2 in the supercritical state. Rev Bras 7–8(1):13–19

Yano K, Kamimura H (1993) Antifeedant activity toward larvae of *Pieris rapae crucivora* of phenolethers related to methyleugenol isolated from *Artemisia capillaris*. Biosci Biotechnol Biochem 57:129–130

Baccharis trimera (Less.) DC.

Daniel Garcia, Marcos Roberto Furlan, and Lin Chau Ming

Baccharis trimera (Less.) DC.
Photo: Gustavo Heiden
Available in: http://floradobrasil.jbrj.gov.br/reflora/floradobrasil/FB26875

D. Garcia (✉)
School of Agriculture and Biology, Shanghai Jiao Tong University (SJTU), Shanghai, China

M. R. Furlan
Agricultural Sciences Department, Universidade de Taubaté (UNITAU), Taubaté, São Paulo, Brazil

L. C. Ming
Laboratory of Medicinal Plants/Department of Horticulture, Universidade Estadual Paulista (UNESP), Agricultural Sciences College (ASC), Botucatu, São Paulo, Brazil

U. P. Albuquerque et al. (eds.), *Medicinal and Aromatic Plants of South America*, Medicinal and Aromatic Plants of the World 5,
https://doi.org/10.1007/978-94-024-1552-0_10

Abstract The producer of medicinal plants can be considered different from others because they need to know the whole steps from cultivation to harvest for each plant, including botanical identification, harvest time, temperature of drying, how to store and, in some cases, the medicinal purposes. Producers of *Baccharis trimera* (Less.) DC., for example, must know its botanical characteristics in order to avoid problems of confusion with *Baccharis coridifolia* DC. (broom), which belongs to the same genus, but it is toxic. *B. trimera*, also known as "Carqueja", is native from Brazil and is among the most important native medicinal plants of Brazil. Furthermore, *B. trimera*, has an ethnopharmacological importance for traditional people. It has many chemical compounds, and among the main are essential oils, sesquiterpene alcohols, resins, vitamins, tannins, flavonoids, lactones and saponin. Fresh or dehydrated *B. trimera* is marketed to produce phytotherapics, teas and is also used in the brewing industry, as well as replacement of hops for flavoring drinks, liqueurs and "cachaça". However, there is only one cultivar of *B. trimera*, called "CPQBA-1". Pioneering agronomic works done with it have shown promising results to cultivate it in the field, but still further studies are needed to ensure the quality and quantity of material.

Keywords Carqueja · Var. CPQBA-1 · Agronomic features · Medicinal purposes · Chemical substances

1 General Aspects

Agronomic research with native medicinal plants in Brazil is rare, as compared to exotic plants (Alonso 1998). This is one of the reasons that hinder the organization of national production of native medicinal plants (Souza et al. 2012). Moreover, the lack of information on the agronomic steps of these plants (Cortés et al. 2007) contributes to obtaining vegetable with the poor quality product (Veiga Jr. 2008), and increases the indiscriminate collection in natural environments (Carvalho 2003). According to Menezes Jr. (2006), about 90% of native medicinal species consumed in Brazil comes from collections without management. Additionally, Reis and Mariot (1998) alert that in Vale do Ribeira do Iguape region (West of São Paulo, Brazil) *Baccharis trimera* (Less.) DC. may be at risk of extinction due to exploration without appropriate management. The *B. trimera* cultivar "CPQBA-1" was the first recorded for a medicinal plant species, in Brazil's Ministry of Agriculture, Livestock and Supply (MAPA), in 2007, under the reference number 21190 (Montanari Jr. et al. 2008). This cultivar has very similar morphological characteristics to the wild type, except by the largest size that can reach up to 1.5 m tall. Another highlight of farming this cultivar is that it is adapted to agricultural environments, has uniform flowering, resistance to environmental factors and high germination rate (Montanari Jr. 2002).

2 Taxonomic Characteristics

B. trimera is native to South and Southeast of Brazil. It is popularly known as carqueja, broom-bitter, bacorida, carque, edge-of-condamine, broom, witches' button sedge-of-frill (Alzugaray and Alzugaray 1988), sweet (Pavan-Fruehauf 2000), and bacanta-Cacalia-bitter (Lorenzi and Matos 2008). These species have two scientific synonyms: *Baccharis genistelloides* var. *trimera* (Less.) Baker and *Molina trimera* Less. (Lorenzi and Matos 2008; Brazilian Pharmacopoeia 2010).

B. trimera belongs to the Asteraceae family. The more than 500 species belonging to the genus *Baccharis* are distributed from the United States of America (Fielding 2001) to the southern tip of Argentina and Chile (Hellwig 1990; Giuliano 2001), much of which is present in South America (Tropicos 2013). In Brazil, the genus *Baccharis* is represented by 120 species, distributed in larger quantities in the southern region (Barroso et al.1991). Some of these species are known for their toxicity, such as *B. coridifolia* (Abreu Matos et al. 2011).

3 Major Chemical Constituents and Bioactive Compounds

The essential oil of *B. trimera* contains monoterpenes (α- and *β*-pinene, nopineno) and sesquiterpene alcohols (carquejol, terpene esters). Soicke and Leng-Peschlow (1987) have investigated the fresh ethanol extract of *B. trimera* and found a mixture of five flavonoids: quercetin, luteolin, nepetina, apigenin and hispidulin. They also found in the same extract: flavones and flavonones; flavonoids, lactones and saponin (Santos et al. 1988; Simões et al. 1998; Pocá 2005), and resin, vitamins, polyphenols, tannins, *α*- and *β*-cadinene, calameno, eledol and eudesmol (Oliveira and Akisue 1997).

The carquejol and carquejila acetate are common in *B. trimera* (Siqueira et al. 1985; Souza et al. 1991), but Palácio et al. (2007) did not detect both chemical compounds in their analysis of essential oil. Lago et al. (2008) also did not notice the carquejila acetate in essential oil of *B. trimera* var. CPQBA-1. Carvalho (2003), evaluating the chemical composition of essential oil from *B. trimera* found great variability in the chemical compounds and in some samples the presence of carquejol and carquejila acetate was not observed, and in another sample was found only carquejila acetate. Morais and Castanha (2011) suggest that the lack of these substances in the analysis may indicate that the species are not *B. trimera*. However, Palácio et al. (2007) confirm that there is the possibility of decomposition of these substances during the extraction process or they may be modified due to the conditions of plant growth. Garcia et al. (2017) did not find both chemical compounds in the analysis of essential oils of *B. trimera* var. CPQBA-1, corroboration with Palácio et al. (2007).

Morais and Castanha (2011) evaluated the chemical composition of two populations of *Baccharis* sp. located in Rio de Janeiro state (Brazil). The authors identified 19 chemical compounds and the main were: *trans*-caryophyllene (22%), spathulenol (13.8%), ledol (13.7%), caryophyllene oxide (8.3%), germacrene-*D* (7%) and bicyclogermacrene (8.5%).

Working with *B. trimera* var. CPQBA-1, Lago et al. (2008) obtained different proportions of chemical compounds in essential oil from male and female plants, but only β-elemene, (*E*)-caryophyllene, aromadendrene, bicyclogermacrene, δ-cadinene, germacrene-*B*, caryophyllene oxide, epi-a-muurolol and α-cadinol were detected in both genders. The main components found in female plants were: (*E*)-caryophyllene, cadinene and α (more than 10%). The main substances found in male plants were: α-humulene and germacrene *D*.

It is known that the terpenoids have protective functions in plants, such as protection against herbivores and microbial activity (Owen and Peñuelas 2005). In work conducted with the cultivation of *B. trimera*, Garcia et al. (2017) identified in whole treatments with escalating doses of organic compost and three harvests the higher accumulation of five chemical compounds: trans-caryophyllene, caryophyllene oxide, spathulenol, bicyclogermacrene and germacrene-*D* (Table 1).

4 Morphological Description

According to the macroscopic analysis described in the Brazilian Pharmacopoeia (2010), *B. trimera* has three wings, cylindrical branches, up to 1 m in length, with rare leafless or sessile and reduced the leaf nodes. Green wings, glabrous, membranous, with 0.5–1.5 cm wide, wards of the flowering branches are narrower than the other. It is dioica plant and when it has flowering branches, these should only be pistillate or only staminate. Inflorescences, when present, the chapter type, yellowish-white, numerous, sessile, arranged along the upper branches. Staminate bracts involucres chapters 0.4–0.5 cm long and gradually the smaller oval and external glabrous, flower with corolla tube form, pentamerous up to 0.4 cm in length. Pistillate chapters up to 0.6 cm long, flowers with filiform corolla, with up to 0.4 cm long; type of fruit achenes, up to 0.2 cm in length with 10 longitudinal grooves.

5 Traditional Use (Part(s) Used) and Common Knowledge

B. trimera is one of the native medicinal plants from Brazil that has a high level of importance in the Brazilian scenario (Furlan 2005). Naiverth and Faria (2007) have emphasized that it is the fourth most widely used medicinal plant in the Pato Branco city (Paraná state, Brazil). Silva Jr. (1997) points out that the region is one of ten medicinal species sold in Brazil. *B. trimera* is sold in the domestic market in dried

Table 1 Biological activities of main chemical compounds of *B. trimera* var. CPQBA-1 and other species that contain the same substances

Chemical compound isolated	Molecule	Vegetal species	Biological activity	Scientific literature
Trans-caryophyllene	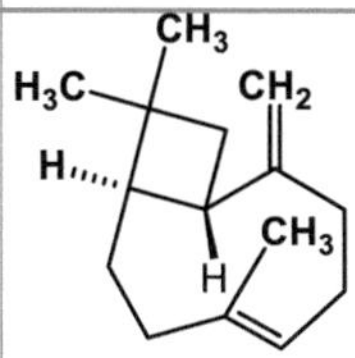	Lippia chevalieri	Antibacterial activity (Staphylococcus aureus and Enterococcus hirae); antifungal (*Saccharomyces cerevisiae*)	Mevy et al. (2007)
Germacrene-D	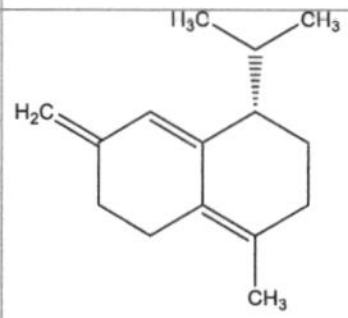	Senecio desiderabilis	Antimicrobial activity	Deuschle (2003)
		S. heterotrichius	Antifungal and antimicrobial activity	Francescato et al. (2007)
		S. bonariensis	Do not have antifungal and antimicrobial activity	Silva et al. (2010)
Bicyclogermacrene	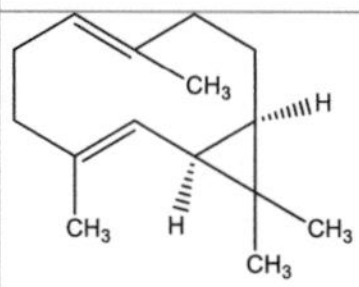	Araucaria columnaris, Agathis moorei, A. ovata, Callitris sulcata, Neocallitropsis pancheri	Natural acaricide	Lebouvier et al. (2013)
Spathulenol	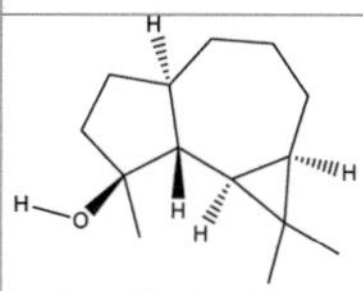	Melaleuca spp.	Antibacterial activity	Amri et al. (2012)
Caryophyllene oxide	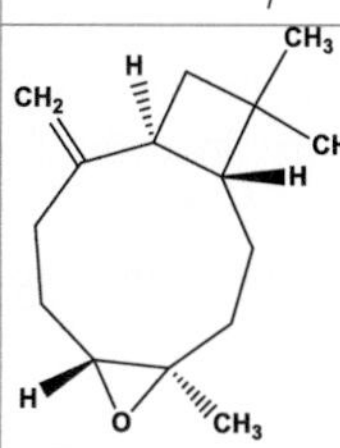	Baccharis trimera	Natural formicide	Marques et al. (2009)

Table 2 Main chemical compounds in the essential oil of *B. trimera*

Main substances	July (%)	August (%)	September (%)	October (%)
Carquejila acetate	68	42,3	60	58,5
β-pinene	5,6	12,6	11,3	12,3
Ledol	5,9	7,2	7,1	7,5
Limonene	3,4	4,2	4,7	4,0

Adapted from Simões-Pires et al. (2005)

form, in capsules, tinctures or tablets (Silva et al. 2006). Pocá (2005) listed some products containing *B. trimera* in its formulation found in the local market of Curitiba city (Paraná state, Brazil), e.g.: capsules, teas in sachets and packets.

B. trimera is known to grow better in full sun (Bona 2002). Is commonly found on roadsides, areas of high slope and wetlands (Correa Jr. et al. 2006). Furthermore, it is considered a weed in fields and pastures (Bona 2002). As for pests, it is usually attacked by aphids, scale insects and chewing (Andrião 2010). With regard to diseases, powdery mildew and some leaf spots (Bona 2002) occur.

The best planting time is from September to October, and culture must be renewed every 3 or 4 years (Correa Jr. et al. 2006; Trani et al. 2007).

The propagation is made of sexual (Castro 1998) and non-sexual form (Biase and Bona 2000; Sousa et al. 2006; Reis et al. 2007; Andrião 2010). Because it is a dioica plant, the agametic propagation of wild species and an except for sexual propagation to *B. trimera* var. CPQBA-1 is recommended (Garcia et al. (2017)).

Seasonality can influence on accumulation of different chemical compounds (Gobbo-Neto and Lopes 2007), as was demonstrated by Simões-Pires et al. (2005), who identified the following proportions of the main chemical compounds in the essential oil of *B. trimera* harvested at four different times in the Guaíba municipality (Rio Grande do Sul state, Brazil) (Table 2).

Regarding the cutting height, Mol et al. (2002) and Bona (2002) suggest leaving 10 cm of aerial part for regrowth, and Palacio et al. (2007) recommend leaving 30 cm.

Regarding the post-harvest of medicinal plants, Correa Jr. et al. (2004) and Reis et al. (2007) suggest that the drying must be done quickly in order to stop the enzyme and microorganisms activity, and consequently, reduce the degradation of their chemical compounds. Andrião (2010) and Garcia et al. (2017) recommend 38 °C as drying temperature of *B. trimera* on the artificial dryer with forced air circulation.

When there is no production of medicinal plants in crops planned, the outcome about the genetic, chemical and sanitary qualities of vegetal material collected is uncertain (Correa Jr. et al. 2004). It should be added that the *B. trimera* has greater genetic variability to be dioica, which also hampers the security of chemical homogeneity of wild plants, those who have not gone for a breeding program.

The aggravating scenario indiscriminate collection of native medicinal plants from Brazil, plus the demand of these plants by industries and population, stimulated the search for development of cultivars. In 2007, *B. trimera* var. CPQBA-1 was registered at the Ministry of Agriculture, Livestock and Supply (MAPA, Brazil) by the Multidisciplinary Center for Chemical, Biological and Agricultural Research

(CPQBA, Brazil) as the first cultivar of the native medicinal plant from Brazil (Montanari Jr. et al. 2008). This cultivar was selected as to dumping, germination dynamic and vigorous growth by the mass process with gametic control for five generations, including parental generation. A voucher specimen was deposited in the CPQBA Herbarium (Brazil) under number 1286.

Davies (1999) has obtained 180 kg ha^{-1} of dry *B. trimera* at 150 DAT. Garcia et al. (2017) obtained 1600 kg ha^{-1} of dry matter at 242 DAT (first regrowth). On the other hand, the results obtained in these studies differ drastically from those obtained by Palacio et al. (2007), who collected data from higher dry matter of *B. trimera* (4600 kg ha^{-1}) at 180 DAT. In this work, the authors used doses and different nitrogen sources (urea and sheep dung containing 4, 8 and 16 g N.$plant^{-1}$) suggesting that this fact may have occurred probably due to initial growth capacity of *B. trimera* as well as influenced by environmental conditions (Pinhais city, Paraná state, Brazil).

Despite there is little information about nutritional aspects of native medicinal plants from Brazil and its development in the field (Cortés et al. 2007), it is known that the availability of nutrients in the soil solution during the life cycle of plants is one of the conditions when wants achieve greater biomass production (Chaves 2002). Thus, it becomes essential to encourage related studies of native medicinal plant, because these lead to understanding and improving the management, thus justifying the production of raw materials with more desirable physicochemical and phytochemical properties industrially marketable.

6 Modern Medicine Based on Its Traditional Medicine Uses

When searching for plants with pharmacological properties in the environment, usually related to the ethnopharmacology studies contribute significantly without having to search for them randomly (Garcia 2009). Some of the main popular uses of *B. trimera* recorded in the scientific literature are to: digestive, diuretic, hepatoprotective, hypoglycemic and combating anemia (Castro and Ferreira 2000), antiemetic and antinauseant (Barbano 2006) and the whole plant as a mild sedative (Garcia et al. 2010).

Many laboratory studies with *B. trimera* has proved its pharmacological potential as: anti-hepatotoxic activity (Soicke and Leng-Peschlow 1987), anti-inflammatory and analgesic (Gené et al. 1996), sedative (Torres et al. 2000), anti-proteolytic and anti-hemorrhagic (Januário et al. 2004), antioxidant (Simões-Pires et al. 2005), antidiabetic (Oliveira et al. 2005) and antisecretory (Biondo et al. 2011). Preliminary studies indicate that some active principles of *B. trimera* act in lowering blood pressure (Saúde 2013). Nevertheless, Grance et al. (2008) observed toxicity activity of the aqueous extract of *B. trimera* cells in the liver and kidneys of pregnant rats; however, a reverse of this toxicity is shown when the extract is used discontinuously.

References

Abreu Matos FJ, Lorenzi H, Dos Santos LFL, Matos MEO, Silva MGV, Sousa MP (2011) Plantas tóxicas: estudos de fitotoxicologia química de plantas brasileiras. Instituto Plantarum de Estudos da Flora, São Paulo 247 p

Alonso JR (1998) Tratado de fitomedicina: bases técnicas y farmacológicas. Isis Ediciones SRL, Bueno Aires 1039 p

Alzugaray D, Alzugaray C (1988) Enciclopédia de plantas brasileiras. Editora Três, São Paulo 431 p

Amri I, Mancini E, De Martino L, Marandino A, Lamia H, Mohsen H, Bassem J, Scognamiglio M, Reverchon E, De Feo V (2012) Chemical composition and biological activities of the essential oils from three *Melaleuca* species grown in Tunisia. Int J Mol Med Sci 13(12):16580–16591

Andrião MA (2010) Marcha de absorção de macronutrientes e acúmulo de Fenólicos totais em [*Baccharis trimera* (Less.) DC.] var. CPQBA-1, sob diferentes podas no plantio. 78 p. Dissertação (Mestrado em Agronomia) – Faculdade de Ciências Agronômicas, UNESP, Botucatu, 2010

Barbano DBA (2006) A fitoterapia no SUS e o Programa de Pesquisas de Plantas Medicinais da Central de Medicamentos. Ministério da Saúde, Brasília 147 p

Barroso GM, Peixoto AL, Costa CG (1991) Sistemática das angiospermas do Brasil. UFV, Viçosa 326 p

Biasi LA, Bona CM (2000) Propagação de Carqueja (*Baccharis trimera* (Less.) A.P. de Candolle) por meio de estaquia. Rev Bras Plant Med 2(2):37–43

Biondo TM, Tanae MM, Coletta ED (2011) Antisecretory actions of *Baccharis trimera* (Less.) DC aqueous extract and isolated compounds: analysis of underlying mechanisms. J Ethnopharmacol 22(2):368–373

Bona CM (2002) Estaquia, calagem e sombreamento de carqueja. 95 p. Dissertação (Mestrado em Agronomia) – Programa de Pós-graduação em Agronomia – Universidade Federal do Paraná, 2002

Brazilian Pharmacopoeia: Brasil (2010) Vol. 2, 5° edição/Agência Nacional de Vigilância Sanitária. Brasília 46 p

Carvalho RIN (2003) Caracterização da exploração de carqueja e espinheira-santa na Região Metropolitana de Curitiba, Relatório de pesquisa. SEAB-PR, Curitiba 60 p

Castro HG (1998) Caracterização isozimática, crescimento e rendimento de tanino em seis acessos de carqueja (*Baccharis myriocephala* DC.) 114 p. Dissertação (Mestrado em Fitotecnia) – Departamento de Fitotecnia, Universidade Federal de Viçosa, Viçosa, 1998

Castro HG, Ferreira FA (2000) Contribuição ao estudo das plantas medicinais: carqueja (*Baccharis genistelloides*). UFV, Viçosa 102 p

Chaves FCM (2002) Produção, Biomassa, Rendimento e Composição de Óleo Essencial de Alfavaca-Cravo (*Ocimum gratissimum* L.) em Função da Adubação Orgânica e Épocas de Corte. Botucatu 144 p. Tese (Doutorado em Horticultura) – Universidade Estadual de São Paulo, 2002

Correa C Jr, Graça LR, Scheffer MC (2004) Complexo agroindustrial das plantas medicinais, aromáticas e condimentares no Estado do Paraná: diagnóstico e perspectivas. Editora Embrapa, Brasil 272 p

Correa C Jr, Ming LC, Scheffer MC (2006) Cultivo agroecológico de plantas medicinais, aromáticas e condimentares. Ministério do Desenvolvimento Agrário, Brasília 75 p

Cortés AMP, Biasi LA, Monte-Serrat B, Nakashima T (2007) Extração de nutrientes pela parte aérea de carqueja sob a influência de fontes e doses de nitrogênio. Ciên Rur 6:1809–1812

Davies P (1999) Experimentation on the propagation of *Baccharis trimera* (Less.) DC., Compositae (Carqueja). Acta Hortic 502:117–120

Deuschle RAN (2003) Atividade antimicrobiana e análise fitoquímica de *Senecio desiderabilis* Vellozo (Asteraceae). Santa Maria p. 124. Dissertação de Mestrado – Programa de Pós-graduação em Ciências Farmacêuticas, Universidade Federal de Santa Maria, 2003

Fapesp. Available online at http://agencia.fapesp.br/14176. Accessed on 01 Dec 2011

Fielding RR (2001) *Baccharis*: a genus of the Asteraceae new to Canada. Proc Nova Scotian Inst Sci 4:214–215

Francescato LN, Deuschle RAN, Mallman NCA (2007) Atividade antimicrobiana de *Senecio heterotrichius* DC. (Asteraceae). Rev Bras Ciên Farma 43(2):239–245

Furlan MR (2005) Cultivo de plantas medicinais, vol 1, 3rd edn. SEBRAE, Cuiabá 137 p

Garcia D (2009) Vozes e Olhares da Cantareira. SP, DVD, Faculdade Cantareira

Garcia D, Domingues MV, Rodrigues E (2010) Ethnopharmacological survey among migrants living in the Southeast Atlantic Forest of Diadema, São Paulo, Brazil. J Ethnobiol Ethnomed 6:29–48

Garcia D, Da Silva PSS, Furlan MR, Isobe MTC, Marques MOM, Ming LC (2017) Effect of organic fertilizer doses on the plant growth, essential oil production and chemical substances of "carqueja" over two harvest moments. J Agri Sci Technol 7:114–124

Gené RM, Cartaña C, Adzet T (1996) Anti-inflammatory and analgesic activity of *Baccharis trimera*: identification of its active constituents. Plant Med 62:232–235

Giuliano DA (2001) Classificación infragenérica de las espécies Argentinas de *Baccharis* (Asteraceae, Astereae). Darwin 39:131–154

Gobbo-Neto L, Lopes NP (2007) Plantas medicinais: fatores de influência no conteúdo de metabólitos secundários. Quím Nova 30:374–381

Grance SEM, Teixeira MA, Leite RS (2008) Baccharis trimera: effect on hematological and biochemical parameters and hepatorenal evaluation in pregnant rats. J Ethnopharmacol 117:28–33

Hellwig F (1990) Die Gattung *Baccharis* sp. (Compositae-Astereae) in Chile. Mitt Botanischen Staatssamml München 29:1–456

Januário AH, Santos SL, Marcussi S (2004) Neo-clerodane diterpenoid, a new metalloprotease snake venom inhibitor from *Baccharis trimera* (Asteraceae): anti-proteolytic and anti-hemorrhagic properties. Chem Biol Interact 7:243–251

Lago JHG, Romoff P, Fávero OA, Souza FO, Soares MG, Baraldi BT (2008) Chemical composition of male and female *Baccharis trimera* (Less.) DC. (Asteraceae) essential oils. Biochem Syst Ecol 36:737–740

Lebouvier N, Hue T, Hnawia E, Lesaffre L, Menut C, Nour M (2013) Acaricidal activity of essential oils from five endemic conifers of New Caledonia on the cattle tick *Rhipicephalus (Boophilus) microplus*. Parasitol Res 112(4):1379–1384

Lorenzi H, Matos FJA (2008) Plantas medicinais no Brasil: nativas e exóticas. Instituto Plantarum de Estudos da Flora, Nova Odessa 512 p

Marques CA, Leitão GG, Bizzo HR, Peixoto AL, Vieira RC (2009) Anatomia e análise de óleo essencial das folhas de *Hennecartia omphalandra* (Monimiaceae). ver Bras Farmacog 19(1):95–105

Menezes A Jr (2006) Aspectos Agronômicos Básicos em Fitoterapia. In: Ferro D (ed) Fitoterapia: conceitos clínicos. Atheneu, São Paulo, pp 67–82

Mevy JPA, Bessiere JM, Dherbomez C, Millogo J, Viano J (2007) Chemical composition and some biological activities of the volatile oils of a chemotype of *Lippia chevalieri* Moldenke. Food Chem 101:682–685

Mol DJS, Silva FG, Pinto JEBP (2002) Acúmulo da biomassa e rendimento do óleo essencial de carqueja em função do sistema de manejo e alturas de poda. In: Congresso Brasileiro de Plericultura, Uberlândia, MG,. Anais. Horticultura Brasileira, Campinas, vol. 20

Montanari I Jr (2002) Aspectos da produção comercial de plantas medicinais nativas. CPQBA-UNICAMP, Campinas Available online at. http://www.cpqba.unicamp.br/plmed/artigos/producao.htm. Accessed on 25 July 2013

Montanari Jr. I, Pereira B, Mello WC (2008) Primeiro registro de cultivar de planta medicinal nativa do Brasil. Horticult. Bras. s/n

Morais LAS, Castanha RF (2011) Composição química do óleo essencial de duas amostras de carqueja (*Baccharis* sp.) coletadas em Paty do Alferes – Rio de Janeiro. Rev Bras Plant Med 13:628–632

Naiverth JA, Faria CMDR (2007) Cultivo de plantas medicinais como alternativa de renda para agricultores familiares do município de Candói-PR. Rev Eletrô Lato Sensu 4:27–32

Oliveira F, Akisue G (1997) Fundamentos de farmacobotânica, 2nd edn. Atheneu, São Paulo, p 178

Oliveira ACP, Endringer DC, Amorim LAS, Brandão MGL, Coelho MM (2005) Effect of the extracts and fractions of *Baccharis trimera* and *Syzygium cumini* on glycaemia of diabetic and non-diabetic mice. J Ethnopharmacol 1:165–169

Owen SM, Peñuelas J (2005) Opportunistic emissions of volatile isoprenoids. Trends Plant Sci 10:420–426

Palácio CPAM, Biasi LA, Nakashima T, Serrat BM (2007) Biomassa e óleo essencial de carqueja [*Baccharis trimera* (Less) DC.] sob influência de fontes e doses de nitrogênio. Rev Bras Plant Med 9(3):58–63

Pavan-Fruehauf S (2000) Plantas medicinais de mata atlântica: manejo sustentado e amostragem. Annablume/Fapesp, São Paulo, p 216

Pocá AMPC (2005) Biomassa, óleo essencial, perfil fitoquímico e nutrientes da carqueja sob influência de fatores e doses de nitrogênio. p 59. Curitiba: Dissertação (Mestrado em Ciências) – Universidade Federal do Paraná, Paraná, 2005

Reis MS, Mariot A (1998) Manejo de populações naturais de plantas medicinais em Santa Catarina. In*:* Jornada catarinense de plantas medicinais, UNISUL – Universidade do Sul de Santa Catarina, p 83–90, 1998

Reis MS, Mariot A, Steenbock W (2007) Diversidade e domesticação de plantas medicinais. In: CMO S, Schenkel EP, Gosmann G, JCP M, De Mentz LA, Petrovick PR (eds) Farmacognosia da planta ao medicamento: da planta ao medicamento. Ed. Universidade UFRGS, Porto Alegre, pp 45–74

Santos CAM, Torres KR, Leonart R (1988) Plantas Medicinais: Herbarium Flora et Scientia, 2nd edn. Ícone, São Paulo 160 p

Saúde (2013) Promessas da Mata. Editora Abril, p.36. Available online at: <http://pharmagistral.blogspot.com.br/2013/06/fitoterapia-quando-vale-pena.html>

Silva AA Jr (1997) Plantas medicinais e aromáticas. Epagri., (CD-ROM), Itajaí

Silva FG, Januário AH, Pinto JEBP, Nascimento VE, Barizan WS, Sales JF, França SC (2006) Teor de flavonóides em populações silvestres e cultivadas de carqueja [*Baccharis trimera* (Less.) DC.] coletadas nas estações seca e úmida. Rev Bras Plant Med 8(2):19–25

Silva CM, Bolzan AA, Mallmann CA, Pozzatti P, Alves SH, Heinzmann BM (2010) Sesquiterpenóides de *Senecio bonariensis* Hook. e Arn., Asteraceae. Rev Bras Farmacog 20(1):87–92

Simões CMO, Mentz LA, Schenkel EP (1998) Plantas da medicina popular no Rio Grande do Sul, 5th edn. Universidade/UFRGS, Porto Alegre, p 173

Simões-Pires CA, Queiroz EF, Henriques AT (2005) Isolation and on-line identification of antioxidant compounds from three *Baccharis* species by HPLC-UV-MS/MS with post-column derivatisation. Phytochem Anal 2005(16):307–314

Siqueira NCS, Silva GAAB, Alice CB, Nitschke M (1985) Análise comparativa dos óleos essenciais de *Baccharis articulata* (Lam) Pers. e *Baccharis trimera* (Less.) DC. (Compositae), espécies espontâneas no Rio Grande do Sul. Rev Bras Farm 3:36–39

Soicke H, Leng-Peschlow E (1987) Characterisation of flavonoids from *Baccharis trimera* and their antihepatotoxic properties. Plant Med 53:37–39

Sousa LA, Sacramento LVS, Ming LC (2006) Propagação por estaquia de três acessos de Baccharis trimera em fenologia reprodutiva. Rev Bras Plant Med 8(4):189–192

Souza MP, Matos MEO, Matos FJA (1991) Constituintes Químicos Ativos de Plantas Medicinais Brasileiras. Edições UFC, Fortaleza

Souza MRMI, Pereira RGF, Fonseca MCMI (2012) Comercialização de plantas medicinais no contexto da cadeia produtiva em Minas Gerais. Rev Bras Plant Med 14(sp. number):242–245

Torres LM, Gamberini MT, Roque NF (2000) Diterpene from *Baccharis trimera* with a relaxant effect on rat vascular smooth muscle. Phytochemistry 55:617–619

Trani PE, Passos FA, Melo AMT (2007) Instruções técnicas para 41 espécies de plantas medicinais. In: Hortaliças e plantas medicinais: manual prático. Campinas: Instituto Agronômico:45–7

Tropicos (2013) Available online at: http://www.tropicos.org/MapsCountry.aspx?maptype=4&lookupid=2728763. Accessed on 25 Jul 2013

Veiga VF Jr (2008) Estudo do consumo de plantas medicinais na Região Centro-Norte do Estado do Rio de Janeiro: aceitação pelos profissionais de saúde e modo de uso pela população. Rev Bras Farmacog 18(2):308–313

Bauhinia forficata Link

Valdir Cechinel Filho

Bauhinia forficata LinkPhoto: Divina Aparicio
Available in: http://www.biodiversidadvirtual.org/herbarium/Bauhinia-forficata-Link-img50266.html

V. Cechinel Filho (✉)
Programa de Pós-Graduação em Ciências Farmacêuticas e Núcleo de Investigações, Químico-Farmacêuticas (NIQFAR), Universidade do Vale do Itajaí (UNIVALI), Itajaí, SC, Brazil
e-mail: cechinel@univali.br

U. P. Albuquerque et al. (eds.), *Medicinal and Aromatic Plants of South America*, Medicinal and Aromatic Plants of the World 5,
https://doi.org/10.1007/978-94-024-1552-0_11

Abstract *Bauhinia forficata* Link (Fabaceae, Leguminosae) is a medicinal plant known in Brazil as "pata de vaca", "mororó", "pé de boi", "casco de vaca" or "unha de boi". It is used in traditional medicine to treat several pathological conditions, especially diabetes. Some studies have confirmed the antidiabetic potential of this plant in preclinical and clinical experiments. Kaempferitrin, the major flavonoid present in the *B. forficata* leaves, appears to be the main active principle, with different kind of medicinal properties, including antidiabetic potential.

Keywords *Bauhinia forficata* · Antidiabetic potential · Flavonoids · Kaempferitrin

1 Taxonomic Characteristics

Bauhinia forficata Link (Fabaceae, Leguminosae), is commonly known as "cow's paw" or "cow's hoof" or mororó. In Brazil it is called "pata de vaca", "mororó", "pé de boi", "casco de vaca" or "unha de boi" (Cechinel Filho 2009, 2015).

2 Crude Drug Used

Bauhinia fortificata leaves and stem-bark are used as a tea or infusion by the population, as a remedy to treat various ailments, especially diabetes (Cechinel Filho 2009). Besides being a popular medicinal plant, it is also considered as ornamental (Gupta 2008).

3 Major Chemical Constituents and Bioactive Compounds

The flavonoid kaempferol-3,7-*O*-(α)-dirhamnoside (kaempferitrin) is the main component present in the leaves and used as a chemotaxonomical marker, but other flavonoids were described, including: kaempferol-3-*O*-(α)-glucoside-(1‴,6″)-rhamnoside-7-*O*-(α)-rhamnoside, kaempferol-7-*O*-(α)-rhamnoside and kaempferol-3-O-(2-rhamnosyl)-rutinoside. Other flavonoids, together with phytosterols glucosides, have also been isolated from this plant (Cechinel-Zanchett et al. 2018; Da Silva and Cechinel Filho 2002; Gupta 2008; Cechinel Filho 2009; Ferreres et al. 2012).

4 Morphological Description

It is a medium-sized tree (from 5 to 9 m) having zigomorphic pentamerous flowers, white color and with wide- linear two petals times longer than the sepals, obtuse at the apex and base contracted in the form of nail. The fruit is a dry vegetable, dehiscent, segmented with the number of seeds varying from 20 to 6 per fruit. The seeds are oval, with husk smooth, greenish brown color (Coutinho et al. 2008; Marques et al. 2013).

5 Geographical Distribution

This species is considered to be native in South America, with an area especially in Argentine, Paraguay, Uruguay, Bolivia and Brazil (Gupta 2008).

6 Ecological Requirements

B. forficata grows mainly in the Ombrofilus Dense Forest (Atlantic Forest) from 50 to 1000 m of altitude and 950 to 2200 mm of rainfall. It is a common plant in riparian vegetation and shows preference for alluvial, deep, permeable, fertile soils, supporting floods (Carvalho 2003).

7 Traditional Uses and Common Knowledge

The leaves and stem-bark are used as a tea or infusion by the population as a remedy to the treatment of diabetes. It is also employed against kidney problems, obesity, diarrhea, skin problems, as a diuretic, etc. (Da Silva and Cechinel Filho 2002; Cechinel Filho 2009; Marques et al. 2013; Pozzobon et al. 2014).

8 Modern Medicine Based on Its Traditional Medicine Uses

Although several experimental studies have confirmed some interesting biological effects for this plant, such as antioxidant, antimicrobial, antitumor and anti-inflammatory properties, the antidiabetic effects are studied most frequently. These experiments have demonstrated efficacy in both animals and humans (Gupta 2008; Cechinel Filho 2009; Marques et al. 2013). With respect to the antidiabetic properties, several experimental studies have confirmed the promising potential of this plant. For example, Lino and co-workers (2004) showed that ethanolic extract of *B. forficata* leaves administered daily for 7 days in diabetic rats at doses of 200 and 400 mg/kg body wt. decreased blood glucose by 42% and 55%, respectively. Dried extracts of *B. forficata* leaves lowered the increased levels of plasma glucose in the STZ-induced diabetic rats. The blood glucose level decreased by 46.42% and 48.17% in the animals treated with oven-dried extract and spray-dried extract respectively, after 7 days of treatment (Cunha et al. 2010). Recently, Curcio and co-workers (2012) concluded that the treatment with the aqueous extract of this plant reduced glucose levels and contributed to weight recovery in treated animals. Some studies have also confirmed the antidiabetic potential of this plant in clinical experiments (Cechinel Filho 2009; Nogueira and Sabino 2012). It was evidenced that kaempferitrin, the major flavonoid present in the *B. forficata* leaves, caused, by oral route, significant hypoglycemic effect in normal and especially in alloxan-induced diabetic rats at all doses tested (50, 100, and 200 mg/kg) (De Souza et al. 2004). More recently, it was demonstrated that kaempferitrin is capable of stimulating the glycolytic enzyme 6-phosphofructo-1-kinase (PFK) in a model of diabetes and that kaempferitrin stimulates glucose-metabolizing enzymes (Da Silva et al. 2014). Recently, Miceli et al. (2015) demonstrated that the flavonoid-rich fraction from the leaves of *B. forficata* showed potent radical-scavenging activity but it did not exert any effect against *Artemia salina* and normal human lymphocytes, indicating that this fraction is not the responsible for the cytotoxic potential exhibited by the extract. Curiously, this plant also have presented several endophytic fungi which produces bioactive compounds with antibacterial properties (Bezerra et al. 2015).

9 Conclusions

B. forficata is a well-known medicinal and ornamental plant in South America. It is known to exhibit various biological effects and has particularly antidiabetic potential that has been confirmed in several experimental models, in both animals and humans. The main components are flavonoids, particularly kaempferitrin. It exhibits antidiabetic properties and occurs only in this species the genus *Bauhinia*. This makes kaempferitrin suitable to serve as a chemical marker for preparations containing this plant.

References

Bezerra JD, Nascimento CC, Barbosa RN, da Silva CC, Svedese VM, Silva-Nogueira EB, Gomes BS, Paiva LM, Souza-Motta CM (2015) Endophytic fungi from medicinal plant *Bauhinia forficata*: diversity and biotechnological potential. Braz J Microbiol 46:49–57

Carvalho PER (2003) *Espécies arbóreas brasileiras*, vol 1. Embrapa Informação Tecnológica, Brasília

Cechinel Filho V (2009) Chemical composition and biological potential of plants from the genus Bauhinia. Phytother Res 23:1347–1354

Cechinel Filho V (2015) Medicamentos de origem vegetal: atualidades, desafios e perspectivas. Ed. UNIVALI, Itajaí 192 p

Cechinel-Zanchett CC, Andrade SF, Cechinel Filho V (2018) Ethnopharmacological, phytochemical, pharmacological aspects of Bauhinia forficata: a mini-review covering the last five years. Nat Prod Comm 13(7):911–916

Coutinho RMA, Bezerra KC, Barbosa VBR, da Silva JVC, Santana JAS, da FCE F (2008) Análise biométrica e morfológica de sementes de uma espécie forrageira: *Bauhinia forficata* Linn (mororó). Encontro científico: 26 a 30 de maio de. PB – UFPB/ABZ, João Pessoa

Curcio SA, Stefan LF, Randi BA, Dias MS, da Silva RE, Caldeira EJ (2012) Hypoglycemic effects of an aqueous extract of Bauhinia forficata on the salivary glands of diabetic mice. Pak J Pharm Sci 25(3):493–499

Da Cunha AM, Menon S, Menon R, Couto AG, Burger C, Biavatti MW (2010) Hypoglycemic activity of dried extracts of Bauhinia forficata Link. Phytomedicine 17(1):37–41

Da Silva KL, Cechinel Filho V (2002) Plantas do gênero Bauhinia: composição química e potencial farmacológico. Quim Nova 25:449–454

Da Silva D, Casanova LM, Marcondes MC, Espindola Netto JM, Paixão LP, De Melo GO, Zancan P, Sola Penna M, Costa SS (2014) Antidiabetic activity of *Sedum dendroideum*: metabolic enzymes as putative targets for the bioactive flavonoid kaempferitrin. IUBMB Life 66(5):361–370

De Sousa E, Zanatta L, Seifriz I, Creczynski-Pasa TB, Pizzolatti MG, Szpoganicz B, Silva FR (2004) Hypoglycemic effect and antioxidant potential of kaempferol-3,7-O-(alpha)-dirhamnoside from Bauhinia forficata leaves. J Nat Prod 67(5):829–832

Ferreres F, Gil-Izquierdo A, Vinholes J, Silva ST, Valentão P, Andrade PB (2012) Bauhinia forficata link authenticity using flavonoids profile: relation with their biological properties. Food Chem 134(2):894–904

Gupta MP (ed) (2008) Plantas medicinales iberoamericanas. Convenio Andrés Bello y CYTED, Bogotá, pp 415–425

Lino CS, Diogenes JPL, Pereira BA, Faria RAPG, Andrade Neto M, Alves RS, de Queiroz MGR, Sousa FCF, Viana GSB (2004) Antidiabetic activity of *Bauhinia forficata* extracts in alloxan-diabetic rats. Biol Pharm Bull 27:125–127

Marques GS, Rolim LA, Alves LDS, Silva CCAR, Soares LAL, Rolim-Neto PJ (2013) Estado da arte de *Bauhinia forficata* Link (Fabaceae) como alternativa terapêutica para o tratamento do Diabetes mellitus. Rev Ciênc Farm Básica Apl 34(3):313–320

Miceli N, Buongiorno LP, Celi MG, Cacciola F, Dugo P, Donato P, Mondello L, Bonacorsi I, Taviano MF (2015) Role of the flavonoid-rich fraction in the antioxidant and cytotoxic activities of *Bauhinia forficata* Link (Fabaceae) leaves extract. Nat Prod Res 30:1229–1239

Nogueira ACO, Sabino CVS (2012) Revisão do gênero *Bauhinia* abordando aspectos científicos das espécies *Bauhinia forficata* Link e *Bauhinia variegata* de interesse para a indústria farmacêutica. Fitosociologia 7(2):77–84

Pozzobon A, Hoerlle J, Carreno J, Strohschoen AG, Rempel C (2014) Verificação do efeito hipoglicemiante da planta medicinal Bauhinia forficata em indivíduos com diabetes mellitus tipo 2. ConScientiae Saúde 13(1):69–75

Byrsonima intermedia A. Juss.

Raquel de Cássia dos Santos, Larissa Lucena Périco, Vinícius Peixoto Rodrigues, Miriam Sannomiya, Lúcia Regina Machado da Rocha, and Clélia Akiko Hiruma-Lima

Byrsonima intermedia A. Juss.Photo: O.M. Montiel
Available in: http://www.tropicos.org/Image/100159675

R. C. dos Santos
Laboratory of Bioactive Compounds, São Francisco University (USF), Bragança Paulista, SP, Brazil

L. L. Périco · V. P. Rodrigues · L. R. M. da Rocha · C. A. Hiruma-Lima (✉)
Department of Physiology, São Paulo State University (UNESP), Institute of Biosciences, Botucatu, SP, Brazil
e-mail: lucia.rocha@unesp.br; clelia.hiruma@unesp.br

M. Sannomiya
School of Arts, Science and Humanities, São Paulo University (USP), São Paulo, SP, Brazil
e-mail: miriamsan@usp.br

U. P. Albuquerque et al. (eds.), *Medicinal and Aromatic Plants of South America*, Medicinal and Aromatic Plants of the World 5,
https://doi.org/10.1007/978-94-024-1552-0_12

Abstract Plants of the genus *Byrsonima*, which is composed of approximately of 150 species, are widely distributed throughout tropical America. In Brazil, these species are known as "murici", and the *Byrsonima* species has a large number of medicinal uses. Byrsonima intermedia A. Juss. is known in Brazilian folk medicine for its popular use of treating diarrhea, dysentery, stomach ache, ulcer and inflammation. Pharmacological pre-clinical studies from this species have proved the anti-inflammatory, antinociceptive, antioxidant, antimicrobial, antidiarrheal and anti-ulcerogenic properties of this plant. Chemical studies of this species have shown the correlation of these activities with the presence of terpenoids, flavonoids and tannins. These evaluations showed the medicinal potential of this plant; however, the presence of the *in vitro* mutagenicity effect requires the careful assessment of the medicine used.

Keywords Murici-pequeno · Murici-do-campo · *Byrsonima intermedia* A. Juss. · Malpighiaceae

1 Taxonomic Characteristics

The *Byrsonima* is a native genus of tropical and subtropical vegetation and is the largest genus in the Malpighiaceae family, with approximately 150 species of trees, shrubs and subshrubs. This genus is widely distributed in Central and South America, Mexico and Florida (Vilas Boas et al. 2013). Approximately 50% of these species are concentrated in Brazil and are an important constituent of the cerrado vegetation (Joly 1998; Davis and Anderson 2010). Species belonging to this genus are known for their use in folk medicine as well as their commercial value in the human diet *in natura* or in the form of juices, jellies and ice creams. One representative species of this genus is *Byrsonima intermedia* A. Juss., a medicinal plant popularly known as "muricido-campo", "murici-anão", "murici-pequeno", "canjica", baga-de-tucano " or "sara-tudo". "Murici" seems to be the most common vernacular name in Brazil for this species but the term is also used for other species, e.g., *Byrsonima cydoniifolia* A. Juss., *B. verbascifolia* Rich ex Juss., *B. basiloba* A. Juss. and *B. crassa* Nied.

Synonyms *B. intermedia* f. latifolia Nied.; *B. intermedia* f. macrobothya Nied.; *B. intermedia* f. parvifolia Nied.; *B. ligustifolia* A. Juss; *Byrsonima bumeliifolia* var. glabrifolia A. Juss. *B. intermedia* f. angustifolia Nied.

2 Crude Drug Used

There are no data about the registration (approval of usage) or commercialization of this species on behalf of the Brazilian drug agency (ANVISA). Traditionally, the leaves and bark of trees are used.

3 Major Chemical Constituents and Bioactive Compounds

The first chemical study in the literature on the species *B. intermedia* resulted in the isolation of gallic acid, pyrogallol, pyrocathecin and β-amyrin from the roots (Silva 1970a, b). A phytochemical screening of the aqueous extract from the stem bark of this species indicated the presence of flavonoids, triterpenes, steroids, tannins and saponins (Orlandi et al. 2011). Phytochemical analysis of methanolic extract from leaves of *B. intermedia* yielded (+)-catechin, (−)-epicatechin, methyl gallate, gallic acid, quercetin, quercetin-3*O*-β-D-galactopyranoside, quercetin-3-*O*-α-L-arabinopyranoside, quercetin-3-*O*-(2″-*O*-galloyl)-β-galactopyranoside, quercetin-3-*O*-(2″-*O*-galloyl)-α-arabinopyranoside, amentoflavone, 3,4di-O-galloylquinic acid, 1,3,5-tri-O-galloylquinic acid and 1,3,4,5-tetra-O-galloylquinic acid (Sannomiya et al. 2007; Santos et al. 2012). According to Rinaldo et al. (2010) the methanolic extract from leaves from *B. intermedia* showed higher amounts of catechin and epicatechin than the infusible form per gram of leaves. Pereira et al. (2015) determined the values of the total phenolic content of ethyl acetate and methanolic extract of leaves and twigs from *B. intermedia.* Their results showed a better correlation between phenolic content and *in vitro* antioxidant activity of methanolic extract from leaves.

4 Morphological Description

B. intermedia is a shrub with upright branches that grows upward, reaching 1–2.5 m height, and forming a clump of up to 3 m in diameter. The opposite leaves are lanceolate with a leathery consistency (similar to leather) and petiole or a very short stem. The shrub flowers with yellow curls that take an orange hue as they age, and drupe fruits are up to 1.2 cm in diameter with a 5–7 mm small seed (Ferri 1969). This species has a flowering season from October to December (Rodrigues and Carvalho 2001). Souto and Oliveira (2005) described the morphology, anatomy, and development of the fruit and seeds. *B. intermedia* has an ovary that is ovate, superior, tricarpellate and trilocular, with one ovule per locule; the outer epidermis is uniseriate and presents a thick cuticle in the apex and a thin cuticle in the base of the ovary, The mesophyll is multiseriate, parenchymatic and vascularized; the inner epidermis is uniseriate with cells that are obliquely elongated. The ovules are subcampilotropous and bitegmic, with the nucleus projecting out of the micropyle; hypostasis and epistasis are observed. During the development of the pericarp, cellular divisions are restricted to the initial phase and occur prior to seminal differentiation. The mature fruit is fleshy, with a fibrous pyrene forming three locules. The exocarp is uniseriate, and the outer mesocarp is parenchymatous. In the apical region of the fruit, sclereids occur that are surrounded by radially arranged parenchyma cells. In the inner mesocarp, some layers of sclereids occur that are elongated in diverse ways. The endocarp is multiseriate with longitudinally elongated sclereids. The meso-endocarpic origin of the lignified regions of the pericarp is not in agreement with the classic definition of the drupoid fruit that only detaches the wood endocarp. The seed presents reduced integuments and endosperm (Souto and Oliveira 2005). The phenology and reproductive biology of *B. intermedia* were studied by Vilas Boas et al. (2013) who described that *B. intermedia* flowered for 9 months (August–April) with a higher intensity at the beginning of the rainy season. The fruit production of *B. intermedia* lasted 8–9 months, principally during the wet season. This species makes oil and pollen available for flower visitors and pollinators through almost the entire year (Vilas Boas et al. 2013).

5 Geographical Distribution

Plant species belonging to the genus *Byrsonima* Rich. Ex. Kunth. are characterized by high phenotypic plasticity, with widespread occurrence in different floristic compositions in South America (Mamede 2011). In Brazil, the occurrence of approximately 300 species in 32 genera has been recorded. These produce edible fruits (Souza and Lorenzi 2005) and oil used by bees of the tribe Centridini (Buchmann 1987). The genus *Byrsonima* is not a unique closed vegetation types; some of them occur in a closed environment, such as São Paulo, Mato Grosso do Sul, Minas Gerais, Tocantins and Goiás states and the north and northeast coast of Brazil (Anderson 1977), including several plant formations, such as fields, closed and salt marshes, rainforests and mesophytic forests (Barroso et al. 1984; Araújo 1994). *B. intermedia* is native to the Brazilian cerrado, the second largest biome in South America (Prevedello and Carvalho 2006).

6 Ecological Requirements

According to Vilas Boas (2009), *B. intermedia* is a keystone species that play a critical role in the maintenance of the cerrado Bioma structure. This is considered to be native species from the cerrado that requires sandy soil. This plant presents with the characteristic of difficult sexual propagation because of the low germination rate and slow seedling emergence in the field. However, the plant is extremely adaptable to climate conditions, and appears on the roadside and in the middle of stones and is the first to sprout when its habitat is burned. The plants can be grown at altitudes from 200 to 1000 m in a sandy or even muddy consistency rather than in soil with well-drained rainwater. It is resistant to frost and drought (Nogueira et al. 2004; Lorenzi 1992).

7 Traditional Use and Common Knowledge

B. intermedia, popularly called 'murici-pequeno', is widely used in folk medicine to treat diarrhea and dysentery and has also been used as an astringent (Lorenzi and Matos 2002).

According to Orlandi et al. (2011), tea made from the stem bark of this species has been popularly used because of its antimicrobial, anti-hemorrhagic, antifungal and anti-inflammatory properties.

The stem bark is prepared with a ratio of 1 teacup of chopped bark to 1L of water and the dosage prescribed is 3–4 cups of tea per day (Rodrigues and Carvalho 2001). The leaves of this species are also used in a tea form with water to treat intestinal infection and diarrhea and to protect from intestinal mucosa. The tea form of the root (1 Tbsp. of root with a half-liter of boiling water) is used externally in compresses for the treatment of wounds and diseases of the mouth and throat. The same tea is also used externally to treat vaginal discharge (Lorenzi and Matos 2002). Moreira et al. (2011), also describe the use of leaves for the treatment of fever, tuberculosis, fungal and bacterial infections and dermal and gastrointestinal diseases.

8 Modern Medicine Based on Its Traditional Medicine

There are no clinical data that support the use of this medicinal plant. However, there are pharmacological studies reporting its therapeutic and toxic effects in a pre-clinical trial. The anti-inflammatory activity of the aqueous extract and fraction from *B. intermedia* leaves and the acute and chronic anti-inflammatory effects were evaluated in rats. This study proved that a combination of several compounds (catechin and flavonoids with their derivatives content) that are present in the aqueous extract and in the aqueous fraction showed greater anti-inflammatory activity compared with isolated catechin that is present in the crude aqueous extract (10%) or the aqueous fraction (18%) (Moreira et al. 2011). Anti-inflammatory and antinociceptive effects in rodents were also observed in the aqueous extract obtained from the stem bark of *B. intermedia* (Orlandi et al. 2011).

The use of *B. intermedia* in traditional medicine, as an anti-ulcerogenic and antidiarrheal substance was studied in a pre-clinical assay. The gastroprotective and healing effect (gastric and duodenal) evaluated with the methanolic extract from leaves of this species by an oral route confirmed this popular use. The extract also showed that *B. intermedia* was able to prevent and reverse diarrhea by decreasing liquid feces and intestinal fluid without changing intestinal motility. This antidiarrheal effect of *B. intermedia* was also accompanied by antimicrobial effects *in vitro* against *Helicobacter pylori, Staphylococcus aureus* and *Escherichia coli* (Santos et al. 2012). The methanolic extract of *B. intermedia* leaves also showed an antimicrobial effect against *Bacillus subtilis* and *Enterococcus faecalis* (Michelin et al. 2008). The effects of a methanolic extract from *B. intermedia* was evaluated on the oxidative burst of *Helicobacter pylori*-stimulated neutrophils, and this extract presents an antioxidant capacity by inhibiting the respiratory burst in a concentration-dependent manner (Bonacorsi et al. 2013). In addition to the antibacterial effect of leaves from *B. intermedia*, a study has also been performed regarding the antiviral effect of the crude aqueous extract against bovine herpesviruses type 1 (BoHV-1) and avian reovirus (Simoni et al. 2007). The ethanolic extract from the aerial parts of *B. intermedia* was assayed for its potential *in vitro* trypanocidal activity against the Y strain of *Trypanosoma cruzi*. However, this species did not present trypanocidal activity in this screening (Cunha et al. 2009). Silva et al. (2014), described the larvicidal activity of the hexanic and remaining fraction obtained from the leaves and bark of *B. intermedia* against *Aedes aegypti*. In addition to the pharmacological studies proving the folk medicine treatment of this medicinal species for diarrhea, inflammation and ulcers, the use of this species requires caution because there are signs of mutagenic activity of methanolic extracts by *in vitro* Ames assay; however, this mutagenic activity was not observed *in vivo* with a micronucleus test (Sannomiya et al. 2007).

9 Conclusions

The pharmacological and toxicological studies of *B. intermedia* have demonstrated its importance in the treatment of diarrhea, ulcer and inflammation and as an antimicrobial species. However, the presence of its *in vitro* mutagenicity effect underlines the importance of the careful assessment of its usage as medicine and calls for the need of further research.

References

Anderson WR (1977) Byrsonimoideae, a new subfamily of the Malpighiaceae. Leandra 7:5–18

Araújo ARB (1994) Morphology of fruits, seeds and seedlings, types and aspects of germination of some species of Malpighiaceae. Universidade Estadual de Campinas, Campinas Portuguese

Barroso GM, Guimarães EF, Ichaso CLF, Costa CG, Peixoto AL (1984) Systematics of angiosperms from Brazil. Imprensa Universitária. Universidade Federal de Viçosa, Viçosa, p 2 Portuguese

Bonacorsi C, Fonseca LM, Raddi MSG, Kitagawa RR, Vilegas W (2013) Comparison of Brazilian plants used to treat gastritis on the oxidative burst of *Helicobacter pylori*-stimulated neutrophil. Evid Based Complement Alternat Med 2013:1–8

Buchmann SL (1987) The ecology of oil flowers and their bees. Annu Rev Ecol Syst 18:343–369

Cunha WR, Santos FM, Peixoto JA, Veneziani RCS, Crotti AEM, Silva MLA, Silva Filho AA, Albuquerque S, Turatti ICC, Bastos JK (2009) Screening of plant extract from the Brazilian Cerrado for their *in vitro* trypanocidal activity. Pharm Biol 47(8):744–749

Davis CC, Anderson WR (2010) A complete generic phylogeny of malpighiaceae inferred from nucleotide sequence data and morphology. Am J Bot 97(12):2031–2048

Ferri MG (1969) Plants of Brazil – species from cerrado. (São Paulo). Portuguese
Joly AB (1998) Botany: introduction to plant taxonomy. Companhia Ed. Nacional, São Paulo Portuguese
Lorenzi H (1992) Brazilian trees: identification manual and cultivation of native tree plants in Brazil. Instituto Plantarum, Nova Odessa Portuguese
Lorenzi H, Matos FJA (2002) Medicinal plants in Brazil: native and exotic. Instituto Plantarum, Nova Odessa, p 324 Portuguese
Mamede MCH (2011) *Byrsonima*. In: REFLORA: list of species of flora from Brazil. Rio de Janeiro. http://floradobrasil.jbrj.gov.br/jabot/floradobrasil/FB8827. Portuguese
Michelin DC, Sannomiya M, Figueiredo ME, Rinaldo D, Santos LC, Souza-Brito ARM, Vilegas W, Salgado HRN (2008) Antimicrobial activity of *Byrsonima* species (Malpighiaceae). Braz J Pharmacogn 18:690–695
Moreira LQ, Vilela FC, Orlandi L, Dias DF, Santos ALA, Silva MA, Paiva R, Alves-da-Silva G, Gisuti-Paiva A (2011) Anti-inflammatory effect of extract and fractions from the leaves of *Byrsonima intermedia* A. Juss. in rats. J Ethnopharmacol 138:610–615
Nogueira RC, Paiva R, Castro AHD, Vieira CV, Abbade LC, Alvarenga AA (2004) *In vitro* germination of murici-pequeno (*Byrsonima intermedia* A. Juss.). Ciência Agrotecnologia 28:1053–1059 Portuguese
Orlandi L, Vilela FC, Santa-Cecília FV, Dias DF, Alves-da-Silva G, Giusti-Paiva A (2011) Anti-inflammatory and antinociceptive effects of the stem bark of *Byrsonima intermedia* A. Juss. J Ethnopharmacol 137:1469–2476
Pereira VV, Borel CR, Silva RR (2015) Phytochemical screening, total phenolic content and antioxidant activity of *Byrsonima* species. Nat Prod Res 21:1–5
Prevedello JA, Carvalho CJ (2006) Conservation of the Brazilian Cerrado: the pan-biogeographic method as a tool for the selection of priority areas. Natureza Conservação 4(1):39–57 Portuguese
Rinaldo D, Batista JM Jr, Rodrigues J, Benfatti AC, Rodrigues CM, Santos LC, Furlan M, Vilegas W (2010) Determination of catechin diastereomers from the leaves of *Byrsonima* species using chiral HPLC-PAD-CD. Chirality 22:726–733
Rodrigues VEG, Carvalho DA (2001) Medicinal plants in the Cerrado area. UFLA, Lavras 69. Portuguese
Sannomiya M, Cardoso CRP, Figueiredo ME, Rodrigues CM, Santos LC, Santos FV, Serpeloni JM, Cólus IMS, Vilegas W, Varanda EA (2007) Mutagenic evaluation and chemical investigation of *Byrsonima intermedia* A. Juss. leaf extracts. J Ethnopharmacol 112(2):319–326
Santos RC, Kushima H, Rodrigues CM, Sannomiya M, Rocha LRM, Bauab TM, Tamashiro J, Vilegas W, Hiruma-Lima CA (2012) *Byrsonima intermedia* A. Juss.: gastric and duodenal antiulcer, antimicrobial and antidiarrheal effects in experimental rodent models. J Ethnopharmacol 140:203–212
Silva JB (1970a) Gallic acid, pyrogallol and pyrocatechol in the roots of *Byrsonima intermedia* Ad. Jussieu, form *Latifolia grisebach*. Rev Farmácia Bioquímica Univ São Paulo 8(2):187–192
Silva JB (1970b) Beta-amyrin in the root of *Byrsonima intermedia* Ad. Jussieu form Latifolia griseback. Rev Farmácia Bioquímica Univ São Paulo 8(1):53–67
Silva CB, Rech KT, Ocampos FMM, Dalarmi L, Jasinski VCG, Dias JFG, Zanin SMW, Kerber VA, Kulik JD, Fujiwara GM, Oliveira M, Miguel OG, Miguel MD (2014) *Byrsonima intermedia* preparations inhibits trypsin and chymotrypsin activities from *Aedes aegypti* larval gut. J Med Plant Res 8(9):416–422
Simoni IC, Manha APS, Sciessere L, Hoe VMH, Takinami VH, Fernandes MJB (2007) Evaluation of the antiviral activity of Brazilian Cerrado plants against animal viruses. Vírus Rev Res 12:1–2
Souto LS, Oliveira DMT (2005) Morphology, anatomy and ontogeny of the fruit and seed of *Byrsonima intermedia* A. Juss. (Malpighiaceae). Braz J Bot 28(4):697–712. Portuguese
Souza VC, Lorenzi H (2005) Systematic botany: illustrated guide to identifying the angiosperm families of Brazilian flora, based on APG II. Instituto Platarum, Nova Odessa, p 639

Vilas Boas JC (2009) Phenology and reproductive biology *Byrsonima intermedia* A. Juss. and B. pachyphylla Griseb (Malpighiaceae): key features in cerrado remaining, Mato Grosso do Sul, Brasil. Dissertation (Masters in Vegetal Biology) Universidade Federal do Mato Grosso do Sul, (Campo Grande). 43. Portuguese

Vilas Boas JC, Fava WS, Laroca S, Sigrist MR (2013) Two sympatric *Byrsonima* species (Malpighiaceae) differ in phenological and reproductive patterns. Flora 208:360–369

Caryocar coriaceum Wittm.

Rogério de Aquino Saraiva, Izabel Cristina Santiago Lemos, Patricia Rosane Leite de Figueiredo, Luiz Jardelino de Lacerda Neto, Cícera Norma Fernandes Lima, Mariana Késsia Andrade Araruna, Renata Evaristo Rodrigues da Silva, Roseli Barbosa, Cícero Francisco Bezerra Felipe, Irwin Rose Alencar de Menezes, and Marta Regina Kerntopf

Caryocar coriaceum Wittm.
Photo source: data bank from Laboratório de Ecologia e Evolução de sistemas socioecológicos

R. de Aquino Saraiva · I. C. S. Lemos · P. R. L. de Figueiredo · L. J. de Lacerda Neto
C. N. F. Lima · M. K. A. Araruna · R. E. R. da Silva · R. Barbosa · C. F. B. Felipe
I. R. A. de Menezes · M. R. Kerntopf (✉)
Biological Chemistry, Regional University of Cariri, Crato, Ceará, Brazil

U. P. Albuquerque et al. (eds.), *Medicinal and Aromatic Plants of South America*, Medicinal and Aromatic Plants of the World 5,
https://doi.org/10.1007/978-94-024-1552-0_13

Abstract *Caryocar coriaceum* Wittm. is an arboreal species of Caryocaraceae family. The fruit pulp is known popularly as *pequi*. The folk knowledge confers to the oil '*pequi*' vast medical applicability. It can be used for treating colds and pulmonary infections, sore throats, rheumatism, external ulcers, muscle pain and skin inflammation. The fruits of *pequi* are known to have aphrodisiac and anti-abortive properties. The leaves are used to treat menstrual disorders. There are few phytochemical studies on *C. coriaceum*, however, pre-clinical tests of the fixed oil of *C. coriaceum* indicated gastro-protective properties and cicatrization in rodents, with a large reduction in ulcers induced by ethanol and aspirin, topical anti-inflammatory effect and efficacy reduction in skin inflammation with chronic treatment in rodents.

Keywords *Caryocar coriaceum* Wittm. · Traditional medicine · Folk knowledge · Ethnopharmacology · Bioprospection

1 Taxonomic Characteristics

Caryocar (souari trees) is a genus of flowering plants, in the family Caryocaraceae described formerly as a genus by Linnaeus, in 1771. Besides *C. coriaceum*, there are other species of the genus *Caryocar* in Brazil: *C. brasiliense*, *C. villosum*, *C. cuneatum* and *C. glabrum*. (Lorenzi and Matos 2008).

Trees of the genus *Caryocar* yield a strong timber. Some of the species have edible fruits, called souari-nuts or sawarri-nuts.

Caryocar coriaceum Wittm., is an endemic species to Brazil, with cultural, alimentary and ethnopharmacological values (term used here to describe medical practices). It is used in traditional medicine systems (see Elisabetzky 2003).

In 2006, *Caryocar coriaceum* was included in the IUCN Red List of Threatened species.

2 Major Chemical Constituents and Bioactive Compounds

Pequi content of the pulp is rich in nutritional compounds, such as fatty acids, carbohydrates, proteins, carotenoids, vitamin E, and retinol. The fruit pulp also has high levels of pectin and tannins, and polyunsaturated oils. In the fixed oil of C. coriaceum, were identified saturated and unsaturated fatty acids, with the major component fatty unsaturated oleic acid. Featuring even in the composition fatty polyunsaturated linoleic acid (Figueiredo 2012).

Regarding the nutritional value for the species C. *coriaceum* Wittm., the study of Oliveira et al. (2010) showed a protein content of 2.09% and 23.19% of lipids. Pequi's pulp is also rich in vitamin A and minerals, especially P, Ca, Cu and Fe (Araújo 1995).

Edible portions of fruit oil are: the pulp and the almond, for its characteristic taste and smell, as well as being a source of lipids and antioxidant vitamins (A and E), they are well used as food in regional food, replacing other sources of fat, such as grease or bacon. Phytochemical analysis of the essential oil obtained from almonds was composed almost exclusively of ethyl hexanoate (Lorenzi and Matos 2008).

Due to the affordable price, pequi is a valuable food source for the low-income population in the region (Figueiredo et al. 1989; Braga 1960; Silva and Medeiros Filho 2006).

Although the fruit is rich in nutrients and has a variety of uses, especially the species *C. coriaceum*, pequi has received inadequate attention in national and international research. There are only a few studies in the special literature that would involve biometry and the chemical and nutritional characterization of fruits of this species (Oliveira et al. 2010; Silva and Medeiros Filho 2006; Oliveira 2009).

As highlighted by Figueiredo (2012), little has been done to preserve the existing germplasm of this species and study its possible domestication with the aim of sustainable utilization.

Pequi has many uses, such as being used to produce oil with high versatility in regional food for sauces and dressings preparation, in cosmetic industry for producing soaps and creams, as well as being used for fuel production and lubricants (Oliveira et al. 2008; Pianovski et al. 2008).

3 Morphological Description

C. coriaceum, popularly known by pequizeiro, pequi, piqui and pequá, is a leafy and branchy tree with trunk coated with dark, thick and furrowed skin and opposite leaves, ternate, oval leaflets, glabrous (hairless), green-glistening, rich in tannin, providing dye substance; more or less leathery. The flowers are large, yellow with red stamens, gathered in terminal bunches (Braga 1960; Figueiredo 2012).

"Pequizeiro" tree reaches an average of 6–8 m high, and its inflorescences produce a varied number of large (5.0–7.5 cm in diameter) and colored from green to white and twilight anthesis flowers (hermaphrodite and actinomorphic) (Araújo 1995).

Studies of C. *brasiliense* and C. *villosum*, indicate that the species of this genus are heavily cross-pollinated, and small nectar bats (socicina geoffroyi and Anoura Glossophaga) are the main pollinators, and the Protandry and herkogamy (spatial separation of anthers and stigmas) work as key mechanisms against autogamous. However, despite the steep allogamy, self-pollination can occur in a small proportion (Gribel and Hay 1993; Martins and Gribel 2007).

The flowering usually occurs between August and November, depending on the region, and the fruit ripening takes from 3 to 4 months after the pollination, with low fruit set rate. However, according to Araújo (1995), one Pequi plant can produce 500–2000 fruits/harvest.

Pequi's fruit is a drupe type with depression-globular shape, leathery and fleshy epicarp, and bright green/slightly yellowish color when ripe, with burry endocarp. Its dimensions range from 4 to 7 cm high and 6–8 cm in diameter, with average mass reaching approximately 120 g, but with variation from 100 to 220 g (Araújo 1995, Lorenzi and Matos 2008).

The pulp is oleaginous, mealy and pasty, varying in color from cream-yellow to intense-yellow and sometimes orange. Generally, the fruit contains only one seed developed (putamen or pyrene), but sometimes it can contain up to three or four seeds (Araújo 1995; Silva and Medeiros Filho 2006; Oliveira 2009).

4 Geographical Distribution

Therefore, *C, coriaceum*, specie found in the northern of Ceará, has an important socio-economic role in Chapada do Araripe, covering the States of Ceará, Pernambuco and Piauí. Can also be found in the states of Tocantins and Maranhão (Saraiva et al. 2011).

5 Collection Practice

Thus, *C. coriaceum* is explored in an extractive way, being seasonal, with flowering from September to November and the season between December and April, period of high rainfall in the region. Then in the off-season, there is the extraction of oil from the pulp and almond, which has greater commercial value (Costa et al. 2004; Lorenzi and Matos 2002).

In pequi's harvest period, the communities near Chapada do Araripe perform extractive activities, collecting fruit for marketing. The fruit is not collected directly from the tree; it is collected after fruit falls to the ground because the taste of the fruit pulp collected from the "floor" is much better (Augusto and Goes 2007; Sousa Junior 2012).

6 Traditional Use (Part(s) Used) and Common Knowledge

However, pequi has a large emphasis in traditional medicine context, highlighted as a relevant ethnopharmacological resource. For example, pequi's bark of the tree and skin of the fruits are used in antipyretic and diuretic infusions (Lorenzi and Matos 2008).

The fruit has anti-abortive and aphrodisiac properties and the leaves are used to treat colds, flu, edemas, menstrual changes and as an antifungal (Vieira and Martins 2000; Batista et al. 2010).

In this context, oil is used in the treatment of burns, colds, broncho-pulmonary infections, skin ulcers, inflammation of the skin and musculoskeletal pain (Saraiva 2009).

At the same time, reports show its use in ophthalmic disorders related to vitamin A deficiency, by its high content of carotenoids with provitamin A activity (Santos 2007; Oliveira et al. 2008).

Furthermore, pequi's fruit is almost fully used, because the skin is consumed by bovine animals, the seed (with pulp) is used in the preparation of dishes – quite appreciated in regional food – and the pulp is still used for extraction and homemade or manufacture edible oil, jellies, jams, liqueurs and animal food (Lorenzi and Matos 2008; Oliveira 2009).

The almond, due to its high nutritional value, shape, size and visual appearance, is also used for fresh consumption in the oil extraction and soap manufacturing, and in cosmetics industry as creams and soaps, being potential as another option in the national market of almonds (Lorenzi and Matos 2008; Oliveira 2009).

Therefore, the therapeutic value of pequizeiro to popular medicine has been researched and some ethnopharmacological and ethnobotanical studies show its real effectiveness, emerging an important bioprospecting research (Batista et al. 2010; Lorenzi and Matos 2008).

7 Modern Medicine Based on Its Traditional Medicine Uses

In the essence of published studies about the medicinal uses of pequi, as well as several other species of traditional use, there is the technique of bioprospecting. Therefore, bioprospecting is basically the identification and evaluation of specific biological material extracted from nature, for its applicability and utility in generating new processes and products. Thus, resources found in nature are experienced, seeking to obtain new resources to be used in everyday life (Palma and Palma 2012).

In the contemporary view of bioprospecting, there are environmental and social aspects associated with new economic paradigms. That is, it is related to biotechnology, with the "biodiversity" and with the agents directly and indirectly involved with the completion of this activity, as entrepreneurs, local communities, indigenous groups, environmental groups, research institutions, international organizations, among others (Palma and Palma 2012).

Bioprospecting also allows the identification of priorities relating to lines of research or for the strengthening of old research. In this sense, some bio-prospective studies nationally and internationally published, corroborate the therapeutic uses of pequi in the Traditional Medicine context (Lorenzi and Matos 2008).

For example, as in some studies in general, there is Passos et al. (2002) research performed with the extract of pequi's leaves, finding antifungal activity by inhibiting the growth of *Cryptococcus neoformans*, *Paracoccidioides brasiliensis* and *Candida albicans*.

In addition, molluscicidal action against *Biomphalaria glabrata* (schistosomiasis vector) was identified in Batista et al. (2010) research, leishmanicide effect by inhibiting the proliferation of the promastigote form of *Leishmania amazonensis* and antimicrobial activity by inhibiting the growth of enterobacteria, according to studies of Paula-Junior et al. (2006).

It is important to mention the research of Alves et al. (2000) showing actions against *Bacillus cereus*, *Pseudomonas aeruginosa* and *Staphylococcus aureus*.

Interference with *T. cruzi parasitemia* curve has been demonstrated in pequizeiro bark extract, thus reducing the number of parasites in the blood (Herzog-Soares et al. 2002).

In this sense, antioxidant and preventive activities for tumors were also observed (Paula-Junior et al. 2006; Khouri et al. 2007), effects against Sarcoma in animals by oleanolic acid content and protease and evidence of hemolytic activity of *C. brasiliense* lectin, as well as in vivo enterotoxic activity in mice (Perez 2004).

It is also noteworthy that pequi's oil is a rich source of vitamin C, with phenolic compounds such as flavonoids, saponins and essential oils in the mesocarp (Miranda-Vilela et al. 2008).

On the other hand, these components have antioxidant properties, mitigating the effects of mutagens and carcinogens agents. In addition, oxidative stress is one of the major risk factors for cardiovascular disease (Alonso 2000; Miranda-Vilela et al. 2008; Tseng et al. 2004).

Furthermore, preclinical studies of C. *coriaceum* fixed oil indicate the gastroprotective activity in rodents with a significant reduction of ulcers induced by ethanol and aspirin, besides to healing activity, topical anti-inflammatory effect in mice, and reduction of skin inflammation, with chronic treatment (Penha 2007; Quirino 2009; Saraiva 2009).

The main compound of fixed oil from the pulp of *C. coriaceum* (OFCC) is the unsaturated fatty oleic acid. The saturated fatty acids increase low-density lipoprotein (LDL) by inhibiting LDL receptor activity and increase the production of apolipoprotein (Aguilar et al. 2012).

In this way, the substitution of saturated fat by poly-unsaturated fat has reduced levels of total cholesterol (TC) and LDL cholesterol levels, decreasing LDL-cholesterol production rates and/or increasing LDL clearance rates (Aguilar et al. 2012).

On the other hand, there is a decreasing in high density lipoprotein (HDL), which together with the reduction in LDL-c, LDL/HDL ratio decreases. Monounsaturated fat also has the same effect on blood cholesterol but the magnitude of the reduction in HDL is lower when compared to poly-unsaturated fats (Aguilar et al. 2012).

In this context, it is important to detail these and some other published studies, such as Saraiva et al. (2011) research assessing the topical effect of *C. coriaceum* against different irritant agents in vivo, in order to verify its effect against dermatoses.

Therefore, Saraiva et al. (2011) research found that the species showed a similar profile of topical anti-inflammatory activity, indicating its potential use against inflammatory skin diseases.

In the study by Oliveira et al. (2010), it was sought to evaluate the effects of the *C. coriaceum* fixed oil (OFCC) on topical inflammation and cutaneous wound healing. In this way, the tests showed that the OFCC was able to reduce inflammation depending on the doses.

Fresh OFCC (100%) inhibited ear edema in 38.01% at the time of 15 min and in 39.20% in 1 h, after induction of the inflammation. Topical administration of OFCC ointment (12%) showed a significant reduction in the unhealed wound area, with the increase in the percentage of wound contraction (96.54%) compared to the other groups. Thus, the conclusion was that *C. coriaceum* inhibits topical inflammation and speeds up the repair of skin wounds.

In Oliveira's research (2013) the antinociceptive activity and anti-inflammatory pequi oil in zymosan-induced arthritis in rats was investigated. The author states that besides the detected anti-inflammatory action, pequi's oil can prevent the inflammatory mechanical hyperalgesia.

On the other hand, in the study of Lacerda Neto (2013), the objective was to verify the gastro-protective activity of a hydroethanolic extract of *C. coriaceum* leaves (EHFCC). Thus, EHFCC gastroprotective activity was evaluated by methods of gastric damage induced by ethanol and a reduction of the lesion area of 69.43% was observed.

Furthermore, quantification of mucus production showed that EHFCC positively influences it and the intestinal motility test reported a decrease in motility under EHFCC action, being as another contribution to its gastro-protective effect (Lacerda Neto 2013).

Thus, the author concludes by highlighting that the described results show the biological potential of EHFCC as a grant for the study of gastro-protection and especially in the formulation of new herbal medicines for the treatment of peptic ulcer (Lacerda Neto 2013).

In *C. coriaceum* influences for cardiovascular diseases, Figueiredo's study (2012) evaluated the toxic effects of *C. coriaceum* fixed oil in biochemical and histopathological parameters of rodents. From the results, the author showed that subchronic toxicity was not revealed at high doses for the evaluated parameters.

Furthermore, it was demonstrated anti-inflammatory and antioxidant activity of *C. coriaceum* fixed oil. The importance of this study was also due to be the first study to report a possible lipid-lowering and hypo-triglyceride activity of *coriaceum* species, showing species with a pharmacological potential related to the management and treatment of cardiovascular diseases, world problems of epidemiological importance (Figueiredo 2012).

Thus, from this information, it can be said that the large and reputable popular use of pequi in Traditional Medicine is supported by the available scientific literature, although further studies to clarify other therapeutic actions of Pequi are necessary for ethnobotanical surveys.

8 Conclusions

The importance of *C. coriaceum* or pequi to traditional communities is not only in the economic context, but also in the context of practices related to Traditional Medicine.

During this chapter, the therapeutic use of pequi was highlighted as being widespread and accepted among the local population of Chapada do Araripe and surrounding regions. Therefore, it is used for a variety of pathologies, suggesting that this species is inserted in a complex set of culturally relevant plants.

However, it is observed that there are a few published studies concerning the applicability of this plant as a viable or complement alternative to conventional pharmacological treatment used for different diseases, ranging from skin diseases to cardiovascular diseases.

In fact, new bioprospective studies should be conducted, addressing chemical, pharmacological characteristics and clinical applicability of *C. coriaceum* species, in order to an efficient use of the properties provided by this plant.

References

Aguilar EC, Jascolka TL, Teixeira LG, Lages PC, Ribeiro ACC, Vieira ELM et al (2012) Paradoxical effect of a pequi oil-rich diet on the development of atherosclerosis: balance between antioxidant and hyperlipidemic properties. Braz J Med Biol Res 45(7):601–609

Alonso JR (2000) Tratado de fitomedicina, bases clínicas e farmacológicas. Isis Ediciones, Buenos Aires

Araujo FD (1995) A review of *Caryocar brasiliense* (Caryocaraceae) – na economically valuable species of the central brazilian cerrados. Econ Bot 49(1):40–48

Augusto LGS, Góes L (2007) Integrated understanding for health surveillance in a forest environment: the case of the Araripe Plateau in Ceará State, Brazil. Cad Saúde Pública 23(4):549–558

Batista JS, Silva AE, Rodrigues CMF, Costa KMFM, Oliveira AF, Paiva ES, Nunes FVA, Olinda RG (2010) Avaliação da atividade cicatrizante do óleo de pequi (*Caryocar Coriaceum* Wittm) em feridas cutâneas produzidas experimentalmente em ratos. Arq Inst Biol 77(3):441–447

Braga R (1960) Plantas do Nordeste, especialmente do Ceará, 2nd edn. Imprensa Oficial, Fortaleza

Costa IR, Araujo FS, Lima-Verde LW (2004) Flora e aspectos auto-ecológicos de um encrave de cerrado na chapada do Araripe, Nordeste do Brasil. Acta Bot Bras 18(4):759–770

de Alves TMA, Silva AF, Brandão M, Grandi TSM, de EFA S, Smânia Junior A et al (2000) Biological screening of Brazilian medicinal plants. Mem Inst Oswaldo Cruz 95(3):367–373

Figueiredo PRL de (2012) Influência do óleo fixo da polpa de *Caryocar Coriaceum* Wittm. sobre o perfil lipídico em modelo animal [dissertation]. Regional University of Cariri, Crato

Lacerda Neto LJ de (2013) Avaliação da atividade antibacteriana e gastroprotetora do extrato hidroetanólico das folhas de Caryocar coriaceum Wittm [dissertation]. Regional University of Cariri, Crato

de Oliveira FFB (2013) Efeito antinociceptivo e anti-inflamatório do óleo da polpa de pequi (*Caryocar coriaceum* Wittm.) na artrite induzida por zymosan em ratos. Federal University of Ceará, Fortaleza

Sousa Júnior JR de (2012) Conhecimento e manejo tradicional de *Caryocar coriaceum* Wittm. (Pequi) na Chapada do Araripe, Nordeste do Brasil [dissertation]. Rural Federal University of Pernambuco, Recife

Elisabetsky E (2003) Etnofarmacologia. Cienc Cult 55(3):35–36

Figueiredo RW, Maia GA, Figueiredo EAT. Propriedades físico-químicas e composição dos ácidos graxos da fração lipídica da polpa e amêndoa do piqui (*Caryocar coriaceum* Wittm). Rev Ciênc Agron 1989; 20(½):135–139

Gribel R, Hay JD (1993) Pollination ecology of *Caryocar brasiliense* (Caryocaraceae) in Central Brazil cerrado vegetation. J Trop Ecol 9:199–211

Herzog-Soares JA, Alves RK, Isac E, Bezerra JCB, Gomes MH, Santos SC et al (2002) Atividade tripanocida in vivo de Stryphnodendron adstringens (barbatimão verdadeiro) e Caryocar brasiliensis (pequi). Rev Bras Farmacognosia 12(1):1–2

Khouri J, Resck IS, Poças-Fonseca M, Sousa TMM, Pereira LO, Oliveira ABB et al (2007) Anticlastogenic potential and antioxidant effects of an aqueous extract of pulp from the pequi tree (*Caryocar brasiliense* Camb). Genet Mol Biol 30(2):442–448

Lorenzi H, Matos FJA (2002) Plantas medicinais no Brasil: nativas e exóticas. 1.ed. Instituto Plantarum, Nova Odessa

Lorenzi H, Matos FJA (2008) Plantas medicinais no Brasil: nativas e exóticas, 2nd edn. Instituto Plantarum, Nova Odessa

Martins RL, Gribel R (2007) Polinização de *Caryocar villosum* (Aubl.) Pers. (Caryocaraceae) uma árvore emergente da Amazônia Central. Rev Bras Bot 30(1):37–45

Miranda-Vilela AL, Resck IS, Grisolia CK (2008) Antigenotoxic activity and antioxidant properties of organic and aqueous extracts of pequi fruit (*Caryocar brasiliense* Camb.) pulp. Genet Mol Biol 31(4):956–963

Oliveira MEB (2009) Características físicas, químicas e compostos bioativos em pequis (Caryocar coriaceum Wittm.) nativos da chapada do Araripe- CE [thesis]. Rural Federal University of Pernambuco, Recife

Oliveira MEB, Guerra NB, Barros LM, Alves RE (2008) Aspectos agronômicos e de qualidade do pequi. Embrapa Agroindústria Tropical, Fortaleza

Oliveira MEB, Guerra NB, de AHN M, Alves RE, dos Matos NMS, Sampaio FGM et al (2010) Características químicas e físico-químicas de pequis da Chapada do Araripe, Ceará. Rev Bras Frutic 32(1):114–125

Palma CM, Palma MS (2012) Bioprospecção no Brasil: análise crítica de alguns conceitos. Cienc Cult 64(3):22–26

Passos XS, Santos SC, Ferri PH, Fernandes OFL, Paula TF, Garcia ACF et al (2002) Atividade antifúngica de *Caryocar brasiliensis* (Caryocaraceae) sobre *Cryptococcus neoformans*. Rev Soc Bras Med Trop 35(6):623–627

Paula-Junior WP, Rocha FH, Donatti L, Fadel-Picheth CMT, Weffort-Santos AM (2006) Observação: a citação encontra-se na lista de referências como. Leishmanicidal, antibacterial, and antioxidant activities of Caryocar brasiliense Cambess leaves hydroethanolic extract. Rev Bras Farmacogn 16(sup):625–630

Penha ARS (2007) Estudo de atividade antiulcerogênica de plantas da chapada do Araripe [monograph]. Regional University of Cariri, Crato

Perez E (2004) Diagnose fitoquímica dos frutos de Caryocar brasiliense Camb. Caryocaraceae [dissertation]. University of Paraná, Curitiba

Pianovski AR, Vilela AFG, Silva AAS, Lima CG, Silva KK, Carvalho VFM et al (2008) Uso do óleo de pequi (*Caryocar brasiliensis*) em emulsões cosméticas: desenvolvimento e avaliação da estabilidade física. Rev Bras Ciênc Farm 44(2):249–259

Quirino GS (2009) Atividade cicatrizante e gastroprotetora de Caryocar coriaceum Wittm [dissertation]. Regional University of Cariri, Crato

Santos RI (2007) Metabolismo básico e origem dos metabólitos secundários. In: CMO S, Schenkel EP, Gosmann G, JCP M, Mentz LA, Petrovick PR (eds) Farmacognosia: da planta ao medicamento, 6th edn. Editora da UFRGS; Florianópolis: Editora da UFSC, Porto Alegre, pp 403–434

Saraiva RA (2009) Efeito do óleo fixo do mesocarpo interno de Caryocar Coriaceum Wittm. em modelos animais de inflamação induzida por agentes flogísticos [dissertation]. Regional University of Cariri, Crato

Saraiva RA, Araruna MKA, Oliveira RC, Menezes KDP, Leite GO, Kerntopf MR et al (2011) Topical anti-inflammatory effect of Caryocar coriaceum Wittm. (Caryocaraceae) fruit pulp fixed oil on mice ear edema induced by different irritant agents. J Ethnopharmacol 136(3):504–510

Silva MAP, Medeiros-Filho S (2006) Morfologia de fruto, semente e plântula de piqui (*Caryocar coriaceum* Wittm.). Rev Ciênc Agron 37(3):320–325

Tseng CF, Lin CC, Huang HY, Liu HC, Mao SJT (2004) Antioxidant role of human haptoglobin. Proteomics 4:2221–2228

Vieira RF, Martins MVM (2000) Recursos genéticos de plantas medicinais do cerrado. Uma compilação de dados. Rev Bras Plantas Med 3(1):13–36

Clinopodium gilliesii (Benth.) Kuntze

Julio Alberto Hurrell

Clinopodium gilliesii (Benth.) Kuntze
L.J Novara
Available in: https://www.sib.gov.ar/ficha/PLANTAE*clinopodium*gilliesii

J. A. Hurrell (✉)
Laboratorio de Etnobotánica y Botánica Aplicada (LEBA), Facultad de Ciencias Naturales y Museo, Universidad Nacional de La Plata, La Plata, Argentina

Consejo Nacional de Investigaciones Científicas y Técnicas, Buenos Aires República, Argentina

U. P. Albuquerque et al. (eds.), *Medicinal and Aromatic Plants of South America*, Medicinal and Aromatic Plants of the World 5,
https://doi.org/10.1007/978-94-024-1552-0_14

Abstract *Clinopodium gilliesii* (Benth.) Kuntze is an aromatic species from the Andean region, from southern Peru to northern-central Argentina. It is mainly known as *muña-muña* and its leaves and tender stems are used as a flavoring and medicinal: stimulant, against mountain sickness, aphrodisiac, digestive, antispasmodic, among others traditional uses. Its bioactive constituents are essential oils, to which the plant owes its aroma and many of its therapeutic properties. The presence of flavonoids and phenolic compounds has also been detected. The essential oil composition of aerial organs is variable according to geographical location and ecological conditions, soil-type, weather-conditions and altitude of the population. Regarding its popular uses, the majority of uses has not been validate by pre-clinical tests, therefore they require experimental founding. Some of its biological activities, e.g.: aphrodisiac (in particular, erectile dysfunction), against some gastrointestinal disorders, antibacterial, antifungal, antiplasmodial, trypanocidal, insect repellent, antioxidant, and cytotoxic activities have already been analyzed. Some data about the similar species: *C. bolivianum* (Benth.) Kuntze and *C. odorun* (Griseb.) Harley is additionally commented.

Keywords *Clinopodium gilliesii* · Lamiaceae · Muña-muña · Andean region · Food and medicinal uses

1 Taxonomic Characteristics

Clinopodium gilliesii (Benth.) Kuntze is an Andean aromatic plant utilized for centuries for medicinal purposes: stimulant, aphrodisiac, digestive, among others. It is also used in different local gastronomies as a food condiment and to flavor milk, infusions, and aperitifs, due to its aroma similar to the mint. The most widespread vernacular name is muña-muña (from Quechua *munay*, 'to love', by referring to its application as an aphrodisiac). It is also called hierba del amor, koa, muiña, mulla-mulla, muña, oreganillo, yerba del amor, yerba del pajarito (Barboza et al. 2009; Hurrell et al. 2008, 2011).

The genus *Clinopodium* belongs to the Family Lamiaceae Martinov, Tribe Mentheae Dumort., and comprises about 100 species, mostly in temperate and tropical New World, and temperate Eurasia, but a few in Africa, tropical Asia and Indomalaysia (Harley et al. 2004). This generic circumscription responds to morphological and molecular studies that defined the boundaries within the complex *Satureja* L./*Calamintha* Mill./*Acinos* Mill./*Micromeria* Benth./*Xenopoma* Willd. (Cantino and Wagstaff 1998; Harley and Granda Paucar 2000; Wood 2011). In this frame, *Clinopodium* includes most of the New World native species of *Satureja* sensu lato (Harley et al. 2004).

Other Andean species of this genus are also utilized for its aromatic and medicinal properties, e.g. *Clinopodium nubigenum* (Kunth) Kuntze [= *Thymus nubigenus* Kunth, *Satureja nubigena* (Kunth) Briq.], sunfillo, from Colombia and Ecuador, *Clinopodium pulchellum* (Kunth) Govaerts [*Gardoquia pulchella* Kunth, *Satureja pulchella* (Kunth) Briq.] and *Clinopodium bolivianum* (Benth.) Kuntze [*Micromeria boliviana* Benth., *Satureja boliviana* (Benth.) Briq.], inca-muña, koa, oregano of the Incas, from Peru, Bolivia and Northwest Argentina, *Clinopodium odorum* (Griseb.) Harley (*Xenopoma odorum* Griseb., and *Satureja odora* (Griseb.) Epling), muña, from Bolivia and Northwest-central Argentina (Pontiroli 1993; Orfila and Farina 1997; Ulloa 2006; Elechosa 2009; Álvarez Sarmiento 2012).

Synonyms *Bystropogon minutus* Briq.; *Micromeria gilliesii* Benth.; *Micromeria eugenioides* Hieron; *Oreosphacus parvifolia* Phil.; *Satureja gilliesii* (Benth.) Briq.; *S. oligantha* Briq.; *S. parvifolia* (Phil.) Epling; *Satureja eugenioides* (Griseb.) Loesener ex R.E.Fries; *Xenopoma eugenioides* Griseb.

2 Crude Drug Used

The drug consists of its leaves and tender stems, sometimes with flowers. Both fresh and dried leaves and stems are used for culinary and therapeutic purposes.

The dry leaves and stems are consumed mostly in infusions or decoctions (20 g per liter of water), two or three cups in daily intakes, also in mother tincture (25 g in 100 cc of 70° alcohol), 25–30 drops in water, three times a day (Burgstaller 1968; Alonso and Desmarchelier 2005; Hurrell et al. 2011).

In the pluricultural urban scenarios, its leaves and tender stems and tincture are commercialized in herb-shops and health food stores, and disseminated by the media, especially the Internet. The dried leaves and stems are sold in bulk or packaged (Hurrell et al. 2011).

3 Major Chemical Constituents and Bioactive Compounds

The essential oil composition from the aerial parts of *C. gilliesii* varies according to geographical areas and its ecological conditions, as soil, weather, and altitude (Viturro et al. 2000). This variable composition is responsible for different scents, defined by olfactory characteristics as mint-like, lemony, fresh, ketonic, phenolic, persistent (Elechosa 2009).

The main essential oils indicated are: carvacrol, carvacryl acetate, carvona, *o*- and *p*-cimene, 1-8-cineol, *cis*-dihydrocarvone, dihydrolippiona, geraniol, *E*-isocitral, isopulegol, limonene, linalool, lippiona, menthol, menthone, methyl nerolate, myrcene, neoisomentol, α- and β-pinene, piperitenone, piperitenone oxide, piperitone, piperitone oxide, pulegone, sabinene, α -thujene. Its flavonoids (e.g. luteolin) and phenolic compounds content have also been studied (Zygadlo et al. 1993; Muschietti et al. 1996; Hernández et al. 2000; Viturro et al. 2000, 2007; Alonso and Desmarchelier 2005; van Baren et al. 2006; Barboza et al. 2009; Dadé et al. 2009; López-Lázaro 2009; Niemeyer 2010; Cabana et al. 2013; Tepe 2015).

4 Morphological Description

C. gilliesii is an aromatic shrub up to 2 m in height, with glabrescent or shortly pubescent branches. Leaves opposite, sub-sessile, simple, oblong, 0.4–2 cm long × 0.1–0.5 cm wide, apex obtuse, margin entire, both faces dotted-glandular and finely pubescent; pubescence is more pronounced in the adaxial face midvein. Axillary verticillasters with three to six flowers or reduced to a single flower, subtended by linear bracts, 1 mm long; pedicels short. Calyx campanulate, pubescent, tube 1–2 mm long, teeth 5, deltoid, acute, subequal, 0.6–1 mm long, somewhat curved. Corolla 2-lipped, white, 2–2.5 mm long, glabrescent, tube exserted, 1.2–1.5 (−2) mm long, upper lip 2-lobed, emarginate, lower lip with three equal lobes. Stamens 4, included, didynamous, the upper ones shorter, thecae divergent. Ovary 2-carpelar, 4-lobed; style enlarged to the base. Fruit formed by four mericarps (nutlets) included in the persistent calyx. Mericarps obovoid, 1.5–1.7 mm long, brown, finely reticulate, apex obtuse or subacute.

Among the species of *Clinopodium* of Bolivia and northeast-central Argentina, *C. odorum* basically differs from *C. gilliesii* by its ovate leaves, 6–20 mm lat., with margins pubescent; meanwhile *C. bolivianum* differs from the two previous by its shorter corolla tube (6–8 mm long.), and its stamens shortly exserted (Pontiroli 1993; Orfila and Farina 1997; Harley et al. 2004; Elechosa 2009).

5 Geographical Distribution

This species is native to the Andean region of southern Peru, Bolivia, Chile and Argentina (Jujuy, Salta, Tucumán, Catamarca, La Rioja, Córdoba, San Juan, San Luis and Mendoza), from 1000 to 4500 m altitude (Pontiroli 1993; Del Vitto et al. 1997; Orfila and Farina 1997; Flores and Ruiz 2006; Hurrell et al. 2011; Wood 2011).

6 Ecological Requirements

C. gilliesii is particularly characteristic of the arid highland Andes. It is more frequent in the upper floor of montane forests, 'ceja de monte' scrub (boundaries of forests), puna vegetation and drier inter-Andean valleys (Orfila and Farina 1997; Wood 2011).

It is a versatile species with a wide range of tolerance to variation in environmental conditions, especially drought and frost, although their growth is optimal in the rainy season when water availability is not a limiting factor. Also tolerates acid soils with moderate moisture (Flores and Ruiz 2006).

7 Collection Practice

As mainly wild plants are collected, the danger of becoming threatened by overexploitation is imminent, in Argentina (Viturro et al. 2007). Branches should be collected when plants are in full bloom (late spring to early autumn). In young plants or second collections make good net cuts at least 10 cm of soil, avoiding uprooting the plants. In older plants, cut branches of smaller diameter 1 cm, leaving 20–30 cm at the bottom. In sustainable harvest, the branches are shaken before bagging, to cause the fall of mature seeds (Elechosa 2009).

The leaves and tender stems that are employed fresh to flavor foods or beverages are harvested just before be used (Hilgert 1999).

In its spontaneous distribution area, it is also cultivated in home gardens (Pochettino et al. 2012), usually for own consumption medicinal purposes. Its cultivation is relatively easy, and it is reproduced by seeds, but is more convenient and simple the multiplication by cuttings (Alonso and Desmarchelier 2005). In vitro propagation was assayed (Díaz et al. 2011).

8 Traditional Use (Part(s) Used) and Common Knowledge

C. gilliesii has a long history of utilization in folk medicine within its spontaneous distribution area. Currently, the dried leaves and tender stems are commercialized in urban herb shops and health food stores to prepare infusions and decoctions; its mother tincture is also marketed (Hurrell et al. 2011).

Its main traditional therapeutic uses include: to treat digestive disorders, and the mountain sickness ('apunamiento', 'mal de puna' or 'soroche': dizziness, headache, nausea, vomiting, lack of appetite, physical exhaustion), aphrodisiac and emmenagogue (Hieronymus 1882; Burgstaller 1968; Orfila 1972; Ratera and Ratera 1980).

Regarding digestive disorders it is consumed as a digestive stimulant, bitter-tonic, stomachic (eupeptic), antacid, antiulcer, to treat stomach aches, and to cure the *empacho* (severe indigestion because many causes, mainly the excessive food intake) mainly in children, antispasmodic, cholagogue, choleretic, carminative, purgative (Bustos et al. 1996; Del Vitto et al. 1997; Hilgert 2001; Villagrán and Castro 2003; Alonso and Desmarchelier 2005; Gupta 2006; Rondina et al. 2008; Campos-Navarro and Scarpa 2013; Ceballos and Perea 2014).

In relation to reproductive medicine, its aphrodisiac properties refer to its use as stimulating libido and to treat male sexual dysfunction (impotence). *C. gilliesii* is utilized also as an emmenagogue, in case of menopausal ailments, to increase fertility, against female infertility, pregnancy and postpartum pains, and facilitating childbirth (Hieronymus 1882; Hilgert and Gil 2007; Barboza et al. 2009; Ceballos and Perea 2014).

Other records of ethnomedical uses include: against colds, anti-catarrhal and febrifuge (León et al. 2003; Villagrán and Castro 2003), in cases of genito-urinary complaints (Martínez and Pochettino 2004), against prolapsed, hernia, bruises, rheumatism (Barboza et al. 2009; Dadé et al. 2009; Hurrell et al. 2011), diuretic (Díaz et al. 2011), hypotensive and to treat heart diseases (Ceballos and Perea 2014).

C. gilliesii is one of the aromatic shrubs (of different families such as Asteraceae, Solanaceae, and Lamiaceae) called *koas* in Andean ritual traditions. These plants are burned and its smoke is an offering to the divinities in ancient ceremonies of the annual cycle. The term *koa* means 'that which is transformed into something else', referring to the transmutation of the plant into smoke (Villagrán and Castro 2003).

In northern Argentina this species is used as a condiment. In the puna region of Jujuy it is utilized for seasoning a traditional food called *pire*, made with corn flour and water. In the Yungas of southern Bolivia and northwestern Argentina, it is used to flavor *diana*, a preparation based on boiled milk, sweetened with sugar or cane honey, to which alcohol and different aromatic herbs are added (Hilgert 1999; Vignale and Gurni 2003; Alonso and Desmarchelier 2005; Giménez and Vignale 2013).

9 Modern Medicine Based on Its Traditional Medicine Uses

Traditional medicinal uses related to gastrointestinal disorders have not been well enough studied from an experimental point of view. However, their effects against these disorders are linked to its content in essential oils, e.g. piperitone has been reported to possess strong enterobactericidal activity, and piperitenone oxide has been reported to be a relaxant of the intestinal smooth muscle (Sousa et al. 1997; Dambolena et al. 2009).

Referring to the traditional use as an aphrodisiac, this term is used to indicate both libido enhancers such as those that increase sexual activity, especially in cases of male sexual dysfunction (erectile dysfunction). This latter use has been supported by an in vitro study about smooth muscle relaxation activity (vasodilatory) on the Guinea pig *corpus cavernosum*, probably due (at least in part) to its phenolic compounds (Hnatyszyn et al. 2003; Singh et al. 2013). Other uses mentioned above related to reproductive medicine have not yet been evaluated.

The trials of antimicrobial, antioxidant and cytotoxic activities of this species are promising for modern medicine. The antibacterial effect of its essential oil and flavonoids was analyzed (Hernández et al. 2000; Feresin et al. 2001; Alonso and Desmarchelier 2005; Luna et al. 2008; Momtaz and Abdollahi 2008; Mattos Cortegana et al. 2013). The antifungal activity of the essential oil was also evaluated (Zygadlo and Grow 1995; Lima et al. 2011). Organic and aqueous extracts showed a trypanocidal effect in vitro (Sülsen et al. 2006; Sülsen 2012; Tepe 2015), in relation to the piperitone and piperitona oxide components. Its antiplasmodial activity was also checked (Debenedetti et al. 2002; van Baren et al. 2006). The essential oil showed properties as an insect repellent, including *Triatoma infestans*, vector of Chagas disease (Tepe 2015), and as anti-head lice (Toloza et al. 2010).

The antioxidant activity has been analyzed by different authors (Desmarchelier et al. 1997; Barboza et al. 2009; Dadé et al. 2009; Cabana et al. 2012, 2013).

Toxicity studies by bioassay of *Artemia salina* from the aqueous extract of the aerial parts of *C. gilliesii* gave a positive result for a concentration of 10 mg/ml, limit value for distinguishing toxic and non-toxic aqueous extracts. On the one hand, this result could be useful in the search for new antitumor compounds (Mongelli et al. 1996). On the other hand, also due to this result the infusion intake for long periods (and preventively during pregnancy and lactation) is not recommended. By contrast, the usual infusion doses are generally well tolerated, except some recorded cases of digestive intolerance and headaches (Alonso and Desmarchelier 2005; Hurrell et al. 2011).

Clinopodium odorum has also been found to show antibacterial action (Mahady 2005; Vazquez et al. 2014), and cytotoxic effect on *Artemia salina* (Mongelli et al. 1996). *Clinopodium bolivianum* have antifungal, anti-inflammatory, and cytoprotective activity (Barboza et al. 2009), anti-*Helicobacter pylori* effect, responsible for gastro-duodenal diseases (Claros et al. 2007), antiviral activity against herpes simplex type I, and vesicular stomatitis virus (Abad et al. 1999; Momtaz and Abdollahi 2008).

10 Conclusions

C. gilliesii, muña muña, is a South American species utilized for centuries in the Andean region for medicinal purposes and as food condiment, mainly due to its essential oil content. Its most widespread traditional medicinal uses are: aphrodisiac, against gastrointestinal disorders, and mountain sickness, among others. Many of these popular applications need scientific validation. Nevertheless, several studies have already checked out some important properties, such as its effect against erectile dysfunction (linked with its aphrodisiac use), enterobactericidal and intestinal smooth muscle relaxant (related with its use in treating gastrointestinal ailments), antibacterial, antifungal, trypanocidal, antiplasmodial, insect repellent (e.g. *Triatoma infestans*, the vector of Chagas disease), anti-head lice, and antioxidant. Its cytotoxic activities have also been studied. These are promising in the search for anticancer compounds.

References

Abad MJ, Bermejo P, Gonzales E, Iglesias I, Irurzun A, Carrasco L (1999) Antiviral activity of Bolivian plant extracts. Gen Pharmacol 32(4):499–503

Alonso J, Desmarchelier C (2005) Plantas medicinales autóctonas de la Argentina. Editorial Lola, Buenos Aires

Álvarez Sarmiento XP (2012) Identificación, historia, características y aplicaciones culinarias de cinco plantas aromáticas endémicas de América. Universidad de Cuenca, Cuenca

Barboza GE, Cantero JJ, Núñez C, Pacciaroni A, Ariza Espinar L (2009) Medicinal plants: a general review and a phytochemical and ethnopharmacological screening of the native Argentine Flora. Kurtziana 34(1–2):7–365

Burgstaller CH (1968) La vuelta a los vegetales. Dinizo, Buenos Aires

Bustos DA, Tapia AA, Feresin GE, Ariza Espinar L (1996) Ethnopharmacobotanical survey of Bauchazeta district, San Juan Province. Argent Fitoterapia 67:411–415

Cabana R, Viturro CI, Heit CI, Saluzzo L, Vinholes J (2012) Avances en el estudio de *Clinopodium gilliesii* (Benth.) Kuntze de la provincia de Jujuy, Argentina. Dominguezia 28(2):71–72

Cabana R, Silva LR, Valentão P, Viturro CI, Andrade PB (2013) Effect of different extraction methodologies on the recovery of bioactive metabolites from Satureja parvifolia (Phil.) Epling (Lamiaceae). Ind Crop Prod 48:49–56

Campos-Navarro R, Scarpa GF (2013) The cultural-bound disease “empacho” in Argentina. A comprehensive botanico-historical and ethnopharmacological review. J Ethnopharmacol 148(2):349–360

Cantino PD, Wagstaff SJ (1998) A reexamination of North American *Satureja* s.l. (Lamiaceae) in light of molecular evidence. Brittonia 50(1):63–70

Ceballos SJ, Perea MC (2014) Plantas medicinales utilizadas por la comunidad indígena de Quilmes (Tucumán, Argentina). Bol Latinoam Caribe Plant Med Aromat 13(1):47–68

Claros PM, Bilbao RP, Damiani ME, Gonzales DE, Estensoro CM, Álvarez AM (2007) Actividad anti-*Helicobacter pylori* de *Plantago major, Clinopodium bolivianum, Calendula officinalis* y *Piper angustifolium* por el método de difusión de disco. BIOFARBO (La Paz) 15(1):37–42

Dadé MM, Fioravanti DE, Schimella GR, Tournier HA (2009) Total antioxidant capacity and polyphenol content of 21 aqueous extracts obtained from native plants of Traslasierra valley (Argentina). Bol Latinoam Caribe Plant Med Aromat 8(6):529–539

Dambolena JS, Zunino MP, Lucini EI, Zygadlo JA, Rotman A, Ahumada O, Biurrun F (2009) Essential oils of plants used in home medicine in North of Argentina. J Essent Oil Res 21:405–409

Debenedetti S, Muschietti L, van Baren C, Clavin M, Broussalis A, Martino V, Houghton PJ, Warhurst D, Steele J (2002) In vitro antiplasmodial activity of extracts of Argentinian plants. J Ethnopharmacol 80(2–3):163–166

Del Vitto LA, Petenatti EM, Petenatti ME (1997) Recursos herbolarios de San Luis (República Arentina) Parte 1. Plantas nativas. Multequina 6:49–66

Desmarchelier C, Repetto M, Coussio J, Liesuy S, Ciccia G (1997) Antioxidant and prooxidant activities in aqueous extracts of Argentine plants. Int J Pharmacogn 35(2):116–120

Díaz MS, Palacio L, Figueroa AC, Goleniowski ME (2011) Propagación in vitro de la especie aromática *Clinopodium gilliesii* (Benth.) Kuntze. Bol Soc Argent Bot 46(Supl):69–70

Elechosa MA (2009) Manual de recolección sustentable de plantas aromáticas nativas de la región central y noroeste de la Argentina. Inst Nac Tecnol Agropecu, Buenos Aires

Feresin GE, Tapia A, López SN, Zacchino SA (2001) Antimicrobial activity of plants used in traditional medicine of San Juan province, Argentine. J Ethnopharmacol 78(1):103–107

Flores PJ, Ruiz MB (2006) Catálogo de vegetación del área de influencia del Parque Nacional San Guillermo, San Juan, Argentina. Inst Nac Tecnol Agropecu, Buenos Aires

Giménez LAS, Vignale ND (2013) Especias y condimentos empleados en la cocina andina tradicional. Agraria (Jujuy) 7(14):33–44

Gupta MP (2006) Medicinal plants originating in the Andean high plateau and central valleys region of Bolivia, Ecuador and Peru. United Nations Industrial Development Organisation

Harley RM, Granda Paucar A (2000) List of species of Tropical American Clinopodium (Labiatae), with new combinations. Kew Bull 55(4):917–927

Harley RM, Atkins S, Budantsev AL, Cantino PD, Conn BJ, Grayer R, Harley MM, de Kok R, Krestovskaja T, Morales R, Paton AJ, Ryding O, Upson T (2004) Labiatae. In: Kubitzki K (ed) The families and genera of vascular plants VII. Springer, Berlin, pp 167–275

Hernández NE, Tereschuk ML, Abdala LR (2000) Antimicrobial activity of flavonoids in medicinal plants from Tafí del Valle (Tucumán, Argentina). J Ethnopharmacol 73(1–2):317–322

Hieronymus J (1882) Plantas diafóricas. Flora Argentina. G. Kraft, Buenos Aires
Hilgert NI (1999) Las plantas comestibles en un sector de las Yungas meridionales (Argentina). An Jard Bot Madrid 51(1):117–138
Hilgert NI (2001) Plants used in Home Medicine in the Zenta River Basin, Northwest Argentina. J Ethnopharmacol 76(1):11–34
Hilgert NI, Gil EG (2007) Reproductive medicine in northwest Argentina: traditional and institutional systems. J Ethnobiol Ethnomed 3:19. https://doi.org/10.1186/1746-4269-3-19
Hnatyszyn O, Moscatelli V, Garcia J, Rondina R, Costa M, Arranz C, Balaszczuk A, Ferraro G, Coussio JD (2003) Argentinian plant extracts with relaxant effect on the smooth muscle of the corpus cavernosum of guinea pig. Phytomedicine 10(8):669–674
Hurrell JA, Ulibarri EA, Delucchi G, Pochettino ML (2008) Plantas aromáticas condimenticias. In: Hurrell JA (ed) Biota Rioplatense XIII. Editorial Lola, Buenos Aires
Hurrell JA, Ulibarri EA, Arenas PM, Pochettino ML (2011) Plantas de Herboristería. Editorial Lola, Buenos Aires
León JF, Sulca Quispe L, Delgado Tello P, Cáceres Musaja C, Bonifacio AA (2003) Diversidad florística medicinal altoandina y propuesta de aprovechamiento de especies endémicas como recurso terapéutico del Departamento de Tacna, Perú. Universidad Nacional Jorge Basadre Grohmann, Tacna
Lima B, López S, Luna L, Agüero MB, Aragón L, Tapia A, Zacchino S, López ML, Zygadlo J, Feresin GE (2011) Essential oils of medicinal plants from the central Andes of Argentina: chemical composition, and antifungal, antibacterial, and insect-repellent activities. Chem Biodivers 8(5):924–936
López-Lázaro M (2009) Distribution and biological activities of the flavonoid luteolin. Mini Rev Med Chem 9(1):31–59
Luna L, Lima B, Tapia A, Egly Feresin G, Duschatzky C, Possetto M, Lampasona M, Schuff C (2008) Chemical composition and antibacterial activity of*Satureja parvifolia* (Phil.) Epling essential oil. J Essent Oil Bearing Plants 11(1):106–111
Mahady GB (2005) Medicinal plants for the prevention and treatment of bacterial infections. Curr Pharm Des 11(19):2405–2427
Martínez MR, Pochettino ML (2004) Microambientes y recursos vegetales terapéuticos. Conocimiento local en Molinos, Salta, Argentina. Zonas Aridas (Lima) 8:18–31
Mattos Cortegana J, Palacios Pinto G, Glorio Paulet P, Morales Cauti S (2013) Efecto de la muña (*Satureja parvifolia*) como aditivo no nutricional sobre el desarrollo de *Lactobacillus* spp. y control de *Salmonella tysphimurium* en cuyes de carne. Científica (Lima) 10(2):123–134
Momtaz S, Abdollahi M (2008) A systematic review of the biological activities of *Satureja* L. species. Pharmacologyonline 2:34–54
Mongelli E, Martino V, Coussio J, Ciccia G (1996) Screening of Argentine medicinal plants using the brine shrimp microwell cytotoxicity assay. Pharm Biol 34(4):249–254
Muschietti L, Van Baren C, Coussio J, Vila R, Clos M, Cañigueral S, Adzet T (1996) Chemical composition of the leaf oil of *Satureja odora* and *Satureja parvifolia*. J Essent Oil Res 8(6):681–684
Niemeyer HM (2010) Composition of essential oils from *Satureja darwinii* (Benth.) Briq. and *S. multiflora* (Ruiz & Pav.) Briq. (Lamiaceae). Relationship between chemotype and oil yield in *Satureja* spp. J Essent Oil Res 22:477–482
Orfila EN (1972) Las especies de la flora medicinal argentina conocidas por "muña-muña". Rev Farm 9-19(114):3–11
Orfila EN, Farina EL (1997) Lamiaceae. In: Novara LR (ed) Flora del Valle de Lerma (Salta, Argentina). Aportes Bot Salta. Ser Flora 4(1):1–62
Pochettino ML, Hurrell JA, Lema VS (2012) Local botanical knowledge and agrobiodiversity: homegardens at rural and periurban contexts in Argentina. In: Luna Maldonado AI (ed) Horticulture. InTech, Rijeka, pp 105–132
Pontiroli A (1993) Labiatae. In: Cabrera AL (ed) Flora del la Provincia de Jujuy. Colecc Cient Inst Nac Tecnol Agropecu 13(9):117–155
Ratera EL, Ratera MO (1980) Plantas de la flora argentina empleadas en medicina popular. Hemisferio Sur, Buenos Aires

Rondina R, Bandoni AL, Coussio JD (2008) Especies medicinales argentinas con potencial actividad analgésica. Dominguezia 24(1):47–69
Singh S, Ali A, Singh R, Kaur R (2013) Sexual abnormalities in males and their herbal therapeutic aspects. Pharmacologia 4(4):165–275
Sousa PJC, Magalhães PJC, Lima CC, Oliveira VS, Leal-Cardoso JH (1997) Effects of piperitenone oxide on the intestinal smooth muscle of the guinea pig. Braz J Med Biol Res 30(6):787–791
Sülsen V (2012) Búsqueda de compuestos antiprotozoarios en especies de la flora medicinal argentina. Dominguezia 28(2):19–27
Sülsen V, Güida C, Coussio J, Paveto C, Muschietti L, Martino V (2006) In vitro evaluation of trypanocidal activity in plants used in Argentine traditional medicine. Parasitol Res 98(4):370–374
Tepe B (2015) Inhibitory effect of *Satureja* on certain types of organisms. Rec Nat Prod 9(1):1–18
Toloza AC, Zygadlo J, Biurrun F, Rotman A, Picollo MI (2010) Bioactivity of Argentinean essential oils against permethrin-resistant head lice, *Pediculus humanus capitis*. J Insect Sci 10:185. https://doi.org/10.1673/031.010.14145
Ulloa C (2006) Aromas y sabores andinos. In: Moraes M, Øllgaard B, Kvist P, Borchsenius F, Balslev H (eds) Botánica Económica de los Andes Centrales. Universidad Mayor de San Andrés, La Paz, pp 313–328
van Baren C, Anao I, Leo Di Lira P, Debenedetti S, Houghton P, Croft S, Martino V (2006) Triterpenic acids and flavonoids from *Satureja parvifolia*. Evaluation of their antiprotozoal activity. Z Naturforsch C 61(3–4):189–192
Vazquez AM, Aimar ML, Demmel GI, Cabalen ME, Decarlini MF, Cantero JJ, Criado SG, Ruiz GM (2014) Identification of volatile compounds of *Clinopodium odorum* (Lamiaceae): a comparison between HS-SPME and classic hydrodistillation. Bol Latinoam Caribe Plant Med Aromat 13(3):285–296
Vignale ND, Gurni AA (2003) Micrografía de plantas medicinales andinas usadas como aditivos alimentarios en la Provincia de Jujuy (Argentina). Bol Soc Argent Bot 38(Suppl):137–142
Villagrán C, Castro V (2003) Ciencia indígena de los Andes del norte de Chile. Editorial Universitaria, Santiago de Chile
Viturro CI, Molina A, Guy I, Charles B, Guinaudeau H, Fournet A (2000) Essential oils of *Satureja boliviana* and *S. parvifolia* growing in the region of Jujuy, Argentina. Flavour Fragr J 15(6):377–382
Viturro CI, Molina A, Heit C, Elechosa MA, Molina AM, Juárez MA (2007) Evaluación de la composición de los aceites esenciales de *Satureja boliviana*, *S. odora* y *S. parvifolia*, obtenidos de colectas en Tucumán, Argentna. Bol Latinoam Caribe Plant Med Aromat 6(5):288–289
Wood JRI (2011) Clinopodium L. (Lamiaceae) in Bolivia. Kew Bull 66(2):199–226
Zygadlo JA, Grow NR (1995) Comparative study of the antifungal activity of essential oils from aromatic plants growing wild in the central region of Argentina. Flavour Fragr J 10(2):113–118
Zygadlo JA, Merino EF, Maestri DM, Guzman CA, Ariza Espinar L (1993) The Essential Oils of *Satureja odora* and *S. parvifolia*from Argentina. J Essent Oil Res 5(5):549–551

Croton zehntneri Pax & K. Hoffm (Euphorbiaceae)

Jackson Roberto Guedes da Silva Almeida, Ana Carolina Murta Ramalho, and Fernanda Guerra da Silveira

Croton zehntneri Pax & K. Hoffm

J. R. G. da Silva Almeida (✉) · A. C. M. Ramalho · F. G. da Silveira
Center for Studies and Research of Medicinal Plants (NEPLAME), Federal University of Vale do São Francisco (UNIVASF), Petrolina, Pernambuco, Brazil
e-mail: jackson.guedes@univasf.edu.br

U. P. Albuquerque et al. (eds.), *Medicinal and Aromatic Plants of South America*, Medicinal and Aromatic Plants of the World 5,
https://doi.org/10.1007/978-94-024-1552-0_15

Abstract The family Euphorbiaceae is comprised of 300 genera and some 7500 species widely distributed over the world, especially in the tropical and subtropical regions of the Americas, Africa and Asia. *Croton* is one of the most important genera of this family that comprises about 1300 species widespread in Africa, Asia and South America. This genus is rich in constituents with biological activities, chiefly diterpenoids such as phorbol esters, clerodane, labdane, kaurane, trachylobane, pimarane, etc. *Croton* is also rich in alkaloids, flavonoids, triterpenoids and steroids. Several species are aromatic, indicating the presence of volatile oils. *Croton zehntneri* Pax & K. Hoffm. is native to Northeastern Brazil, where it is often used in folk medicine to treat anxiety, as sedative, appetite stimulating, antianorexigen and for the relief of gastrointestinal disturbances. In view of its popular uses in treating various diseases, this chapter reviews scientific studies on the chemical and biological properties of this species.

Keywords *Croton zehntneri* · Euphorbiaceae · Essential oils · Biological activity

1 Taxonomic Characteristics

The family Euphorbiaceae is comprised of 300 genera and some 7500 species widely distributed over the world, especially in tropical and subtropical regions of the Americas, Africa and Asia. The most important genera are: *Euphorbia*, *Croton*, *Phyllantus*, *Acalypha*, *Macaranga*, *Antidesma*, *Drypetes*, *Jatropha*, *Manihot* and *Tragia*. *Croton* is one of the largest genera that comprise about 1300 species of trees, shrubs and herbs distributed in tropical and subtropical regions of both hemispheres, widespread in Africa, Asia and South America (Webster 1994; Salatino et al. 2007). Species of this genus are ecologically prominent and often important elements of secondary vegetation in the tropics and subtropics worldwide (Simionatto et al. 2007).

Synonyms *Croton grewioides* Baill. (Cordeiro et al. 2015)

2 Crude Drug Used

The main parts of the plant used in folk medicine are the leaves and stems. The crude drug has an aroma reminiscent of a mixture of star anise (*Illicium verum*) and clove India (*Eugenia caryophyllata*) which is due to the presence of essential oils. However, this aroma has been shown to vary according to copies of this plant collected in different locations in Northeast. This is due to variation in the concentration of the chemical constituents more abundant in its essential oils (Morais et al. 2006).

The leaves contain usually 2–4% essential oil. Its production is variable and undergoes changes according to ecological conditions. Diurnal changes during the day have also been recorded. All these factors determines the optimal time of collection, including the season of the year and the region within Northeast Brazil (Cavalcanti et al. 2012).

As most Euphorbiaceae, *Croton* species may contain latex, which is red-colored in some species, a characteristic usually associated with medicinal properties (Salatino et al. 2007).

3 Major Chemical Constituents and Bioactive Compounds

The genus *Croton* is rich in constituents with biological activities, chiefly diterpenoids such as phorbol esters, clerodane, labdane, kaurane, trachylobane, pimarane, etc. *Croton* is also rich in biologically active alkaloids, flavonoids, triterpenoids and steroids. Several species of the genus are aromatic, indicating the presence of volatile oil constituents (Salatino et al. 2007).

The richness of the chemical composition of the plants of the genus *Croton* has been subject of a comprehensive review of the special literature, covering its phytochemistry, and the use of some species in folk medicine (Medeiros et al. 2012).

The phytochemistry of the species *C. zehntneri* is characterized by the presence of aliphatics, monoterpenes, phenylpropanoids and sesquiterpenes (Medeiros et al. 2012). Remarkably, the triterpenoid acetyl aleuritolic acid also was isolated from *C. zehntneri*. Its structure was characterized by NMR spectroscopy (Melo et al. 2014).

The first studies carried out with the essential oils from the stems and leaves of *C. zehntneri* showed the presence of chemical constituents such as *n*-eicosane, *n*-heptadecane, isoborneol, camphor, 1, 8-cineole, myrcene, α-pinene, β-pinene, estragole, eugenol methyl ether, safrole, anethole, caryophyllene and γ-elemene (Craveiro et al. 1978), as well as the presence of *p*-cymene, geranial, linalool, neral, eugenol, β-farnesene, β-guayene, γ-muurolene and α-bergamotene (Craveiro et al. 1981).

Since then, other chemical studies have been realized. For a more complete list of compounds see Table 1. The chemical composition of the essential oil was analyzed by gas chromatography coupled to mass spectrometry (GC-MS), and its inclusion complex with β-cyclodextrin (β-CD) was characterized by both vibrational spectroscopy and differential scanning calorimetry (DSC). Estragol was the major component identified in the essential oil (Aguiar et al. 2014). In another study, the chemical composition was analyzed by GC-MS. This method permitted to identify a total of 97.40% of the components, with a major presence of estragole (76.80%). Estragol was previously reported as being responsible for antibacterial activities (Costa et al. 2008). *trans*-anethole was the major constituent found in the essential oil of *C. zehntneri* and it is closely implicated with the pharmacologic activity attributed to essential oil (Cavalcanti et al. 2012).

Table 1 Presence of compounds in *Croton zehntneri*

Compound type	Chemical structure	Reference
Triterpene		
Acetyl aleuritolic acid	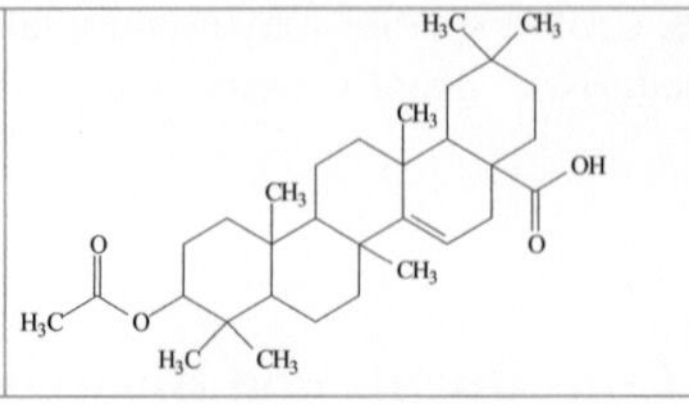	Melo et al. (2014)
Tropone derivative		
Crototropone	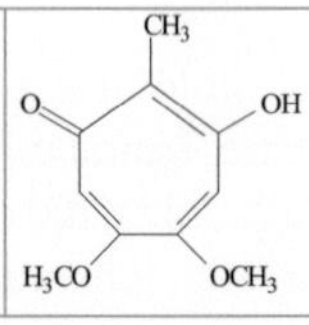	Bracher et al. (2008)
Volatile constituents		
Anethole	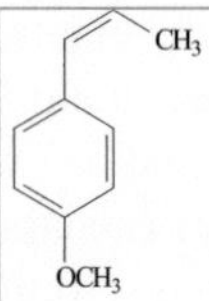	Fontenelle et al. (2008)
para-Anisaldehyde	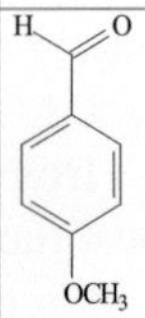	Morais et al. (2006)
Anisil formiate	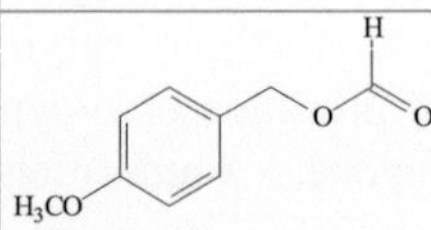	Morais et al. (2006)
Camphene	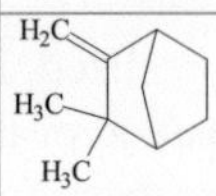	Morais et al. (2006)
Camphor	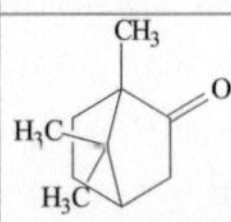	Morais et al. (2006)
Caryophyllene	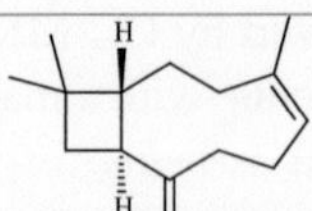	Morais et al. (2006)
Caryophyllene oxide	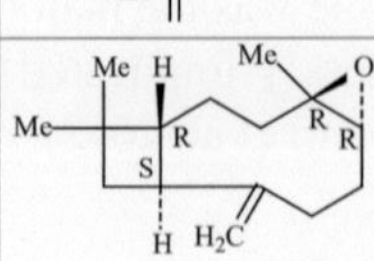	Morais et al. (2006)

(continued)

Table 1 (continued)

Compound type	Chemical structure	Reference
1,8-Cineole	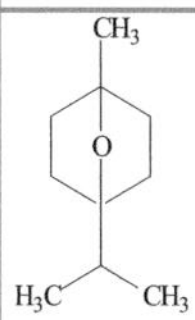	Morais et al. (2006)
β-Elemene	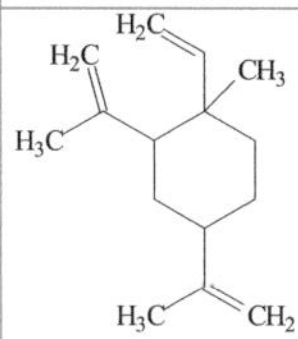	Morais et al. (2006)
Estragole	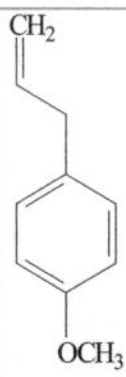	Fontenelle et al. (2008)
Guaiene	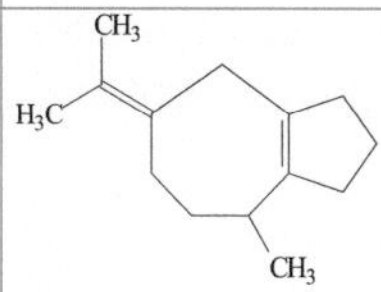	Morais et al. (2006)
α-Pinene	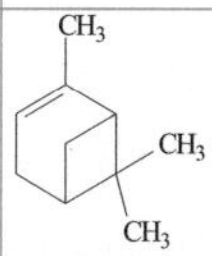	Morais et al. (2006)
β-Selinene	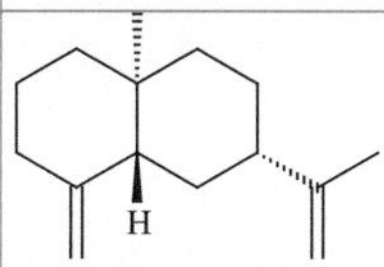	Morais et al. (2006)

Crototropone (3-hydroxy-5, 6-dimethoxy-2-methylcyclohepta-2,4,6-trien-1-one) was isolated from roots of *C. zehntneri*. The structure was established by spectroscopic methods (Bracher et al. 2008).

4 Morphological Description

Croton zehntneri is an aromatic bush. It has induments of star trichomes, petioles with sessile glands. Leaves are alternate, oval, with short petioles. Pseudoracemes with unisexual summits. Flowers have generally 11 stamens. Fruits are capsules containing three seeds, 4–5 cm (Fernandes et al. 1978).

5 Geographical Distribution

The genus *Croton* is widespread in the Northeast region of Brazil, mainly in the Caatinga (semi-arid vegetation). Generally, the genus presents a consistent profile of biological activities and folk use (Ramos et al. 2013). *C. zehntneri* is a bush native to the drought ecosystem of the Caatinga from the Northeast Brazil (Cavalcanti et al. 2012).

6 Ecological Requirements

There are interesting aspects of the ecology of C. zehntneri that are not yet well elucidated. Although the species is well adapted to the Caatinga biome, some rural populations, who know well the distribution of the species in the native forest, report that this plant does not distribute evenly over the forest land, but forms groups in certain places. Remarkably, the essential oil of the species collected in different regions of Northeast Brazil presents different chemical composition/constituents. This variation in the chemical composition seems to refer to strong influence of the different ecological conditions on the major constituents of the essential oil (Leal-Cardoso et al. 2013).

7 Collection Practice

The aerial parts of the plant are collected and crushed. In this process the plants release a characteristic odor, the smell of its essential oil. The bark and leaves of the plant are collected and widely used, in the Northeast region of Brazil, as sweeteners, as well as for medicinal use (Leal-Cardoso et al. 2013).

8 Traditional Use (Part(s) Used) and Common Knowledge

Many species from the genus *Croton* have been used in traditional medicine and its pharmacological activities have been demonstrated. Popular uses include treatment of cancer, constipation, diabetes, digestive problems, dysentery, external wounds, fever, hypercholesterolemia, hypertension, inflammation, intestinal worms, malaria, pain, ulcers, and weight-loss (Salatino et al. 2007).

In Northeastern Brazil *C. zehntneri* is popularly called "canela de cunhã", "canelinha", "canelinha brava" and "canela brava". In folk medicine, infusions or decoctions of leaves from *C. zehntneri* are used mainly to treat anxiety. It is also used as sedative, appetite stimulating, antianorexigen and for the relief of gastrointestinal disturbances (Cunha et al. 2012; Salatino et al. 2007; Oliveira et al. 2001). Because *C. zehntneri* is characterized by a strong and pleasant odor reminiscent of anise and clove, extracts of its bark and leaves are used in perfumes and as sweeteners in foods and in drinks (Siqueira et al. 2006).

A herbal tea prepared by pouring water over dried leaves or over branches of *C. zehntneri* is one of the most popular remedies in Brazilian folk medicine for treating "nervous disturbances" such as irritability, anxiety and seizures (Bernardi et al. 1991).

9 Modern Medicine Based on Its Traditional Medicine Uses

Plants of the genus *Croton* have been used extensively in the Northeast of Brazil for treating various clinical conditions. Previous studies have demonstrated that the essential oils are responsible for the pharmacologic effects.

The effects of the essential oil of *C. zehntneri* (EOCz) and its main constituent anethole on several models of gastric lesions were studied in mice and rats. Oral treatment with EOCz and anethole, both at doses of 30–300 mg/kg, caused similar and dose-dependent gastroprotection against ethanol- and indomethacin induced gastric damage, but did not change cold-restraint stress-induced ulcers in rats. Furthermore, EOCz and anethole (both at 30 and 300 mg/kg) similarly and significantly increased the mucus production by the gastric mucosa, measured by Alcian blue binding, in ethanol-induced ulcer model. The results of this study showed for the first time that EOCz possesses a gastroprotective potential, an effect mostly attributed to the action of anethole. This activity is related predominantly to the ability of EOCz and anethole to enhance the production of gastric wall mucus, an important gastroprotective factor (Coelho-de-Souza et al. 2013).

The cardiovascular effects of the EOCz in deoxycorticosterone-acetate (DOCA)-salt hypertensive rats was evaluated. Furthermore, in vitro experiments using isolated thoracic aortic rings were performed to assess the vascular effects of the EOCz. The data showed that i.v. administration of EOCz in DOCA-salt hypertensive rats induces a vago-vagal reflex decreases in heart rate and blood pressure (phase 1). EOCz may induce a second and delayed hypotension due to its direct endothelium-independent vasorelaxant effects, but it seems to be buffered by the pressor component (subsequent to phase 1) of EOCz (Siqueira et al. 2013). Cardiovascular effects of the essential oil of *C. zehntneri* leaves and its main constituents, anethole and estragole, in normotensive conscious rats were investigated. The administration of EOCz induces an initial hypotension followed by a pressor response, two effects that appear mainly attributed to the actions of anethole and estragole (Siqueira et al. 2006).

Antifungal activity of essential oils of several *Croton* species from the Brazilian Caatinga biome was evaluated against *Candida albicans*, *Candida tropicalis*, and *Microsporum canis* by the agar-well diffusion method and the minimum inhibitory concentration (MIC) by the broth microdilution method. The main constituents for

C. zehntneri were estragole and anethole. The essential oil demonstrated better activity against *M. canis* (Fontenelle et al. 2008).

The anti-nociceptive effects of EOCz were evaluated in mice using chemical and thermal models of nociception. EOCz was administered orally at doses of 100 and 300 mg/kg, and reduced paw licking time in the second phase of the formalin test. During the first phase of the formalin test only 300 mg/kg induced a significant alteration. The number of contortions in response to intraperitoneal injections of acetic acid did not differ significantly between controls and experimental animals. In the hot-plate test, EOCz at doses > or = 100 mg/kg significantly increased the latency time with respect to controls. The data showed that EOCz is effective as an antinociceptive agent (Oliveira et al. 2001).

C. zehntneri is a popular plant used to treat nervous disturbance. It contains a complex mixture of compounds, including substances exhibiting central nervous system activity. The effects of EOCz administration (p.o.) on the rat's central nervous system were studied in behavioral models used to evaluate anxiety and antidepressive drugs. The results showed that administration of EOCz: (1) increased the immobility duration measured in the forced swimming test as compared to control group; (2) reduced the locomotion frequency observed in the open field; (3) had no effect on the experimental group (1 μl) observed in open field; (4) had no effect on animals tested in social interactions, plus-maze and hole-board tests. These data suggested that EOCz produced central depressor effects in rats without any anxiety alterations. These results may explain the popular use of this plant in Brazilian folk medicine for treating nervous disturbances (Lazarini et al. 2000).

The effects of essential oil of *C. zehntneri*, orally administered, were studied on behavioral parameters using rats and mice. The oil suspension did not modify pentobarbital induced hypnosis, stereotypic behavior, catalepsy and amphetamine-induced hypermotility. The open-field behaviors were decreased and the minimal convulsant dose of pentylenetetrazole was increased (Batatinha et al. 1995). The effects of aqueous *C. zehntneri* leaf and branch extracts, orally administered, on some dopaminergic- and cholinergic-related behaviours were studied in rats and mice. The leaf extract did not modify apomorphine-induced stereotypic behavior, haloperidol-induced catalepsy and active avoidance/escape responses. The branch extract reduced stereotypy but did not interfere with catalepsy and active avoidance behavior. Both extracts were capable of increasing the tremor induced by oxotremorine (Giorgi et al. 1991).

10 Conclusions

The species *C. zehntneri* is rich in essential oils. The presence of compounds such as α-pinene and β-pinene has been reported in essential oils of also other species of *Croton*, indicating that this species is typical representative of the Euphorbiaceae family. Anethole and estragole, the main constituents of this species could be considered responsible for several of the biological activities presented by the plant.

Despite of its extensive use in folk medicine, the phytochemical studies of this species are restricted to the identification of chemical constituents of the essential oils. Some of these studies seem to support also the popular use of the plant in traditional medicine.

References

Aguiar UN, Lima SG, Rocha MS, Freitas RM, Oliveira TM, Silva RM, Moura LCB, Almeida LTG (2014) Preparation and characterization of the inclusion complex essential oil of *Croton zehntneri* with β-cyclodextrin. Quim Nova 37(1):50–55

Batatinha MJM, Souza-Spinosa H, Bernardi MM (1995) *Croton zehntneri*: possible central nervous system effects of the essential oil in rodents. J Ethnopharmacol 45(1):53–57

Bernardi MM, Souza-Spinosa H, Batatinha MJM, Giorgi R (1991) *Croton zehntneri*: possible central nervous system effects in rodents. J Ethnopharmacol 33(3):285–287

Bracher F, Randau KP, Lerche H (2008) Crototropone, a new tropone derivative from *Croton zehntneri*. Fitoterapia 79(3):236–237

Cavalcanti JM, Leal-Cardoso JH, Diniz LR, Portella VG, Costa CO, Linard CF, Alves K, Rocha MV, Lima CC, Cecatto VM, Coelho-de-Souza AN (2012) The essential oil of *Croton zehntneri* and *trans*-anethole improves cutaneous wound healing. J Ethnopharmacol 144(2):240–247

Coelho-de-Souza AN, Lahlou S, Barreto JE, Yum ME, Oliveira AC, Oliveira HD, Celedônio NR, Feitosa RG, Duarte GP, Santos CF, de Albuquerque AA, Leal-Cardoso JH (2013) Essential oil of *Croton zehntneri* and its major constituent anethole display gastroprotective effect by increasing the surface mucous layer. Fundam Clin Pharmacol 27(3):288–298

Cordeiro I, Secco R, Carneiro-Torres DS, Lima LR de, Caruzo MBR, Berry P, Riina R, Silva OLM, Silva MJ da, Sodré RC (2015) Croton in Lista de Espécies da Flora do Brasil. Jardim Botânico do Rio de Janeiro. Disponível em: http://floradobrasil.jbrj.gov.br/jabot/floradobrasil/FB29246. Acesso em: 29 Mai. 2015

Costa JGM, Rodrigues FFG, Angélico EC, Pereira CKB, Souza EO, Caldas GFR, Silva MR, Santos NKA, Mota ML, Santos PF (2008) Chemical composition and evaluation of the antibacterial activity and toxicity of the essential oil of *Croton zehntneri* (variety estragol). Rev Bras Farmacogn 18(4):583–586

Craveiro AA, Andrade CHS, Matos FJA, Alencar JW (1978) Anise-like flavor *Croton aff zehntneri*. J Agric Food Chem 26(3):772–773

Craveiro AA, Rodrigues AS, Andrade CHS, Matos FJA, Alencar JW, Machado MIL (1981) Volatile constituents of Brazillian Euphorbiaceae. Genus *Croton*. J Nat Prod 44(5):602–608

Cunha CSM, Maia SSS, Coelho MFB (2012) Cuttings of *Croton zehntneri* Pax et Hoffm. at different concentrations of indole butyric acid. Cienc Rural 42(4):621–626

Fernandes AA, Alencar IW, Matos FJA, Craveiro AA, Andrade CHS, Fonteles MC, Viana GSB, Capelo LR, Matos FF (1978). Canelas silvestres nordestinas: aspectos botânicos, químicos e farmacológicos. In: Simpósio de Plantas Medicinais do Brasil, vol 5, 1978. Ciência e Cultura, São Paulo 197, 32:26–36

Fontenelle RO, Morais SM, Brito EH, Brilhante RS, Cordeiro RA, Nascimento NR, Kerntopf MR, Sidrim JJ, Rocha MF (2008) Antifungal activity of essential oils of *Croton* species from the Brazilian Caatinga biome. J Appl Microbiol 104(5):1383–1390

Giorgi R, Batatinha MJ, Bernardi MM, Souza-Spinosa H, Spinosa FR, Palermo-Neto J (1991) Effects of *Croton zehntneri* aqueous extracts on some cholinergic- and dopaminergic-related behaviours of laboratory rodents. J Ethnopharmacol 34(2–3):189–193

Lazarini CA, Uema AH, Brandão GMS, Guimarães APC, Bernardi MM (2000) *Croton zehntneri* essential oil: effects on behavioral models related to depression and anxiety. Phytomedicine 7(6):477–481

Leal-Cardoso JH, Albuquerque AAC, Cecatto VM, Souza ANC (2013). *Croton zehntneri* Pax et Hoffm. (canela de cunhã). In: Viana GSB, Leal LKAM, Vasconcelos SMM (eds) Plantas medicinais da Caatinga: atividades biológicas e potencial terapêutico. Ed. UFC, 131, 2013

Medeiros VM, Tavares JF, Almeida JRGS, Araujo-Junior VT, Athayde-Filho PF, Cunha EVL, Barbosa-Filho JM, Silva MS (2012). Phytochemistry of the genus *Croton*. In: Natural products: research reviews, vol 1. Daya Publishing House, New Delhi, pp 217–366

Melo IRS, Teixeira AMR, Sena-Junior DM, Santos HS, Albuquerque MRJR, Bandeira PN, Rodrigues AS, Braz-Filho R, Gusmão GOM, Silva JH, Fariad JLB, Bento RRF (2014 Jan) FT-Raman and FTIR-ATR spectroscopies and DFT calculations of triterpene acetyl aleuritolic acid. J Mol Struct 1058:221–227

Morais SM, Catunda-Junior FEA, Silva ARA, Martins-Neto JS, Rondina D, Leal-Cardoso JH (2006) Antioxidant activity of essential oils from Northeastern Brazilian *Croton* species. Quim Nova 29(5):907–910

Oliveira AC, Leal-Cardoso JH, Santos CF, Morais SM, Coelho-de-Souza AN (2001 Nov) Antinociceptive effects of the essential oil of *Croton zehntneri* in mice. Braz J Med Biol Res 34(11):1471–1474

Ramos JMO, Santos CA, Santana DG, Santos DA, Alves PB, Thomazzi SM (2013) Chemical constituents and potential antiinflammatory activity of the essential oil from the leaves of *Croton argyrophyllus*. Rev Bras Farmacogn 23(4):644–650

Salatino A, Salatino MLF, Negri G (2007) Traditional uses, chemistry and pharmacology of croton species (Euphorbiaceae). J Braz Chem Soc 18(1):11–33

Simionatto E, Bonani VFL, Morel AF, Poppi NR, Raposo-Junior JL, Stuker CZ, Peruzzo GM, Peres MTLP, Hess SC (2007) Chemical composition and evaluation of antibacterial and antioxidant activities of the essential oil of Croton urucurana Baillon (Euphorbiaceae) stem bark. J Braz Chem Soc 18(5):879–885

Siqueira RJ, Magalhães PJ, Leal-Cardoso JH, Duarte GP, Lahlou S (2006) Cardiovascular effects of the essential oil of *Croton zehntneri* leaves and its main constituents, anethole and estragole, in normotensive conscious rats. Life Sci 78(20):2365–2372

Siqueira RJ, Duarte GP, Magalhães PJ, Lahlou S (2013) Cardiovascular effects of the essential oil of *Croton zehntneri* leaves in DOCA-salt hypertensive, conscious rats. Nat Prod Commun 8(8):1167–1170

Webster GL (1994) Systematics of the euphorbiaceae. In: Annals of the Missouri Botanic Garden, vol 1. EUA, California, n. 81, p 44

Cymbopogon citratus (DC.) Stapf

Wendy Marisol Torres-Avilez, Flávia dos Santos Silva, and Ulysses Paulino Albuquerque

Cymbopogon citratus (DC.) Stapf
Photo: David Stang
Available in: http://www.tropicos.org/Image/100111485

W. M. Torres-Avilez · F. S. Silva
Laboratório de Ecologia e Evolução de Sistemas Socioecológicos, Departmento de Botânica, Universidade Federal de Pernambuco, Recife, PE, Brazil

U. P. Albuquerque (✉)
Departamento de Botânica, Centro de Biociências, Universidade Federal de Pernambuco, Recife, Brazil

U. P. Albuquerque et al. (eds.), *Medicinal and Aromatic Plants of South America*, Medicinal and Aromatic Plants of the World 5,
https://doi.org/10.1007/978-94-024-1552-0_16

Abstract *Cymbopogon citratus* (DC.) Stapf is an herbaceous species native to tropical Asia that has been introduced to several South American countries, including Brazil, where it is known by several local names. This species is recognized in the pharmaceutical, food and cosmetic industries for the chemical properties of its essential oil, which is composed of citral, as main component. The essential oil of this plant is in high demand in countries such as the United States, Japan, France and Switzerland and must be used with caution because large doses can damage the body.

Keywords Medicinal plants · Lemon grass · Essential oil · Crop · Production

1 Part I: General Aspects

1.1 Description of the Plant

Cymbopogon citratus (DC.) Stapf is an herb that can measure up to 2 m in height. Its glabrous leaves, up to 70 cm long and 18 mm wide, are light green, rough, basal in their vegetative form, highly aromatic and elongated like strips, which sprout from the ground forming dense clumps (Stevens et al. 2001). In communities in South American countries such as Colombia, Venezuela and Brazil, the plant is known locally as malojillo, malojillo criollo, capim-limão, capim-santo, capim-cidró, erva-cidreira, limonaria and limocillo (TRAMIL 2014; Bermúdez and Velázquez 2002; Almeida et al. 2010; Miranda et al. 2011; Zucchi et al. 2013; Toscano 2006). It has also been recognized for its medicinal use to calm the nerves and to treat gastrointestinal problems (poor digestion and stomach pain), fever, headache, tonsillitis and sores (Bermúdez and Velázquez 2002, Toscano 2006; Zank and Hanazaki 2012; Albuquerque 2006).

There is a demand for *C. citratus* in the international market for the compounds in its essential oil, which is used in the pharmaceutical, food and perfumery industries. Major consumers are the United States, Japan, Canada, Switzerland, Great Britain and France (Department of Agriculture, Forestry and Fisheries 2009). In South America, Brazil is one of the producers of *C. citratus*, where Paraná State is the largest producer (Secretary of State of Agriculture and Supply 2002). In Brazil, plants are used for the same purposes as at the international level, i.e. for the extraction of the essential oil and as its dry mass for the production of tea. However, production in Brazil is only for the regional and domestic market (Gomes 2003).

1.2 Classification and Synonyms

The genus *Cymbopogon* comprises 30 species native to the Old World and has a broad distribution (Stevens et al. 2001).

The species *C. citratus* has eight synonyms (*Andropogon cerifer* Hack, *Andropogon ceriferus* Hack, *Andropogon citratus* DC, *Andropogon citriodorus* Desf., *Andropogon fragrans* C. Cordem, *Andropogon nardus* subsp. *ceriferus* (Hack.) Hack, *Andropogon nardus* var. *ceriferus* (Hack.) Hack and *Andropogon roxburghii* Nees ex Steud) (Tropics, Missouri Botanical Garden 2015; The Plant List 2015).

1.3 Origin and Distribution

This species of tropical Asian origin is a widely cultivated in the tropics. It is not known to grow in the wild. The plant are known to bare flowers rarely and have a strong lemon-like smell (Stevens et al. 2001).

1.4 Soil Requirements

Producers of *C. citratus* in Brazil recommend a soil pH of approximately 5.5 (Gomes 2003). In Colombia, it has been documented that this plant is resistant to acidity (Chemonics Foundation Colombia 2003), and the Department of Agriculture of South Africa recommends that plantations of C. *citratus* maintain a soil pH of 5.0–8.4. More alkaline soils are associated with greater quantities of citral in the oil, without leaving aside the good soil drainage required (Department of Agriculture, Forestry and Fisheries 2009).

1.5 Climatic Requirements

Tropical and subtropical climates with abundant rainfall (2000 mm or more) are optimal for growing *C. citratus* because the leaves are sensitive to frost, in cold climates (Castro and Ramos 2002; Chemonics Foundation Colombia 2003). In Colombia, crops have been reported to grow in temperatures between 20 and 32 °C at an altitude of 0–1500 m above sea level (Chemonics Foundation Colombia 2003). In Cuba, Soto et al. (2002) report that the greatest root growth occurs when the soil temperature ranges between 21 and 23 °C and that development slows below 21 °C during the months of December, January and February.

1.6 Leaf Production

It has been reported that in the 1st year of harvest, 10,000 kg/ha of green mass can be obtained, reduced to 60% as dry weight (the yield of essential oil varies from 0.4% to 0.6% of green mass) (Castro and Ramos 2002). In their study in Cuba, Soto et al. (2002) indicate that under conditions of fertilization and irrigation and depending on the number of cuts, the yield of green mass ranges from 50 to 60 t/ha/year. The producers of Paraná in Brazil report a production between 7.5 and 19 t/ha/year (Gomes 2003). *C. citratus* plantations can be maintained in economic production between 4 and 5 years, after which the production decreases, so the renewal of the plantation is recommended (Gomes 2003; Chemonics Foundation, Colombia 2003).

1.7 Cultivars

A study conducted at the Central Institute of Medicinal and Aromatic Plants of India confirmed the production of ten varieties of lemon grass (Pragati, Krishna, Cauvery, Nima, YEL-1 and LMH-4 of *Cymbopogon flexuosus*, Praman of *Cymbopogon pendulus*, T-1 of *C. citratus*, and CIMAP Suwarana and parent-1 of *Cymbopogon khasianus*) and recognizes the cultivar 'T-1' of *C. citratus.* As a result of genetic selection, this cultivar differs from the species *C. citratus* and guarantees an increased yield in the production of the essential oil (Lal 2012).

2 Part II: Cultivation Practices

2.1 Propagation

C. citratus flowers rarely and propagates through propagules (Chemonics Foundation Colombia 2003; Castro and Ramos 2002). Soto et al. (2002) recommend that propagules come from seed banks that have not been cut during a period from 10 months up to 1 year. A propagule must also produce between 40 and 70 useful shoots at 10/12 months after being planted. After obtaining the shoots, packets of 100 units were made, tied gently and placed vertically in a cool place until planting time. It is recommended leave the shoots in a place for 3–5 days where water is in constant circulation to stimulate the production of root primordia (Soto et al. 2002).

2.2 Soil Preparation

The preparation of the soil, according to Soto et al. (2002), must not be less than 40 days and must be carried out in the following sequence: (a) plowing, which consists of plowing the soil to a depth of 10–12 cm; (b) 12–15 cm average grade to break the thicker soil structures (lumps), eliminate undesirable grasses and split the crop residues to accelerate decomposition; (c) irrigating partially with 250 m^3/ha after grading, to accelerate the germination of undesirable herbs, favor the decomposition of the organic matter incorporated as green manure and restore soil fitness; (d) 15 days after irrigation, crossing by plowing the soil to a depth of 25–30 cm regardless of the soil type, to eliminate any undesirable weeds that have sprouted. This process leaves the weeds deeper so that the decomposition process continues, which increases the percentage of organic matter, and many of the seeds of undesirable grasses are moved deeper, preventing their germination; (e) immediately after performing the crossing, passing the Tiller perpendicularly, in such a way that the soil remains soft on the surface, avoiding excessive superficial pulverization that would be produced with another grading pass and thus counters wind erosion; and (f) after the cross is completed, furrowing at the same depth of the cross because, for *C. citratus,* a larger furrow ensures a greater yield of green mass in the plantation.

2.3 Planting

The best time to plant *C. citratus* are the months from March to May, allowing for the first harvest to occur after 9 months with a minimum yield of 18–22 t/ha of green mass and an essential oil concentration of 0.3–0.5% rich in citral, reaching a plantation height of 1.10–1.20 m due to the prevailing weather conditions during that period (Soto et al. 2002). In their study conducted in Brazil, Gomes (2003) observe that the producers of Paraná plant between August and November because the climatic situation is more suitable for propagule establishment during that time of the year. Castro and Ramos (2002) mention that in other places in Brazil, propagation is carried out from the end of August until October, and in warmer areas, propagation can be performed between March and April.

2.4 Fertilization

Cultivation of *C. citratus* requires a supply of nitrogen, phosphorus, potassium and organic matter to obtain good yields. Studies of fertilization recommend 100 kg/ha of nitrogen per year in two applications: the first at 2 months after sowing and the second after the harvest. Urea at 46% and 50 kg/ha of potassium and phosphorus per year applied as base fertilization before planting is recommended (Soto et al. 2002). Organic fertilization in the cultivation of *C. citratus* is demanding because fertilizer must be locally applied in the furrow before the planting at a dose of 20 t/ha. Organic fertilizers that can be used include *cachaça*, manure and others. The use of arbuscular mycorrhizal fungi was also studied by planting 10 g of commercial inoculum with 62% root colonization and strains *Glomus muscae* güira 8, or Fasaculatum-1, *G. amarillo* Topes 7 and *G. pelu* Topes-5, which increases crop yield from 3% to 10%. If the essential oil is to be used in the manufacture of drugs, the only authorized fertilizer is organic (Soto et al. 2002).

2.5 Irrigation

Soil humidity must be 85% until plants reach the tillering period, after which the humidity needs to be 80%. In the event the irrigation system is not able to maintain soil moisture, it is recommended to sow in the rainy season. Water deficit in the crop manifests in the leaves as accelerated necrosis in old leaves, starting at the apex and covering the total leaf area (Soto et al. 2002).

2.6 Weed Control

Depending on their abundance, some small-scale producers from Paraná in Brazil manually eliminate weeds (Gomes 2003). Soto et al. (2002) recommend the first elimination of weeds at 20 or 25 days after planting. Weeding with or without leaving the dead cover improves the production yield of the essential oil of organic crops (Lemos et al. 2013).

2.7 Pest Control

Studies from Colombia, Brazil and Cuba document that there are no pests that cause significant damage to the crop (Soto et al. 2002; Chemonics Foundation Colombia 2003; Gomes 2003). The presence of pests of *Chilotrea* larvae that perforate the stem and feed on the strands has been observed in crops from Southeast Asia (Department of Agriculture, Forestry and Fisheries 2009). The presence of nematodes that affect crops has also been reported, such as *Tylenchorhynchus vulgaris* (Stunt), *Rotylenchulus reniformis* (Reniform), *Helicotylenchus* (Spiral) spp. and *Pratylenchus* (Lesion) spp. To control them, organic fertilizer is recommended as well as sunning the soil for a few days so that the heat from the sun kills the nematodes. Marigolds can also be used, and only as a last resort should chemical control be used (Department of Agriculture, Forestry and Fisheries 2009).

2.8 Disease Control

The Department of Agriculture of South Africa reports four diseases in *C. citratus* crops (Table 1) (Department of Agriculture, Forestry and Fisheries 2009).

In commercial crops of *C. citratus* in India, the presence of rust disease generated by the fungus *Puccinia nakanishikii* has been documented, which caused major losses in the green mass and the essential oil extracted (Boruah et al. 1995). In a study in Venezuela, Antolinez et al. (2008) report the presence of *Puccinia sp.* in crops; however, this presence does not affect the quality and quantity of essential oil extracted. In Brazil, *Puccinia cymbopogonis* was reported for the first time in crops of Paraná (Vida et al. 2006), and another study has reported the presence of the species *Puccinia nakanishikii* in Brazil (Melo et al. 2010). The presence of these two species in Colombia has also been reported in crops of *C. citratus* (Álvarez and Salazar 2014). Fungicide is recommended for the control of this disease (Lorenzetti et al. 2012).

2.9 Harvesting

In Cuba, after 9–11 months, the crop is ready to be harvested. One of the characteristics of mature plants, which have an optimal amount of essential oil, is the yellowish brown color at the leaf apex (Soto et al. 2002). The producers of Paraná Brazil

Table 1 Diseases observed in *Cymbopogon citratus* (DC.) Stapf (Department of Agriculture, Forestry and Fisheries 2009)

Disease	Disease characteristics	Control
Long smut	Inflorescences are thin tubular with rust color cream that comes off at maturity from the tip and hang in pieces	Fumigation with fungicide before flowering. To prevent this disease, it is recommended to treat the seeds with fungicide
Red leaf spot	On the underside of the leaf are brown spots with concentric rings in the center	Application of fungicide
Leaf blight	Circular reddish-brown spots on the margins and the tips of the leaves; when the spots unite, they form an elongated reddish and brownish necrotic lesions which dry the leaves. Old leaves are most susceptible to infection	Application of fungicides
Rust	Brown linear uredinia on the underside of the leaves associated with chlorotic stripes	Application of fungicide

harvest between 6 and 8 months after planting, without the use of an agricultural calendar, collecting according to market demand, which can be up to five times a year (Gomes 2003). The harvest can be manual or mechanized depending on the size of the planted area. Mechanized harvesting is carried out with a silage harvester. Manual harvest is performed with a sickle or machete; the leaves are cut at the height of leaf overlap at 20–25 cm (Soto et al. 2002; Gomes 2003).

3 Part III: Post-Harvest Management

3.1 Part Harvested and Harvesting Techniques

The parts always harvested are the leaves, and the technique is the same regardless of the final product. What varies depending on the final product is the post-harvest handling. When the green mass is going to be used for essential oil extraction, it is better to transfer the leaves quickly after harvest because it ensures a better quality and yield of oil. When the transfer cannot be made immediately, it is recommended that the leaves be left in areas with low light, good ventilation and a surface that allows the leaves to spread and not form lumps that favor the appearance of microorganisms (Soto et al. 2002). Martinazzo et al. (2013) demonstrate that cutting the leaves in 2 cm fragments contributes to a better extraction of essential oil. When the final product to be marketed is natural, the green mass is subjected to a post-harvest drying process that is carried out naturally by the sun, naturally in the shade at room temperature or with a hot air dryer (Gomes 2003).

3.2 Packaging

The dry green plant mass is sold in double paper sacks with 15–50 kg, which may be smaller or larger depending on the buyer. The packages also have ingredient information, batch and origin identification, expiration date and content labels (Gomes 2003). The essential oil is marketed in dark glass bottles labeled with the same marketing information as the dry green mass (Gomes 2003; Department of Agriculture, Forestry and Fisheries 2009).

3.3 Storage

Leaves of *C. citratus* need to be firmly packed to avoid vapor channels. If the leaf is very large, it is recommended to cut it into pieces to ensure firm packaging (Department of Agriculture, Forestry and Fisheries 2009).

The essential oil of *C. citratus* needs to be stored in sealed dark glass bottles. Once opened, the bottle should be refrigerated. Deterioration is detected when the oil is darker or more viscous than normal (Department of Agriculture, Forestry and Fisheries 2009).

3.4 Marketing

The essential oil of *C. citratus* is marketed internationally for cosmetic, food and pharmaceutical purposes, and its price on the market varies depending on demand and foreign currency exchange. Countries with the highest demand are the United States for the soft drink industry; Japan and France for perfumery; Switzerland for pharmaceutical purposes; and Britain and India for the flavoring market (Department of Agriculture, Forestry and Fisheries 2009). Marketing of these products in Brazil has the same purpose as at the international level; however, at regional and national levels, the product produced in Paraná is purchased by industries in São Paulo (Gomes 2003).

4 Part IV: Utilization

4.1 Beauty

The essential oil of *C. citratus* (citral) is used in pharmaceuticals; perfumery for the fragrance of soaps and detergents; and in the cosmetic industry specifically for the synthesis of vitamin A and ionones (Dawson 1994; Gomes and Negrelle 2015).

4.2 Pharmaceutical and Therapeutic

C. citratus is used in different cities across the world for cough, elephantiasis, malaria, pneumonia and ophthalmic and vascular disorders (Poonpaiboonpipat et al. 2013) as well as for diarrhea, stomach pain, fever, flatulence, flu, cold and cough (TRAMIL 2014). Table 2 shows some of the biological activities of *C. citratus* that have been studied.

Studies have reported that 96.9% of the chemical compounds of the essential oil of *C. citratus* belong to the monoterpenes chemical group and 0.6% to sesquiterpenes (Table 3) (Kpoviessi et al. 2014). Other studies have also reported a high percentage of geranial and neral in the essential oil of *C. citratus* (Blanco et al. 2009; Bassolé et al. 2011). The combination of these two geometric isomer compounds constitutes citral: $C_{10}H_{16}O$.

Table 2 Biological activity of the essential oil of *Cymbopogon citratus* (DC.) Stapf

Extract	Activity	Organisms	References
Essential oil	Antibacterial	*Bacillus cereus*, Bacillus subtilis, Enterococcus faecalis, Escherichia coli, Klebsiella pneumoniae, Listeria monocytogenes, Pseudomonas aeruginosa, Salmonella enterica, Salmonella typhimurium, Shigella dysenteriae, Staphylococcus aureus, Staphylococcus mutans, Staphylococcus epidermis	Onawunmi et al. (1984), Naik et al. (2010), Bassolé et al. (2011), Almeida et al. (2013), and Lucena et al. (2015)
Essential oil	Antifungical activity	Aspergillus ochraceus, *Candida albicans*, *Candida tropicalis*, Candida glabrata, Penicillium expansum, Penicillium verrucosum	Nguefack et al. (2009), Tyagi and Malik (2010), and Almeida et al. (2013)
Essential oil	Antiprotozoal activity	Trypanosoma brucei brucei	Kpoviessi et al. (2014)
Essential oil	Antiprotozoal activity	Trypanosoma cruzi, Leishmania amazonensis	Santoro et al. (2007), Santi et al. (2009), and Rojas et al. (2012)
Essential oil	Hypertension	–	Moreira et al. (2010)
Essential oil	Anxiolytic hypnotic and anticonvulsant	–	Blanco et al. (2009)
Standardized hexanic extract of Cymbopogon citratus	Anti-allergic asthma	Mite Blomia tropicalis	Machado et al. (2015)
Extract rich in polyphenols	Antiinflammatory	–	Vera et al. (2013)

4.3 *Food and Flavoring*

The leaves and stems of lemon grass are consumed fresh in Asian cuisine (Department of Agriculture, Forestry and Fisheries 2009), and fresh or dried leaves are used to make tea (Martinazzo et al. 2013).

4.4 *Industrial*

The essential oil of *C. citratus* possesses repellent activity (Oyedele et al. 2002) and is used in industry for the manufacture of repellents for insects, candles and waxes (Department of Agriculture, Forestry and Fisheries 2009). This oil is also a component of organic pesticides (Gomes and Negrelle 2015).

Table 3 Chemical composition of the essential oil of *Cymbopogon citratus* (DC.) Stapf using the Kovats index (KI) on HP-5 MS columns (Kpoviessi et al. 2014)

Components	Mean ± Standard deviation
β-Pinene	10.0 ± 0.04
p-Cymene	0.5 ± 0.00
(Z)-β-ocimene	0.4 ± 0.00
(E)-β- ocimene	0.2 ± 0.00
α – Terpinolene	0.2 ± 0.00
Myrcenol	0.4 ± 0.00
β – Linalool	0.9 ± 0.00
trans-3(10)-caren-2-ol	0.1 ± 0.00
Cis-p-mentha-2,8-dienol	0.1 ± 0.00
α – Phellandren -8-ol	0.5 ± 0.00
β-Citronellol	0.4 ± 0.00
Neral	35.5 ± 0.15
cis-geraniol	4.3 ± 0.02
p- Mentha-1(7), 8 (10)-dien-9-ol	0.1 ± 0.00
Geranial	39.5 ± 0.00
Nopol	0.4 ± 0.00
β- Bourbonene	0.5 ± 0.00
Geranyl acetate	1.0 ± 0.00
2-Undecanone	0.1 ± 0.00
β- Caryophyllene	0.2 ± 0.00
Neric acid	0.1 ± 0.00
Geranic acid	0.1 ± 0.00
τ-Gurjenene	0.1 ± 0.00
α-Bergamotene	0.1 ± 0.00
β-Caryophyllene oxide	0.1 ± 0.00
Eudesm-7(11)-en-4-ol	0.1 ± 0.00

4.5 Safety Data

One study of *C. citratus* shows that the essential oil is not toxic when taken as infusion tested on rats at a dose of 3.4 g/kg (Costa et al. 2011). However, caution needs to be taken with the concentrations consumed (Fandohan et al. 2008; Sinha et al. 2014). Regarding the quality of the products which contain *C. citratus* offered on the market, a study conducted of herbal products shows that these products do not have sufficient information on quality, and in addition, purity is very low (Melo et al. 2007).

Acknowledgments We are especially grateful to the National Institute of Science and Technology in Ethnobiology, Bioprospecting and Nature Conservation, certified by CNPq, with financial support from FACEPE (Foundation for the Support of Science and Technology of the State of Pernambuco).

References

Albuquerque UP (2006) Re-examining hypotheses concerning the use and knowledge of medicinal plants: a study in the Caatinga vegetation of NE Brazil. J Ethnobiol Ethnomed 2(30). https://doi.org/10.1186/1746-4269-2-30.

Almeida CFCBR, Ramos MA, Amorim ELC et al (2010) A comparison of knowledge about medicinal plants for three rural communities in the semi-arid region of northeast of Brazil. J Ethnopharmacol 127(3):674–684

Almeida RBA, Akisue G, Cardoso LML, Junqueira JC, Jorge AOC (2013). Antimicrobial activity of the esencial oil of *Cymbopogon citratus* (DC) Stapf. on *Staphylococcus spp., Streptococcus mutas* and *Candida Spp*. rev. Brav. Planta Med 15(4):474–482

Álvarez ML, Salazar YM (2014) Caracterización morfológica de las royas (Pucciniales) que afectan el limoncillo (*Cymbopogon citratus* (DC.) Stapf) en Colombia. Bioagro 26(3):171–176

Antolinez GJC, de Colmenares NG, Usubillaga A, Darghan E, Linares S (2008) Evalución de variables agronómicas em el cultivo de Limonaria *(Cymbopogon citratus* Stapf) para la producción de aceite essencial. Interciencia 33(9):693–699

Bassolé IHN, Lamien-Meda A, Bayala B et al (2011) Chemical composition and antimicrobial activity of *Cymbopogon citratus* and Cymb*opogon giganteus* essential oils alone and in combination. Phytomedicine 18:1070–1074

Bermúdez A, Velázquez D (2002) Etnobotánica médica de una comunidad campesina del estado de Trujillos, Venezuela: un estudio preliminar usando técnicas cuantitativas. Rev Fac Farm 44:2–6

Blanco MM, Costa CARA, Freire AO et al (2009) Short Communication. Neubehavioral effect of essential oil of *Cymbopogon citratus* in mice. Phytomedicine 16:265–270

Boruah P, Misra B, Pathac M, Ghosh A (1995) Dynamics of essential oil of *Cymbopogon citratus* (DC) Stapf under rust disease indices. J Essent Oil Res 7:337–338

Castro LO, Ramos RLD (2002) Principais gramíneas produtoras de óleos essenciais: *Cymbopogon citratus* (DC) Stapf., capim-cidró, *Cymbopogon martinii* (Rox.) J.F. Watson, palma-rosa, *Cymbopogon nardus* (L.) Rendle, citronela, *Elyonurus candidus* (Trin.) Hack., capim-limão, *Vetiveria zizanioides* (L.) Nash, vetiver. Porto Alegre, FEPAGRO. 31p

Costa CARA, Bidinotto LT, Takahira RK, Salvadori DMF, Barbisan LF, Costa M (2011) Cholesterol reduction and lack of genotoxic or toxic effects in mice after repeated 21-day oral intake of lemongrass (*Cymbopogon citratus*) essencial oil. Food Chem Toxicol 49:2268–2272

Dawson FA (1994) The amazing terpenes. Nav Stores Rev:6–12

Departament of Agricultura Forestry and Fisheries (2009) Essential oil crops, production guidelines for lemongrass. Lemograss production. Departament: Agricultura and Fisheries, Republic of South Africa 19 p

Fandohan P, Gnonlonfin B, Laleye A, Gbenou JD, Darboux R, Moudachirou M (2008) Toxicity and gastric tolerance of essential oils from *Cymbopogon citratus*, *Ocimum gratissimum* and *Ocimum basilicum* in Wistar rats. Food Chem Toxicol 46:2493–2497

Fundación Chenomics Colombia. Manual de fitoprotección y análisis de plaguicidad (estrído de PERSUAP) Cultivo (2003) Plantas medicinales y aromáticas (Curcuma- *Curcuma longa*, estevia- *Stevia rebaudiana*, jengibre- *Zingiber officinale*, anamú- *Petiveria alliacea*, limonaria – *Cymbopopgon citratus*, ruda –*Ruta gravealens*). Colomb Alternat Dev (CAD) Proj:75

Gomes EC (2003) Capim-Limão- Cymbopogon citratus (D.C.) Stapf: subsídios para melhoria de qualidade do cultivo, industrialização e comercialização no estado do Paraná [dessertação]. Universidade Federal do Paraná.

Gomes EC, Negrelle RRB (2015) Análises da cadeia produtiva do capim limão: estudo de caso. Rev Bras Pl Med 17(2):201–209

Kpoviessi S, Bero J, Agbani P et al (2014) Chemical composition cytoxicity and in vitro antitrypanosomal and antiplasmodial activity of the essencial oils of four Cymbopogon species from Benin. J Ethnopharmacol 15:1652–1659

Lal RK (2012) Stability for oil yield and variety recommendations' using AMMI (additive main effects and multiplicative interactions) model in Lemongrass (Cymbopogon species). Ind Crops Prod 40:296–301

Lemos GCS, Santos AD, Freitas SP, Gravina GA (2013) Controle de plantas invasoras em cultivo orgánico e convencional de capim-limão (*Cymbopogon citratus* (DC) Stapf.). Rev Bras Pl med 15(3):405–414

Lorenzetti ER, Conceição DM, Sacramento LVS, Furtado EL (2012) Controle da ferrugem do capim-limão (*Cymbopogon citratus* (DC.) Stapf) com produtos naturais. Rev Bras Pl Med 14(4):571–578

Lucena FFB, Tintino RS, Figueredo GF, Oliveira DMCD, Aguiar DSJJ, Cardoso DNE et al (2015) Avalição da atividade antibacteriana e moduladora de aminoglicosideos do óleo esencial de Cymbopogon citratus (DC.) Stapf. Acta biol Colomb 20(1):39–45

Machado MSS, Silva HBF, Rios R, de Oliveira AP, Carneiro NVQ, Costa RS et al (2015) The anti-allergic activity of *Cymbopogon citratus* is mediated via inhibition of nuclear factor kappa B (Nf-Hb) activation. BMC Complement Altern Med 15:168. https://doi.org/10.1186/s12906-015-0702-8

Martinazzo AP, De Castro ME, Demuner AJ, Amorim BP (2013) Avaliação do óleo essencial folhas de *Cymbopogon citratus* (DC.) Stapf após o processo de secagem. Bol Latinoam Caribe Plant Med Aromat 12(5):523–536

Melo JG, Járisson Martins DGR, Amorim ELC, Albuquerque UP (2007) Qualidade de produtos a base de plantas medicinais comercializados no Brasil: castanha-da-índia (*Aesculus hippocastanum* L.), capim-limão (*Cymbopogon citratus* (DC.) Stapf) e centelha (*Centella asiática* (L.) Urban). Acta Bot Bras 21(1):27–36

Melo MP, Araújo JSP, Carvalho Junior AA, Tostes GO, Arêas MS (2010) *Puccinia nakanishikii*, nova ocorrência de ferrugem em capim-limão (*Cymbopogon citratus*) no Brasil. Short comunication. Tripical Plant Pathol 35(2):129–130

Miranda TM, Hanazaki N, Govone JS et al (2011) Is there effective resources utilization among Cardoso Island population ("caiçaras"), São Paulo State, Brazil? Rodriguesia 62(1):153–169

Moreira FV, Bastos JFA, Blank AF, Alves PB, Santos MR (2010) Chemical composition and cardiovascular effects induced by the essencial of *Cymbopogon citratus* DC. Stapf, Poaceae, in rats. Braz J Pharmacog 20(6):904–909

Naik MI, Ahmad FB, Jaykumar E et al (2010) Antibacterial activity of lemongrass (*Cymbopogon citratus*) oil against some selected pathogenic bacterias. Asian Pac J Trop Med:535–538

Nguefack J, Lekagne Dongmo LB, Dakole CD (2009) et la. Food preservative potential of essential oils and fractions from *Cymbopogon citratus*, *Ocimum gratissimum* and Thymus vulgaris against mycotoxigenic fungi. Int J F Microbiol 131:151–156

Onawunmi GO, Yisak W, Ogunlana EO (1984) Antibacterial constituents in the essential oil of *Cymbopogon citratus* (DC.) Stapf. J Ethnopharmacol 12:274–286

Oyedele AO, Gbolade AA, Sosan MB, Adewoyin FB, Soyelu OL, Orafidiya OO (2002) Formulation of an effective mosquito-repellent topical product from Lemongrass oil. Phytomedicine 9(3):259–262

Poonpaiboonpipat T, Pagnakorn U, Suvunnamek U et al (2013) Phytotoxic effects of essential oil from *Cymbopogon citratus* and its physiological mechanisms on barnyardgrass (Echinochloa crus-galli). Ind Crop Prod 41:403–407

Rojas J, Roceros S, Palacios O, Sevilla C et al (2012) Efecto anti-Trypanosoma cruzi del aceite esencial de *Cymbopogon citratus* (DC) Stapf (hierba luisa) en ratones Balb/c. An Fac Med 73(1):7–12

Santi RM, Dos Santos AO, Nakamura CV et al (2009) In vitro activity of the essential oil of *Cymbopogon citratus* and its major component (citral) on Leishmania amazonensis. Parasitol Res 105:1489–1496

Santoro GF, Cardoso MG, Guimarães LGL et al (2007) Anti-proliferative effect of the essential oil of *Cymbopogon citratus* (DC) Stapf (lemongrass) on intracellular amastigotes, bloodstream trypomastigotes and culture epimastigotes of *Trypanosoma cruzi* (Protozoa: Kinetoplastida). Parasitology 134:1649–1656

Secretaria de Estado da Agricultura e do Abastecimento. Departamento de Economia Rural (2002) Levantamento do valor bruto da produção agropecuária. produtos especiais: safra 2000/2001. PARANÁ Curitiba

Sinha S, Jothiramajayam M, Ghosh M, Mukherjee A (2014) Evaluation of toxicity of essential oils palmorosa, citronela, lemongrass and vetiver in human lymphocytes. Food Chem Toxicol 68:71–77

Soto OR, Vega M, Tamajón AL (2002) Instructivo técnico del cultivo de *Cymbopogon citratus* (D.C) Stapf (caña santa). Rev Cubana Med 7(2):89–95

Stevens WD, Ulloa C, Pool A, Montiel OM (2001) Flora de Nicaragua. Monogr Syst Bot Missouri Bot Gard 85:i–xlii

The Plant list.org [Internet] (2015) Updated 2013; cited 2015 May 15. Available from: http://www.theplantlist.org

Toscano JYG (2006) Uso tradicional de plantas medicinales en la vereda San Isidro, municipio de San José de Pare-Boyacá: un estudio preliminar usando técnicas cuantitativas. Acta Biol Colomb 11(2):1–10

TRAMIL (2014) Farmacopea vegetal caribeña. tercera edición. CICY, México 400 p

Tropicos, Missouri Botanical Garden.org [Internet] (2015) Update 2015 January 25; cited 2015 May 15. Available from: http://www.tropicos.org

Tyagi AK, Malik A (2010) Liquid and vapour-phase antifungal activities of selected essential oils against Candida albicans: microscopic observations and chemical characterization of *Cymbopon citratus*. BMC Complement Altern Med 10:65

Vera F, Costa G, Figueirinha A, Marques C, Pereira P, Neves BM et al (2013) Anti-inflammatory activity of *Cymbopogon citratus* leaves infusión via proteasome and nuclear factor-kB pathway inhibition: contribution of chlorogenic acid. J Ethnopharmacol 148(1):126–134

Vida JB, Carvalho Junior AA, Verzignassi JR (2006) Primeira ocorrência de ferrugem em capim-limão causada por *Puccinia cymbopogonis* no Brasil. Summa Phytopathol 32(1):86–91

Zank S, Hanazaki N (2012) Exploring the links between ethnobotany, local therapeutic practices, and protected areas in Santa Catarina Coastline, Brazil. Evid Based Complement Alternat Med. https://doi.org/10.1155/2012/563570

Zucchi MR, Oliveira Júnior VF, Gussoni MA et al (2013) Levantamento etnobotânico de plantas medicinais na cidade de Ipameri – GO. Rev Bras Pl Med 15(2):273–279

Dysphania ambrosioides (L.) Mosyakin & Clemants

Julio Alberto Hurrell

Dysphania ambrosioides (L.) Mosyakin & Clemants
Photo: David G. Smith
Available in: http://www.delawarewildflowers.org/plant.php?id=0478

J. A. Hurrell (✉)
Laboratorio de Etnobotánica y Botánica Aplicada (LEBA), Facultad de Ciencias Naturales y Museo, Universidad Nacional de La Plata, La Plata, Argentina

Consejo Nacional de Investigaciones Científicas y Técnicas, Buenos Aires República, Argentina

U. P. Albuquerque et al. (eds.), *Medicinal and Aromatic Plants of South America*, Medicinal and Aromatic Plants of the World 5,
https://doi.org/10.1007/978-94-024-1552-0_17

Abstract *Dysphania ambrosioides* (L.) Mosyakin & Clemants (= *Chenopodium ambrosioides* L.) is an American aromatic species used for medicinal and culinary purposes, since pre-Columbian times by Aztecs and Mayans in Mesoamerica (where is called *epazote*) and Andean communities and many others in South America (where it is known as *paico*). Currently, it is globally known by a wide diversity of cultures around the world, due to its cultivation and naturalization. Its uses are currently widespread in pluricultural contexts, by means of the commercial circuits and mass media, especially the Internet. The main active constituents of the plant are essential oils, to which it owes its aroma and flavor. It is toxic in high doses, but safe when consumed in appropriated concentrations. The most widespread folk therapeutic use is as antiparasitic (anthelmintic, antimicrobial), and it is also employed against gastrointestinal disorders, as hypotensive, antipyretic, vulnerary, analgesic, anti-inflammatory, antitumor, sedative and anxiolytic, among others, many of which have been evaluated scientifically. Recent research results on its anticancer activity are very promising.

Keywords *Dysphania ambrosioides* · Chenopodiaceae · Paico · Epazote · Food and medicinal uses

1 Taxonomic Characteristics

Dysphania ambrosioides (L.) Mosyakin & Clemants (= *Chenopodium ambrosioides* L.) is an aromatic plant used in America since pre-Hispanic times for medicinal purposes, mainly as anthelmintic, and it is also widespread in different local culinary traditions as food condiment and beverage flavoring. In Mexico and Central America, this species is called epazote (from Náhuatl *epatl*, 'stench', 'skunk', and *tzotl*, 'sweat', 'grime', referring to the unpleasant aroma of its leaves. In South America it is commonly known as paico (from *páykko* o *payqu*, Quechua name of this plants). The name Guaraní is *ka'arẽ* (from *ka'a*, 'planta' and *-arẽ*, one Tupi Guaraní tribe). In Spanish, it is also known as: hierba de Santa María, quenopodio, té de los jesuitas, té de México, among others. In English: wormseed and Mewican tea (Pinedo et al. 1997; Mejía and Rengifo 2000; Barboza et al. 2009; Hurrell et al. 2011).

The genus *Dysphania* R.Br. is now accepted in an expanded circumscription (Mosyakin and Clemants 2002), including taxa previously treated in *Chenopodium* L. subg. *Ambrosia* A. J. Scott, or segregated in genera such as *Roubieva* Moq., *Teloxys* Moq., and *Neobotrydium* Moldenke. The most indicative trait of *Dysphania* is the presence of glandular hairs, glands and (or) simple hairs on the stem, leaves or perianth which often impart an aromatic smell to the plant. In its strict sense, this genus included the only species 7–10 species from Australia, *Dysphania* sensu lato includes about 32–40 species worldwide, from tropics to warm-temperate zones (Clemants and Mosyakin 2003; Zhu et al. 2003; Sukhorukov and Zhang 2013).

This genus is usually placed in the Family Chenopodiaceae Vent. (Kühn 1993; Giusti 1997; Clemants and Mosyakin 2003; Kadereit et al. 2003; Zhu et al. 2003). However, other authors consider that Chenopodiaceae and Amaranthaceae Juss. should be considered together (Chenopodiaceae-Amaranthaceae alliance), as a single family under the name Amaranthaceae, for being the oldest (Judd et al. 2002; Pratt 2003; Culham 2007). Morphological, molecular and phylogenetic evidence supports both positions according to the interpretations, so the issue is still controversial. The Amaranthaceae sensu stricto, with ca. 69 genera and 1000 species, are most diverse in the tropics. Meanwhile, the Chenopodiaceae included ca. 100 genera and 1400 species, and are most diverse in temperate regions (Pratt 2003).

Synonyms *Ambrina ambrosioides* (L.) Spach, *A. parvula* Phil., *A. spathulata* Moq., *Atriplex ambrosioides* (L.) Crantz, *Blitum ambrosioides* (L.) Beck, *Chenopodium ambrosioides* L., *C. ambrosioides* L. var. *suffruticosum* (Willd.) Graebn., *C. anthelminticum* L., *C. spathulatum* (Moq.) Sieber ex Moq., *C. suffruticosum* Willd., *Teloxys ambrosioides* (L.) W.A. Weber.

2 Crude Drug: The Crude Drug Used

The drug consists of its dried aerial parts: leaves, stems, inflorescences, and fruits (*Herba Chenopodii ambrosioides*), used to make therapeutic preparations. Sometimes the root is also used in rural areas. Fresh leaves are consumed as condiment and infusion-flavoring. The dried leaves are used less often, mostly in feeding and mostly for therapeutic purposes.

The dried aerial parts are consumed mostly in infusions or decoctions: 30 g per liter of water in adults, four cups per day (4–5 g per cup in children), also in tincture: 20 g in 100 cc of 70° alcohol, a teaspoon diluted in water, tea or mate, in fasting and before lunch and dinner (Burgstaller 1968).

In appropriated doses its consumption is safe, but in high doses, it causes various disorders and death (Duke et al. 2002; Gadano et al. 2006; Monzote et al. 2009). It is not indicated during pregnancy and lactation, for children up to 3 years old and adult patients debilitated or with hepatic, renal, and hearing diseases. The essential oil (*Oleum Chenopodii*) is included in different editions of the pharmacopoeias of various countries of the World, as Argentina, Brazil, Mexico, United States, France, Italy, Portugal, Spain, India, Turkey, and Vietnam (Alonso and Desmarchelier 2005; Hurrell et al. 2008, 2011).

In pluricultural contexts, dried aerial parts are commercialized bulk or packaged, both in traditional markets in urban areas (Macía et al. 2005; Pochettino et al. 2012), as well as in herb shops and health food stores (Hurrell et al. 2011).

The fragmented dry plant material that is marketed as a herbal product in Argentina is sometimes adulterated or substituted by *Dysphania multifida* (L.) Mosyakin & Clemants (= *Chenopodium multifidum* L.), from Bolivia, Chile, Argentina and Uruguay, which can be distinguished by morpho-histological characteristics of its trichomes, epidermis, mesophyll, leaf margin, and stem growth type (Bonzani et al. 2003).

3 Major Chemical Constituents and Bioactive Compounds

The essential oil of paico or epazote is responsible for its aroma and most of its therapeutic properties. The leaves, stems and inflorescences containing up to 0.35% essential oil; the fruits contain between 0.6% and 3%. The essential oil is a colorless or slightly yellow liquid, not very viscous, with sharp and pungent camphor-like odor, and a slightly bitter taste. It is extracted from the whole plant, especially seeds and fruits, by steam distillation (Gadano et al. 2006).

The ascaridole is the main component (42–90% of the essence). Its concentration varies with the season of collection, temperature and humidity (Alonso and Desmarchelier 2005; Dembitskya et al. 2008; Gómez Castellanos 2008). Also contains aritasone, camphor, β-carophyllene, p-cimol, p-cymene, n-docosane, geraniol, γ-gurjunene, n-hentriacontane, n-heptacosane, limonene, myrcene, n-octacosane, phellandrene, α- and β-pinene, pinocarvone, safrol, spinasterol, α-terpinene, α- and γ-terpineol, terpynil-acetate, terpynil-salicylate, thymol, triacontyl-alcohol, among others (Alonso and Desmarchelier 2005; Potawale et al. 2008; Barboza et al. 2009; Kokanova-Nedialkova et al. 2009; Alitonou et al. 2012; Zhu et al. 2012).

Also contains saponins (entire plant), organic acids (butyric, citric, ferulic, malic, succinic, tartaric, vanillic), tannins (aerial parts), anethole, kaempferol, quercetin, santonin (fruits), betain, *chenopodioside*s, heterosides (roots), among others (Pinedo et al. 1997; Alonso and Desmarchelier 2005; Kokanova-Nedialkova et al. 2009; Okhale et al. 2012).

4 Morphological Description

D. ambrosioides is a strongly scented annual or biennial herb, 30–80 cm tall, stems erect to ascending, much branched, striated, ± glandular-pubescent. Leaves alternate, sessile (distal) to petiolate, petiole to 18 mm long; blade ovate-elliptic, oblong-elliptic to elliptic, the upper ones gradually reduced, 2–8(−15) cm long × 0.5–4(−5.5) cm wide, apex acute to acuminate, margins entire, sparsely and irregularly coarsely dentate, base cuneate or attenuate, abaxially with scattered glands, slightly hairy around veins, adaxially subglabrous. Inflorescences in axillary glomerules, globose, 1.5–2.3 mm diam, with three to five flowers, gathered in terminal spikelike arrays; bracts absent but glomerules often subtended by reduced leaves ('leaflike bracts'), elliptic, spatulate, or linear, 0.3–2.5 cm long. Perianth segments (3−) 4–5, membranous, connate for ca. 1/2 their length, segments ovate, 0.7–1 mm long, apex obtuse, glandular-pubescent, persistent in maturity. Stamens 4–5, anthers ca. 0.5 mm long. Ovary superior, 1-locular, 1-ovulate; stigmas 3(−4), filiform, exserted from perianth. Fruit utricule, enclosed in the perianth, ovoid to depressed globose, pericarp membranous, non-adherent, rugose to smooth. Seeds in a horizontal position, lenticular, 0.6–1 mm long, ×0.4–0.5 mm wide, glabrous, black to dark reddish brown. $2n = 32$ (Grozeva and Stoeva 2006).

5 Geographical Distribution

This species is distributed in warm and warm-temperate America, from the southern United States and Mexico to austral South America: Brazil, Paraguay, Chile, Uruguay, and Argentina. It was introduced in Spain in the sixteenth century and spread under cultivation since the seventeenth century in Eurasia, and since the nineteenth century in the United States. At present, it is wide naturalized in tropical, subtropical, and warm-temperate regions around the world (Uotila 1990; Giusti 1997; Clemants and Mosyakin 2003; Zhu et al. 2003; Randall 2005).

6 Ecological Requirements

D. ambrosioides usually grows from sea-level up to about 3000 m altitudes, in disturbed soils, waste areas, embankments, roadsides, edges of ditches, orchards and gardens, rivers and dry lake beds, sandy soils, and nitrophilous grasslands. It is a secondary weed of field-crops and fruit tree orchards. It has a long flowering and fruiting period, from between spring and autumn (Giusti 1997; Pinedo et al. 1997; Clemants and Mosyakin 2003; Hurrell et al. 2008).

7 Collection Practice

Collection is done in wild and cultivated specimens. Occasionally, some people protect plants growing near their homes. It is grown from seed mainly in spring, preferably in shaded locations in the tropics, in sandy-loamy, fertile, and well-drained soils. Tolerates shade, but in full sun acquires a loose and wispy habitus. Germination occurs within 7–10 days after seeding. When cultivated, 10–12 cm tall seedlings are transplanted, at the age of 30 days. Its has a 9 months' vegetation period. Leaf harvest begins 80 days after seeding and subsequent cuts are made at 30 days intervals, at height of 10 or 12 cm from soil surface in order to facilitate re-growth. The harvested parts should preferably be dried under shade for conservation (Pinedo et al. 1997).

When the crop is destined for seed-production, it should be harvested just before the apexes turn brown. The plants are cut and left to dry, after which the grains are separated and cleaned using sieves. For essential oil harvest takes place when most of the seeds have turned dark: all aerial parts of the plant are cut and subjected to steam distillation. The essential oil yield is about 0.02% of dried matter (Alonso and Desmarchelier 2005).

8 Traditional Use (Part(s) Used) and Common Knowledge

D. ambrosioides have a long history of utilization as an aromatic plant in America, both in folk medicine as gastronomy. It has been a medicinal plant used as traditional anthelmintic by Aztecs and Mayans (Kliks 1985). The first recorded use as parasitic, antidysenteric and anti-inflammatory in Mexico corresponds to the Spanish physician Francisco Hernández (1517–1578) by the end of the sixteenth century, published in 1651 (Micheli-Serra 2001; Carballo et al. 2005). For the Andean region its medicinal use was reported by the Spanish chronicler Bernabé Cobo (1582–1657, who comments in 1654 (*History of the New World*) that it was used as emollient in patches, and its decoction against gout in topical use (Alonso and Desmarchelier 2005).

The infusions and decoctions are widespread as a vermifuge traditional remedy in Latin America and the Caribbean. Until the early decades of twentieth century was one of the most anthelmintic used in ethnomedicine and ethnoveterinary. Towards the 1940s its use declined with the discovery of less toxic products (Quinlan et al. 2002; Gómez Castellanos 2008).

Among the most widespread popular uses in America are found: vermifuge, tonic, digestive, stomachic, antispasmodic, to cure the *empacho* (severe indigestion), anti-ulcers, appetizer, hepatic, carminative, laxative, antidiarrheal, antidysenteric, antiemetic, antihaemorrhoidal, antidiabetic, antipyretic, pectoral, anticatarrhal, antitussive, antiasthmatic, anti-tuberculosis, hypotensive, haemostatic, anti-inflammatory, antiarthritic, antirheumatic, analgesic, diuretic, antiseptic, emollient, antitumor, vulnerary, to treat skin diseases and urinary infections, anti-asthenia, nervous affections, sedative, mnemonic, emmenagogue, against uterine fibroids and haemorrhaging, contraceptive, abortifacient, anti-head lice, insecticide (Hieronymus 1882; Conway and Slocumb 1979; Kliks 1985; Hurrell 1991; Berlin et al. 1996; Barrett and Kiefer 1997; Pinedo et al. 1997; Heinrich et al. 1998; Ruffa et al. 2002; González Torres 2005; Gupta 2006; Adams et al. 2007; Mendes and Carlini 2007; Yadav et al. 2007; Potawale et al. 2008; Volpato et al. 2009; Mejía and Rengifo 2000; Hurrell et al. 2011).

Some uses mainly registered in the Old World: galactogogue, hypoglycaemic, anti-headache in Morocco (Bnouham et al. 2002; Abouri et al. 2012; Montanari 2014), anxiolytic, antiepileptic, and hypnotic in Cameroon (Bum et al. 2011), to treat oedema in Nigeria (Kayode et al. 2008), against Cryptococcal meningitis and Herpes simplex in Tanzania (Kisangau et al. 2007), mosquito repellent with antimalarial applications in South Africa (Maharaj et al. 2010), against toothache in India (Kala 2005), anti-scabies in Philippines (Balangcod and Balangcod 2011).

Regarding its culinary uses, the paico or epazote is used since pre-Hispanic times as condiment for soups, stews, sauces, salads, *tamales* (corn dough stuffed with meat, cheese, vegetables and condiments, which is steamed or boiled in a leaf wrapper), *chupes* (stew generally made with chicken, meat, fish or shellfish, and vegetables), and many bean dishes, because of its carminative effect, also for flavouring infusions and boiled milk. The leaves can be consumed like potherb (Horkheimer 1973; Pinedo et al. 1997; Pöll 2005; Ulloa 2006; Hurrell et al. 2008).

9 Modern Medicine Based on Its Traditional Medicine Uses

The folk use as antiparasitic has been well studied, being ascaridole the mainly responsible, but not the only (MacDonald et al. 2004; Kokanova-Nedialkova et al. 2009). Studies with parasitized patients tested its anthelmintic effect against nematodes (Kliks 1985; Giove Nakazawa 1996; Navone et al. 2014). Same results were obtained in goats and lambs (Kato et al. 2000; Ketzis et al. 2002). Its action against *Schistosoma mansoni* (trematode that causes schistosomiasis) was checked in infected mice (Kamel et al. 2011). The anthelmintic action was also evaluated in vitro (Eguale and Giday 2009; Wabo Poné et al. 2011).

D. ambrosioides essential oil exhibited in vitro and in vivo antifungal activity (Kishore et al. 1996; Kumar et al. 2007; Goka Chekem et al. 2010; Shah 2014), antileishmanial activity in vivo (Monzote et al. 2014), anti-*Entamoeba histolytica* in vitro and in vivo (Ávila-Blanco et al. 2014). Monoterpene hydroperoxides from aerial parts showed activity in vitro against *Trypanosoma cruzi*, etiologic agent of Chagas disease (Kiuchi et al. 2002). The antimalarial activity was tested: the ascaridole found to be a potent inhibitor on the growth of *Plasmodium falciparum* (Pollack et al. 1990), and anti-*Plasmodium berghei* (Misra et al. 1991). The essential oil showed a promising activity against *Trichomonas vaginalis* that parasitizes the urogenital tract of both men and women (Kokanova-Nedialkova et al. 2009). Antibacterial activities were studied, including anti-*Helicobacter pylori* (cause of gastritis and ulcer), against *Mycobacterium tuberculosis* and skin pathogen bacteria (Lall and Meyer 1999; Larhsini et al. 2001; Liu et al. 2013; Shah 2014).

Its antiviral activity against influenza type A has been tested (Kokanova-Nedialkova et al. 2009).

Insect repellent, insecticide and acaricidal effects were studied, including human head lice and mosquitoes that transmit malaria (Chiasson et al. 2004; Gillij et al. 2008; Fekadu et al. 2009; Toloza et al. 2010; Zhu et al. 2012; Bigoga et al. 2013).

Regarding its uses in treating gastrointestinal disorders, some activities have been evaluated: antispasmodic (Toso and Boeris 2010), antidiarrheal, antidysenteric (Velázquez et al. 2006), digestive, against indigestion, and laxative (Florian et al. 2013). Regarding its folk use as abortifacient and contraceptive, the aqueous extract did not promote maternal or fetal toxicity, nor did it impair reproductive performance and fertility in rats (Medeiros et al. 2011).

D. ambrosioides also shows the following tested effects: antioxidant (Speiky et al. 2006; Kumar et al. 2007), immunomodulatory (Rossi-Bergmann et al. 1997), cardio-depressant, muscle relaxant (Alonso and Desmarchelier 2005), hypotensive (Assaidi et al. 2014), antipyretic (Hallal et al. 2010; Bum et al. 2011), vulnerary (Trivellato Grassi et al. 2013), anti-inflammatory (Ibironke and Ajiboye 2007; Trivellato Grassi et al. 2013), analgesic/antinociceptive (Okuyama et al. 1993; Amole and Yusuf 2002; Ibironke and Ajiboye 2007; Hallal et al. 2010; Trivellato Grassi et al. 2013), sedative (Okuyama et al. 1993), and anxiolytic (Bum et al. 2011). Its potential action on Alzheimer's disease treatment has been suggested (Carpinella et al. 2010).

Regarding its anticancer activity, this species exerts antitumor activity against different tumor cell lines studied in vitro and in vivo (Ruffa et al. 2002; Nascimento et al. 2006; Potawale et al. 2008; Kokanova-Nedialkova et al. 2009; Barros et al. 2013; Wu et al. 2013).

10 Conclusions

D. ambrosioides is utilized as therapeutic and as condiment in the New World since pre-Columbian times. Its early introduction and its subsequent naturalization in the Old World make it a species whose uses are globally known for a very wide diversity of cultures. Its uses against very diverse parasites are well validated in numerous essays. The same applies to its effects on gastrointestinal disorders, antioxidant, hypotensive, anti-inflammatory, analgesic, anxiolytic, among others. Its use as an adaptogen (e.g. anti-asthenia) and cognitive enhancer (e.g. hypnotic, mnemonic), require validation studies. It has also been established that consumption is toxic in high concentrations. The research on its anticancer activity is promising.

References

Abouri M, El Mousadik A, Msanda F, Boubaker H, Saadi B, Cherifi K (2012) An ethnobotanical survey of medicinal plants used in the Tata Province, Morocco. Int J Med Plant Res 1(7):99–123

Adams M, Gmünder F, Hamburger M (2007) Plants traditionally used in age related brain disorders. A survey of ethnobotanical literature. J Ethnopharmacol 113(3):363–381

Alitonou GA, Sessou P, Tchobo FP, Noudogbessi JP, Avlessi F, Yehouenou B, Menut C, Villeneuve P, Sohounhloue DCK (2012) Chemical composition and biological activities of essential oils of *Chenopodium ambrosioides* L. collected in two areas of Benin. Int J Biosci 2(8):58–66

Alonso J, Desmarchelier C (2005) Plantas medicinales autóctonas de la Argentina. Editorial Lola, Buenos Aires

Amole OO, Yusuf OG (2002) The analgesic effects of *Chenopodium ambrosioides*. Nig J Nat Prod Med 6:36–38

Assaidi A, Legssyer A, Berrichi A, Aziz M, Mekhfi H, Bnouham M, Ziyyat A (2014) Hypotensive property of *Chenopodium ambrosioides* in anesthetized normotensive rats. J Complement Integr Med 11(1):1–7

Ávila-Blanco ME, Rodríguez MG, Moreno Duque JL, Muñoz-Ortega M, Ventura-Juárez J (2014) Amoebicidal activity of essential oil of *Dysphania ambrosioides* (L.) Mosyakin & Clemants in an amoebic liver abscess hamster model. Evid-Based Complement Alternat Med. https://doi.org/10.1155/2014/930208

Balangcod TD, Balangcod AK (2011) Ethnomedical knowledge of plants and healthcare practices among the Kalanguya tribe in Tinoc, Ifugao, Luzon Philippines. Indian J Tradit Knowl 10(2):227–238

Barboza GE, Cantero JJ, Núñez C, Pacciaroni A, Ariza EL (2009) Medicinal plants: a general review and a phytochemical and ethnopharmacological screening of the native Argentine Flora. Kurtziana 34(1–2):7–365

Barrett B, Kiefer D (1997) Ethnomedical, biological, and clinical support for medicinal plant use on Nicaragua's Atlantic Coast. J Herbs Spices Med Plants 4(3):77–108

Barros L, Pereira E, Calhelha RC, Dueñas M, Carvalho AM, Santos-Buelga C, Ferreira ICFR (2013) Bioactivity and chemical characterization in hydrophilic and lipophilic compounds of *Chenopodium ambrosioides* L. J Funct Foods 5(4):1732–1740

Berlin EA, Berlin B, Lozoya X, Meckes M, Tortoriello J, Villareal ML (1996) The scientific basis of gastrointestinal herbal medicine among the highland Maya of Chiapas, Mexico. In: Nader L (ed) Naked science: anthropological inquiry into boundaries, power, and knowledge. Routledge, London, pp 43–68

Bigoga JD, Saahkem PA, Ndindeng SA, Ngondi JL, Nyegue M, Oben JE, Leke RGF (2013) Larvicidal and repellent potential of *Chenopodium ambrosioides* L. essential oil against *Anopheles gambiae* Giles (Diptera: Culicidae). The Open Entomol J 7:16–22

Bnouham M, Mekhfi H, Legssyer A, Ziyyat A (2002) Medicinal plants used in the treatment of diabetes in Morocco. Int J Diabetes Metab 10:33–50

Bonzani NF, Barboza GE, Bugatti MA, Ariza Espinar L (2003) Morpho-histological studies in the aromatic species of *Chenopodium* from Argentina. Fitoterapia 74(3):207–205

Bum EN, Soudi S, Ayissi ER, Dong C, Lakoulo NH, Maidawa F, Seke PF, Nanga LD, Taiwe GS, Dimo T, Njikam N, Rakotonirina A, Rakotonirina SV, Kamanyi A (2011) Anxiolytic activity evaluation of four medicinal plants from Cameroon. Afr J Tradit Complement Altern Med 8(5 Suppl):130–139

Burgstaller CH (1968) La vuelta a los vegetales. Dinizo, Buenos Aires

Carballo MA, Cortada CM, Gadano AB (2005) Riesgos y beneficios en el consumo de plantas medicinales. Theoria 14(2):95–108

Carpinella MC, Andrione DG, Ruiz G, Palacios SM (2010) Screening for acetylcholinesterase inhibitory activity in plant extracts from Argentina. Phytother Res 24:259–263

Chiasson H, Bostanian NJ, Vincent C (2004) Acaricidal properties of a *Chenopodium*-based botanical. J Econ Entomol 97(4):1373–1377

Clemants SE, Mosyakin SL (2003) Dysphania. In: Flora of North America Editorial Committee (ed) Flora of North America North of Mexico, vol 4. Oxford University Press, New York, pp 267–275

Conway GA, Slocumb JC (1979) Plants used as abortifacients and emmenagogues by Spanish New Mexicans. J Ethnopharmacol 1(3):241–261

Culham A (2007) Amaranthaceae. In: Heywood VH, Brummitt RK, Culham A, Seberg O (eds) Flowering plant families of the world. Royal Botanic Gardens, Kew, pp 28–29

Dembitskya V, Shkrobb I, Hanusa LO (2008) Ascaridole and related peroxides from the Genus *Chenopodium*. Biomed Pap Med Fac Univ Palacky Olomouc Czech Repub 152(2):209–215

Duke J, Bogenshutz M, Du-Cellier J, Duke A (2002) Handbook of medicinal herbs, 2nd edn. CRC Press, Boca Raton

Eguale T, Giday M (2009) In vitro anthelmintic activity of three medicinal plants agains *Haemonchus contortus*. Int J Green Pharm 3(1):29–34

Fekadu M, Mekuria T, Tesfaye B, Meshesha B, Teshome GM (2009) Evaluation on larvicidal effects of essential oils of some local plants against *Anopheles arabiensis* Patton and *Aedes aegypti* L. (Diptera, Culicidae) in Ethiopia. Afr J Biotechnol 8(17):4183–4188

Florian DD, Attindehou S, Takin MC, Gbangboche AB, Gnancadja A, Salifou S (2013) Preliminary study of the digestive effects of *Chenopodium ambrosioides* L. (Chenopodiaceae) leaves extracts in goats. Indian J Pharm Biol Res 1(4):55–58

Gadano AB, Gurni AA, Carballo MA (2006) Argentine folk medicine: genotoxic effects of Chenopodiaceae family. J Ethnopharmacol 103(2):246–251

Gillij YG, Gleiser RM, Zygadlo JA (2008) Mosquito repellent activity of essential oils of aromatic plants growing in Argentina. Bioresour Technol 99(7):2507–2515

Giove Nakazawa RA (1996) Medicina tradicional en el tratamiento de la enteroparasitosis. Rev Gastroenterol Peru 16(3):197–202

Giusti L (1997) Chenopodiaceae. In: Hunziker AT (ed) Flora Fanerogámica Argentina, vol 40. CONICET-Froflora, Córdoba, pp 1–52

Goka Chekem MS, Keilah Lunga P, De Dieu Tamokou J, Kuiate JR, Tane P, Vilarem G, Cerny M (2010) Antifungal properties of *Chenopodium ambrosioides* essential oil against Candida species. Pharmaceuticals 3(9):2900–2909

Gómez Castellanos JR (2008) Epazote (*Chenopodium ambrosioides*). Revisión a sus características morfológicas, actividad farmacológica, y biogénesis de su principal principio activo, ascaridol. Bol Latinoam Caribe Plant Med Aromat 7(1):3–9

González Torres DM (2005) Catálogo de plantas medicinales (y alimenticias y útiles) usadas en Paraguay, 2nd edn. Servilibro, Asunción

Grozeva N, Stoeva M (2006) Reports 1466–1472. In: Kamari G, Blanché C, Garbari F (eds) Mediterranean chromosome number reports 16. Fl Medit 16, pp 400–408

Gupta MP (2006) Medicinal plants originating in the Andean high plateau and central valleys region of Bolivia, Ecuador and Peru. United Nations Industrial Development Organisation

Hallal A, Benali S, Markouk M, Bekkouche K, Larhsini M, Chait A, Romane A, Abbad A, El Abdouni MK (2010) Evaluation of the analgesic and antipyretic activities of *Chenopodium ambrosioides* L. Asian J Exp Biol Sci 1(1):189–192

Heinrich M, Aankli A, Frei B, Weinmann C, Sticher O (1998) Medicinal plants in Mexico: healer's consensus and cultural importance. Soc Sci Med 47(11):1859–1871

Hieronymus J (1882) Plantas diafóricas. Flora Argentina. G. Kraft, Buenos Aires

Horkheimer H (1973) Alimentación y obtención de alimentos en el Perú prehispánico. Universidad Nacional Mayor de San Marcos, Lima

Hurrell JA (1991) Etnomedicina: enfermedad y adaptación en Iruya y Santa Victoria (Salta, Argentina). Rev Mus La Plata (n.s.) Antropol 9(69):109–124

Hurrell JA, Ulibarri EA, Delucchi G, Pochettino ML (2008) Plantas aromáticas condimenticias. In: Hurrell JA (ed) Biota Rioplatense XIII. Editorial Lola, Buenos Aires

Hurrell JA, Ulibarri EA, Arenas PM, Pochettino ML (2011) Plantas de Herboristería. Editorial Lola, Buenos Aires

Ibironke GF, Ajiboye KI (2007) Studies of anti-inflammatory and analgesic properties of *Chenopodium ambrosioides* leaf extract in rats. Int J Pharmacol 3(1):111–115

Judd WS, Campbell CS, Kellogg EA, Stevens PF (2002) Amaranthaceae. In: Plant systematics: a phylogenetic approach. Sinauer Associates, Saunderland, pp 245–246

Kadereit G, Borsch T, Weising K, Freitag H (2003) Phylogeny of Amaranthaceae and Chenopodiaceae and the evolution of C4 photosynthesis. Int J Plant Sci 164(6):959–986

Kala CP (2005) Ethnomedicinal botany of the Apatani in the Eastern Himalayan region of India. J Ethnobiol Ethnomed 1:11. https://doi.org/10.1186/1746-4269-1-11

Kamel EG, El-Emam MA, Mahmoud SS, Fouda FM, Bayaumy FE (2011) Parasitological and biochemical parameters in *Schistosoma mansoni*-infected mice treated with methanol extract from the plants *Chenopodium ambrosioides*, *Conyza dioscorides* and *Sesbania sesban*. Parasitol Int 60(4):388–392

Kato S, Bowman DD, Brown DL (2000) Efficacy of *Chenopodium ambrosioides* as an antihelmintic for treatment of gastrointestinal nematodes in lambs. J Herbs Spices Med Plants 7(2):11–25

Kayode J, Aleshinloye L, Ige OE (2008) Ethnomedicinal use of plant species in Ijesa Land of Osun State, Nigeria. Ethnobot Leafl 12:164–170

Ketzis JK, Taylor A, Bowman DD, Brown DL, Warnick LD, Erb HN (2002) *Chenopodium ambrosioides* and its essential oil as treatments for *Haemonchus contortus* and mixed adult-nematode infections in goats. Small Rumin Res 44(3):193–200

Kisangau DP, Lyaruu HVM, Hosea KM, Joseph CC (2007) Use of traditional medicines in the management of HIV/AIDS opportunistic infections in Tanzania: a case in the Bukoba rural district. J Ethnobiol Ethnomed 3:29. https://doi.org/10.1186/1746-4269-3-29

Kishore N, Chansouria JPN, Dubey NK (1996) Antidermatophytic action of the essential oil of *Chenopodium ambrosioides* and an ointment prepared from it. Phytother Res 10(5):453–455

Kiuchi F, Itano Y, Uchiyama N, Honda G, Tsubouchi A, Nakajima-Shimada J, Aoki T (2002) Monoterpene hydroperoxides with trypanocidal activity from *Chenopodium ambrosioides*. J Nat Prod 65(4):509–512

Kliks MM (1985) Studies on the traditional herbal anthelmintic *Chenopodium ambrosioides* L.: ethnopharmacological evaluation and clinical field trials. Soc Sci Med 21(8):879–886

Kokanova-Nedialkova Z, Nedialkov PT, Nikolov SD (2009) The genus *Chenopodium*: phytochemistry, ethnopharmacology and pharmacology. Pharm Rev 3(6):280–306

Kühn U (1993) Chenopodiaceae. In: Kubitzki K (ed) The families and genera of vascular plants II. Springer, Berlin, pp 253–281

Kumar R, Mishra AK, Dubey NK, Tripathi YB (2007) Evaluation of *Chenopodium ambrosioides* oil as a potential source of antifungal, antiaflatoxigenic and antioxidant activity. Int J Food Micobiol 115(2):159–164

Lall N, Meyer JJM (1999) In vitro inhibition of drug-resistant and drug-sensitive strains of *Mycobacterium tuberculosis* by ethnobotanically selected South African plants. J Ethnopharmacol 66(3):347–354

Larhsini M, Oumoulid L, Lazrek HB, Wataleb S, Bousaid M, Bekkouche K, Jana M (2001) Antibacterial activity of some Moroccan medicinal plants. Phytother Res 15:250–252

Liu W, Liu Y, Zhang X, Li N, Cheng H (2013) In vitro bactericidal activity of jinghua weikang capsule and its individual herb *Chenopodium ambrosioides* L. against antibiotic-resistant *Helicobacter pylori*. Chin J Int Med 19(1):54–57

MacDonald D, VanCrey K, Harrison P, Rangachari PK, Rosenfeldt J, Warren C, Sorger G (2004) Ascaridole-less infusions of *Chenopodium ambrosioides* contain a nematocide(s) that is(are) not toxic to mammalian smooth muscle. J Ethnopharmacol 92(2–3):215–221

Macía MJ, García E, Vidaurre P (2005) An ethnobotanical survey of medicinal plants commercialized in markets of La Paz and El Alto, Bolivia. J Ethnopharmacol 97(2):337–350

Maharaj R, Maharaj V, Newmarch M, Crouch NR, Bhagwandin N, Folb PI, Pillay P, Gayaram R (2010) Evaluation of selected South African ethnomedicinal plants as mosquito repellents against the *Anopheles arabiensis* mosquito in a rodent model. Malar J 9:301. https://doi.org/10.1186/1475-2875-9-301

Medeiros IU, Figueiredo IMF, Junior VFM, Oliveira CN, Schwarz A (2011) Reproductive study of *Chenopodium ambrosioides* aqueous extract in rats. In: Riet-Correa F, Pfister J, Schild AL, Wierenga T (eds) Poisoning by plants, mycotoxins and related toxins. CABI, Wallingford, pp 655–659

Mejía K, Rengifo E (2000) Plantas medicinales de uso popular en la Amazonia Peruana, 2nd edn. Agencia española de Cooperación Internacional, Lima

Mendes FR, Carlini EA (2007) Brazilian plants as possible adaptogens: an ethnopharmacological survey of books edited in Brazil. J Ethnopharmacol 109(3):493–500

Micheli-Serra A (2001) Médicos y medicina en la Nueva España del siglo XVI. Gac Méd Méx 137(3):257–263

Misra P, Pal NL, Guru PY, Katiyar JC, Tandon JS (1991) Antimalarial activity of traditional plants against erythrocytic stages of *Plasmodium berghei*. Pharm Biol 29(1):19–23

Montanari B (2014) Aromatic, medicinal plants and vulnerability of traditional herbal knowledge in a Berber community of the High Atlas Mountains of Morocco. Plant Div Resour 36(3):388–402

Monzote L, Stamberg W, Staniek K, Gille L (2009) Toxic effects of carvacrol, caryophyllene oxide, and ascaridole from essential oil of *Chenopodium ambrosioides* on mitochondria. Toxicol Appl Pharmacol 240(3):337–347

Monzote L, Pastor J, Scull R, Gille L (2014) Antileishmanial activity of essential oil from *Chenopodium ambrosioides* and its main components against experimental cutaneous leishmaniasis in BALB/c mice. Phytomedicine 21(8–9):1048–1052

Mosyakin SL, Clemants SE (2002) New nomenclatural combinations in *Dysphania* R. Br. (Chenopodiaceae): taxa occurring in North America. Ukrayins'k Bot Zhurn (n. s.) 59(4):380–385

Nascimento FRF, Cruz GV, Pereira PV, Maciel MC, Silva LA, Azevedo AP, Barroqueiro ES, Guerra RN (2006) Ascitic and solid Ehrlich tumor inhibition by *Chenopodium ambrosioides* L. treatment. Life Sci 78(22):2650–2653

Navone GT, Zonta ML, Gamboa MI (2014) Fitoterapia Mbyá-Guaraní en el control de las parasitosis intestinales. Un estudio exploratorio con *Chenopodium abrosioides* var. *anthelminticum* en cinco comunidades de Misiones, Argentina. Polibotánica 37:135–151

Okhale SE, Egharevba HO, Ona EC, Kunle OF (2012) Phytochemicaland proximate analyses and thin layer chromatography fingerprinting of the aerial part of *Chenopodium ambrosioides* L. (Chenopodiaceae). J Med Plant Res 6(12):2289–2294

Okuyama E, Umeyama K, Saito Y, Yamazaki M, Satake M (1993) Ascaridole as a pharmacologically active principle of "paico", a medicinal Peruvian plant. Chem Pharm Bull (Tokyo) 41(7):1309–1311

Pinedo M, Rengifo E, Cerruti T (1997) Plantas Medicinales de la Amazonia Peruana. Estudio de su uso y cultivo. Instituto de Investigaciones de la Amazonia Peruana (IIAP), Iquitos

Pochettino ML, Puentes JP, Buet Costantino F, Arenas PM, Ulibarri EA, Hurrell JA (2012) Functional foods and nutraceuticals in a market of Bolivian immigrants in Buenos Aires (Argentina). Evid-Based Complement Alternat Med. https://doi.org/10.1155/2012/320193

Pöll E (2005) Medicinal and aromatic plants of Guatemala and the need for their conservation. Acta Hortic 676:167–170

Pollack Y, Segal R, Golenser J (1990) The effect of ascaridole on the in vitro development of *Plasmodium falciparum*. Parasitol Res 76(7):570–572

Potawale SE, Luniya KP, Mantri RA, Mehta UK, Sadiq W, Vetal YD, Deshmukh RS (2008) *Chenopodium ambrosioides*. An ethnopharmacological review. Pharmacologyonline 2:272–286

Pratt DB (2003) Phylogeny and morphological evolution of the Chenopodiaceae-Amaranthaceae alliance. Restrospective Theses and Dissertartions. Paper 613, pp 1–116

Quinlan MB, Quinlan RJ, Nolan JM (2002) Ethnophysiology and herbal treatments of intestinal worms in Dominica, West Indies. J Ethnopharmacol 80(1):75–83

Randall RP (2005) A global compendium of weeds, 2nd edn. Department of Agriculture and Food of Western Australia, Perth

Rossi-Bergmann S, Costa S, VLG DM (1997) Brazilian medicinal plants: a rich source of immunomodulatory substances. Ciênc Cult (São Paulo) 49(5/6):395–401

Ruffa MJ, Ferraro G, Wagner ML, Calcagno ML, Campos RH, Cavallaro L (2002) Cytotoxic effect of Argentine medicinal plant extracts on human hepatocellular carcinoma cell line. J Ethnopharmacol 79(3):335–339

Shah H (2014) Antibacterial and antifungal activities of the crude extracts from the stem of *Chenopodium ambrosioides* L., an indiginous medicinal plant. Afr J Pharm Pharmacol 8(8):231–234

Speiky H, Rocco C, Carrasco C, Lissi EA, López-Alarcón C (2006) Antioxidant screening of medicinal herbal teas. Phytother Res 20(6):462–467

Sukhorukov AP, Zhang M (2013) Fruit and seed anatomy of *Chenopodium* and related genera (Chenopodioideae, Chenopodiaceae/Amaranthaceae): implications for evolution and taxonomy. PLos One 8(4):e61906. https://doi.org/10.1371/journal.pone.0061906

Toloza AC, Zygadlo J, Biurrun F, Rotman A, Picollo MI (2010) Bioactivity of Argentinean essential oils against permethrin-resistant head lice, *Pediculus humanus capitis*. J Insect Sci 10:185. https://doi.org/10.1673/031.010.14145

Toso RE, Boeris MA (2010) Validación de la actividad antiespasmódica de *Sida rhombifolia, Baccharis articulata, Chenopodium ambrosioides* y *Conyza bonariensis*. Ciencia Veterinaria (La Pampa) 12(1):20–24

Trivellato Grassi L, Malheiros A, Meyre-Silva C, Buss Zda S, Monguilhott ED, Fröde TS, da Silva KA, de Souza MM (2013) From popular use to pharmacological validation: a study of the antiinflammatory, antinociceptive and healing effects of *Chenopodium ambrosioides* extract. J Ethnopharmacol 145(1):127–138

Ulloa C (2006) Aromas y sabores andinos. In: Moraes M, Øllgaard B, Kvist P, Borchsenius F, Balslev H (eds) Botánica Económica de los Andes Centrales. Universidad Mayor de San Andrés, La Paz, pp 313–328

Uotila PJ (1990) Chenopodium. In: Castroviejo S (ed) Flora Iberica, vol 2. Real Jardín Botánico-CSIC, Madrid, Madrid, pp 484–500

Velázquez C, Calzada F, Torres J, Gonzalez F, Ceballos G (2006) Antisecretory activity of plants used to treat gastrointestinal disorders in Mexico. J Ethnopharmacol 103(1):66–70

Volpato G, Godínez D, Beyra A, Barreto A (2009) Uses of medicinal plants by Haitian immigrants and their descendants in the Province of Camagüey, Cuba. J Ethnobiol Ethnomed 5:16. https://doi.org/10.1186/1746-4269-5-16

Wabo Poné J, Jeannette Y, Olivia Fossi T, Komtangi MC, Bilong Bilong CF, Mpoame M (2011) In vitro effects of *Chenopodium ambrosioides* extracts on the parasitic nematode *Heligmosomoides bakeri* (Nematoda, Heligmosomatidae). J Pharmacogn Phytother 3(4):56–62

Wu JL, Ma DW, Wang YN, Zhang H, He B, Li Q, Zou ZY, Feng J (2013) Cytotoxicity of essential oil of *Chenopodium ambrosioides* L against human breast cancer MCF-7 cells. Trop J Pharm Res 12(6):929–933

Yadav N, Vasudeva N, Singh S, Sharma SK (2007) Medicinal properties of the genus *Chenopodium* L. Nat Prod Radiance 6(2):131–134

Zhu G, Mosyakin SL, Clemants SE (2003) Chenopodiaceae. In: Wu ZY, Raven PH, Hong DY (eds) Flora of China, vol 5. Science Press/Missouri Botanical Garden Press, Beijing/St. Louis, pp 351–414

Zhu WX, Zhao K, Chu SS, Liu ZL (2012) Evaluation of esential oil and its three main active ingredients of Chinese *Chenopodium ambrosioides* (Chenopodiaceae) against *Blatela germanica*. J Arthropod-Borne Dis 6(2):90–97

Echinodorus macrophyllus (Kunth) Micheli

Maria Izabela Ferreira, Gabriela Granghelli Gonçalves, and Lin Chau Ming

Echinodorus macrophyllus (Kunth) Micheli
Rich Hoyer
Available in: https://www.flickr.com/photos/birdernaturalist/38989275251/

M. I. Ferreira · G. G. Gonçalves · L. C. Ming (✉)
Horticulture Department, School of Agronomic Sciences,
Universidade Estadual Paulista (UNESP), Botucatu, São Paulo, Brazil
e-mail: linming@fca.unesp.br

U. P. Albuquerque et al. (eds.), *Medicinal and Aromatic Plants of South America*, Medicinal and Aromatic Plants of the World 5,
https://doi.org/10.1007/978-94-024-1552-0_18

Abstract *Echinodorus macrophyllus* (Kunth) Micheli pertaining to Alismataceae family, is a perennial, rhizomatous and aquatic herb that also occurs in wetlands and flooding areas, in several countries of South America. In Brazil, it is known as chapéu-de-couro, chá-mineiro, erva-de-pântano, erva-de-bugre, congonha-do-brejo e erva-do-brejo. Other species of the genus, such as *E. grandiflorus* is also used in folk medicine, with very similar indications for use. *E. macrophyllus* leaves have been used in folk medicine in the form of decoction, infusion, or bottled, considered a reputed remedy for the treatment of infections, respiratory diseases, inflammatory conditions, kidney dysfunctions, diuretic, anti-hypertensive and against pains of the genito-urinary system. Although many studies have shown positive results in pre-clinical trials and this herb seems to be safe to human organism, it is important to be careful with its indiscriminate use to avoid side effects and health damage, as well as with collection practices. In Brazil, this species is still wild-crafted and extensively extracted. These are conditions that make it a high-priority species for conservation.

Keywords *Chapéu-de-couro* · Aquatic herbs · Antihypertensive drugs · Diuretic plants

1 Taxonomic Characteristics

Echinodorus macrophyllus (Kunth) Micheli belongs to the Alistaceae family and *Echinodorus* genus (Tropicos 2015). *Echinodorus* is the second largest genus in the aquatic plant family *Alismataceae* (Lehtonen and Myllys 2008) and comprises 14 genera (Haynes et al. 1998). The species-level classifications are typically conflicting among different authors. The taxonomy of the genus has been partially revised in a recent phylogenetic relationships study that has shown that genus *Echinodorus* has 28 species (Lehtonen 2008).

E. macrophyllus was first described by Kunth, a German botanist, under the name *Alisma macrophyllum* Kunth, (basionym) published in *Enumeratio Plantarum Omnium Hucusque Cognitarum,* in 1841. Later, in 1881, the Swiss botanist Micheli listed the species as *E. macrophyllus,* published in *Monographiae Phanerogamarum* (Tropicos 2015).

Synonyms *Alisma macrophyllum* Kunth and *Echinodorus scaber* Rataj.

2 Crude Drug Used

The part of *E. macrophyllus* described in the two first editions of the Brazilian Pharmacopoeia (1924, 1959) used as a drug, is the leaf, which is odorless and has slightly bitter flavor (Leite et al. 2007).

3 Major Chemical Constituents and Bioactive Compounds

Phytochemical analysis of the leaves revealed the presence of triterpenoids, steroids, flavones, flavonols, and xanthones (Tanus-Rangel et al. 2010). Farther important compounds isolated from the leaves are: echinophyllins A, B, C, and F, chapecoderins A and C (Kobayashi et al. 2000; Kobayashi and Ohsaki 2000) echinodolides A and B (Shigemori et al. 2002) isovitexinandvitexin (Tanus-Rangel et al. 2010).

4 Morphological Description

It is a perennial herb, robust, estyliform, pubescent. Takes root in the soil and maintains its lower portions immersed, while exposing its petioles, leaves and inflorescences. Cylindrical petiole with emerged leaves rough and eatery, dark green color, with prominent veins. Leaf generally oval, rarely oval-lanceolate, obtuse to the acute apex, cordate to truncate base, absent translucent marks. Panicle inflorescences are composed of numerous cylindrical, hermaphroditic flowers, with around 5 cm in diameter, white and yellow petals in the basal part and with bracts lanceolate (Lorenzi and Matos 2002; Pansarin and Amaral 2005). They have rounded infructescences of brown color when ripe, fruit achene type, with only one seed.

5 Geographical Distribution

Echinodorus genus has a sub-cosmopolitan distribution. It occurs in the Western hemisphere, mostly in the tropics. The native of tropical America, but some species reaching temperate climates, occurring from the Northern United States of America to Argentina and Chile (Haynes and Holm-Nielsen 1994; Lehtonen and Myllys 2008).

E.macrophyllus occurs in several countries in South America: Nicaragua (Chontales); Guiana; Suriname; Venezuela (Guarico, Monagas) Brazil (Amapá, Bahia, Goiás, Mato Grosso, Mato Grosso do Sul, Minas Gerais, Pará, Paraná, Pernambuco, Piauí, Rio de Janeiro, Roraima, São Paulo); Bolivia (Beni); Columbia (Antioquia); Paraguay (Pansarin and Amaral 2005; USDA).

6 Ecological Requirements

E. macrophyllus is an aquatic herb that also occurs in swamps, wetlands and flooded areas (Haynes and Holm-Nielsen 1994; Pio-Correa and Pena 1984), but it is able to survive fully immersed for a certain period, although not bloom as well as being able to tolerate short periods of drought. In the state of São Paulo, it blooms during the period from October to January and fructifies from November to July (Pansarin and Amaral 2005).

Plants are vegetatively propagated from rhizome runners, from adventitious plantlets developed at the nodes of the scape, or by divisions of the rhizome. Sexual propagation is reported to be difficult and germination temperatures range between 25 and 30 °C (Castro and Chemale 1995; Haynes and Holm-Nielsen 1994).

It is a rustic species that propagates and grows quickly; it has been used in landscaping projects, around lakes, due to its decorative, beautiful foliage and inflorescences. Although less cultivated, it can be grown in damp, shady locations, such as floodplain, river banks, lakes and drainage ditches, at a spacing of 50×70 cm between plants (Corrêa Junior et al. 1994).

7 Collection Practice

E. macrophyllus occurs in wet and marshy areas, therefore harvests should be done before the dry season. During this period, the plant loses its aerial parts (leaves and petioles) but keeps the rhizomes in the ground to sprout in favorable terms. Commercially available plant material is still mostly collected from the naturally occurring populations since there is no established technology for growing species (Ming et al. 2012). Fresh leaves are cut to facilitate drying and then stored away from light and heat, in tightly closed containers.

In Brazil, this species has a priority for germplasm collection and conservation, because it is extensively extracted (Vieira 1999), and being a native species with scarce cultivation, management and commercialization. *E. macrophyllus* requires registration at IBAMA (Brazilian Institute of Environment and Renewable Natural Resources).

8 Traditional Use (Part(s) Used) and Common Knowledge

In Brazil, it is known as *chapéu-de-couro, chá-mineiro, erva-de-pântano, erva-de-bugre, congonha-do-campo* and *erva-do-brejo* (Leite et al. 2007; Nunes et al. 2003). Other species of the genus, such as *E. grandiflorus* is also used in folk medicine, with very similar instructions for use.

With a long tradition of use in Brazil, *E. macrophyllus* is referred in the 1st Edition of the Brazilian Pharmacopoeia, published in 1929, as a medicinal plant, extensively used in traditional medicine. The use of *E. macrophyllus* leaves is documented in the first edition of the Pharmacopoeia as fluid extract and the second edition as a vegetable drug. Its use is cited historically in ancient bibliographies as anti-inflammatory, depurative, diuretic, to treat arthritis, urinary disorders, hydrops, liver disorders, rheumatism, cutaneous affections, venereal diseases (Brandão and Cosenza 2009). Beyond that, it is well known in Brazil as a diuretic and antihypertensive but also is regionally used against pains of the genitourinary system (Nunes et al. 2003).

Its leaf has been used in Brazilian folk medicine in decoction, infusion, or bottled. It is a reputed remedy for the treatment of infections, respiratory diseases, inflammatory conditions and kidney dysfunctions (De la Cruz 2008; Tanus-Rangel et al. 2010).

Its extract is also used for the manufacture of soft drinks traditionally associated with the city of Niterói in Rio de Janeiro state and called as Mineirinho®. It was first made in the state of Minas Gerais, so it is called "Mineirinho", but more recently the soda is called Mate Couro® and has regional popularity as a cultural and industrial reference of Rio de Janeiro. By having *E. macrophyllus* extract in its composition, due to the diuretic properties of the plant, the product is also very much in demand. Farther uses include hot or ice tea, made from the leaves and petioles, dehydrated or fresh, as well as beers and semi-sparkling wine (Pio-Correa and Pena 1984).

9 Modern Medicine Based on Its Traditional Medicine Uses

In order to evaluate the bioactivity of *E. macrophyllus* leaves aqueous extracts many preclinical studies have been conducted. As it is utilized in a large range of diseases, it is important to know if its use can cause health damage.

Infusion of dried leaves is regulated by National Agency for Sanitary Surveillance of Brazil (ANVISA), indicated for therapeutic use in the treatment of edema (swelling) by fluid retention and inflammation process. So, it is indicated as weak diuretic and anti-inflammatory, and must be prepared with 1 g of leaf in 150 ml of water, and should be consumed immediately after preparation, three times a day. But it should not be used by children under 12 years of age, people with kidney or heart insufficiency and using antihypertensive drugs (Brasil 2010, 2011).

Preclinical studies show that aqueous extract of *E. macrophyllus* leaves has antioxidant and renoprotective effect (Nascimento et al. 2014), showed no mutagenic activity (Rivera et al. 1994), no cytotoxicity effects, as well as show a reduction of body weight (Costa Lopes et al. 2000). A modest immunosuppressive effect of aqueous extract supports a potential therapeutic use to control exacerbated humoral and/or cellular immune response, as in autoimmune rheumatic diseases (Pinto et al. 2007). Ethanolic leaf extract shows anti-inflammatory action in acute and subchronic models of inflammation (Tanus-Rangel et al. 2010).

But the presence of kidney cells alterations in mice exposed to subchronic treatment, in the highest dose tested, point to the presence of substances potentially genotoxic to the kidney. On the other hand, exposure dose equivalent to the daily dose recommended to humans (23 mg/kg) did not reveal any genotoxic effect (Costa Lopes et al. 2000).

So, although many studies have shown positive results in preclinical trials and this herb seems to be safe to the human organism, it is important to be careful with the indiscriminate use of this plant drug to avoid side effects and health damage.

10 Conclusions

E. macrophyllus is a species of therapeutic and commercial importance with a wide range of reported ethnomedicinal uses, as well as many biological activities studies and industrial potential uses.

This species is the priority for conservation because it is extensively extracted and widely used in traditional medicine in Brazil, with preclinical studies proving its therapeutic action. However, the scarcity of research on plants cultivation shows the difficulties for its utilization as raw material for industry, particularly *E. macrophyllus,* is a hygrophilous species. This life-form does not favor the development of very specific agronomic practices for its large-scale cultivation.

References

Brandão M, Cosenza G (2009) Traditional uses of American plant species from the 1st edition of Brazilian Official Pharmacopoeia. Braz J Pharmacogn 18(2A):478–487

Brasil (2010) Resolução – RDC No 10, de 9 de Março de 2010. Diario Oficial da União, Brasilia, pp 52–59

Brasil (2011) In: Agência Nacional de Vigilância Sanitária (ed) Formulário de Fitoterápicos Farmacopeia Brasileira, 1st edn. Agência Nacional de Vigilância Sanitária, Brasília, p 126

Castro L, Chemale V (1995) Plantas medicinais, condimentares e aromáticas: descrição e cultivo. Agropecuária, Guaíba

Corrêa Junior C, Ming L, Scheffer M (1994) Cultivo de plantas medicinais, condimentares e aromáticas, 2nd edn. FUNEP, Jaboticabal

Costa Lopes L, Albano F, Laranja GAT, Alves LM, LFM S, Souza GP et al (2000) Toxicological evaluation by in vitro and in vivo assays of an aqueous extract prepared from *Echinodorus macrophyllus* leaves. Toxicol Lett 116(3):189–198

De la Cruz MG (2008) Plantas utilizadas por raizeiros na medicina popular em Cuiabá, Mato Grosso, Plantas Med. Mato Grosso A Farm. Pop. dos Raizeiros. Carline & Carliato, Cuiabá, pp 63–128

Haynes RR, Holm-Nielsen LB (1994) The Alismataceae. Flora Neotrop. New York Botanical Garden Press, New York

Haynes R, Les D, Holm-Nielsen L (1998) Flowering plants monocotyledons. In: Kubitzki K (ed) The families of genera and vascular plants. Springer, Berlin/Heidelberg, pp 11–18

Kobayashi J, Ohsaki A (2000) Echinophyllins C-F, new nitrogen-containing clerodane diterpenoids from *Echinodorus macrophyllus*. J Nat Prod 63:1576–1579

Kobayashi J, Sekiguchi M, Shigemori H, Ohsaki A, Chapecoderins A-C (2000) New labdane-derived diterpenoids from *Echinodorus macrophyllus*. J Nat Prod 63:375–377

Lehtonen S (2008) An integrative approach to species delimitation in Echinodorus (Alismataceae) and the description of two new species. Kew Bull 63(4):525–563

Lehtonen S, Myllys L (2008) Cladistic analysis of *Echinodorus* (Alismataceae): simultaneous analysis of molecular and morphological data. Cladistics. Blackwell Publishing Ltd 24(2):218–239

Leite JPV, Pimenta DS, Gomes RSDL, Dantas-Barros AM (2007) Contribuição ao estudo farmacobotânico da *Echinodorus macrophyllus* (Kunth) Micheli (chapéu-de-couro) – Alismataceae. Rev Bras Farmacogn. Soc Bras Farmacognosia 17(2):242–248

Lorenzi H, Matos F d A (2002) Plantas medicinais no Brasil: nativas e exóticas, 2nd edn. Plantarum, Nova Odessa

Ming LC, Ferreira MI, Gonçalves GG (2012) Pesquisas agronômicas das plantas medicinais da Mata Atlântica regulamentadas pela ANVISA. Rev Bras Plantas Med Soc Bras Plantas Medicinais 14(spe):131–137

Nascimento E, Watanabe M, Dezoti da Fonseca C, Schlottfeldt F, Vattimo M (2014) Efeito renoprotetor do Echinodorus macrophyllus na lesão renal induzida. Acta Paul Enferm 27(1):12–17

Nunes GP, Da Silva MF, Resende UM, De Siqueira JM (2003) Plantas medicinais comercializadas por raizeiros no Centro de Campo Grande, Mato Grosso do Sul. Rev Bras Farmacogn 13(2):83–92

Pansarin E, Amaral M (2005) Alismataceae. In: Wanderley MGL, Shepherd GGAM (eds) Flora fanerogâmica do estado São Paulo. Rima, São Paulo, pp 1–10

Pinto AC, Rego GCG, Siqueira AM, Cardoso CC, Reis PA, Marques EA et al (2007) Immunosuppressive effects of *Echinodorus macrophyllus* aqueous extract. J Ethnopharmacol 111(2):435–439

Pio-Correa M, de Pena LA (1984) Dicionário de plantas úteis do Brasil e das exóticas cultivadas. Ministério da Agricultura, Instituto Brasileiro de Desenvolvimento Florestal, Rio de Janeiro

Rivera IG, Martins MT, Sanchez PS, Sato MIZ, Coelho MCL, Akisue M et al (1994) Genotoxicity assessment through the Ames test of medicinal plants commonly used in Brazil. Environ Toxicol Water Qual 9(2):87–93

Shigemori H, Shimamoto S, Sekiguchi M, Ohsaki A, Kobayashi J (2002) Echinodolides A and B, new cembrane diterpenoids with an eight-membered lactone ring from the leaves of *Echinodorus macrophyllus*. J Nat Prod Am Chem Soc 65(1):82–84

Tanus-Rangel E, Santos SR, Lima JCS, Lopes L, Noldin V, Monache FD et al (2010) Topical and systemic anti-inflammatory effects of *Echinodorus macrophyllus* (Kunth) Micheli (Alismataceae). J Med Food Mary Ann Liebert 13(5):1161–1166

Tropicos (2015) !Echinodorus macrophyllus (Kunth) Micheli [Internet]. Missouri Bot. Gard. Available from: http://www.tropicos.org/Name/900028

USDA. *Echinodorus macrophyllus* [Internet]. Information from Natl. Genet. Resour. Program. Germplasm Resour. Inf. Available from: http://www.ars-grin.gov/cgi-bin/npgs/html/taxon.pl?402575

Vieira R (1999) Conservation of medicinal and aromatic plants in Brazil. In: Janick J (ed) Perspectives on new crops and new uses. ASHS Press, Alexandria, pp 152–159

Equisetum giganteum L.

Ivanilda Soares Feitosa, Rafael Corrêa Prota dos Santos Reinaldo, Augusto César Pessôa Santiago, and Ulysses Paulino Albuquerque

Equisetum giganteum L.
Photo courtesy of Dr. Vinícius Antônio de Oliveira Dittrich

I. S. Feitosa · R. C. P. d. S. Reinaldo
Laboratório de Ecologia e Evolução de Sistemas Socioecológicos, Departamento de Botânica, Universidade Federal de Pernambuco, Cidade Universitária, Recife, PE, Brazil
e-mail: ulysses@pq.cnpq.br

A. C. P. Santiago
Laboratory of Biodiversity (Laboratório de Biodiversidade). Biology Nucleus (Núcleo de Biologia), Federal University of Pernambuco (Universidade Federal de Pernambuco), Vitória Academic Center (Centro Acadêmico de Vitória), Bela Vista, Vitória de Santo Antão, PE, Brazil

U. P. Albuquerque et al. (eds.), *Medicinal and Aromatic Plants of South America*, Medicinal and Aromatic Plants of the World 5,
https://doi.org/10.1007/978-94-024-1552-0_19

Abstract *Equisetum giganteum* L. is a fern with numerous uses in popular medicine in Latin and Central America. In particular, it is used as a diuretic, anti-inflammatory, and astringent and to treat ophthalmologic and renal disorders. *E. giganteum* is also used for pest control and as a fertilizer in agriculture due to its high silicon content. It possesses antidiabetic and antifungal activities. Several compounds have been identified from this species, namely tannins, flavonoids and alkaloids.

Keywords Medicinal ferns · Traditional use · Equisetaceae · Flavonoids · Oleoresin

1 Taxonomic Characteristics

Equisetum L. is the only living genus of the family Equisetaceae (Equisetales and Equisetopsida). The species of this genus are commonly known as "horsetails" (Smith et al. 2008). Due to its peculiar morphology, this group was formerly considered a separate Pteridophyta division (Equisetophyta) or class (Tryon and Tryon 1982; Christenhusz and Chase 2014). However, recent molecular studies have included the Equisetaceae and Psilotaceae (whisk ferns) in the evolutionary line of ferns (Pryer et al. 2001; Smith et al. 2006).

Equisetum (Equisetaceae) is monophyletic, and recent studies suggest that it is the basal clade of ferns (Knie et al. 2015). The genus *Equisetum* is usually subdivided into two subgenera, *Equisetum* and *Hippochaete*, with *E. giganteum* included in the second subgenus (Tryon and Tryon 1982; Guillon 2004).

In addition to being commonly known as "horsetails" in English, they are also known as "cavalinhas" in Portuguese or "cola de caballo," "limpia plata," "yerba del platero" and "rabo de mula" in some Spanish-speaking countries.

E. giganteum has the following synonyms: *Equisetum bolivianum* Gand., *E. martii* Milde, *E. pyramidale* Goldm., *E. ramosissimum* Kunth, *E. schaffneri* Milde and *E. xylochaetum* Mett (Mobot Tropicos 2015).

2 Crude Drug Used

Although the therapeutic potential of *E. giganteum* is well known in the traditional communities living along its geographical distribution area and there is evidence of its pharmacological potential (Farinon et al. 2013), the safety and efficacy of its use in humans have not been confirmed. However, the pharmacological properties of *Equisetum arvense* L., which belongs to the same genus, have

U. P. Albuquerque (✉)
Departamento de Botânica, Centro de Biociências, Universidade Federal de Pernambuco, Recife, Brazil

been studied in humans, and *E. arvense* L. has been included in the official Brazilian Pharmacopoeia (Brasil 2011). *E. giganteum* shoots and occasionally all plant parts are commonly used in the form of a decoction (Macía 2004; Kloucek et al. 2005; Quiroga et al. 2001).

3 Major Chemical Constituents and Bioactive Compounds

E. giganteum contains tannins, flavonoids, saponins and alkaloids (Santos et al. 2010; Mir et al. 2013). The following compounds also occur in this species: *dodecanoic acid, 3-Nonynoic acid methyl ester, 3,6-Dimethyl decane, n-Heneicosane, 6-Hydroxicholesterol, Ergosta-4,7,22-trien-3-one, 8,12-Dimethyl-4Z,8E,12E-octadecatriene, Methenolone, Gorgost-5-en-3-ol, 2,6,10,14-Hexadecatetraen-1-ol, 3,7,11,15-*Tetramethyl-acetate *(E,E,E), Z-13-Octadecenal, bufa-20,22-dienolide* and *3,14-dihydroxy* (Michielin et al. 2005).

4 Morphological Description

Horsetails display a short or long creeping underground stem and a hollow aerial stem that is branched or non-branched, articulated, and impregnated with silicon and that contains nodes. There are small, verticillate, teeth-shaped leaves at each node, with free apices, and the sporangia are in the apical strobili. Horsetails are the only ferns that exhibit alete spores with elaters (Tryon and Tryon 1982; Hauke 1995).

The subgenera are separated based on the characteristics of their gametophytes, chromosome size, and morphological characteristics of their sporophytes (Guillon 2004). *Hipppochaete* sporophytes exhibit perennial non-branched shoots, sunken stomata arranged in long regular lines, and an apiculate strobilus, whereas *Equisetum* exhibits deciduous, branched aerial stems, superficial stomata arranged irregularly, and a blunt or non-apiculate strobilus (Tryon and Tryon 1982; Guillon 2004).

E. giganteum differs from the neotropical species because it exhibits branched aerial stems, with regular whorls of branches, persistent leaf apices (teeth), stem crests with tubers that are nearly square in profile, and an apiculate strobilus (Tryon and Tryon 1982; Hauke 1995).

5 Geographical Distribution

E. giganteum is a native plant from Central and South America (Farinon et al. 2013), that is found in Guatemala, El Salvador, Costa Rica, Great Antilles, Colombia, Venezuela, Ecuador, Peru, Bolivia, Chile, Argentina, Paraguai, Uruguai and Brazil

(Hauke 1995). In Brazil, it occurs in the South, Southeast and West-Central regions and is cultivated in some states, in the North and Northeast (Tryon and Tryon 1982; Salino and Almeida 2015).

6 Ecological Requirements

E. giganteum plant individuals can colonize a wide variety of habitats with different salinity gradients. Its salinity tolerance is believed to depend on sodium extrusion from the cells and potassium accumulation at the root (Husby 2009).

E. giganteum grows in areas with a substantial underground water source, often along rivers and in swamps (Hauke 1963). Therefore, it is always found in humid places, such as humid wood and road fills, where there is a sufficient underground water supply. *E. giganteum* exhibits clonal growth via rhizomes, which is very important for its ability to use underground water sources, and the deep growth of the rhizomes confers resistance to severe environmental variations, such as fire and drought (Husby 2009).

7 Traditional Use (Part(s) Used) and Common Knowledge

E. giganteum is popularly used as a medicinal resource by traditional communities (Bussman et al. 2007). It has been reportedly used as a diuretic, digestive, antianemic, and for the treatment of gastrointestinal problems (Barros et al. 2007). It has also been reported to be used as an anti-inflammatory agent, to treat urinary tract infections (Estomba et al. 2006) and hemorrhoids, as an astringent (Kloucek et al. 2005), to treat ophthalmologic and renal disorders (Nunes et al. 2003), to treat hypertension (Mello and Budel 2013), as an antifungal (Mir et al. 2013; Farinon et al. 2013), and to treat male impotence and female sterility.

E. giganteum is also used as an alternative insecticide for agricultural pests and as a fertilizer, likely due to its high silicon content (Bertalot et al. 2012).

The shoot is the plant part that is most commonly used by traditional communities to treat the different diseases listed above (Gorzalczany et al. 1999; Portillo et al. 2001; Martinez et al. 2004; Kloucek et al. 2005; Rodrigues et al. 2012); however, there are report according to also the whole plant is used (Quiroga et al. 2001; Fenner et al. 2006).

8 Modern Medicine Based on Its Uses in Traditional Medicine

Studies assessing the therapeutic potential of ferns are still scarce. In the case of *E. giganteum,* the compounds extracted from the oleoresins have been studied (Michielin et al. 2005). The oleoresins may contain compounds such as triterpenes, steroids and alkanes (Farinon et al. 2013).

Kloucek et al. (2005) tested ethanol extracts of *E. giganteum*, and reported that these extracts exhibited biological activity against six species of Gram-positive bacteria (*Bacillus cereus, Bacillus subtilis, Enterococcus faecalis, Staphylococcus aureus, Staphylococcus epidermidis,* and *Streptococcus pyogenes*) and one species of Gram-negative bacteria (*Bacteroides fragilis*). *Equisetum* had no activity against two other species of Gram-negative bacteria (*Escherichia coli* and *Pseudomonas aeruginosa*). Rodrigues et al. (2012) surveyed the antidiabetic activity of the plants used by the traditional communities and confirmed the reported activities by conducting a series of laboratory studies on these plants. Although *E. giganteum* is popularly referred to as an antidiabetic, the authors reported that this activity was not confirmed in the laboratory.

Portillo et al. (2001) investigated the use of *E. giganteum* as an antifungal in popular medicine in Paraguay, and demonstrated that none of the 11 different fungal species tested were sensitive to the extracts from *E. giganteum.* Quiroga et al. (2001) also demonstrated that the *E. giganteum* alcohol extracts did not exhibit antifungal activity against two species of fungi.

9 Conclusions

Based on the aforementioned studies, *E. giganteum* exhibits significant therapeutic potential, and further studies may lead to its use for the development of new phytotherapeutic drugs. The use of this species by traditional communities and the pharmacological studies of its biological activities indicate that it may also exhibit additional unknown activities. Therefore, we recommend that additional ethnodirected studies be performed to identify the potential additional activities of this species, based on popular indications.

Acknowledgments We are especially grateful to the National Institute of Science and Technology in Ethnobiology, Bioprospecting and Nature Conservation, certified by CNPq, with financial support from FACEPE (Foundation for the Support of Science and Technology of the State of Pernambuco).

References

Barros FMC, Pereira K, Zanetti GD, Heinzmann BM (2007) Plantas do município de São Luiz Gonzaga, RS, Brasil. Lat Am J Pharm 26(5):652–662

Bertalot MJA, Carvalho-Pupatto JG, Rodrigues E, Mendes RD, Buso D (2012) Controle alternativo de doenças no morango. Ass Bras Agric Biod 7(2):170–177

Brasil: ANVISA (2011) Formulário de Fitoterápicos Farmacopeia Brasileira

Bussman RW, Sharon D, Vandebroek I, Jones I, Revene Z (2007) Health for sale: the medicinal plant markets in Trujillo and Chiclayo, northern Peru. J Ethnobiol Ethnomed 3:1–9

Christenhusz MJM, Chase MW (2014) Trends and concepts in fern classification. Ann Bot 113:571–594

Estomba D, Ladio A, Lozada M (2006) Medicinal wild plant knowledge and gathering patterns in a Mapuche community from North-Western Patagonia. J Ethnofarmacol 103:109–119

Farinon M, Lora PS, Francescato LN, Bassani VL, Henriques AT, Xavier RM, Oliveira PG (2013) Effect of aqueous extract for giant horsetail (*Equisetum giganteum* L.) in antigen-induced arthritis. Open Rheumatol J 7:129–133

Fenner R, Betti AH, Mentz LA, Rates SMK (2006) Plantas utilizadas na medicina popular brasileira com potencial atividade antifúngica. Rev Bras Cienc Farm 42(3):369–394

Gorzalczany S, Rojo A, Rondina R, Debenedetti S, Acevedo C (1999) Estudio de toxicidade Aguda por via oral de plantas medicinales Argentinas. Acta Farm Bonaer 18(3):221–224

Guillon JM (2004) Phylogeny of horsetails (*Equisetum*) based on the chloroplast rps4 gene and adjacent noncoding sequences. Syst Bot 29:251–259

Hauke RL (1963) A taxonomic monograph of the genus Equisetum subgenus Hippochaete. Nova Hedwigia 8:1–123

Hauke RL (1995) Equisetaceae, pp. 4–5. In: G. Davidse, Sánchez MS, Knapp S (eds) Vol. 1: Psilotaceae a Salviniaceae. In: Davidse G, Sánchez MS, Chater AO (eds). Flora Mesoamericana. Universidad Nacional Autónoma de México, Mexico

Husby CE (2009) Ecophysiology and biomechanics of *Equisetum giganteum* in South America. Miami. Thesis [Doctor of philosophy in Biology]-Florida Internacional University

Kloucek P, Polesny Z, Svobodova B, Vlkova E, Kokoska L (2005) Antibacterial screening of some Peruvian medicinal plants used in Calleria District. J Ethnofarm 99:309–312

Knie N, Fischer S, Grewe F, Polsakiewicz M, Knoop V (2015) Horsetails are the sister group to all other monilophytes and Marattiales are sister to leptosporangiate ferns. Mol Phylogenet Evol 90(2015):140–149

Macía MJ (2004) A comparison of useful Pteridophytes between two Amerindian groups from Amazonian Bolivia and Ecuador. Am Fern Soc 94(1):39–46

Martinez MR, Pochettino ML, Cortella AR (2004) Environment and illness in the Calchaqui Valley (Salta, Argentina): phytotherapy for osteo-articular and cardio-circulatory diseases. J Ethnofarm 95(2004):317–327

Mello M, Budel JM (2013) *Equisitum* L. (Equisetaceae): uma Revisão. Cadernos da Escola de Saúde 9:1–15

Michielin EMZ, Bresciani LFV, Danielski L, Yunes RA, Ferreira SRS (2005) Composition profile of horsetail (*Equisetum giganteum* L.) Oleoresin: comparing SFE and organic solventes extraction. J Supercrit Fluids 33(2005):131–138

Mobot Trópicos (2015) Tropicos.org. Missouri Botanical Garden. Accessed 20 Jul 2015

Mir AS, Mishira AK, Reshi ZA, Sharma MP (2013) Preliminary phytochemical screening of some pteridophytes from District Shopian (J&K). Internacional J Pharm Sci 5(4):632–637

Nunes GP, Silva MF, Resende UM, Siqueira JM (2003) Plantas medicinais comercializadas por raizeiros no Centro de Campo Grande, Mato Grosso do Sul. Rev Bras Farmacogn 3(2):83–92

Portillo A, Vila R, Freixa B, Adzet T, Canigueral S (2001) Antifungal activity of Paraguayan plants used in tradicional medicine. J Ethnofarm 76(2001):93–98

Pryer KM, Schneider H, Smith AR, Cranfill R, Wolf PG, Hunt JS, Sipes SD (2001) Horsetails and ferns are a monophyletic group and the closest living relatives to seed plants. Nature 409:618–622

Quiroga EM, Sampietro AR, Vattuone MA (2001) Screening antifungical activities of selected medicinal plants. J Ethnofarm 74(2001):89–96

Rodrigues MT, Alves TLS, Soares GLG, Ritter MR (2012) Plants used as antidiabetics in popular medicine in Rio Grande do Sul, Southern, Brasil. J Ethnofarm 139(2012):155–163

Salino A, Almeida TE (2015) Equisetaceae in Lista de Espécies da Flora do Brasil. Jardim Botânico do Rio de Janeiro. Available at: HYPERLINK http://floradobrasil.jbrj.gov.br/jabot/floradobrasil/FB91157. Accessed 15 Sept 2015

Santos MG, Kelecom A, Paiva SR, Moraes MG, Rocha L, Garret R (2010) Phytochemical studies in Pteridophytes growing in Brazil: a review. Am J Plant Sci Biotechnol 4(1):113–125

Smith AR, Pryer KM, Schuettpelz E, Korall P, Schneider H, Wolf PC (2006) A classification for extant ferns. Taxon 55:705–731

Smith AR, Pryer KM, Schuettpelz E, Korall P, Schneider H, Wolf PG (2008) Fern classification. In: Ranker TA, Haufler CH (eds) The biology and evolution of ferns and lycophytes. Cambridge University Press, Cambridge, pp 417–467

Tryon RM, Tryon AF (1982) Ferns and allied plants, with special reference to tropical America. Springer, New York

Heteropterys tomentosa A. Juss.

Fúlvio Rieli Mendes and Eliana Rodrigues

Heteropterys tomentosa A. Juss.
Photo: Maria de Fátima Barbosa Coelho
Available in: http://www.scielo.br/scielo.php?script=sci_arttext&pid=S1516-05722011000400013

F. R. Mendes (✉)
Centro de Ciências Naturais e Humanas, Universidade Federal do ABC,
São Bernardo do Campo, SP, Brazil

E. Rodrigues
Centro de Estudos Etnobotânicos e Etnofarmacológicos, Universidade Federal de São Paulo,
Rua Arthur Riedel, 275, Diadema, SP, CEP 09972-270, Brazil

U. P. Albuquerque et al. (eds.), *Medicinal and Aromatic Plants of South America*, Medicinal and Aromatic Plants of the World 5,
https://doi.org/10.1007/978-94-024-1552-0_20

Abstract *Heteropterys tomentosa* A. Juss. (Syn. *Heteropterys aphrodisiaca*, Malpighiaceae) is a species that occurs in Brazilian Cerrado, where it is known as *nó-de-cachorro* and used as a tonic, aphrodisiac, depurative, a treatment for nervous debility, and other uses. Preparations in traditional medicine use the leaves and, mainly, the roots, as a decoction or macerated in alcoholic beverages. The roots of *H. tomentosa* are obtained mainly by extractivism, which has contributed to the decline of populations of the plant. Agronomic studies indicate that the species is easy to cultivate. There are reports indicating the sale of botanical material from other species as nó-de-cachorro, which indicates the importance of quality control studies. The *H. tomentosa* roots have flavonoids, tannins, saponins, terpenoids, and other phytochemical classes. The flavonoids astilbin, neoastilbin, and isoastilbin were identified in the plant and an aliphatic nitro compound with antimicrobial activity suggested as a possible marker. Pharmacological tests indicate that the hydroalcoholic extract of *H. tomentosa* has antioxidant activity and a beneficial effect on memory, especially in aged rats. The aqueous infusion produced positive effects on spermatogenesis and on the reproductive tract of male rats. Preclinical toxicological data are conflicting, indicating that the toxicity depends on the route of administration, dose, and animal species used.

Keywords Heteropterys tomentosa · Heteropterys aphrodisiaca · Malpighiaceae · nó-de-cachorro · Memory · Astilbin · Aliphatic nitro compound

Abbreviations

TLC	Thin layer chromatography
HPLC	High pressure liquid chromatography
CO_2	Carbon dioxide
ACTH	Adrenocorticotropic hormone
DPPH	2,2-diphenyl-1-picrylhydrazyl
LD_{50}	Median lethal dose

1 Introduction

Heteropterys tomentosa A. Juss. is a species native to Cerrado (Central Brazilian Savanna). It is used medicinally as a tonic and aphrodisiac, among other uses. Most studies on this species used the name *Heteropterys aphrodisiaca* O. Mach, given by Othon Machado (1949). However, *H. tomentosa* is currently considered as the correct name and will, therefore, be used in this chapter including reference to studies using *H. aphrodisiaca*, considered a botanical synonym (Amorin 2015).

2 Taxonomic Characteristics

The genus *Heteropterys* Kunth belongs to the taxonomic family Malpighiaceae, which is embedded in the Malpighiales order, Magnoliidae subclass, within the class Equisetopsida (Tropicos 2015). According to the website of the Brazilian Flora Checklist (Lista de Espécies da Flora do Brasil) (Amorin 2015) *H. aphrodisiaca* is a synonym accepted, as well as *H. spectabilis* A. Juss. and *H. verbascifolia* Grisep. The Missouri Botanical Garden gives *H. brachiata* (L.) DC. as an accepted name (Tropicos 2015), but according to Plant List website (Plant List 2013) the plant name *Heteropterys aphrodisiaca* is unresolved and currently is not accepted.

In Latin, the term *hetero* means uneven and *pterys* means wing, referring to the winged fruit with an asymmetric shape. *Tomentosa* comes from the Latin *tomentum* meaning hairy, in reference to the trichomes that cover some new leaves. The aphrodisiaca term used as specific epithet in its synonym refers to Aphrodite, goddess of love, because of its use as a sexual stimulant.

The *H. tomentosa* is commonly known as nó-de-cachorro, nó-de-porco, guaco, jasmim-amarelo, quaró, resedá-amarelo, tintureiro, coração-de-são-francisco, cordão-de-são-francisco, and raiz-de-santo-antônio (Sangirardi 1981; Corrêa 1984; Pott and Pott 1994; Coelho et al. 2011). It is called ocinanta-sá-caá by Karajá Indians. The most notably used name is nó-de-cachorro (dog's knot), due to the appearance of its roots, with thickened parts and knots, which resemble canine penises during intercourse (Pott and Pott 1994). According to Corrêa (1984), the name nó-de-cachorro is also used for other species of the genus: *H. anceps* NDZ.

3 Major Chemical Constituents and Bioactive Compounds

Qualitative phytochemical analysis indicated the presence of the following chemical groups in the hydroalcoholic extract of the *H. tomentosa* roots: flavonic glycosides, simple aromatic glycosides, anthracene compounds, polyphenols, condensed and hydrolysable tannins, alkaloids, cardiac glycosides, and saponins (by foam test) (Galvão 1997; Galvão et al. 2002). The presence of polyphenols, flavonoids, tannins, saponins, and anthracene steroidal substances in the plant roots was confirmed by Marques et al. (2007), including the isolation of the flavonoids astilbin, isoastilbin, and neoastilbin, which were used in the quality control study of the species. The presence of flavonoids and terpenoids was also confirmed using thin-layer chromatography (TLC) by Veggi et al. (2014). An aliphatic nitro compound with antimicrobial activity was isolated from a fraction of the *H. tomentosa* roots' acetone extract (Roman Júnior et al. 2005; Melo et al. 2008) and, according to the authors, it could be used as a marker for the species.

Paula-Freire et al. (2013) compared the chemical composition of the hydroalcoholic extract of roots, branches, and leaves of *H. tomentosa* by TLC and HPLC. The qualitative assessment indicated the presence of hydrolysable tannins, flavonoids, triterpenes, and saponins (foam test) in the three parts of the plant, whereas alkaloids, coumarins, lignans, iridoids, and naphthoquinones were not observed. The HPLC analysis showed distinct chromatographic profiles for the three parts of the plant: the main aglycones found were taxifolin (in the roots and branches), catechin (roots and leaves), rutin and chlorogenic acid (only in the extract of the leaves).

In a pharmacognostic study the authors tested the extraction with water or water mixed with ethanol, methanol, acetone (1:1), and acetone (7:3), and they showed that the aqueous extract yielded higher extractive content, indicating that the root is rich in polar compounds (Marques et al. 2007). Veggi et al. (2014) compared the supercritical and subcritical fluid extraction of *H. tomentosa* roots using pure CO_2 or combined with ethanol or water. The extraction with only CO_2 was the one that produced the highest extraction of phenolic compounds, but the extraction made with CO_2 + water proved to be the most economically viable due its high phenolic content and low cost.

The characterization of plant material and its extracts is important, as there are reports of adulteration with the use of other species in place of *H. tomentosa* (Marques et al. 2007). Marques et al. (2007) performed the morphological, anatomical and physical chemical characterization of *H. tomentosa*, with a description of various characteristics of the whole material and its powder. The use of TLC with astilbin as a marker and physical chemical tests have been proposed as simple and inexpensive methods for quality control of the botanical drug for this species (Braz et al. 2012).

4 Morphological Description

Heteropterys tomentosa is a 1–2 m high shrub (Fig. 1), with subscandent; rust-colored, reddish branches with internodes from 11 to 14 cm long. Opposite leaves with canaliculated petiole, thick, puberulent, sometimes granular, elliptic-ovate or nearly oval, with ciliate margin, acute apex, base rounded, slightly contracted, entire and flat margin; when new the leaves are tomentose on both sides, glabrous on the upper side and tomentose-velutinous on the lower, 4.5–23 cm in length (Corrêa 1984; Coelho et al. 2011). It has odorless flowers with yellow corolla, assembled into inflorescences, becomes rosy after the period of fertilization and subsequent red; produces fruit type samara (Fig. 2), with seed in the basal portion and a wing on the terminal (Corrêa 1984; Barata et al. 2009; Coelho et al. 2011). It displays cylindrical and irregular tuberous roots (Fig. 3), with dimensions in the adult plant between 0.5 and 2 cm in diameter and lengths between 3 and 30 cm, with thicker parts and others with marked narrowing (Marques et al. 2007).

A key based on leaf anatomy was proposed to distinguish 16 species of Malpighiaceae, including *H. tomentosa* and 3 species of the genus (Araújo et al. 2010).

Fig. 1 General aspect of *Heteropterys tomentosa*

Fig. 2 Fruits type-samara of *Heteropterys tomentosa*

5 Geographical Distribution

H. tomentosa is native to Brazil, occurs in dystrophic soils of Cerrado, especially in the states of São Paulo, Mato Grosso, and Goiás (Corrêa 1984; Guarin-Neto 1987; Pott and Pott 1994; Coelho et al. 2011). Besides Brazil, the species has been found in Paraguay, Bolivia, and Peru (Tropicos 2015).

Fig. 3 Roots of *Heteropterys tomentosa*

6 Ecological Requirements

The species tolerates fire; its spread is promoted by deforestation (Pott and Pott 1994), with the budding of new branches from the underground tuberous structure. The flowers have spontaneous self-pollination, but pollinating bees are needed to help break the cuticle that covers the stigma (Coelho et al. 2011).

According to Coelho et al. (2011) the reproduction of *H. tomentosa* is exclusively sexual, the plant being propagated by its seeds scattered by the wind. It presents deciduous behavior, with leaf fall and budding occurring at the same time. It produces flowers and fruit in the dry season, between April and August (Coelho and Spiller 2008; Coelho et al. 2011). The common practice of burning during the months of reproduction of the species, the substitution of the Cerrado areas for cultivation of grain or pasture for cattle, and removal of plants to obtain its roots as medicinal drug have threatened populations of *H. tomentosa*, therefore studies on the sustainable cultivation and management of the species are important.

7 Cultivation and Agronomic Aspects

Studies indicate that *H. tomentosa* is easy to cultivate. The seeds can be stored for 2 years and still have good germination rate (Coelho et al. 2011). The best temperature for seed germination is around 30 °C and the lighting conditions do not appear to influence the germination (Arruda et al. 2003; Hernandez et al. 2011). The seedlings develop slowly in the first months and may be transplanted after 6–8 months, preferably at the beginning of the rainy season. The species grows well in poor soils (Arruda 2001) and after 24 months produces about 80 g of dried roots, making it viable for commercial exploitation. Coelho et al. (2011) suggest that cultivation can be associated with other cultures, optimizing the cultivation of crops and providing an alternative income to farmers. A germplasm collection is maintained in an Experimental Station of the Universidade Federal do Mato Grosso, with samples collected in various regions of the state.

8 Ethnopharmacology

The leaves are employed in teas and baths, and the roots can be prepared as a tea decoction or macerated in alcoholic beverages (Macedo and Ferreira 2000). Preparations known as "garrafadas", made with the roots macerated in wine or "cachaça", a local spirit made of sugar cane, are used as sexual stimulants and aphrodisiacs (Mendes and Carlini 2007; Barata et al. 2009). It is also used as a depurative, dysenteric, tonic, uterine, for uric acid problems, nerve weakness, venereal diseases, ophthalmic ailments, and others (Corrêa 1984; Guarin-Neto 1987; López-Palacios 1983; Pott and Pott 1994; Macedo and Ferreira 2004). There are also reports of the use of crushed roots macerated in water as a tonic, against diarrhea or to heal dermal ulcers (Coelho et al. 2011). The tea, prepared by decoction, is used to treat diabetes, flu, diarrhea, and intestinal and kidney infections; the leaves can be used in baths applied to the legs for strengthening the muscles of children and the elderly or to wash the eyes in the treatment of cataracts (Coelho et al. 2011). We also obtained information that in addition to the root some communities in Mato Grosso use the skin of the roots to prepare a reddish spirit and to strengthen the nerves of children who have difficulty walking and to facilitate labor, as well as an aphrodisiac (personal communication). A review carried by Coelho et al. (2011) mentions more than 30 uses in studies with traditional communities for the nó-de-cachorro.

H. tomentosa was classified by Rizzini (1983) as a psychoactive plant with a stimulating effect. The diversity of popular uses for the nó-de-cachorro also allows it to be included in the category of adaptogenic plants, which are often used chronically to improve the general functions of the body, such as the *Panax ginseng* C.A. Meyer and *Eleutherococcus senticosus* (Rupr. & Maxim.) Maxim., among other classic adaptogens (Mendes 2011). In fact, in an ethnopharmacological survey carried out with practitioners of Umbanda, the chronic use of the root macerated on "cachaça" was nominated for three simultaneous therapeutic purposes: as an aphrodisiac, to thin the blood, and to improve memory (Rodrigues and Carlini 2004).

9 Pharmacology and Toxicology

There are several studies evaluating the biological properties of *H. tomentosa*, but more detailed studies and especially clinical studies are needed to confirm the effects alleged by the population. The early pharmacological studies with nó-de-cachorro were carried out by Galvão (1997) and were the basis for further studies evaluating the effects of the species on the central nervous system. This initial study evaluated the acute and chronic effects of the hydroalcoholic extract of *H. tomentosa* in rodents. The acute oral treatment in mice showed a stimulating effect and did not affect motor coordination and the sleep time of animals, demonstrating a possible absence of toxic effects. Oral chronic treatment of aged rats with 50 mg/kg of the lyophilized extract produced positive effects on memory in rats (Galvão 1997; Galvão et al. 2002). Further studies were carried out by the same group with different doses and times of treatment using experimental models of learning and memory. Aged rats that received doses of 25 and 50 mg/kg orally for 45 days learned a discriminative task in a T-maze in a shorter time than did animals of the same age without treatment (Galvão et al. 2004–2005, 2011). The flavonoids astilbin, isoastilbin, and neoastilbin were identified in the hydroalcoholic extract and it was suggested that they may be involved with the positive effects on memory observed for the plant (Galvão et al. 2011). Moreover, aged rats treated with the extract at a dose of 50 mg/kg for 26 or 7 days and tested in a passive avoidance test had a moderate improvement of memory, which was not observed after acute treatments (Galvão et al. 2002, 2004–2005). However, the acute treatments for 7 or 21 days at doses of 100–400 mg/kg did not reverse the scopolamine-induced amnesia in mice (Galvão et al. 2004–2005, 2011). To assess whether the stimulating effect of the extract was due to dopaminergic action, young and aged rats treated with *H. tomentosa* were challenged with a moderate dose of apomorphine, a dopaminergic drug that induces stereotypy. Pretreatment with the hydroalcoholic extract of nó-de-cachorro for 7 days did not alter the stereotypy of young animals but increased the degree of stereotypy of aged animals at 20 and 30 min after apomorphine (Galvão et al. 2004–2005). The stereotypy of aged rats treated for 120 days does not differ from the control group, indicating that chronic treatment may induce tolerance. All these studies used a patented standardized extract called BST 0298 (Biosintética/UNIFESP 2000), although so far it has not led to the development of a medicine.

In contrast to earlier results, a study by Paula-Freire et al. (2013) showed no improvement in memory of aged rats treated orally for 80 days with extracts of roots or stems of *H. tomentosa* (75 mg/kg). This study employed hydroalcoholic extracts of different plant parts (roots, branches and leaves) to assess the possible adaptogen action of nó-de-cachorro. However, the treatment for 14 days at doses of 100 and 300 mg/kg did not protect the rats from cold and restraint stress (measured by stomach ulcerations, organ weights, and ACTH and corticosterone levels), and the same doses administered for 7 days did not change the response of mice in a test of self-analgesia induced by stress (Paula-Freire et al. 2013). Although the extract used in this study is different from that employed by Galvão and colleagues, the phenolic content of both extracts was quite similar.

Regarding the possible sexual stimulant/aphrodisiac effect of nó-de-cachorro, which is the main popular use described for the plant, we found only one study in which the oral administration of hydroalcoholic extract for 7 days showed a stimulating effect on the sexual behavior of rats 12 months old, without producing these effects in younger animals (Santos and Carlini 2000). However, the effect was not maintained with continued treatment for another week and the authors conclude that more experiments are needed to confirm the sexual stimulant effect of *H. tomentosa*. In another study, an infusion prepared from the roots macerated in hot water (proportions of 12.5 or 25 g dried root for each 100 ml of boiling water) was administered orally to rats for 56 days and at the end of the treatment the animals' weight and the weight of testis was higher than those of control animals, although the gonadosomatic index (which considers body weight) did not significantly change (Chieregatto 2005). The analysis of the testis showed that the treatment with *H. tomentosa* increased the thickness of the seminiferous epithelium and tubule diameter while decreasing its length. There was also an increase in the volume of Leydig cells, which the author attributes to a possible increase in the production of testosterone (Chieregatto 2005).

Based on data suggesting that nó-de-cachorro has an androgenic effect, the effect of its treatment on the male reproductive tract organs of healthy mice and animals treated with other drugs was investigated. The oral administration of *H. tomentosa* (infusion prepared with 25 g of root in 100 ml of boiling water) for 56 days induced no significant morphological changes in the testis or prostate epithelium (Monteiro et al. 2008; Freitas et al. 2012). The effect of treatment with the same extract, dose, and duration was also evaluated in rats that received cyclosporin A, an immunosuppressive agent that induces various side effects. Cyclosporin A caused several changes in testis tissue, as seminiferous epithelium degeneration and Sertoli cell vacuolization, among other damages, and most of the changes were decreased or prevented with the concomitant use of *H. tomentosa* (Monteiro et al. 2008). Cyclosporin A also caused changes in the ventral prostate tissue and increased the levels of glutamic oxalacetic transaminase, cholesterol, triglycerides, and glucose, but animals that received the *H. tomentosa* infusion did not show these changes (Freitas et al. 2013).

Other studies from the same group evaluated the effect of treatment with the same preparation of nó-de-cachorro on animals subjected to forced exercise (Gomes et al. 2011; Monteiro et al. 2011). The exercise protocol did not affect spermatogenesis and the biometric data of animals, but the treatment with the infusion of *H. tomentosa* was enough to increase the secretion of testosterone, promoting increased cell division in the germ cells, and increased spermatogenesis (Gomes et al. 2011). The treatment with the infusion (104 mg/day) for 8 weeks induced anabolic-like effects with the significant increase in stress and maximum load capacity on the tendons of animals, which was attributed to more organized collagen bands and positive modulation on biochemical parameters involved with physical activity (Monteiro et al. 2011). Moreover, Gomes et al. (2011) have found a reduction in the number of apoptotic cells in the testis of rats treated with the aqueous infusion of *H. tomentosa*. These results contrast with a study where young and aged rats treated for

30 days with a hydroalcoholic extract of the plant did not present difference in the number of apoptotic cells in the hippocampus when compared to control animals of the same age (Bezerra et al. 2013). Gomes et al. (2011) suggest that the antioxidant effect of the extract, among other mechanisms, may be responsible for the protective effect observed.

In fact, the antioxidant action of *H. tomentosa* has been well documented (Mattei et al. 2001; Galvão et al. 2004–2005, 2011; Veggi et al. 2014). The hydroalcoholic extract of the plant showed antioxidant activity in vitro in homogenates of rat brain and increased the activity of total superoxide dismutase, as well the manganese- and copper-zinc-dependent isoforms in the brain tissue of aged rats treated with the extract (50 mg/kg, orally) for 90 days (Mattei et al. 2001). Although the treatment of young rats for the same period did not change the activity of antioxidant enzymes, a reduction of free iron levels (25%) and thiobarbituric acid reactive substances (30%) was observed in the brain of animals. The antioxidant activity of several lots of roots collected in the same region, but in different seasons and years, proved to be relatively similar, with the best activity observed for the lot collected in the summer. It was also observed that the fraction rich in astilbin and neoastilbin showed a weak antioxidant effect, while the nitrogen fraction did not exhibit antioxidant activity (Galvão et al. 2004/2005). A study which used the supercritical extraction with CO_2 showed that the extraction using CO_2 + ethanol presented higher scavenging activity of DPPH free radical than CO_2 alone or CO_2 + water extraction (Veggi et al. 2014).

An aliphatic nitro compound isolated from *H. tomentosa* was tested for its antifungal activity (*Candida albicans*, *C. krusei*, *C. parapsilosis* and *C. tropicalis*) and bactericidal (*Bacillus subtilis*, *Staphylococcus aureus*) and was effective in different concentrations against all strains investigated (Roman Júnior et al. 2005). The nitro compound was moderately active against poliovirus type 1 and type 1 bovine herpes virus in cell cultures, with 50% inhibitory concentration of 22 and 21 μg/ml, respectively (Melo et al. 2008). However, treatment of the cells before infection did not inhibit virus replication.

Micronucleus and Ames tests were performed with the hydroalcoholic extract and did not indicate signs of genotoxicity and mutagenicity, respectively (Galvão 2003). With regard to toxicological studies in animals, the data are somewhat controversial. Most studies with *H. tomentosa* extracts found no signs of toxicology, but initial studies conducted by Galvão (1997, 2003) employing the hydroalcoholic extract showed some toxicity, depending on the dose, route, and animal species used.

Several lots of nó-de-cachorro were evaluated for acute oral and ip toxicity at increasing doses. The median lethal dose (LD_{50}) for oral administration was >5000 mg/kg, and by via ip the LD_{50} ranged between 380 and 1047 mg/kg for the different lots tested (Galvão 2003). The oral chronic treatment of young rats at the dose of 100 mg/kg resulted in lower levels of glucose, cholesterol, and triglyceride in rats (Galvão 1997). The oral administration of doses from 200 to 800 mg/kg for 30 days to guinea pigs did not change the general condition of the animals, food consumption, and weight gain. However, there was a reduction in weight gain in rats treated orally for 30 days with doses of 100 and 200 mg/kg or with 800 mg/kg for 7 days. There were some changes in biochemical and hematological parameters evaluated, but apparently not related to the doses (Galvão 2003). The extract administration to female rats did not affect the estrous cycle or pregnancy or the weight gain and development of offspring (Galvão 2003).

Toxicological studies using lots of the same hydroalcoholic extract of *H. tomentosa* were also conducted with three breeds of dogs (Galvão 2003). Mongrels were treated for 90 days at doses of 50 and 100 mg/kg and showed no signs of toxicity. However, Beagles treated with doses of 100 and 200 mg/kg showed motor incoordination, ataxia, and muscle rigidity after the first days of administration and some animals presented seizures and death. The anatomopathological evaluation indicates the presence of microhemorrhages, neurodegenerative processes, with demyelination, among other findings. Boxers were also treated for 90 days with doses up to 400 mg/kg. There were no deaths among these animals, but some dogs showed signs of sedation and somnolence after receiving the extract, and histopathological examination at the end of treatment showed some changes similar to those observed for Beagles (Galvão 2003).

A phase I clinical toxicology study with BST 0298 extract was initiated at the Federal University of São Paulo, but the results are not available in the literature.

10 Conclusions

The ethnobotanical surveys, phytochemical, and pharmacological studies with *H. tomentosa* indicate that the species has a great potential for medical use and economic exploitation. However, sustainable management and plant cultivation are necessary to ensure the availability of the raw material, as well as quality control tests to certify the authenticity of the botanical material. In addition, more preclinical toxicological studies and clinical trials are essential to ensure safety and to validate the alleged popular uses for nó-de-cachorro (*H. tomentosa*).

Acknowledgments The authors thank Prof. Wayne Losano for the grammar review.

References

Amorin AMA (2015) [Internet]. *Heteropterys* in Lista de Espécies da Flora do Brasil. Jardim Botânico do Rio de Janeiro. Available from: http://floradobrasil.jbrj.gov.br/. Accessed 2 Mar 2015

Araújo JS, Azevedo AA, Silva LC, Meira RMSA (2010) Leaf anatomy as an additional taxonomy tool for 16 species of Malpighiaceae found in the Cerrado area (Brazil). Plant Syst Evol 286(1–2):117–131

Arruda JB (2001) Aspectos da germinação e cultivo do nó-de-cachorro (*Heteropteris aphrodisiaca* O. Mach.). [dissertation]. Universidade Federal do Mato Grosso, Cuiabá

Arruda JB, Camargo IP, Albuquerque MCF, Coelho MFB, Ferronato A (2003) Efeito da luminosidade na germinação de sementes de nó-de-cachorro (*Heteropteris aphrodisiaca* O. Mach.). Rev Bras Pl Med 5(2):55–59. Portuguese

Barata LES, Alencar AAJ, Tascone M, Tamashiro J (2009) Plantas medicinais brasileiras III. *Heteropterys aphrodisiaca* Machado (nó-de-cachorro). Rev Fitos 4(1):129–131. Portuguese

Bezerra AG, Smaili SS, Lopes GS, Carlini EA (2013) Effects of *Panax ginseng*, *Turnera diffusa* and *Heteropterys tomentosa* extracts on hippocampal apoptosis of aged rats. Einstein (São Paulo) 11(2):163–167

Biosintética/UNIFESP, inventors; Extratos hidroalcoólicos da *Heteropteris aphrodisiaca* (nó-de-cachorro). Brazilian Patent protocol INPI # 98035518-A. 2000 Feb 8

Braz R, Wolf LG, Lopes GC, Mello JCP (2012) Quality control and TLC profile data on selected plant species commonly found in the Brazilian market. Rev Bras Farmacog 22(5):1111–1118

Chieregatto LC (2005) Efeito do tratamento crônico com extratos de *Heteropterys aphrodisiaca* O. Mach. e *Anemopaegma arvense* (Vell.) Stellf. no testículo de ratos Wistar adultos. [dissertation]. Universidade Federal de Viçosa, Viçosa

Coelho MFB, Jorge AS, Macedo M, Nogueira Borges HB, Spiller C (2011) Nó-de-cachorro (*Heteropterys tomentosa* A. Juss.): espécie de uso medicinal em Mato Grosso, Brasil. Rev Bras Pl Med 13(4):475–485. Portuguese

Coelho MFB, Spiller C (2008) Phenology of *Heteropterys aphrodisiaca* O. Mach. Malpighiaceae in Mato Grosso. Rev Bras Pl Med 10(1):1–7

Corrêa MP (1984) Dicionário das plantas úteis do Brasil e das exóticas cultivadas, vol V. Ministério da Agricultura, Instituto Brasileiro de Desenvolvimento Florestal, Brasília, p 687

Freitas KM, Monteiro JC, Gomes MLM, Taboga SR, Dolder H (2012) Study of the ventral prostate of Wistar rats treated with *Heteropterys tomentosa* (A. Juss.). J Med Pl Res 6(44):5640–5646

Freitas KM, Monteiro JC, Gomes MLM, Taboga SR, Dolder H (2013) *Heteropterys tomentosa* (A. Juss.) infusion counteracts Cyclosporin A side effects on the ventral prostate. BMC Complement Altern Med 13:30

Galvão SMP (1997) Estudos farmacológicos e toxicológicos de *Heteropteris aphrodisiaca* O. Mach. – Malpighiaceae (nó-de-cachorro) em roedores jovens e idosos. [dissertation – PhD Thesis]. Universidade Federal de São Paulo, São Paulo

Galvão SMP (2003) *Heteropterys aphrodisiaca* O. Mach. (EXTRATO BST 0298): estudos pré-clínicos farmacológicos e toxicólogicos. [dissertation]. Universidade Federal de São Paulo, São Paulo

Galvão SMP, Marques LC, Oliveira MGM, Carlini EA (2002) *Heteropterys aphrodisiaca* (extract BST0298): a Brazilian plant that improves memory in aged rats. J Ethnopharmacol 79(3):305–311

Galvão SMP, Mendes FR, Oliveira MGM, Mattei R, Carlini EA (2004–2005) Possíveis efeitos adaptógenos da *Heteropterys aphrodisiaca* O. Mach. – extrato BST 0298: uma planta da área do Pantanal brasileiro. Arq Bras Fitomed Cient 2(1): 41–55. Portuguese

Galvão SMP, Mendes FR, Oliveira MGM, Mattei R, Mello JCP, Roman Junior WA, Carlini EA (2011) Memory retrieval improvement by *Heteropterys aphrodisiaca* in aging rats. Braz J Pharm Sci 47(4):825–832

Gomes MLM, Monteiro JC, Freitas KM, Sbervelheri MM, Dolder H (2011) Association of the infusion of *Heteropterys aphrodisiaca* and endurance training brings spermatogenetic advantages. Biol Res 44(3):235–241

Guarin-Neto G (1987) Plantas utilizadas na medicina popular do Estado de Mato Grosso. Ministério da Ciência e Tecnologia e CNPq, Brasília, p 58

Hernandez FMP, Coelho MFB, Maia SSS, Albuquerque MCF (2011) Germinação de sementes de *Heteropteris tomentosa* A. Juss. sob diferentes temperaturas e períodos de armazenamento. Rev Bras Ciên Agrár 6(4):617–621. Portuguese

López-Palacios S (1983) Enumeracion de algunos afrodisiacos vegetales. Rev Fac Farm ULA 23:5–63. Spanish

Macedo M, Ferreira AR (2000) *Heteropteris aphrodisiaca* O. Mach. "nó-de-cachorr": uma espécie usada na medicina popular em Mato Grosso, Brasil. Anais do XVI Simpósio de Plantas Medicinais do Brasil, Recife, p 90

Macedo M, Ferreira AR (2004) Plantas hipoglicemiantes utilizadas por comunidades tradicionais da Bacia do Alto Paraguai e Vale do Guaporé, Mato Grosso, Brasil. Rev Bras Farmacog 14(Suppl 1):45–47. Portuguese

Machado OXB (1949) Nova espécie do gênero Heteropterys Kunth. Rodriguésia 11–12:113–119. Portuguese

Marques LC, Pieri C, Roman Junior WA, Cardoso MLC, Milaneze-Gutierre MA, Mello JCP (2007) Controle farmacognóstico das raízes de *Heteropteris aphrodisiaca* O. Mach. (Malpighiaceae). Rev Bras Farmacog 17(4):604–615. Portuguese

Mattei R, Barros MP, Galvão SMP, Bechara EJH, Carlini ELA (2001) *Heteropteris aphrodisiaca* O. Machado: effects of extract BST 0298 on the oxidative stress of young and old rat brains. Phytother Res 15(7):604–607

Melo FL, Benati FJ, Roman Junior WA, Mello JCP, Nozawa C, Linhares REC (2008) The in vitro antiviral activity of an aliphatic nitro compound from *Heteropteris aphrodisiaca*. Microbiol Res 163(2):136–139

Mendes FR (2011) Tonic, fortifier and aphrodisiac: adaptogens in the Brazilian folk medicine. Rev Bras Farmacog 21(4):754–763

Mendes FR, Carlini EA (2007) Brazilian plants as possible adaptogens: an ethnopharmacological survey of books edited in Brazil. J Ethnopharmacol 109(3):493–500

Monteiro JC, Gomes MLM, Tomiosso TC, Nakagi WR, Sbervelheri MM, Ferrucci DL, Pimentel ER, Dolder H (2011) More resistant tendons obtained from the association of *Heteropterys aphrodisiaca* and endurance training. BMC Complement Altern Med 11:51

Monteiro JC, Predes FS, Matta SLP, Dolder H (2008) *Heteropterys aphrodisiaca* infusion reduces the collateral effects of cyclosporine A on the testis. Anat Rec Adv Integr Anat Evolut Biol 291(7):809–817

Paula-Freire LIG, Mendes FR, Molska GR, Duarte-Almeida JM, Carlini EA (2013) Comparison of the chemical composition and biological effects of the roots, branches and leaves of *Heteropterys tomentosa* A. Juss J Ethnopharmacol 145(2):647–652

Pott A, Pott VJ (1994) Plantas do Pantanal. EMBRAPA-SPI, Corumbá. 320 p

Rizzini CT (1983) Efeitos psicotrópicos de plantas brasileiras. Parte II: aspectos botânicos. Ciên Cult 35(4):434–438. Portuguese

Rodrigues E, Carlini EA (2004) Plants used by a Quilombola group in Brazil with potential central nervous system effects. Phytother Res 18(9):748–753

Roman Júnior WA, Cardoso MLC, Vilegas W, Nakamura CV, Dias Filho BP, Mello JCP (2005) 2,3,4,6-Tetra-*O*-(3-nitropropanoyl)-*O*-β-D-glucopyranoside, a new antimicrobial from the roots of *Heteropteris aphrodisiaca*. Acta Farm Bonaer 24(4):543–545

Sangirardi JR (1981) Plantas eróticas. Codecri, Rio de Janeiro, pp 222–223

Santos R, Carlini E (2000) Efeitos da *Heteropteris aphrodisiaca* sobre o comportamento sexual de ratos. Anais do XVI Simpósio de Plantas Medicinais do Brasil, Recife, pp 260–261

The Plant List (2013) Version 1.1. [Internet]. Available from: http://www.theplantlist.org/. Accessed 18 Jul 2018

Tropicos.org (2015) [Internet]. Missouri Botanical Garden. Available from: http://www.tropicos.org/. Accessed 10 Mar 2015

Veggi PC, Cavalcanti RN, Meireles MAA (2014) Production of phenolic-rich extracts from Brazilian plants using supercritical and subcritical fluid extraction: experimental data and economic evaluation. J Food Eng 131:96–109

Himatanthus drasticus (Mart.) Plumel

André Sobral, Alessandro Rapini, and Ulysses Paulino Albuquerque

Himatanthus drasticus
Photo: data bank from Laboratório de Ecologia e Evolução de sistemas socioecológicos

A. Sobral
Laboratório de Ecologia e Evolução de Sistemas Socioecológicos, Departamento de Botânica, Universidade Federal de Pernambuco, Recife, PE, Brazil

A. Rapini
Biological Sciences Department, State University of Feira de Santana (Universidade Estadual de Feira de Santana), Feira de Santana, Bahia, Brazil

U. P. Albuquerque (✉)
Departamento de Botânica, Centro de Biociências, Universidade Federal de Pernambuco, Recife, Brazil

U. P. Albuquerque et al. (eds.), *Medicinal and Aromatic Plants of South America*, Medicinal and Aromatic Plants of the World 5,
https://doi.org/10.1007/978-94-024-1552-0_21

Abstract The South American genus *Himatanthus* (Apocynaceae) includes nine species. *Himatanthus drasticus*, known in Brazil as "janaguba," is used in popular medicine to treat inflammation, gastric ulcers and tumors. Scientific studies have confirmed that its latex, popularly referred to as "leite de janaguba" (janaguba milk), exhibits some important therapeutic activities.

Keywords Apocynaceae · Janaguba · Latex · Cerrado · Brazil

1 Taxonomic Characteristics

Himatanthus Willd. ex Schult. is a South American genus belonging to the family Apocynaceae. This family is widely distributed in tropical and subtropical regions (Spina et al. 2013) and includes 366 genera (Endress et al. 2014). It belongs to the tribe Plumerieae E. Mey. and the subtribe Plumeiriinae Pichon & Leeuwenb., together with the monotypic genus *Mortoniella* Woodson and the ornamental genus *Plumeria* L. The taxonomy of the genus *Himatanthus* was revised by Woodson (1938), Plumel (1991) and Spina (2004). Spina et al. (2013), based on Spina (2004), recognized nine species in this genus, six of which are primarily found in the Amazon: *Himatanthus articulatus* (Vahl) Woodson, *Himatanthus attenuatus* (Benth.) Woodson, *Himatanthus phagedaenicus* (Mart.) Woodson, *Himatanthus revolutus* (Huber) Spina & Kin. Gouv., *Himatanthus semilunatus* (Markgr.) and *Himatanthus tarapotensis* (K.Schum. ex Markgr.) Plumel. Of the three species that are primarily extra-Amazonian, *Himatanthus obovatus* (Müll. Arg.) Woodson occurs in the Cerrado (Brazilian savanna) areas of Brazil and Bolivia, and *Himatanthus drasticus* (Mart.) Plumel and *Himatanthus bracteatus* (A.DC.) Woodson occurs exclusively in Brazil, mostly in the Northeast region.

In Northeast Brazil, *H. drasticus* occurs most frequently in the state of Ceará, more specifically in Chapada do Araripe in the Southernmost point of the state (Colares et al. 2008). This species was originally described by Martius, based on his own collection performed in Caetité, state of Bahia. Its basionym is *Plumeria drastica* Mart., and *Himatanthus fallax* Müll. Arg. is a heterotypic synonym (Spina et al. 2013). It is popularly known as "janaguba" in Ceará and is highly valued as a medicinal plant. In the states of Minas Gerais and Bahia, it is popularly known as "tiborna," "jasmim-manga" and "raivosa." It is also known as "pau-de-leite" in Piauí, "joanaguba" in Rio Grande do Norte, "sucuuba" in the Amazon region (Plumel 1991), and "janaúba" in Maranhão (Linhares and Pinheiro 2013). Its distribution also extends to Guiana, French Guiana and Suriname (Amaro et al. 2006), where it is popularly known as "caterpillar tree" (Moragas 2006).

Another extra-Amazonian species that is used as a medicinal plant is *H. obovatus*. It is primarily distributed in the Cerrado phytogeographic domain, particularly in savannah vegetation (Morokawa et al. 2013), but it also occurs in the Amazon and Caatinga (Spina 2014). Its basionym is *Plumeria obovata* Müll. Arg. (Spina et al. 2013). It has been observed in North (Pará, Rondônia and Tocantins), Northeast (Alagoas, Bahia, Maranhão and Piauí), West-Central (Distrito Federal, Goiás, Mato

Grosso do Sul and Mato Grosso), and Southeast (Minas Gerais and São Paulo) Brazil, and extends to Bolivia (Spina 2014; Morokawa et al. 2013), where it is popularly known as "mangava brava" (Plumel 1991).

2 Crude Drug and Its Uses

The latex of *H. drasticus* is the main product used for medicinal purposes. It is extracted by removing the plant's bark. The latex is popularly known as "leite de janaguba" (janaguba milk) and is extracted from the janaguba populations at the Araripe National Forest (Floresta Nacional do Araripe) in the state of Ceará (Baldauf et al. 2014). The latex is mixed with water and sold in local markets, and it is indicated for the treatment of different diseases, such as gastritis, anemia, and inflammations, as well as several types of tumors (Baldauf and Santos 2013).

3 Major Chemical Constituents and Bioactive Compounds

A distinctive characteristic of family Apocynaceae is the presence of laticifers, which produce a latex that is rich in alkaloids related to the plant's defense against herbivory (Linhares et al. 2013). The latex contains depsides, terpenes and iridoids, such as fulvoplumierin, isoplumericin and plumericin (Colares et al. 2008). These iridoids possess confirmed antineoplastic, antiphlogistic and antimicrobial activities (Colares et al. 2008). The latex of several species from genus *Himatanthus*, including *H. drasticus*, is also rich in triterpenes. Pentacyclic triterpenes, including lupeol, are promising plant secondary metabolites (Laszczyk 2009). Recently, lupeol acetate was isolated from the latex of *H. drasticus*. This compound exhibited anti-inflammatory activity, which likely prevents the production of pro-inflammatory mediators such as TNF-α and IL-1β (Lucetti et al. 2010).

4 Morphological Description

H. drasticus is a medium sized lactescent tree that can reach up to 7 m in height. It possesses large leaves, which are more dense at the end of the branches, with short petioles, usually one pair of colleters on the leaf axil, and an oblanceolate to elliptic, glabrous, sub-coriaceous leaf blade, with a dark green upper side and a light green lower side. The flowers are relatively large, pentameric, actinomorphic, with approximately 3 cm in length, with a soft odor, a green calyx and a white corolla, yellowish fauces, hypocrateriformis, convolute and sinistrorse, and lobes that are slightly longer than the tube. The tube has a glabrous exterior and pubescent interior. The flowers are arranged in terminal articulated cymes, with a pair of large, deciduous bracts that are up to 2 cm in length and cover each pair of floral buds.

These large bracts are exclusive to *Himathanthus,* and leave scars in the axis of the inflorescence when they fall. There are five stamens, and the two carpels of the gynoecium are fused at the apex, forming an obconical style head, but are free at the ovary level. The ovaries are semi-inferior. The fruits are formed by a pair of slightly curved, divergent, fusiform follicles that are usually 15–20 cm in length, each with numerous round seeds with concentric wings. (Spina 2004).

5 Geographical Distribution

The genus *Himatanthus* is widely distributed in South America, occurring from Southeast Brazil to French Guiana, Guiana and Suriname. In Brazil, *H. drasticus* has been recorded in the states of Minas Gerais, Espírito Santo, Bahia, Ceará, Maranhão, Sergipe, Alagoas, Pernambuco, Rio Grande do Norte, Paraíba, Piauí, Pará, Roraima, Goiás, Mato Grosso and Mato Grosso do Sul. However, it occurs predominantly in the Caatinga domain, in Northeast Brazil (Sousa et al. 2010; Spina 2014). It is abundant in the Chapada do Araripe (Sousa et al. 2010), which is at an altitude of 900 m and is located between the states of Ceará, Pernambuco and Piauí (Costa et al. 2004). The annual rainfall in this region varies between 600 and 2200 mm, and the average temperature is 23 °C. The Chapada do Araripe includes different types of vegetation, with transition zones between Cerrado (wooded savannah) and Cerradão (densely wooded savannah), tropical forest, and carrasco (xerophytic scrubland) (IBAMA 2004).

6 Ecological Requirements

H. drasticus is restricted to the tropical and subtropical areas in Brazil, particularly the areas in the Northeast, which are characterized by high temperatures and low annual rainfall, with marked seasonality. It is considered a pioneer species in the Cerrado and Caatinga areas, grows well in open vegetation with high sunlight incidence, and is resistant to fire. Its seeds are wind dispersed, and both their germination and seedling establishment require open areas with a high sunlight incidence (Baldauf et al. 2014; Baldauf and Santos 2013).

7 Collection Practice

The medicinal value of *H. drasticus* is widely acknowledged in traditional medicine and has been confirmed by pharmacological studies. This confirmation has led to an increase in the extraction of the bark and latex from the natural populations of *H.*

drasticus to meet the increasing demand for leite de janaguba (Baldauf et al. 2011; Baldauf and Santos 2014).

According to Baldauf and Santos (2014) the collection sites are selected based on the density of the janagubas. Trees that are between 7 and 40 cm diameter at breast height (DBH) are selected for extraction. Trees of this size have reached the reproductive phase and are more resilient and better able to recover after bark removal (Baldauf and Santos 2014), and the latex is extracted by performing a vertical cut of approximately 2 m in the tree bark down to the base of the plant using a machete or scythe (Baldauf and Santos 2014). Bark removal causes latex exudation, which is then collected using a water-soaked sponge or a spoon (Linhares et al. 2013).

The regeneration capacity and the time needed for plant regeneration depend on the number of sides exploited. The more sides that are exploited, the longer the time that is required for plant regeneration and until the next extraction, and between 6 and 18 months are estimated to be required for full bark regeneration (Baldauf and Santos 2014).

At the Chapada do Araripe, the latex collection from *H. drasticus* starts before 5:00 am because the harvesters believe that the plants produce more latex at that time, and the collection lasts for 5 h until 10:00 am, because heat decreases exudation (see Baldauf and Santos 2013, 2014). The extraction is more intense in the rainy season, between December and May, in the Northeast region. The harvesters consider this to be the period of highest latex production per plant (see Baldauf and Santos 2013, 2014). However, many harvesters also collect the latex during the dry season, as they consider that the latex is of higher quality during this season (Linhares et al. 2013).

8 Traditional Use (Part(s) Used) and Common Knowledge

H. drasticus is used in popular medicine for the treatment of inflammatory processes (Lucetti et al. 2010), ulcers (Colares et al. 2008), gastritis and tumors (Ribeiro et al. 2014; Souza et al. 2014) and is used as an immunostimulant (Mousinho et al. 2011) and antimicrobial agent (Luz et al. 2014). In addition to these uses for human health, it has also been used to feed goats because it is known to help control worm infestations (Luz et al. 2014). The main plant parts that are used for these purposes are the bark and latex (Ribeiro et al. 2014). The bark infusions are used to treat tumors, gastritis, arthritis and hemorrhoids, and the fresh leaves are crushed and used as compresses against herpes, mycoses and warts. There are also records of the use of leaf infusions or decoctions (Ribeiro et al. 2014) to treat urethra irritation and uterus inflammation (Colares et al. 2008; Sousa et al. 2010).

The latex, a milky white juice extracted from the trunk and branches, is one of the most commercialized products of *H. drasticus* and is used in popular medicine for the treatment of tumors, worm infestations, gastritis, arthritis and cancer (Colares et al. 2008; Ribeiro et al. 2014; Sousa et al. 2010). The ethanol extracts from the leaves and roots exhibited cytotoxic activity against cerebrovascular diseases, colon carcinoma, melanoma and leukemia cells in vitro (Melo et al. 2011a).

H. obovatus is used in popular medicine for the treatment of several infection-related afflictions, such as wound cicatrization, which indicates that it may exhibit antibiotic activity (Bieski et al. 2012). Mesquita et al. (2005) observed that the extracts from the *H. obovatus* leaves exhibited activity against the promastigote form of *Leishmania donovani*. Bieski et al. (2012) considered that the use of *H. obovatus* in popular medicine may be related to its immune system modulating capacity, which increases the activity of the physiological mechanisms involved in the resolution of inflammation and pain and in wound cicatrization. Moragas (2006) observed that the extracts from the leaves and latex of *H. obovatus* contained the same chemical substances that are present in *H. drasticus*.

Himatanthus phagedaenicus, known as "leiteiro," "banana-de-papagaio" and "angelica-do-mato" (Plumel 1991), is popularly used in Northeast Brazil for the treatment of ulcers, diabetes, inflammations, hepatic diseases, and warts, in addition to as an anthelmintic agent (Agra et al. 2007). Brandão et al. (2011) observed that extracts from the bark and leaves of *H. phagedaenicus* exhibited antiviral activity against the human herpes simplex virus type-1 (HSV-1).

Himatanthus articulatus, which is popularly known as "sucuuba" in the Brazilian Amazon, is also used in popular medicine to treat ulcers, tumors, inflammation (Agra et al. 2007), syphilis (Barreto et al. 1998), and malaria (Milliken 1997).

Studies have shown that *H. lancifolius* (Muell. Arg.) Woodson, a heterotypic synonym of *H. bracteatus* (Spina et al. 2013), also exhibits pharmacological potential (Baratto et al. 2010). Its stem bark is traditionally used to treat asthma, skin diseases, syphilis and menstrual disturbances (Côrrea 1926). The latex extracted from the stem is used as an anthelmintic agent (Côrrea 1926), and the latex from the roots is used to treat problems with the uterus and ovaries (Plumel 1991).

9 Modern Medicine Based on Its Uses in Traditional Medicine

Colares et al. (2008) tested the gastroprotective activity of *H. drasticus* and showed that the latex prevented the gastric lesions induced by ethanol and indometacin in mice. The latex from janaguba was shown to be rich in triterpenes, compounds that possess antioxidant and cytoprotective properties and have confirmed anti-ulcerogenic actions.

Mousinho et al. (2011) tested the popular indication of the antitumor activity of the latex from janaguba using in vitro and in vivo experimental models and observed that the latex extracts had no cytotoxic effects in vitro but had antitumor activity on both of the systems tested in vivo (sarcoma 180 and carcinosarcoma Walker 256), which may be associated with the stimulation of the immune system (Mousinho et al. 2011). To date, the pharmacological studies that have been performed to test the therapeutic activities of *H. drasticus* have used animal (mice) or in vitro cell-based models, and there are no reports of tests with human subjects.

Pharmacological analyses of the extracts from other species, such as *H. phagedaenicus* (Brandão et al. 2011) and *H. articulatus* (Rebouças et al. 2011), showed that the presence of iridoids is associated with its antineoplastic, antiphlogistic, antimicrobial (Colares et al. 2008) and antiviral actions (Brandão et al. 2011). Iridoids are present in different species of *Himatanthus* (Rebouças et al. 2011). A recent study (Rebouças et al. 2011) tested the genotoxic and mutagenic activity of a bark extract from *H. articulatus* and found no antitumor activity. However, the authors demonstrated that it had a protective effect against hydrogen peroxide-induced DNA damage (Rebouças et al. 2013).

Extracts from the bark of *H. lancifolius* contain indole alkaloids (Nardin et al. 2010; Souza et al. 2007), and these compounds exhibit gastroprotective (Baggio et al. 2005), antimicrobial (Morel et al. 2006; Souza et al. 2004), antispasmodic (Rattmann et al. 2005) and anti-inflammatory activities (Nardin et al. 2009). Jiménez et al. (2001) reported that the triterpenoids and flavonoids present in the extracts of *H. attenuatus* were associated with decreased blood pressure in rats, without changes in their heart rates (Jiménez et al. 2001).

10 Conclusions

H. drasticus is primarily distributed in Northeast Brazil and is known in traditional (popular) medicine to possess substances that can treat diseases, such as inflammation, gastric ulcers and tumors. Its therapeutic action has been confirmed by pharmacological studies, which has led to increased demand and increased extraction of the *H. drasticus* latex, popularly known as "leite de janaguba," and bark (Mousinho et al. 2011; Lucetti et al. 2010, etc.). Remarkably, this confirmation has led to the increased commercialization of latex, and consequently, to an increased latex harvesting, which – ultimately – may have a negative impacts on the natural populations of this species (Baldauf and Santos 2013).

Although pharmacological studies have indicated the therapeutic efficacy of *H. drasticus* products, no tests have been performed on human subjects, and there are also not-known patents or drugs made from janaguba.

References

Agra MF, Freitas PF, Barbosa-Filho JM (2007) Synopsis of the plants known as medicinal and poisonous in Northeast of Brazil. Braz J Pharmacogn 17(1):114–140

Amaro MS, Medeiros Filho S, Guimarães RM, Teófilo EM (2006) Morfologia de frutos, sementes e de plântulas de janaguba (*Himatanthus drasticus* (Mart.) Plumel – Apocynaceae). J Seed Sci 28(1):63–71

Baggio CH, Otofuji GD, Souza WM, Santos CAD, Torres LMB, Rieck L, Marques MCD, Mesia-Vela S (2005) Gastroprotective mechanisms of indole alkaloids from *Himatanthus lancifolius*. Planta Med 71:733–738

Baldauf C, Santos FAM (2013) Ethnobotany, traditional knowledge and diachronic changes in non-timber forest products management: a case study of Himatanthus drasticus (Apocynaceae) in the Brazilian savanna. Econ Bot 67(2):110–120

Baldauf C, dos Santos FAM (2014) The effect of managem ent systems and ecosystem types on bark regeneration in Himatanthus drasticus (Apocynaceae): recommendations for sustainable harvesting. Environ Monit Assess 186(1):349–359

Baldauf C, Ciampi MB, Vigna BBZ, Mori GM, Guedes JPP, Souza AP et al (2011) Characterization of microsatellite *loci* in *Himatanthus drasticus* (Apocynaceae), a medicinal plant from the brazilian savanna. Am J Bot 98(9):244–246

Baldauf C, Silva AS, Sfair JC, Ferreira R, Santos FAM (2014) Harvesting increases reproductive activity in *Himatanthus drasticus* (Mart.) Plumel (Apocynaceae), a non-timber forest product of the Brazilian savanna. Biotropica 46(3):341–349

Barreto A, Carvalho M, Almeida Nery I, Gonzaga L, Kaplan MAC (1998) Chemical constituents from *Himatanthus articulata*. J Braz Chem Soc 9(5):430–434

Bieski IGC, Santos FR, Oliveira RM, Espinosa MM, Macedo M, Albuquerque UP, Martin DTO (2012) Ethnopharmacology of medicinal plants of the Pantanal Region (Mato Grosso, Brazil). Evid Based Complement Alternat Med 2012:1–36

Brandão GC, Kroon EG, Santos JR, Stehmann JR, Lombardi JL, Oliveira AB (2011) Antiviral activity of plants occurring in the State of Minas Gerais (Brazil): part III. J Chem Pharm Res 3:223–236

Colares AV, Cordeiro LN, Costa JGM, Cardoso AH, Campos AR (2008) Efeito gastroprotetor do látex de *Himatanthus drasticus* (Mart.) Plumel (Janaguba). Inf Dent 20:34–36

Côrrea MP (1926) Dicionário das plantas úteis do Brasil e das exóticas cultivadas. Imprensa Nacional, Rio de Janeiro

Costa IR, Araújo FS, Lima-Verde LW (2004) Flora e aspectos auto-ecológicos de um encrave de cerrado na chapada do Araripe, Nordeste do Brasil. Acta Bot Bras 18(4):759–770

Endress M, Liede-Schumann S, Meve U (2014) An updated classification for Apocynaceae. Phytotaxa 159:175–194

IBAMA (2004) Plano de Manejo da Floresta Nacional do Araripe. Brasília, Instituto Brasileiro do Meio Ambiente e dos Recursos Naturais Renováveis

Jiménez G, Hasegawa M, Rodríguez M, Estrada O, Méndez J, Castillo A, Gonzalez-Mujica F, Motta N, Vásquez J, Romero-Vecchione E (2001) Biological screening of plants of the Venezuelan Amazon. J Ethnopharmacol 77:77–83

Laszczyk MN (2009) Pentacyclic triterpenes of the lupane, oleanane and ursane group as tools in cancer therapy. Planta Med 75:1549–1560

Linhares JFP, Pinheiro CUB (2013) Caracterização do sistema de extração de látex de janaúba (*Himatanthus* Willd. ex Schult. – Apocynaceae), no município de Alcântara, Estado do Maranhão, Brasil. Pan-Amazonian J Health 4(1):23–31

Lucetti DL, Lucetti EC, Bandeira MAM, Veras HN, Silva AH, Leal LKA et al (2010) Anti-inflammatory effects and possible mechanism of action of lupeol acetate isolated from *Himatanthus drasticus* (Mart.) Plumel. J Inflamm 7(60):1–11

Luz HS, Santos ACG, Machado KRG (2014) Prospecção fitoquímica de *Himatanthus drasticus* Plumel (Apocynaceae), da mesorregião leste maranhense. Braz J Med Plant 16(3):657–662

Mesquita ML, Desrivot J, Bories C, Fournet A, Paula JE, Grellier P, Espindola LS (2005) Antileishmanial and trypanocidal activity of Brazilian Cerrado plants. Mem Inst Oswaldo Cruz 100(7):783–787

Milliken W (1997) Traditional anti-malarial medicine in Roraima. Brazil Econ Bot 51:212–237

Moragas CJ (2006) Estudo do gênero *Himatanthus*: anatomia vegetal, fitoquímica, farmacologia e biotransformação [thesis]. Universidade Federal do Rio de Janeiro, Rio de Janeiro

Morel AF, Graebner IB, Porto C, Dalcol II (2006) Study on the antimicrobial activity of *Himatanthus sucuuba*. Fitoterapia 77:50–53

Morokawa R, Simões AO, Kinoshita LS (2013) Apocynaceae s. str. of the Serra da Canastra National Park, State of Minas Gerais, Brazil. Rodriguésia 64(1):179–199 http://rodriguesia.jbrj.gov.br. Accessed 20 Mar 2015

Mousinho KC, Oliveira CDC, Ferreira JRDO, Carvalho AA, Magalhães HIF, Bezerra D et al (2011) Antitumor effect of laticifer proteins of *Himatanthus drasticus* (Mart.) Plumel – Apocynaceae. J Ethnopharmacol 137(1):421–426

Nardin JM, Souza WM, Lopes JF, Florão A, Santos CAM, Weffort-Santos AM (2009) Effects of *Himatanthus lancifolius* on human leukocyte chemotaxis and their adhesion to integrins. Planta Med 74:1253–1258

Nardin JM, Lima MP, Machado JCJ, Hilst LF, Santos CAM, Weffort-Santos AM (2010) The uleine-rich fraction of *Himatanthus lancifolius* blocks proliferative responses of human lymphoid cells. Planta Med 76(7):697–700

Plumel MM (1991) Le genre *Himatanthus* (Apocynaceae) révision taxonomique. Bradea 5(suplemento):1–118

Rattmann YD, Terluk MR, Souza WM, Santos CA, Biavatti MW, Torres LB, Mesia-Vela S, Rieck L, Silva-Santos JE, Marques MC (2005) Effects of alkaloids of *Himatanthus lancifolius* (Muell. Arg.) Woodson, Apocynaceae, on smooth muscle responsiveness. J Ethnopharmacol 100:268–275

Rebouças SDO, Grivicich I, Santos MS, Rodriguez P, Gomes MD, Oliveira SQ, Silva J, Ferraz ADBF (2011) Antiproliferative effect of a traditional remedy, *Himatanthus articulatus* bark, on human cancer cell lines. J Ethnopharmacol 137:926–929

Rebouças SDO, Silva J, Bertoni RS, Decker N, Santos MS, Rossatto RR, Corrêa DS, Ferraz ABF (2013) Assessment of the genotoxic and mutagenic properties of *Himatanthus articulatus* bark extracts used as phytotherapeutic drug in the Amazon. J Ethnopharmacol 147:474–480

Ribeiro DA, Oliveira LGS, Macêdo DG, Menezes IRA, Costa JGM, Silva MAP et al (2014) Promising medicinal plants for bioprospection in a Cerrado area of Chapada do Araripe, Northeastern Brazil. J Ethnopharmacol 155:1522–1533

Sousa EL, Grangeiro ARS, Bastos IVGA, Rodrigues GCR, Silva MJ, Anjos FBR et al (2010) Antitumor activity of leaves of *Himatanthus drasticus* (Mart.) Plumel-Apocynaceae (janaguba) in the treatment of Sarcoma 180 tumor. Braz J Pharm Sci 46(2):199–203

Souza WM, Stinghen AE, Santos CAM (2004) Antimicrobial activity of alkaloidal fraction from barks of *Himatanthus lancifolius*. Fitoterapia 75:750–753

Souza WM, Brehmer F, Nakao LS, Stinghen AEM, Santos CAM (2007) Ação da uleína sobre a produção de óxido nítrico em células RAEC e B16F10. Rev Bras Farmacogn 17:191–196

Souza RKD, Silva MAP, Menezes IRA, Ribeiro DA, Bezerra LR, Souza MMDA (2014) Ethnopharmacology of medicinal plants of carrasco, northeastern Brazil. J Ethnopharmacol 157:99–104

Spina AP (2004) Estudos taxonômicos, micro-morfológico e filogenético do gênero *Himatanthus* Willd. ex Schult. (Apocynaceae: Rauvolfioideae – Plumerieae). [thesis]. Universidade Estadual de Campinas, Campinas

Spina AP (2014) *Himatanthus* in Lista de Espécies da Flora do Brasil. http://reflora.jbrj.gov.br/jabot/floradobrasil/FB4621. Accessed 5 Dec 2014

Spina AP, Bittrich V, Kinoshita LS (2013) Typifications, new synonyms and a new combination in *Himatanthus* (Apocynaceae). Taxon 62(6):1304–1307

Woodson RE (1938) Studies in the Apocynaceae. VII- an evaluation of the genera *Plumeria* L. and *Himatanthus* Willd. Ann Missouri Bot Gard 25:189–224

Justicia pectoralis Jacq.

Carles Roersch

Justicia pectoralis Jacq.
David Neill
Available in: http://www.tropicos.org/Image/100222854

C. Roersch (✉)
Herbario "Dr. Henri Alain Liogier", Universidad Nacional Pedro Henriquez Ureña (UNPHU), Santo Domingo, Dominican Republic
e-mail: croersch@unphu.edu.do; croersch@imd-medicina-dominicana.org

U. P. Albuquerque et al. (eds.), *Medicinal and Aromatic Plants of South America*, Medicinal and Aromatic Plants of the World 5,
https://doi.org/10.1007/978-94-024-1552-0_22

Abstract *Justicia pectoralis* is used as a medicinal plant in Central America, the Caribbean and the tropical parts of South America. It has a longstanding history, being already mentioned in the sixteenth century in nowadays the Dominican Republic. The species is included in the pharmacopeia of Brazil and Cuba for its applications as an expectorant and sedative for nervous affections, respectively. In traditional medicine, the most frequent application is for illnesses of the respiratory tract. The scientific research concentrates on the anti-inflammatory, analgesic and sedative effects of the plant, with positive results that confirm traditional uses. Toxicity has hardly been reported. It would be recommended that more research should be done on the pharmacokinetics of the extracts and the clinical aspects.

Keywords *Justicia pectoralis* · Traditional uses · Chemical constituents · Expectorant · Sedative

1 Taxonomic Characteristics

Synonyms *Dianthera pectoralis* (Jacq.) Murray, *Dianthera pectoralis* (Jacq.) J.F. Gmel, *Ecbolium pectorale* (Jacq.) Kuntze, *Justicia pectoralis* var. *latifolia* Bremek, *Justicia stuebelii* Lindau, *Psacadocalymma pectorale* (Jacq.) Bremek, *Rhytiglossa pectoralis* (Jacq.) Nees, *Stethoma pectoralis* (Jacq.) Raf. (The Plant List 2014, Tropicos n.d.).

Justicia is the largest genus of the family Acanthaceae, with approximately 400 species that are distributed in pantropical and tropical regions (The Plant List 2014). In the Amazon basis and in Cuba a variety of *J. pectoralis* is described, *J. pectoralis* var. *stenophylla*, as an erect herb of 15–20 cm. with linear-lanceolate leaves of a dark green color (Fuentes et al. 2000). However this variety is considered more to be a growth form than a genetic variant (MacRae and Towers 1984). In The Plant List (2014) and Tropicos (n.d.), this variety is not mentioned.

The hydroalcoholic extracts from both varieties of *J. pectoralis* turned to be very similar according to the results of the chemical, toxicological and pharmacological studies performed on equal footing (Rodríguez et al. 2008). As a result, we have included the literature data on *J. pectoralis* var. *Stenophylla*, without mentioning it separately.

Common Names
English: carpenter's grass, garden balsem, death-angel
Spanish: carpintero, curía, tila, tilo
French: herbe à charpentier, charpentier
Portuguese: chamb'a, anador

Beside these widely used common names, *J. pectoralis* is locally known under a great variety of names. In Peru, it is called azul, cuya-cuya and lluichu (Egg 1999; Rutter 1990; Duke et al. 2009), in French Guyana as carmentin (DeFilipps et al. 2008), in Jamaica as fresh cut (Facey et al. 1999; Picking et al. 2011), in Venezuela as hierba de San Antonio and ancú (Gupta 1995), in Panama as mojo bren (Joly et al. 1990), in Ecuador as moradilla blanca (Tene et al. 2007), in Surinam as papawiwiri and tonkawiwiri (van Andel and Ruysschaert 2011), in the Virgin Islands as rock balsam and sweet mint (Thomas 1997; Cabi no date), in Guyana as toyeau (van Andel 2000), and in Colombia as curibano and mejorana (Cabi no date).

Several common names refer to the fragrance of one of its principal constituents, coumarin. Garden balsam is a common name in Barbados (Honychurch 1986), Trinidad and Tobago (Seaforth et al. 1983) and Montserrat (Brussell 1997, 2004). In Montserrat, the plant is also called bitter balsam (Brussell 1997, 2004), in Belize, balsam vine (Balick et al. 2000; Duke et al. 2009) and as balsam and rock balsam in the Virgin Islands (Thomas 1997). In Cuba, the plant is generally known as Tilo or Tila. This name refers to the European Tilo, *Tilia europea* L. This European tilo was imported to Cuba and widely used as a sedative. During the Second World War (1940–1945) it was hardly possible to import tilo, so that a substitute was found in *J. pectoralis* (Roig and Mesa 1965).

2 Crude Drug Used

All parts of *J. pectoralis* are used as a drug. These can be fresh as well as dried. The most common application form is as a tea (infusion or decoction). Externally the crushed leaves are used as a poultice. In Costa Rica, dried ethanol extracts of the aerial parts of *J. pectoralis* are commonly sold as an over-the-counter sleep aid under the name of Estilo© (Locklear et al. 2010).

3 Major Chemical Constituents and Bioactive Compounds

The major chemical constituents of *J. pectoralis* are coumarin (1,2-benzopirona), and umbelliferone (7-hydroxycoumarin) (MacRae and Towers 1984; Oliveira et al. 2000; Fonseca et al. 2010). The presence of betaine was also confirmed (MacRae and Towers 1984). Small amounts of ortho-hydroxy-transcinnamic acid (acetylated coumaric acid), ortho-hydroxydihydrocinnamic acetylated acid (acetylated melilotic acid) and β-sitosterol were found in a Brazilian *J. pectoralis* (Taveira 1993; Lino et al. 1997). Furthermore, Oliveira et al. (2000) identified O-glycosides (quercetin and kaempferol) and stigmasterol and De Vries et al. (1988) also identified dihydroxycoumarin and dihydrocoumarin. Joseph et al. (1988a) detected the flavonoids, swertisin and swertiajaponin and the O-methylated C-glycosylflavones 2′-O-rhamnosylswertisin and 2′-O-rhamnosylswertiajaponin. Also the lignan, Justicidin B was described (Joseph et al. 1988b). Alkaloids were not detected (MacRae and Towers 1984; Oliveira et al. 2000). Coumarin, for its fragrance, has been commonly incorporated into cosmetics and detergents (Opdyke 1974). Justicidin B, 1-aryl-2,3-naphthalide lignan, is active in NCI murine P-388 lymphocytic leukemia (Joseph et al. 1988b). Coumarin and umbelliferone that was isolated from the aerial parts of *J. pectoralis* showed anti-inflammatory activity in rats (2.5–5 mg/kg, orally) (Lino et al. 1997). Umbelliferone (30, 60, and 90 mg/kg, orally) attenuates airway inflammation in a murine model of asthma (Vasconcelos et al. 2009). It also has an antihyperglycemic effect in Streptozotocin-diabetic rats (30 mg/kg body weight) comparable with glibenclamide (Ramesh and Pugalendi 2006). Ramalingam and Vaiyapuri (2013) found a possible protective action of umbelliferone against liver damage, lipid peroxidation and the antioxidant defense system in N-Nitrosodiethylamine – induced liver carcinogenesis in rats.

Natural as well as synthetic coumarin-derived compounds demonstrate very promising anti-inflammatory activity. However, no such compound has yet been developed as a commercial drug (Bansal et al. 2013).

4 Morphological Description

Justicia pectoralis is an ascendant or decumbent herb that grows up to a height of 1.5 m. Thin, often rooting at lower nodes, leaves lanceolate to ovate – lanceolate, acuminate at apex, acute or obtuse at the base, glabrous, inflorescence terminal in panicle with few to many flowers, alternate branches 2–12 cm, more or less glandular, bracts and bracteoles subulate up to 3 mm.; calyx segments 5, subulate 2 mm, the posterior somewhat shorter; pink corolla puberula 8–15 mm, upper lip 4 mm, 2 lobed, lower lip 7 mm purple with white stripes; capsule of 5–6 mm, puberula (Liogier 1995).

5 Geographical Distribution

J. pectoralis is a fairly common tropical plant in various states in Mexico, Central and South America (tropical regions) and the Caribbean (USDA, ARS no date).

6 Ecological Requirements

In Cuba, the plant is cultivated in rows of 1 m in width. The plant needs sufficient water to develop, but resist some periods of drought. *J. pectoralis* needs sufficient sunlight to produce enough coumarins that are greatly responsible for its medicinal use (Fuentes et al. 2000). In the wild, *J. pectoralis* can be found along roadsides, riverbanks, streams and waste places. It grows well in moist to wet forests (Cabi no date).

7 Collection Practice

In general, the plant is collected in the wild-state. Due to its medicinal uses, many families are reported to grow the plant in their gardens. In Surinam, *J. pectoralis, although* originally a native species, is now sources almost exclusively from home gardens (van Andel and Havinga 2008). It is also present in 21% of the stalls at the market in Paramaribo and Albina in Surinam (van Andel et al. 2007). It can also be bought in herb stores from Surinam in Amsterdam, The Netherlands (van Andel and Ruysschaert 2011).

8 Traditional Use (Part(s) Used) and Common Knowledge

The first mention of *J. pectoralis* was made by Gonzalo Fernández de Oviedo, considered as the first author of the Americas, who described in an organized form the flora and the fauna of the New World. In his famous 'Historia general y natural de las Indias' (General and Natural History of the Indies), he describes a plant called Curía by the Taino people in now called the Dominican Republic, which almost certainly corresponds to *Justicia pectoralis*. The plant was used by the Tainos as an aphrodisiac and for wound healing (Fernández De Oviedo 1851). The latter use is still present in Venezuela (Gupta 1995), Trinidad and Tobago (Seaforth et al. 1983), Virgin Islands (Thomas 1997), Haíti (Beauvoir et al. 2001; Duke et al. 2009), Puerto Rico (Nuñez 1992) and the Dominican Republic (Cordero 1986). The use as an aphrodisiac has been lost.

In the literature, we have found a total of 126 recipes describing the traditional uses of *J. pectoralis* in 19 countries. By far, most recipes refer to ailments, illnesses of the respiratory tract (29%), followed by the digestive tract (12%), wounds, bruises and sprains (10%), Nerves (9%) and Pain (9%).

In the Respiratory tract the whole plant or leaves are used for *Influenza* (Guianas, DeFilipps et al. 2008), *Whooping cough* (Guianas, DeFilipps et al. 2008; van Andel 2000), *Cough* (Guianas, DeFilipps et al. 2008; van Andel 2000; Ecuador, Tene et al. 2007; Brazil, Agra et al. 2008; Albuquerque et al. 2007, Surinam, van Andel and Ruysschaert 2011; Trinidad and Tobago, Seaforth et al. 1983; Morton 1977; Wong (1976); Virgin Islands, Thomas 1997), *Colds* (Guianas, DeFilipps et al. 2008; van Andel 2000; Ecuador, Tene et al. 2007; Brazil, Albuquerque et al. 2007; Dominican Republic, Cordero 1986; Montserrat, Brussell 1997, 2004; Surinam, van Andel and Ruysschaert 2011; Costa Rica, Gupta 1995; Jamaica, Facey et al. 1999; Trinidad and Tobago, Seaforth et al. 1983; Wong (1976); Virgin Islands, Thomas 1997; Martinique, Honychurch 1986), *Chills* (Dominican Republic, Beauvoir et al. 2001, in Duke et al. 2009), *Pneumonia* (Trinidad, Wong (1976); Brazil, Albuquerque et al. 2007), *Asthma* (Surinam, Ruysschaert et al. 2009; Brazil, Agra et al. 2008; Albuquerque et al. 2007), *Bronchitis* (Brazil, Agra et al. 2008; Surinam, van Andel and Ruysschaert 2011) and *Expectorant* (Brazil, Agra et al. 2008; Puerto Rico, Nuñez 1992).

In the digestive tract the whole plant and the leaves are used to cure: *Stomach(ache)* (Guianas, DeFilipps et al. 2008; Surinam, van Andel and Ruysschaert 2011; Haíti, Germosén-Robineau 2005; Gupta 1995; Beauvoir et al. 2001, in Duke et al. 2009; Weniger et al. 1986; Martinique, Honychurch 1986; Panama, Morton 1977), *Dyspepsia* (Dominican Republic, Beauvoir et al. 2001, in Duke et al. 2009), *Antiemetic* (Guianas, DeFilipps et al. 2008), *Dysentery* (Surinam, van Andel and Ruysschaert 2011), *Intestines* (Surinam, van Andel and Ruysschaert 2011) and *Flatulence* (Martinique, Longuefosse and Nossin 1996).

The next category is wounds, bruises and sprains. Recipes are described for: *Hematoma* (Guianas, DeFilipps et al. 2008), *Bruises and sprains* (Dominica, Germosén-Robineau 2005; Martinique, Germosén-Robineau 2005; Puerto Rico, Nuñez 1992), *Sprains*, *Fracture* (Dominican Republic, Beauvoir et al. 2001, in Duke et al. 2009), *Sprains* (Martinique, Longuefosse and Nossin 1996), *Cuts* (Jamaica (Gupta 1995; Trinidad and Tobago, Seaforth et al. 1983; Virgin Islands, Thomas 1997), *Wounds* (Trinidad, Wong (1976); Venezuela, Gupta 1995; Haíti, Beauvoir et al. 2001, in Duke et al. 2009; Puerto Rico, Nuñez 1992) and *Vulnerary* (Dominican Republic, Cordero 1986).

Nerves are calmed down by *J. pectoralis* in a limited amount of countries (5). Most recipes come from Cuba. Remedies are described for: *Sedative* (*nerves*) (Cuba, Roig and Mesa 1928, 1965; Beyra et al. 2004; Virgin Islands, Thomas 1997), *Calmative* (Guianas, DeFilipps et al. 2008; Costa Rica, Gupta 1995), *Anxiety* (Cuba, Macias-Peacok et al. 2009), *Tension* (Costa Rica, Garcia Gonzalez et al. 2002), *Nerves* (Costa Rica, Garcia Gonzalez et al. 2002; Germosén-Robineau 2005; Cuba, Germosén-Robineau 2005; Morton 1977; Puerto Rico, Nuñez 1992).

Analgesic effects of *J. pectoralis* form the fifth category. It is used against: *Headache* (Guianas, DeFilipps et al. 2008; Brazil, Coelho-Ferreira 2009; Surinam, van Andel and Ruysschaert 2011), *Pains* (Brazil, Albuquerque et al. 2007; Surinam, van Andel and Ruysschaert 2011; Panama, Caballero-George and Gupta 2011), *Legs and pain* (Brazil, Coelho-Ferreira 2009; Surinam, van Andel and Ruysschaert 2011; Panama, Morton 1977). In women diseases, we have five recipes which almost all concern *menstruation ailments* (Ecuador, Tene et al. 2007; Surinam, van Andel and Ruysschaert 2011; Venezuela, Gupta 1995; Costa Rica, Locklear et al. 2010).

Heart problems count the following four remedies: *Heart problems* (Surinam, van Andel and Ruysschaert 2011), *Thoracic pain* (Martinique, Longuefosse and Nossin 1996), *Hypertension* (Surinam, van Andel and Ruysschaert 2011; Seaforth et al. 1983), equal to *Fever* (Guianas, DeFilipps et al. 2008; Brazil, Albuquerque et al. 2007; Dominican Republic, Cordero 1986), Surinam, van Andel and Ruysschaert 2011).

Finally, there is a wide range of other ailments, in which cure is attributed to *J. pectoralis*. To name a few: *Rheumatism* (Brazil, Coelho-Ferreira 2009; Venezuela, Gupta 1995; Martinique, Longuefosse and Nossin 1996), '*Tranga wiwiri*' (leaves that make you strong) (Surinam, Ruysschaert et al. 2009), *Prostrate problems* (Trinidad and Tobago, Lans 2007), *Inflammation and infection of the ear* (Venezuela, Meléndez et al. 2012), *Antiinflammatory* (Brazil, Aversi-Ferreira et al. 2013) and *Hepatic disorders* (Panama, Joly et al. 1990; Panama, Caballero-George and Gupta 2011).

In the markets of Costa Rica, the dried plant material is widely advertised as a treatment for menopause and other menstrual ailments (Locklear et al. 2010).

In Surinam *J. pectoralis* is considered as an important ritual plant. It is used in baths, in combination with several other plants, to calm down enemies, to resolve problems with the police, to get lucky, to eliminate nightmares, bad spirits, to reinforce one's own soul; after giving birth, the placenta is buried together with *J. pectoralis* and other strong aromatic herbs (van Andel and Ruysschaert 2011).

The dried leaves of *J. pectoralis* are used as an ingredient in a hallucinogenic *Virola* snuff prepared by the Yanomami Indians in the Amazonas (Schultes and Holmstedt 1968). The plant usually is described as *J. pectoralis* var. *stenoplylla*. However, this variety is considered more to be a growth form than a genetic variant (MacRae and Towers 1984). In The Plant List (2014) and Tropicos (n.d.), this variety is not mentioned. As *J. pectoralis* does not contain any chemical compound with hallucinogen activity, it is thought that the plant is added for its flavor (Agm 1985). Recently, Khan et al. (2012) mention that *J. pectoralis* contain DMT (N,N – dimethyltryptamine), without any literature references.

9 Modern Medicine Based on Its Traditional Medicine Uses

In Cuba the sedative action in nervous affections of *J. pectoralis* was recognized by the health authorities in 1992 and hence the plant was included in the list of therapeutic agents used by the national Cuban health system (MINSAP 1992). More recently in Brazil, the Health Ministry included *J. pectoralis* as an expectorant in a list of 71 medicinal plants within its National Program of Medicinal Plants and Phytotherapeutics (Ministério da Saúde 2008; Formulario de Fitoterápicos 2011).

Some clinical experiments have been performed with *J. pectoralis*. A syrup of the plant was given to asthmatic patients with mild to moderate asthma. Within a week an increase in maximum expiratory flow, forced vital capacity and forced expiratory volume was noted. Also reduced obstruction of the airways was observed in the patients (Nobre et al. 2006; Fonseca et al. 2010). A double-blind clinical test was performed giving one group of patients a capsule of the water extract of *J. pectoralis* and the other group Diazepam. The sedative effect was confirmed and no adverse effect was noticed (Gupta 1995). In another experiment, the decoction of the aerial parts (2% and 6%) was given orally in normal adults (25–35 years) in a clinically controlled study showing significant electroencephalographic modifications, demonstrated in Broad Band Spectral Parameters (BBSPs), revealing neurotropic activity (Rodriguez et al. 1989; Germosén-Robineau 2005).

The pharmacokinetics of one of the principal constituents of *J. pectoralis*, coumarin, has been studied in man by Ritschel et al. (1977, 1979 in De Smet 1985). Given orally, the compound is absorbed completely, but only 2–6% reaches intact the systemic circulation because of extensive first-pass metabolism. The major metabolite is 7-hydroxycoumarin, which in its turn undergoes glucuronidation. This could mean that coumarin acts like a prodrug and is active as 7-hydroxycoumarin or its 7-hydroxyglucuronide form.

In a study on gerbils, intraperitoneally administered coumarin distributed rapidly into the cerebral tissue, whereas its metabolites 7-hydroxycoumarin and 7-hydroxycoumarin glucuronide entered the brain only to a small extent, if at all. The used dose of 40 mg/kg produced transient sedation, and this effect corresponded rather well with the time of maximal coumarin brain concentration and with the subsequent rapid removal of coumarin from the brain (Ritschel and Hardt 1983; De Smet 1985). The same intraperitoneal dose of 40 mg/kg of coumarin was found to cause a longer and deeper level of sedation in the rat, but this species is a poor 7-hydroxylator of coumarin (Hardt and Ritschel 1983, in De Smet 1985).

Several animal studies have been performed to elucidate the possible mechanism of the sedative effect. Male rats treated with a hydro-alcoholic extract of *J. pectoralis* (100 mg/ml, orally) did not show a depressive action on the Central Nervous System (Fica 2005). Fernandez et al. (1987, in Germosén-Robineau 2005) found a significant sedative effect in mice of the decoction of the fresh aerial parts ((10%) in doses of 0.1 ml/g) or the dry aerial parts (10%) (7.5, 15, 75, 400 and 700 mg/kg, via intraperitonally), which showed a comparable dose-dependent curve as for the controls diazepam (0.1, 0.5, 1 and 5 mg/kg), chlorpromazine (0.2, 2 and 7.5 mg/kg) and haloperidol (0.1, 0.3, 1 and 5 mg/kg)). The decoction of the green and dry leaves and stems (1.4% and 10%) produced a decrease in the aggressive conduct and exploratory activity. It was shown that this activity did not correspond to the pharmacological profile of antipsychotic drugs, tricyclic antidepressants and anxiolytic benzodiazepines (Fernandez et al. 1989; Gupta 1995). The decoction of the leaves (75 mg/ml) in mice (via oral, 1 g/kg/day/5 days) did not produce any sedative effect or introduced sleep (Germosén-Robineau 2005).

The water, ethyl acetate and diethyl ether extracts of the leaves, when administered orally to mice (250 mg/kg) reduced spontaneous activity (with the ethyl acetate fraction with the strongest effect) but did not demonstrate any psychotomimetic activity (MacRae and Towers 1984). The behavioral effects in animal models like the elevated plus maze (EPM), light/dark, open field, rotarod and pentobarbital sleep time of the aqueous standardized extract of the aerial parts of *J. pectoralis* (50, 100 and 200 mg/kg, intragastrically) was investigated by Venâncio et al. (2011). Diazepam and flumazenil were used to determinate the interference of benzodiazepinic receptors. The outcome was that the extract showed anxiolytic effects but no sedative effects. In addition, the decoction of the aerial parts of *J. pectoralis*, dry or fresh, do not block the convulsions produced by pentylenetetrazole, unlike diazepam, which suggests that the sedative action does not follow the mechanism of action of benzodiazepines (Perez et al. 1987; Germosén-Robineau 2005).

Despite the fact that more than 500 years ago *J. pectoralis* was mentioned for its wound-healing properties, only one experiment is described in the literature. Mills et al. (1986) tested the dried aqueous and organic extract of the leaves and twigs of *J. pectoralis* and the isolated coumarin (2H-1-Benzopyran-2-one) on wounds (0.5 mg each) in rats. Coumarin attenuated the inflammation and significantly enhanced the healing process. The other two extracts did not significantly improve the healing process, but they nevertheless showed less inflammation and the healing process was better compared with the controls. The anti-inflammatory activity has been demonstrated in several experiments. The standardized extract of the aerial parts of *J. pectoralis* has anti-inflammatory actions that prevent the development of tracheal hyperresponsiveness after antigen challenge in rats (Moura et al. 2013). In the carrageenan-induced rat hind paw edema test the hydroalcoholic extract of the leaves (400 mg/kg, orally) showed anti-inflammatory activity (Leal et al. 2000; Lino et al. 1997). This effect increased when administered intraperitoneally (68%inhibition at 200 mg/kg) (Leal et al. 2000).

The analgesic effect of the hydroalcoholic extract of the leaves of *J. pectoralis* has been established (Fica 2005). Antinociceptive activity was exhibited in the formalin-induced nociception test in mice (Leal et al. 2000) and it also possesses analgesic activity using the writhing test and formalin test in mice (Lino et al. 1997).

Bronchodilator activity (EC_{50} 1.5 ± 0.18 mg/ml of hydro-alcoholic extract of the leaves) was established in carbachol – treated trachea from guinea-pigs (Leal et al. 2000). The dried powder of *J. pastoralis* showed an antioxidant activity in both spontaneous and nonspontaneous self-oxidation of phospholipids in brain tissue of rats (Perez et al. 2001).

The juice and decoction of the leaves and stem (1 mg/ml), in vitro, did not show activity against *Salmonella typhi, Shigella flexneri, S. dysenteriae, Pseudomonas aeruginosa* and *Staphylococcus aureus* (Germosén-Robineau 2005). Also, the extract of the aerial parts (concentration not very clear) of *J. pectoralis* does not show anti-bacterial activity. Against *S.aureus, E. coli, P. mirabilis, P. aeruginosa* and *Streptococcus A* (Facey et al. 1999). However, Chariandy et al. (1999) found an antibacterial effect against *E. coli* and *S. epidermides* (extract of the aerial parts (1000 μg/ml)). They also found a high insecticidal activity of *J. pectoralis* (0.50 mg/ml ethyl-acetate extract of the leaves) against *Aedes aegypti* (Chariandy et al. 1999).

The metanolic extract of the aerial parts of *J. pectoralis* show estrogenic, progestogenic and anti-inflammatory effects (IC_{50} between 4.8 and 50 μg/ml) which give a plausible mechanism of action for its traditional use for menopause and PMS (Locklear et al. 2010). The use against stomachache and nerves, sprains and bruises is recommended by the TRAMIL group (Germosén-Robineau 2005).

The hydroalcoholic extract of the aerial parts of *J. pectoralis* was orally administered to male and female Wistar rats at 10, 100, and 1000 mg/kg/day, 5 days a week during 90 days. The results showed that organs and tissues abnormalities were not observed and only slight variation in blood clotting time and biochemical parameters was present. In the acute toxicity test, rats of both sexes were given an orally single dose of *J. pectoralis* extract at 2000 mg/kg. After 14 days no mortality was observed. The autopsy revealed no signs of toxicity (Lagarto et al. 2009). The alcoholic extract of the aerial parts of *J. pectoralis* given orally to mice resulted in an LD_{50} of 3531.11 mg/kg (Lagarto et al. 2001).

The water extract of fresh aerial parts (2.889 kg in 7.850 l distilled water) was applied to mice via oral (5 g/kg/day/5 days). It did not show any death or toxic signs (Germosén-Robineau 2005).

The water extract (decoction 30%) of the aerial parts, using the Draize Model in rabbit (patch of 0.6 ml/6 cm^2 during 4 h on the shaved skin), did not produce any clinical signs of edema or erythema after 1, 24, 48 and 72 h. The extract of *J. pectoralis* can be classified as not irritating (Germosén-Robineau 2005).

Fresh, crushed aerial parts (0.6 g) were placed on the skin of Wistar rats to determine the acute toxicity by topical application. During 14 days the animals were daily observed. No death or any other signs of adverse effect were noticed. Necropsy revealed no damage to any organ (Germosén-Robineau 2005).

The water extract of *J. pectoralis* applied intravenously in mice resulted in an LD_{50} of 1.344,00 mg/kg. The highest technical administered dose in rats via intraperitoneal was 4.000 mg/kg without giving morbidity whatsoever (Palacios et al. 1989; Gupta 1995).

10 Conclusions

J. pectoralis is widely used as a medicinal plant in Central America, the Caribbean and the tropical parts of South America. It has a longstanding history, being already mentioned in the sixteenth century in nowadays the Dominican Republic. The plant is incorporated in the pharmacopeia of Brazil and Cuba for its applications as an expectorant and sedative for nervous affections respectively. In traditional medicine, the most frequent application is for illnesses of the respiratory tract. On the contrary, in the laboratory studies on *J. pectoralis* there is just one experiment concerning this category. Scientific research with a focus on the anti-inflammatory, analgesic and sedative effects of the plant, have produced positive results that confirm traditional uses. Toxicity has hardly been reported. It would be recommended that more research should be aimed both at the pharmacokinetics of the extracts and the clinical aspects.

References

AGM DS (1985) Ritual enemas and snuffs in the Americas, Latin America Studies no. 33. CEDLA, Amsterdam

Agra M, Silva K, Basílio I, Freitas P, Barbosa-Filho JM (2008) Survey of medicinal plants used in the region Northeast of Brazil. Braz J Pharmacogn 18(3):472–508

Albuquerque UP, Monteiro JM, Ramos MA, Amorim ELC (2007) Medicinal and magic plants from a public market in northeastern Brazil. J Ethnopharmacol 110:76–91

Aversi-Ferreira TA, Ribeiro PP, Silva NC, Brandão LD, Gratão LH, Nyamdavaa E, Aversi-Ferreira RA, Nishijo H, Nascimento GN (2013) Confrontation between ethnopharmacology and scientific results of the herbal medicaments from Brazil to be applied in primary health care. J Med Plant Res 7(14):845–856

Balick MJ, Nee MH, Atha DE (2000) Checklist of the vascular plants of Belize. Mem NY Bot Gard 85:1–246

Bansal Y, Sethi P, Bansal G (2013) Coumarin: a potential nucleus for anti-inflammatory molecules. Med Chem Res 22:3049–3060

Beauvoir MG, DeFillips RA, Wolpert BJ, Crepin J (2001) Selected medicinal plants of Haitian Vodou. typescript. Smithsonian Institution, Washington, DC

Beyra A, León M, Iglesias E, Ferrándiz D, Herrera R, Volpato G, Godínez D, Guimarais M, Álvarez R (2004) Estudios etnobotánicos sobre plantas medicinales en la provincia de Camagüey (Cuba). An Jard Bot Madr 61(2):185–203

Brussell DE (1997) Potions, poisons, and panaceas: an ethnobotanical study of Montserrat. Southern Illinois University Press, Carbondale/Edwardsville

Brussell DE (2004) A medicinal plant collection from Montserrat, West Indies. Econ Bot 58(Supplement):S203–S220

Caballero-George C, Gupta MP (2011) A quarter century of pharmacognostic research on Panamanian flora: a review. Planta Med 77:1189–1202

Cabi (n.d.) Invasive species compendium. http://www.cabi.org/iscbeta/datasheet/29291. Accessed 3 Apr 2014

Chariandy CM, Seaforth CE, Phelps RH, Pollard GV, Khambay BPS (1999) Screening of medicinal plants from Trinidad and Tobago for antimicrobial and insecticidal properties. J Ethnopharmacol 64:265–270

Coelho-Ferreira M (2009) Medicinal knowledge and plant utilization in an Amazonian coastal community of Marudá, Pará State (Brazil). J Ethnopharmacol 126:159–175

Cordero AB (1986) Manual de Medicina Domestica; Plantas Medicinales Dominicanas, Publicaciones de la Universidad Autónoma de Santo Domingo, vol CCLII, 2nd edn. Colección Ciencia y Tecnología no. 7, Santo Domingo

De Smet AGM (1985) A multidisciplinary overview of intoxicating snuff rituals in the western Hemisphere. J Ethnopharmacol 13:3–49

De Vries JX, Tauscher B, Wurzel G (1988) Constituents of *Justicia pectoralis* Jacq. 2. Gas chromatograph/mass spectrometry of simple coumarins, 3-phenylpropionic acids and their hydroxy and methoxy derivates. Biol Mass Spectrom 15:413–417

Defilipps RA, Maina SL, Crepin J (2008) Medicinal plants of the Guianas (Guyana, Surinam, French Guiana). Smithsonian national museum of natural history. Smithsonian Inst, Washington, DC

Duke JA, Bogenschutz-Godwin MJ, Ottensen AR (2009) Duke's handbook of medicinal plants of Latin America. CRC Press/Taylor & Francis Group, Boca Raton/Florida

Egg AB (1999) Diccionario Enciclopédico de Plantas Utiles del Perú. Centro de Estudios Regionales Andina Bartolomé de Las Casas. Cuzco, Peru

Facey PC, Pascoe KO, Porter RB, Jones AD (1999) Investigation of plants used in Jamaican folk medicine for anti-bacterial activity. J Pharm Pharmacol 51:1455–1460

Fernández de Oviedo G (1851–1855) Historia general y natural de las Indias, Islas y Tierra Firme del Mar Océano. Publíca la Real Academia de la Historia. Ed. De José Amador de los Ríos, Madrid, Spain. In: fascimile: www.cervantesvirtual.com

Fernandez L, Perez H, Mas R, Rodriguez L, Galan L, Bisca YR (1987) Efecto de *Justicia pectoralis* sobre la conducta exploratoria en ratones. Centro Nacional de Investigaciones Científicas (CENIC) Ed. Estudios Avanzados en Neurociencias, La Habana, pp 254–264

Fernandez L, Mas R, Perez H, Biscay R, Galan L (1989) Evaluación preliminar de los efectos nuerofarmacológicos de *Justicia pectoralis*. Rev Cuba Farm 23(1/2):161–166

Fica SE (2005) Efecto de la administración del extracto de *Justicia pectoralis* sobre la conducta de ratas sometidas a pruebas de comportamiento. Thesis, Institute of Pharmacology, Faculty of Veterinary Sciences, Austral University of Chile, Chile

Fonseca FN, Silva AH, Leal LKAM (2010) *Justicia pectoralis* Jacq., Acanthaceae: preparation and characterization of the plant drug including chromatographic analysis by HPLC-PDA. Braz J Pharmacogn 20(6):871–877

Formulário de Fitoterápicos. Farmacopeia Brasileira (2011) 1a edição, Agência Nacional de Vigilância Sanitária, Brasilia, Brazil

Fuentes V, Lems C, Rodriguez C, Germosen-Robineau L (2000) Manual de cultivo y conservación de plantas medicinales. Enda-caribe, UAG, INIFAT & MINSAP, Santo Domingo, Dominican Republic

Garcia Gonzalez M, Saenz Campos D, Rojas Mora L, Tinoco Mora Z, Bonilla PJ (2002) Exploración del uso de plantas medicinales en zonas urbana de Costa Rica. Fármacos 15(2):53–64

Germosén-Robineau L (2005) Farmacopea vegetal caribeña. Segunda edición actualizada. Editorial Universitaria, Unan, Leon, Nicaragua

Gupta M (1995) 270 Plantas medicinales Iberoamericanas. CYTED-SECAB, Bogota

Hardt TJ, Ritschel WA (1983) Dose-related pharmacokinetics of coumarin, 7-hydroxycoumarin and 7-hydroxycoumarin glucuronide upon intraperitoneal administration in the rat. Arzeimittel-Forschung 33:1442–1446

Honychurch PN (1986) Caribbean wild plants and their uses: an illustrated guide to some medicinal and wild ornamental plants of the West Indies. Macmillan Publishers, Ltd., London

Joly LG, Guerra S, Séptimo R, Solis PN, Correa MD, Gupta MP, Levy S, Sandberg F, Perera P (1990) Ethnobotanical inventory of medicinal plants used by the Guaymi Indians in Western Panama. Part II. J Ethnopharmacol 28:191–206

Joseph H, Gleye J, Fouraste MI, Stanislas E (1988a) O-methoxylated C-glycosylflavones from *Justicia pectoralis*. J Nat Prod 51(4):804–805

Joseph H, Gleye J, Moulis C, Mensah LJ, Roussakis C, Gratas C (1988b) Justicidin B, a cytotoxic principle from *Justicia pectoralis*. J Nat Prod 51(3):599–600

Khan JI, Kennedy TJ, Christian DR Jr (2012) Basic principles of forensic chemistry. Humana Press, Springer/New York

Lagarto A, Silva Yhebra R, Guerra Sardiñas I, Iglesias Buel L (2001) Comparative study of the assay of *Artemia salina* L. and the estimate of the medium lethal dose (LD50 value) in mice, to determine oral acute toxicity of plant extracts. Phytomedicine 8(5):395–400

Lagarto A, Bueno V, Guerra I, Valdés O, Gabilondo T, Rodríguez J (2009) Acute and subchronic oral toxicities of *Justicia pectoralis J.* extract in Wistar Rats. Open Nat Prod J 2:53–58

Lans C (2007) Ethnomedicines used in Trinidad and Tobago for reproductive problems. J Ethnobiol Ethnomed. https://doi.org/10.1186/1746-4269-3-13

Leal LKAM, Ferreira AAG, Bezerra GA, Matos FJA, Viana GSB (2000) Antinociceptive, anti-inflammatory and bronchodilator activities of Brazilian medicinal plants containing coumarin: a comparative study. J Ethnopharmacol 70:151–159

Lino CS, Taveira ML, Viana GSB, Matos FJA (1997) Analgesic and antiinflammatory activities of *Justicia pectoralis* Jacq. and its main constituents: coumarin and umbelliferone. Phytother Res 11:211–215

Liogier AH (1995) La Flora de la Española. VII. Universidad Central del Este, Volumen LXXI, Serie Científica 28, San Pedro de Macorís, Dominican Republic

Locklear TD, Huang Y, Frasor J, Doyle BJ, Perez A, Gomez-Laurito J, Mahady GB (2010) Estrogenic and progestagenic effects of extracts of *Justicia pectoralis* Jacq., an herbal medicine from Costa Rica used for the treatment of menopause and PMS. Maturitas 66:315–322

Longuefosse J-L, Nossin E (1996) Medical ethnobotany survey in Martinique. J Ethnopharmacol 53:117–142

Macias-Peacok B, Perez-Jackson L, Suarez-Crespo MF, Fong-Dominguez CO, Pupo-Perera E (2009) Consumo de plantas medicinales por mujeres embarazadas. Use of medicinal plants during pregnancy. Rev Med Inst Mex Seguro Soc 47(3):331–334

MacRac WD, Towers GHN (1984) *Justicia pectoralis*: a study of the basis for its use as a hallucinogenic snuff ingredient. J Ethnopharmacol 12:93–111

Meléndez M, Alvarado S, Castro de Rojas L (2012) Identificación y conocimiento de las plantas medicinales expedidas en los mercados principal y libre de Maracay, estado Aragua, Venezuela. Rev Fac Agron (UCV) 38(2):64–70

Mills J, Pascoe KO, Chambers J, Melville GN (1986) Preliminary investigations of the wound-healing properties of a Jamaican folk medicinal plant (*Justicia pectoralis*). W Indian Med J 35:190–193

Ministério da Saúde (2008) Relação Nacional de Plantas Medicinais de interesse para o SUS (Sistema unica da saude). http://portalsaude.saude.gov.br/index.php/cidadao/principal/agencia-saude/noticias-anteriores-agencia-saude/3487. Accessed Sept 2014

MINSAP (1992) Guía Terapéutica y Dispensarial de Fitofármacos y Apifármacos. ECIMED, La Habana, pp 109–113

Morton JF (1977) Some folk-medicine plants of Central American markets. Quart J Crude Drug Res 15:165–192

Moura CT, Lima FJ, Vasconcelos TB, de Siqueira RJ, Leal LK, Havt A, Magalhães PJ (2013) The anti-inflammatory effects of a standardized extract of *Justicia pectoralis* (SEJP) on the antigen-elicited rat airway hyperresponsiveness involve changes in gene expression of canonical transient receptor proteins (TRPC). Planta Med 79:PN56

Nobre MEP, Leite GL, Barbosa MAC, Sousa LR, Sólon PCD, De Brito SPC, Viana GSB (2006) Avaliação da eficácia do xarope de chambá (*Justicia pectoralis* Jacq.) na função pulmonar da pacientes asmáticos. J Bras Fitomedicina 4:4–10

Nuñez E (1992) Plantas Medicinales de Puerto Rico. Reimpresión. Editorial de la Universidad de Puerto Rico, Puerto Rico

Oliveira AFM, Xavier HS, Silva NH, Andrade LHC (2000) Screening Cromatográfico de Acanthaceae Medicinais: *Justicia pectoralis* Jacq. e *J. gendarussa* Burm. Rev Bras Plant Med, Botucatu 3(1):37–41

Opdyke DLJ (1974) Monographs on fragrance raw materials – coumarin. Food Cosmet Toxicol 12:385–388

Palacios M, Tillan J, Garcia G, Mas R, Cabrera Y (1989) Reporte toxicológico del extracto liofilizado de *Justicia pectoralis* (Tilo). Informe Técnico Laboratorio de Control Biológico. IMEFA. Ministerio de Salud de Cuba

Perez H, Mas R, Fernandez L, Rodriguez L (1987) *Justicia pectoralis* no previene las convulsiones inducidas por PTZ y PTX. Centro Nacional de Investigaciones Científicas (CENIC) Ed. Estudios Avanzados en Neurociencias, La Habana, pp 273–283

Perez G, Rivero R, Pardo Z, Rodriguez J (2001) Evaluación de la actividad antioxidante de *Justicia pectoralis* Jacq. Rev Cubana Invest Bioméd 20(1):30–33. Accessed 9 April 2014. http://scielo.sld.cu/scielo.php?script=sci_arttext&pid=S0864-03002001000100006&lng=es&nrm=iso. ISSN 1561-3011

Picking D, Younger N, Mitchell S, Delgoda R (2011) The prevalence of herbal medicine home use and concomitant use with pharmaceutical medicines in Jamaica. J Ethnopharmacol 137:305–311

Ramalingam R, Vaiyapuri M (2013) Effects of umbelliferone on lipid peroxidation and antioxidant status in diethylnitrosamine-induced hepatocellular carcinoma. JACME 3:73–82

Ramesh B, Pugalendi KV (2006) Antihyperglycemic effect of umbelliferone in streptozotocin-diabetic rats. J Med Food 9(4):562–566

Ritschel WA, Hardt TJ (1983) Pharmacokinetics of coumarin, 7-hydroxycoumarin and 7-hydroxycoumarin glucuronide in the blood and brain of gerbils following intraperitoneal administration of coumarin. Arzneimittelforschung 33:1254–1258

Ritschel WA, Brady ME, Tan HSI, Hoffmann KA, Yiu IM, Grummich KW (1977) Pharmacokinetics of coumarin and its 7-hydroxy-metabolites upon intravenous and peroral administration of coumarin in man. Eur J Clin Pharmacol 12:457–461

Ritschel WA, Brady ME, Tan HSI (1979) First-pass effect of coumarin in man. Int J Clin Pharmacol Biopharm 17:99–103

Rodriguez E, Virnes A, Aleman J (1989) Estudio preliminar del efecto de *Justicia pectoralis* sobre el EEG de adultos normales. Rev Cuba Farm 23(3):302–308

Rodríguez JE, Roche A, Vega R, Rodríguez C, Carballo C, Guerra I. et al (2008) Estudios preliminares de extractos fluidos 30% de *Justicia pectoralis* Jacq. var. *stenophylla* Leonard. Rev Cubana Plant Med [journal in the Internet]. Dic [assessed 2014 Apr 09]; 13(4):. Disponible en: http://scielo.sld.cu/scielo.php?script=sci_arttext&pid=S1028-47962008000400012&lng=es

Roig, Mesa JT (1928) Diccionario Botánico de Nombres Vulgares Cubanos. Imprenta y papeleria Ramela Bauzá y Cia, La Habana (On line by Ann Arbor Michigan, University of Michigan Library, 2005)

Roig y Mesa JT (1965) Diccionario Botánico de Nombres Vulgares Cubanos. Tercera Edición. Editora Nacional de Cuba, Editora del Consejo Nacional de Universidades. La Habana, Cuba

Rutter RA (1990) Catalogo de Plantas Utiles de la Amazonia Peruana. Instituto Linguistico del Verano, Yarinacocha

Ruysschaert S, van Andel T, Van de Putte K, Van Damme P (2009) Bathe the baby to make it strong and healthy: plant use and child care among Saramaccan Maroons in Suriname. J Ethnopharmacol 121:148–170

Schultes RE, Holmstedt B (1968) De plantis toxicariis e Mundo Novo tropicale commentationes II. The vegetal ingredients of the myristicaceous snuffs of the northwest Amazon. Rhodora 70:113–160

Seaforth CE, Adams CD, Sylvester YA (1983) Guide to the medicinal plants of Trinidad & Tobago. Commonwealth Secretariat, London

Taveira ML (1993) Contribuição ao conhecimento químico de plantas do Nordeste: *J. pectoralis* Jacq. var. *Stenophylla* Leon. (Acanthaceae) e *Bombax cearensis* Ducke (Bombaceae). Fortaleza. Dissertação (Mestrado em Química Orgãnica). Universidade Federal do Ceará, Brazil

Tene V, Malagón O, Finzi PV, Vidari G, Armijos C, Zaragoza T (2007) An ethnobotanical survey of medicinal plants used in Loja and Zamora-Chinchipe, Ecuador. J Ethnopharmacol 111:63–81

The Plant List (2014) Version 1.1 Published on the Internet: http://www.theplantlist.org. Accessed 5 Nov 2014

Thomas T (1997) Traditional medicinal plants of St. Croix. St. Thomas and St. John. University of the Virgin Islands, St. Thomas

Tropicos (n.d.) Missouri botanical garden. http://www.tropicos.org. Accessed 5 Nov 2014

USDA, ARS, National Genetic Resources Program (n.d.) *Germplasm Resources Information Network – (GRIN)* [Base de Datos en Línea]. National Germplasm Resources Laboratory, Beltsville, Maryland. URL: http://www.ars-grin.gov/cgi-bin/npgs/html/index.pl. Accessed 03 Apr 2014

van Andel T (2000) Non-timber forest products of the north-West District of Guyana. Part II a field guide, Tropenbos-Guyana Series 8b. Tropenbos-Guyana programme-Georgetown, Guyana

van Andel T, Havinga R (2008) Sustainability aspects of commercial medicinal plant harvesting in Suriname. For Ecol Manag 256:1540–1545

van Andel T, Ruysschaert S (2011) Medicinale en rituele planten van Suriname. KIT Publishers, Amsterdam

van Andel T, Behari-Ramdas J, Havinga R, Groenendijk S (2007) The medicinal plant trade in Surinam. Ethnobot Res Appl 5:351–372

Vasconcelos JF, Teixeira MM, Barbosa-Filho JM, Agra MF, Nunes XP, Giulietti AM, Ribeiro-dos-Santos R, Soares MBP (2009) Effects of umbelliferone in a murine model of allergic airway inflammation. Eur J Pharmacol 609:126–131

Venâncio ET, Rocha NFM, Rios ERV, Feitosa ML, Linhares MI, Melo FHC, Matias MS, Fonseca FN, Sousa FCF, Leal LKAM, Fonteles MMF (2011) Anxiolytic-like effects of standardized extract of *Justicia pectoralis* (SEJP) in mice: involvement of GABA/Benzodiazepine in receptor. Phytother Res 25:444–450

Weniger B, Rouzier M, Daguilh R, Henrys D, Henrys JH, Anton R (1986) Popular medicine of the Central Plateau of Haiti. 2. Ethnopharmacological inventory. J Ethnopharmacol 17(1):13–30

Wong W (1976) Some folk medicinal plants from Trinidad. Econ Bot 30:103–142

Kalanchoe brasiliensis Camb. and *Kalanchoe pinnata* (Lamk.) Pers.

Rosilene Gomes da Silva Ferreira, Nilma de Souza Fernandes, and Valdir Florêncio da Veiga-Junior

Kalanchoe pinnata (Lamk.) Pers
Photo: Source: data bank from Laboratório de Ecologia e Evolução de sistemas socioecológicos

R. G. da Silva Ferreira
Pharmaceutical Sciences College, Amazonas Federal University, Manaus, AM, Brazil

N. de Souza Fernandes · V. F. da Veiga-Junior (✉)
Chemistry Department, Amazonas Federal University, Manaus, AM, Brazil
e-mail: nilmafernandes@ufam.edu.br; valdirveiga@ufam.edu.br

U. P. Albuquerque et al. (eds.), *Medicinal and Aromatic Plants of South America*, Medicinal and Aromatic Plants of the World 5,
https://doi.org/10.1007/978-94-024-1552-0_23

Abstract *Kalanchoe brasiliensis* Cambs. and *Kalanchoe pinnata* (Lamk.) Pers species belong to *Crassulaceae* family. *Kalanchoe brasiliensis* is popularly known as '*saião*', 'white coirama', 'thick leaf', 'leaf of luck' and 'leaf of the coast', and *Kalanchoe pinnata* as '*saião-roxo*', 'leaf-of-fortune', 'leaf of the coast', 'yellow flower of fortune' and '*para-tudo*'. In ethnopharmacology, there are reports of the use of the extract of the leaves of *Kalanchoe brasiliensis* for skin infections and oral mucosa, bronchitis, nasal congestion, chest infections, yellow fever, gastric ulcers and arthritis. Leaves and stalks are the most commonly used parts. The leaves of *K. brasiliensis* contain high concentrations of flavonoids; while fatty acids, acyclic and aromatic organic acids, amino acids, bufadienolides, α-β unsaturated acyclic ketones, fenantrenic derivatives, sterols, long-chain hydrocarbons and triterpenoids are found mainly in the leaves of *K. pinnata*. Analgesic, anti-inflammatory, anti-leishmaniotic, antimalarial, antipyretic, antimicrobial, antithyroidal, antitumor, antiulcer, hepatoprotective, immunosuppressive, pesticide, inhibition in uterine contractions, neuropsicofarmacologic and hypoglycemic properties of these species have already been evaluated in experimental pharmacology.

Keywords Saião · Flavonoids · *Kalanchoe* · *Crassulaceae*

1 Taxonomic Characteristics

Kalanchoe pinnata and *Kalanchoe brasiliensis* belong to the genus *Kalanchoe* (synonym *Bryophyllum* and *Cotyledon*), family *Crassulaceae* (Maurice 1993). Popular names include plant of life, air plant, plant of love, canterbury bells, cathedral bells, green love, curtain plant, parnabija, white coirama, coirama-brava, leaf of the coast and *saião* (Anjoo and Kumar 2000).

Synonyms Both species have botanical synonyms: *Kalanchoe brasiliensis* syn *Cotyledon brasilica* Vell, *Kalanchoe pinnata* syn *Bryophyllum pinnatum* (Lamk.) Oken; *Bryophyllum pinnatum* Kurz., *Cotyledon pinnata* Lamk, among others.

2 Crude Drug Used

The aqueous extract of the leaves of *K. pinnata* has been used for the treatment of cutaneous leishmaniasis and to decrease acute anaphylactic reactions (Cruz et al. 2008, 2012). Investigating anti-tumor action of *Kalanchoe brasiliensis*, an aqueous solution containing 50 mg/kg of the raw extract diluted in saline was administered intraperitoneally in mice, showing that it could be used for treatment of sarcoma 180 (Machado and Melo-Junior 2009).

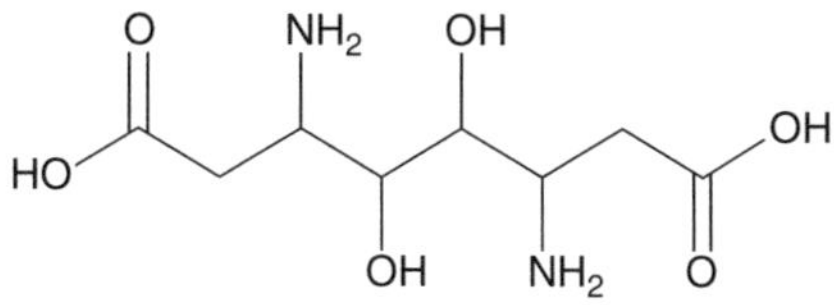

Kalanchosine

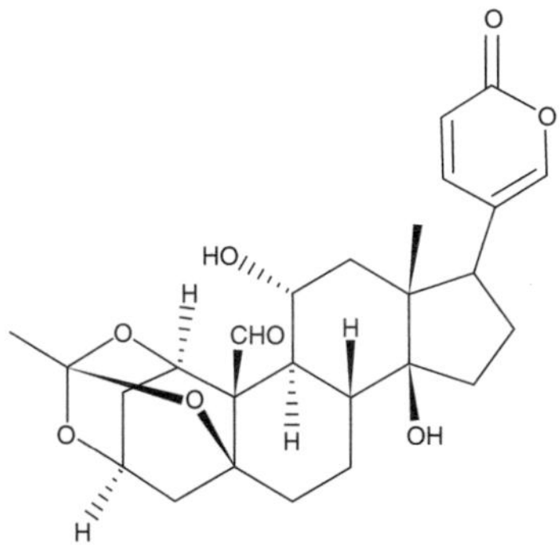

Bryophyllin A

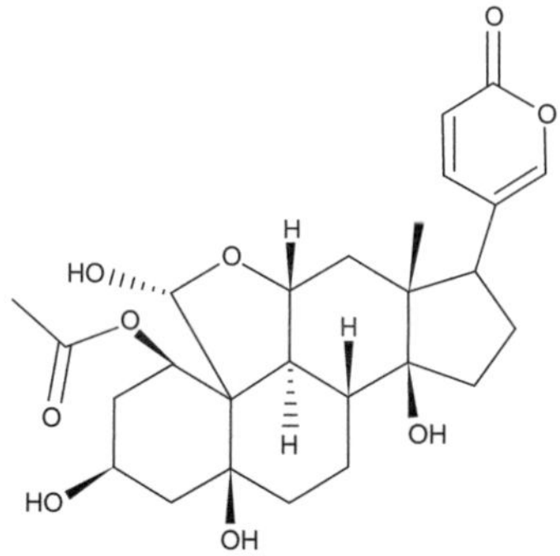

Bryophyllin B

Fig. 1 Chemical structures of the main compounds isolated from *Kalanchoe*. (**a**) Kalanchosine (1), 3,6-diamino-4,5-dihydroxyoctanedioic acid and (**b**) Bryophillin A and B

3 Major Chemical Constituents and Bioactive Compounds

Species of the genus *Kalanchoe* contain a wide variety of secondary substances, mainly terpenes (Anjoo and Kumar 2000; Siddiqui et al. 1989), flavonoids (Gaind and Gupta 1972; Muzitano et al. 2006), alkaloids (Biswas 2011; Okwu and Josiah 2006), bufadienolides (Anjoo and Kumar 2000; Milad et al. 2014; Supratman et al. 2001) glycosides, steroids, saponins, tannins, reduced sugars and aminoacids (Biswas 2011; Matthew et al. 2013; Pattewar 2012) (Fig. 1).

Flavonoid glycosides derived from patuletin were isolated from the leaves and branches of *K. brasiliensis*, as 8-methoxykaempferol-3,7-di-O- rhamnopyranoside, as 8-methoxyquercetin, 3,7-di-O-rhamnopyranoside and quercetin (Trevisan et al. 2006; Veiga-Junior 2005). Malic acid and an organic salt – kalanchosin dimalate (KMC), belonging to a new class of metabolites, called kalanchosine – were isolated from extracts of areal parts of *K. brasiliensis* (Costa et al. 2006).

From *K. pinnata*, triterpenes and sterols were identified such as α and β-amyrin, taraxerol, acetylated derivatives of cycloartan-3-ol, ψ-taraxasterol; (24R) – stigmast-5, 25-dien-3β-ol (24 epiclerosterol); (24R) – 5α- stigmast-7, 25-dien-3β-ol; 5α- stigmast- 24-en-3β-ol; 25 methyl-5α-ergost-24 (28) – en-3β-ol, and others. The bufadienolides isolated from *K. pinnata* were identified as bryophillin A and B (Supratman et al. 2001). The presence of bufadienolids suggests a potential antitumor and bactericidal ability (Pattewar 2012; Supratman et al. 2001).

The compounds α-ramnoisorobin, kaempferitrine (Tatsimo et al. 2012) and quercetin (Muzitano et al. 2006) are among flavonoids isolated from *K. pinnata*. Due to the restricted occurrence and great abundance of flavonoids in *K. pinnata*, it has been suggested that this class of metabolites may be responsible for the high therapeutic potential of the species (Pattewar 2012).

4 Morphological Description

K. brasiliensis has herbaceous features and grows to a height of 30 cm to 1 m. Leaves are sparsely branched, oval or obovate oppositely succulent, peciolated and crenated. A characteristic feature that facilitates differentiation between *K. pinnata* and *K. brasiliensis* species is the appearance of the leaf, since the latter has a corrugated subcrenated edge, whereas *K. pinnata* has a crenated leaf. *K. brasiliensis* has a yellow-orange inflorescence with small flowers (Lorenzi and Matos 2008).

5 Geographical Distribution

The *Kalanchoe* genus includes native species from Africa and Brazil (Boulos 1999). In Brazil, *K. brasiliensis* is a native species, with an area from the southeast to the northeast. It is common in the coastal zone. *K. pinnata* has a pantropical distribution, both continental and insular (Veiga-Junior 2005).

6 Ecological Requirements

Species of the genus *Kalanchoe* inhabit different regions, ranging from rainforests to arid environments (Rauh 1973).

K. pinnata is intolerant to long periods of drought. As invasive species, it adapts and colonizes different areas, are abundant in sandy soils and rocky coastal regions in different countries, such as Madagascar, the United States, Brazil and Australia. It is still found in areas with human disturbance. It adapts to humid and semi-humid climates, with a precipitation between 1000 and 2000 mm (Smith 1985).

7 Collection Practice

The special literature relating to the harvesting of *K. pinnata* and *K. brasiliensis* is either scarce or unavailable. In general, during collection of medicinal plants, one should take into account population survival and maintenance of the ecosystem.

Furthermore, the species should be collected during the appropriate season, climate and time, as the secondary metabolites are variable according to different periods (World Health Organization 2003).

8 Traditional Use and Common Knowledge

Different species of the genus *Kalanchoe* are traditionally used in folk medicine in many parts of the world, particularly in South America. In Guyana, the leaves of *K. pinnata* are traditionally used as an anti-inflammatory and antiseptic to treat coughs, ulcers and wounds (El Abdellaoui et al. 2010). In Brazil, the most studied and used species are *K. pinnata* and *K. brasiliensis*.

K. brasiliensis is widely used in the treatment of boils. The pure juice is used orally in cases of ovarian and uterine inflammation or mixed with other plants such as *malvarisco*, used in the preparation of cough syrups. *K. pinnata* is used in inflammatory diseases, gastric ulcers, burns, diarrhea, vomiting, insect bites, body aches, and as an antifungal and antibacterial (Almeida et al. 2000; Anjoo and Kumar 2000; Okwu and Josiah 2006).

9 Modern Medicine Based on Uses Its Traditional Medicine

K. brasiliensis and *K. pinnata* are extensively used in traditional medicine. There are a significant number of studies that describe their biological effects, especially for *K. pinnata*. However, the evaluation of the active chemical compounds and their biological activity is far from being complete. There is also a need for more detailed studies looking on large scale production and economic viability. Preclinical studies of pharmacological activities in vitro and in vivo are also described in the special literature on these species.

In vitro assays using extracts of the leaves of different species of the genus Kalanchoe (including *K. brasiliensis*) in ethyl acetate, hexane and methanol acetate, identified larvicidal activity effective against *Aedes aegypti* at concentrations of 500, 250 and 100 ppm (Salles Trevisan et al. 2006). Two bufadienolids isolated from *K. pinnata* demonstrated a high degree of effectiveness against the third larval stage of the silkworm (Supratman et al. 2001).

In vitro assays showed that the raw extract of *Kalanchoe brasiliensis* contains active substances with antitumor effects against Sarcoma 180 cells. The results indicated an inhibitory effect of the growth of this kind of tumor, with 52.8% reduction ($p < 0.05$) of tumor mass (Machado and Melo-Junior 2009). Raw extract and fractions of *K. pinnata* also exhibited dose-dependent cytotoxic activity, with IC_{50} 550.0 μg/mL and 91.0 μg/mL, respectively; against cervical cancer (Mahata et al. 2012). Additionally, it exhibited cytotoxic activity against KB cells (Yamagishi et al. 1989). Finally, leaves of the species have been shown to have anti-mutagenic properties (Obaseiki-Ebor et al. 1993).

The leaf extract of *Kalanchoe pinnata* in dichloromethane (DCM), chloroform, petroleum ether and aqueous fractions have been evaluated in an oral glucose tolerance test (OGTT) at a concentration of 10 mg/kg in rats. In this study, the fraction in DCM produced an improved hypoglycemic action. In addition, the dose-dependent effects of the same fraction of *Kalanchoe pinnata* were evaluated. It was concluded that the DCM fraction demonstrated antihyperglycemic activity in a dose-dependent pattern, which is comparable to the glibenclamide (with the same dose of 2.5 mg/kg body weight). According to researchers, among four concentrations tested, the maximum concentration used (10 mg/kg body weight) showed prominent hypoglycemic activity (Patil et al. 2013). The study by Ojewole (2005) demonstrated significant hypoglycemia in mice when treated with aqueous extract of *K. pinnata*.

In models of severe anaphylactic reaction, the aqueous extract of leaves of *Kalanchoe pinnata* was effective. In studies by Cruz et al. (2012) the effect of *K. pinnata* flavonoids quercetin (QE) and quercitrin (IQ) was evaluated in the activation of mast cells in vitro in a model of the allergic disease in vivo. The study showed that this extract and QE prevented mast cell degranulation and lessened the action of TNF and IL-6 released in vitro and in vivo. These findings demonstrate that treatment with *K. pinnata* or QE is effective in the treatment of allergic respiratory diseases, providing new perspectives on the immunomodulatory functions of this plant.

The leaf extract of *K. pinnata* in DCM/methanol (1:1) and hexane/DCM reduced at least 30% acetic acid-induced pain and also increased the latency period between seizures (Nguelefack et al. 2006). The effect was greater with higher doses per kilogram (between 200 and 300 mg/kg) (Veiga-Junior 2005).

Cruz et al. (2008) identified a protective effect of aqueous extract of *K. pinnata* in fatal anaphylactic shock, an immune-mediated Th2 pathology, and also identified the active component. Mice oral treated daily with the extract survived during sensitization with ovalbumin when tested with this allergen, while there was a 100% mortality rate in the untreated group. The intraperitoneal single dose 3 h before the test was partially effective. Oral protection was accompanied by a decreased production of anti-OVA IgE antibodies, eosinophilia and decreased the production of cytokines IL-5, IL-10 and TNF-α. In vitro, these extract prevented mast cell degranulation and histamine release induced by antigens. Oral treatment with the flavonoid quercitrin from *K. pinnata* prevented fatal anaphylaxis in 75% of animals. These results indicate that oral treatment effectively attenuates anaphylactic pro-immune responses. The protection obtained with quercitrin, although not maximal, suggests that the flavonoid is a critical component of *K. pinnata* extract against this extreme allergic reaction.

Studies by Biswas et al. (2011) evaluated ethanol extracts of leaves and stems of *K. pinnata*. The ethanolic extract demonstrated significant antimicrobial activity against gram-positive (*B. subtilis, S. aureus*) and gram-negative (*E. coli, P. aeruginosa, S. dysenteriae*) bacteria, with zones of inhibition of 6.0 ± 0.35 to 8.2 ± 0.22 mm.

Yadav and Dixit (2003) observed that the juice of the fresh leaves of *K. pinnata* was used as a treatment for jaundice, and the ethanolic extract was tested on rats against tetrachloride-induced hepatotoxicity. The test was effective in vivo and in vitro, based on the histological analysis. The juice was more effective than the ethanol extract. In another study using the aqueous extract in mice, showed that this extract protect the gentamicin-induced nephrotoxicity. A significant antioxidant activity of the aqueous extract was observed in the same study (Harlalka et al. 2007).

There was a reduction in blood pressure in rats after administration of the aqueous leaf extract of *K. pinnata*. In rabbits, this extract protected the kidneys and the liver (Ghasi et al. 2011). The alcoholic extract of the leaves administered orally and intraperitoneally in rats showed significant diuretic action, especially with the intraperitoneal administration (Patil et al. 2013).

According to tests in a murine model of cutaneous leishmaniasis, where different flavonoids were used, the glycosides were defined as active compounds with evident action against *Leishmania amazonensis* (Muzitano et al. 2006).

10 Conclusions

The widespread use of the genus *Kalanchoe*, and specifically of the species *K. brasiliensis* and *K. pinata* in traditional medicine, as well as their acceptance by many researchers, is strong evidence that these species can be effective for treating the conditions described and can be considered as a possible source for healing the pathological cases investigated. Extracts of *K. pinnata* and *K. brasiliensis* have been reported to possess anti-inflammatory, antihypertensive, antimicrobial, antifungal, antidiabetic and antitumor effects. Several active compounds have been identified in *K. pinnata,* such as glycosides, organic acids, steroids and bufadienolides. These compounds have also a variety of demonstrated effects including antibacterial and antitumor effects.

References

Almeida AP, Da Silva SAG, Souza MLM, Lima LMTR, Rossi-Bergmann B, Gonçalves de Moraes VL, Costa SS (2000) Isolation chemicals analysis of a fatty acid fraction of *Kalanchoe pinnata* with a potent lymphocyte suppressive activity. Planta Med 66:134–137

Anjoo K, Kumar SA (2000) Microscopical and preliminary phytochemical studies on aerial part (leaves and stem) of *Bryophyllum pinnotum* Kurz. Pharm J 2:254–259. https://doi.org/10.1016/S0975-3575(10)80113-0

Biswas K (2011) Literature review on pharmacological potentials of *Kalanchoe pinnata* (Crassulaceae). Afr J Pharm Pharmacol 5:1258–1262. https://doi.org/10.5897/AJPP11.273

Biswas SK, Chowdhury A, Das J, Karmakar UK, Shill MC (2011) Assessment of cytotoxicity and antibacterial activities of ethanolic extracts of *Kalanchoe pinnata linn.* (family: crassulaceae) leaves and stems. Int J Pharm Sci Res 2:2605–2609

Boulos L (1999) Flora of Egypt. Vol. 1 (Azollaceae – Oxalidaceae), nordic. J Bot 19(3):328. https://doi.org/10.1111/j.1756-1051.1999.tb01119.x

Costa SS, de Souza MDLM, Ibrahim T, de Melo GO, de Almeida AP, Guette C, Férézou J-P, Koatz VLG (2006) Kalanchosine dimalate, an anti-inflammatory salt from *Kalanchoe brasiliensis*. J Nat Prod 69:815–818. https://doi.org/10.1021/np050475+

Cruz E, Da-Silva S, Muzitano MF, Silva PMR, Costa SS, Rossi-Bergmann B (2008) Immunomodulatory pretreatment with *Kalanchoe pinnata* extract and its quercitrin flavonoid effectively protects mice against fatal anaphylactic shock. Int Immunopharmacol 8:1616–1621. https://doi.org/10.1016/j.intimp.2008.07.006

Cruz E, Reuter S, Martin H, Dehzad N, Muzitano MF, Costa SS, Rossi-Bergmann B, Buhl R, Stassen M, Taube C (2012) *Kalanchoe pinnata* inhibits mast cell activation and prevents allergic airway disease. Phytomedicine 19:115–121. https://doi.org/10.1016/j.phymed.2011.06.030

El Abdellaoui S, Destandau E, Toribio A, Elfakir C, Lafosse M, Renimel I, André P, Cancellieri P, Landemarre L (2010) Bioactive molecules in *Kalanchoe pinnata* leaves: extraction, purification, and identification. Anal Bioanal Chem 398:1329–1338. https://doi.org/10.1007/s00216-010-4047-3

Gaind KN, Gupta RL (1972) Alkanes, alkanols, triterpenes and sterols of *Kalanchoe pinnata*. Phytochemistry 11:1500–1502. https://doi.org/10.1016/S0031-9422(00)90117-1

Ghasi S, Egwuibe C, Achukwu PU, Onyeanusi JC (2011) Assessment of the medical benefit in the folkloric use of *Bryophyllum pinnatum* leaf among the igbos of Nigeria for the treatment of hypertension. Afr J Pharm Pharmacol 5:83–92. https://doi.org/10.5897/AJPP10.309

Harlalka GV, Patil CR, Patil MR (2007) Protective effect of *Kalanchoe pinnata* pers. (Crassulaceae) on gentamicin-induced nephrotoxicity in rats. Indian J Pharmacol 39(4):201–205. https://doi.org/10.4103/0253-7613.36540

Lorenzi H, Matos FJA (2008) Plantas Medicinais no Brasil – Nativas e Exóticas, 2nd edn. Instituto Plantarum, Nova Odessa

Machado MCF, Melo-Junior MR (2009) Evaluation of antitumoral effect of the *Kalanchoe brasiliensis* on the Sarcoma 180 on rats. Rev Eletrônica Farmácia VI:1–6

Mahata S, Maru S, Shukla S, Pandey A, Mugesh G, Das BC, Bharti AC (2012) Anticancer property of *Bryophyllum pinnata* (Lam.) Oken. leaf on human cervical cancer cells. BMC Complement Altern Med 12:15–15. https://doi.org/10.1186/1472-6882-12-15

Matthew S, Jain AK, James M, Matthew C, Bhowmik D (2013) Analgesic and anti-inflammatory activity of *Kalanchoe pinnata* (Lam.) Pers material. J Med Plants Stud Analg 1:23–28

Maurice M (1993) Handbook of African medicinal plant. CRC Press, London

Milad R, El-Ahmady S, Singab AN (2014) Genus *Kalanchoe* (Crassulaceae): a review of its ethnomedicinal, botanical, chemical and pharmacological properties. Eur J Med Plants 4(1):86–104

Muzitano MF, Cruz EA, De Almeida AP, Da Silva SAG, Kaiser CR, Guette C, Rossi-Bergmann B, Costa SS (2006) Quercitrin: an antileishmanial flavonoid glycoside from Kalanchoe pinnata. Planta Med 72:81–83. https://doi.org/10.1055/s-2005-873183

Nguelefack TB, Nana P, Atsamo AD, Dimo T, Watcho P, Dongmo AB, Tapondjou LA, Njamen D, Wansi SL, Kamanyi A (2006) Analgesic and anticonvulsant effects of extracts from the leaves of Kalanchoe crenata (Andrews) Haworth (Crassulaceae). J Ethnopharmacol 106:70–75. https://doi.org/10.1016/j.jep.2005.12.003

Obaseiki-Ebor EE, Odukoya K, Telikepalli H, Mitscher LA, Shankel DM (1993) Antimutagenic activity of extracts of leaves of four common edible vegetable plants in Nigeria (West Africa). Mutat Res Lett 302:109–117. https://doi.org/10.1016/0165-7992(93)90012-K

Ojewole JAO (2005) Antinociceptive, anti-inflammatory and antidiabetic effects of *Bryophyllum pinnatum* (Crassulaceae) leaf aqueous extract. J Ethnopharmacol 99:13–19. https://doi.org/10.1016/j.jep.2005.01.025

Okwu DE, Josiah C (2006) Evaluation of the chemical composition of two Nigerian medicinal plants. Afr J Biotechnol 5:357–361

Patil S, Dongare V, Kulkarni C, Joglekar M, Arvindekar A (2013) Antidiabetic activity of Kalanchoe pinnata in streptozotocin-induced diabetic rats by glucose independent insulin secretagogue action. Pharm Biol 51:1411–1418. https://doi.org/10.3109/13880209

Pattewar SV (2012) *Kalanchoe pinnata*: phytochemical and pharmacological profile. Int J Phytopharm 1:1–8. https://doi.org/10.7439/ijpp.v2i1.223

Rauh W (1973) Über die Zonierung und Differenzierung der Vegetation Madagaskars. Akad Wiss Mainz. Trop Subtrop Pflanzenwelt 1:146

Salles Trevisan MT, Barbosa Bezerra MZ, Pinheiro Santiago GM, Feitosa CM, Verpoorte R, Braz Filho R (2006) Atividades larvicida e anticolinesterésica de plantas do gênero *Kalanchoe*. Quim Nova 29:415–418. https://doi.org/10.1590/S0100-40422006000300002

Siddiqui S, Faizi S, Siddiqui BS, Sultana N (1989) Triterpenoids and phenanthrenes from leaves of *Bryophyllum pinnatum*. Phytochemistry 28:2433–2438. https://doi.org/10.1016/S0031-9422(00)97999-8

Smith AC (1985) Flora Vitiensis nova: a new flora of Fiji. Natl Trop Bot Gard Lawai 3:624–625. https://doi.org/10.5962/bhl.title.44033

Supratman U, Fujita T, Akiyama K, Hayashi H, Murakami A, Sakai H, Koshimizu K, Ohigashi H (2001) Anti-tumor promoting activity of bufadienolides from *Kalanchoe pinnata* and *K. daigremontiana x tubiflora*. Biosci Biotechnol Biochem 65:947–949. https://doi.org/10.1271/bbb.65.947

Tatsimo S, Tamokou J, Havyarimana L, Csupor D, Forgo P, Hohmann J, Kuiate J-R, Tane P (2012) Antimicrobial and antioxidant activity of kaempferol rhamnoside derivatives from *Bryophyllum pinnatum*. BMC Res Notes 5:158. https://doi.org/10.1186/1756-0500-5-158

Trevisan MTS, Zeneide M, Bezerra B, Maria G, Santiago P (2006) Atividades larvicida e anticolinesterásica de plantas do gênero *Kalanchoe*. Quim Nova 29:415–418

Veiga-Junior VF (2005) *Kalanchoe brasiliensis* Camb. *Kalanchoe pinnata* (Lamk.), In: Amaral A (Org.), Coletânea científica de plantas de uso medicinal. Editora FioCruz, Rio de Janeiro

World Health Organization (2003) WHO guidelines on good agricultural and collection practices (GACP) for medicinal plants. World Health 99:67–73

Yadav NP, Dixit VK (2003) Hepatoprotective activity of leaves of *Kalanchoe pinnata* Pers. J Ethnopharmacol 86:197–202. https://doi.org/10.1016/S0378-8741(03)00074-6

Yamagishi T, Haruna M, Yan X-Z, Chang J-J, Lee K-H (1989) Antitumor agents, 110, bryophyllin B, a novel potent cytotoxic bufadienolide from *Bryophyllum pinnatum*. J Nat Prod 52:1071–1079. https://doi.org/10.1021/np50065a025

Lantana camara L. and *Lantana montevidensis* (Spreng.) Briq.

Erlânio O. de Sousa, Sheyla C. X. de Almeida, Sarah S. Damasceno, Camila B. Nobre, and José Galberto M. da Costa

Lantana camara L.
Jean Hivert
Available in: http://www.tropicos.org/Image/100543679

E. O. de Sousa (✉) · S. C. X. de Almeida · S. S. Damasceno
C. B. Nobre · J. G. M. da Costa (✉)
Department of Biological Chemistry, Laboratory of Research in Natural Products, Regional University of Cariri, Crato, CE, Brazil

U. P. Albuquerque et al. (eds.), *Medicinal and Aromatic Plants of South America*, Medicinal and Aromatic Plants of the World 5,
https://doi.org/10.1007/978-94-024-1552-0_24

Abstract Species of the genus *Lantana* (Verbenaceae) are among the species studied scientifically. *Lantana camara* L. and *Lantana montevidensis* (Spreng.) Briq. are shrubs present in various countries, where they are often grown as ornamental plants. They have been used in many parts of the world to treat a variety of diseases. For decades, these species have been widely studied with regards to their phytochemical components, among them terpenoids, flavonoids and phenylpropanoids being the more commonly isolated secondary metabolites. Ethnopharmacological information, isolated constituents, as well as the activities of their different phytochemicals are the focus of this chapter. All these aspects allow an evaluation of the ethnopharmacological potential of these species for the utilization of the large biomass of these plants.

Keywords *Lantana camara* L. · *Lantana montevidensis* (Spreng.) Briq. · Chemical constituents · Biological activities

1 Taxonomic Characteristics

Lantana camara L. has long been reported popularly as "wild sage" and *Lantana montevidensis* (Spreng.) Brinq. as "cambara". They have been introduced to many countries as ornamental plants (Nagão et al. 2002). The term *Lantana* probably comes from the old Latin name of the genus *Viburnum*, which resemble a little in leaves and inflorescence. Taxonomically, the *Lantana* genus is divided into four sections based on floral and carpological features: *Lantana*, *Callioreas*, *Rhytidocamara* and *Sarcolippia*. The taxonomy of the species is however difficult, normally are not stable, hybridization is very widespread, the shape of the inflorescence changes with age, and color of the flowers varies with age and maturity (Ghisalberti 2000).

2 Major Chemical Constituents and Bioactive Compounds

Due to the medicinal properties exhibited by these species, a large number of studies have had the goal to identify and isolate their volatile and non-volatile chemical constituents. Various constituents with varied structural patterns belonging to triterpenoids **(1–64)**, flavonoids **(65–87)**, phenylethanoid glycosides **(88–94)** furanonaphthoquinones **(95–104)**, iridoid glycosides **(105–110),** steroids **(111–119)** and other compounds **(120–134)** have been elucidated over several years, specially *L. camara*, as shown in Table 1 and Fig. 1.

Studies revealed the chemical composition of essential oils of these species that were collected in different locations and ecological conditions. Several mono- and sesquiterpenes were identified, but with a greater predominance of the latter (Dambolena et al. 2010). Cited among the common major constituents identified are the sesquiterpenes, α and ß-caryophillene, isocaryophillene, caryophyllene oxide, caryophyllene epoxide, germacrene D and bicyclogermacrene (Sena Filho et al. 2010).

Table 1 Chemical constituents of *L. camara* and *L. montevidensis*

Compound	Species (parts used)	References
Lantadene A (**1**), lantadene B (**2**), lantadene D (**3**), 22β-angeloyloxy-3β-hydroxyolean-12-en-28-oic acid (**4**), 22β-dimethylacryloyloxy-3 B-hydroxyolean-12-en-28-oic acid (**5**), 22b-hydroxyoleanonic acid (**6**)	*L. camara* (leaves, stems, roots)	Hart et al. (1976), Sharma and Dawra (1991), Pan et al. (1993), Begum et al. (1995), Sharma et al. (2000), and Litaudon et al. (2009)
Oleanonic acid (**7**), oleanolic acid (**8**)	*L. camara* (aerial parts, stems, roots)	Begum et al. (1995), Misra et al. (2007), and Sousa (2014)
22β-hydroxy-3-oxoolean-12-en-28-oic acid (**9**), 24-hydroxy-3-oxoolean-12-en-28-oic acid (**10**), icterogenin (**11**), 22β-dimethylacryloyloxy-24-hydroxy-3-oxo-olean-12-en-28-oic acid (**12**), 22β-*o*-angeloyl-oleanoic acid (**13**), 22b-*o*-senecioyl-oleanoic acid (**14**), hederagenin (**15**), 25-hydroxy-3-oxoolean-12-en-28-oic acid (**16**), 21,22b-epoxy-3β-hydroxyolean-12-en-28-oic methyl ester (**17**), camarin (**18**), lantanone (**19**), 22β-tigloyloxylantanolic acid (**20**)	*L. camara* (leaves, stems, roots)	Hart et al. (1976), Pan et al. (1993), Mahato and Kundu (1994), Siddiqui et al. (1995), Singh et al. (1996), Lai et al. (1998), Misra and Laatsch (2000), Begum et al. (2008b), and Litaudon et al. (2009)
Camarilic acid (**21**)	*L. camara* (aerial parts)	Begum et al. (1995)
Lantanilic acid (**22**)	*L. camara* (leaves, stems, roots)	Pan et al. (1993) and Siddiqui et al. (1995)
Lantanolic acid (**23**), camaric acid (**24**)	*L. camara* (aerial parts, roots)	Pan et al. (1993) and Siddiqui et al. (1995)
Camarolic acid (**25**), lantrigloylic acid (**26**), 22β-dimethylacryloyloxy-lantanolic acid (**27**)	*L. camara* (leaves)	Barre et al. (1997) and Begum et al. (2008b)
Ursangilic acid (**28**), lancamaric acid (**29**), camangeloyl acid (**30**), camarinin (**31**), lantadienone (**32**), camaradienone (**33**), pomonic acid (**34**), 3β,19α-dihydroxy-ursan-28-oic acid (**35**), 19α-hydroxy ursolic (**36**), lantaiursolic acid (**37**)	*L. camara* (aerial parts, roots)	Pan et al. (1993), Misra and Laatsch (2000), and Begum et al. (2003, 2008a)
Ursonic acid (**38**), lantacin (**39**), pomolic acid (**40**), 3,24-dioxo-urs-12-en-28-oic acid (**41**), α-amyrin (**42**) methyl 3-oxours-late (**43**), camaranoic acid (**44**), lantoic acid (**45**), camarinic acid (**46**), 22β-dimethylacryloyloxylantic acid (**47**), lantic acid (**48**), camaracinic acid (**49**), methyl ursoxylate (**50**), ursoxy acid (**51**), ursethoxy acid (**52**), methylcamaralate (**53**), camariolic acid (**54**), camarolide (**55**), betulinic acid (**56**), betulonic acid (**57**), betulonol (**58**), lantabetulic acid (**59**), euphane lactone B (**60–61**), euphane lactone C (**62–63**), euphane lactone A (**64**)	*L. camara* (aerial parts, leaves, stems, roots)	Barua et al. (1969), Ahmed et al. (1972), Hart et al. (1976), Mahato and Kundu (1994), Begum et al. (1995, 2003, 2008b), Barre et al. (1997), O'Neill et al. (1998), Saleh et al. (1999), and Yadav and Tripathi (2003)

(continued)

Table 1 (continued)

Compound	Species (parts used)	References
Luteolin (**65**), 7,3′,4′-trimethoxyluteolin (**66**), 7,3′-dimethyxyluteolin (**67**), 5,6-dihydroxy-7,3′,4′-trimethoxyflavone (**68**), 5,6,3′-trihydroxy-7,4′-dimethoxyflavone (**69**)	*L. montevidensis* (leaves)	Wollenweber et al. (1997)
3-methoxy-quercetin (**70**), 3-methoxy-3,7-dimethoxy-quercetin (**71**), 3,7,4′-trimethoxy-quercetin (**72**)	*L. camara* (leaves)	
Apigenin (**73**), cirsilineol (**74**), eupatorin (**75**), hispidulin (**76**), 5,4′-dihydroxy-6,7,3′,5′-tetramethoxyflavone (**77**), 5,3′,4′-trihydroxy-6,7,5′-trimethoxyflavone (**78**), 5,6,4′-trihydroxy-7,3′,5′-trimethoxyflavone (**79**), cirsiliol (**80**), Eupafolin (**81**)	*L. montevidensis* (leaves)	Nagão et al. (2002)
Pectolinarigenin (**82**), pectolinarin (**83**), camaroside (**84**) camaraside (**85**), lantanoside (**86**), linaroside (**87**), calceolarioside E (**88**), isonuomioside A (**89**), isoverbascoside (**90**), derhamnosylverbascoside (**91**), lantanaside (**92**), verbascoside (**93**), martynoside (**94**)	*L. camara* (aerial parts, stems)	Pan et al. (1993), Mahato and Kundu (1994), Taoubi et al. (1997), Syah et al. (1998), Begum et al. (2000), and Juang et al. (2005)
6-methoxydiodantunezone (**95**), 6-methoxy-8-hydroxy-diodantunezone (**96**), 7-methoxydiodantunezone (**97**), 7-methoxy-5-hydroxy-isodiodantunezone (**98**), 7-methoxy-8-hydroxy-diodantunezone (**99**), 6-methoxy-7-hydroxy-diodantunezone (**100**), 8-hydroxy-13-(methyl-dimethyl-hydroxy)-diodantunezone (**101**), 5-hydroxy-13-(methyl-dimethyl-hydroxy)-diodantunezone (**102**) diodantunezone (**103**), isodiodantunezone (**104**), geniposide (**105**), theviridoside (**106**)	*L. camara* (roots)	Abeygunawardena et al. (1991) and Pan et al. (1992)
Theveside (**107**), 8-epiloganin (**108**), lamiridoside (**109**), shanzhiside methyl ester (**110**)	*L. camara* (leaves, stems, roots)	Ford and Bcndal (1980) and Pan et al. (1992)
β-sitosterol (**111**), β-sitosterol-3-*O*-β-D-glucopiranoside (**112**), β-sitosterol-3-*O*-β-D-glicoside (**113**), β-sitosterol acetate (**114**), stigmasterol acetate (**115**), stigmasterol (**117**), 3β-hydroxystigmast-5-en-7-one (**117**), campesterol (**118**), lancamarone (**119**), *p*-coumaric acid (**120**), ethyl-β-D-galactoside (**121**), octanoic acid (**122**), cotriacontanoic acid (**123**), tetracosanoic acid (**124**), palmitic acid (**125**), docosanoic acid (**126**), octadecanoic acid (**127**)	*L. camara* (aerial parts, stems)	Ahmed et al. (1972), Jain et al. (1989), Siddiqui et al. (1995), Misra et al. (1997), Begum et al. (2003, 2008b)
Arachidic acid (**128**), 1-triacontanol (**129**)	*L. camara* (leaves, stems)	Ahmed et al. (1972)
Ajugose (**130**), verbascose (**131**), verbascotetrose (**132**), lantanose A and B (**133**), stachyose (**134**)	*L. camara* (roots)	Pan et al. (1992)
Pheophorbide A (**135**)	*L. camara* (leaves)	Sousa (2014)

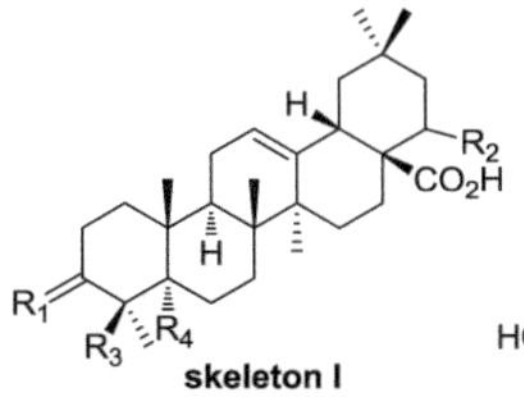

skeleton I

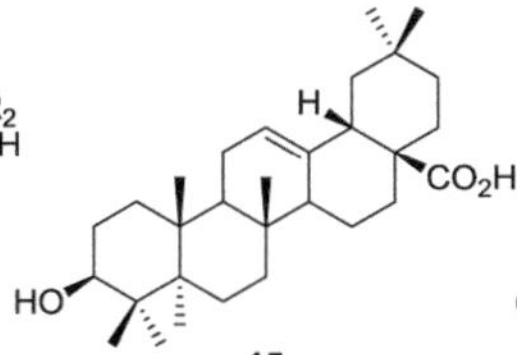

15

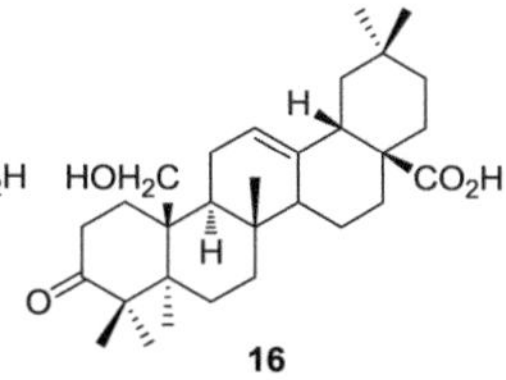

16

1 R_1=O; R_2=β-OCOC(CH_3)=CHCH_3; R_3=CH_3; R_4=H
2 R_1=O; R_2=β-OCOCH=C$(CH_3)_2$; R_3=CH_3; R_4=H
3 R_1=O; R_2=β-OCOCH$(CH_3)_2$; R_3=CH_3; R_4=H
4 R_1=H, β-OH; R_2=β-OCOC(CH_3)=CHCH_3; R_3=CH_3; R_4=H
5 R_1=H, β-OH; R_2=β-OCOCH=C$(CH_3)_2$; R_3=CH_3; R_4=H
6 R_1=O; R_2=β-OH; R_3=CH_3; R_4=H
7 R_1=O; R_2=R_4=H; R_3=CH_3
8 R_1=H, β-OH; R_2=R_4=H; R_3=CH_3
9 R_1=O; R_2=β-OH; R_3=R_4=CH_3
10 R_1=O; R_2=H, R_3=CH_2OH, R_4=CH_3
11 R_1=O; R_2=β-OCOC(CH_3)=CHCH_3; R_3=CH_2OH; R_4=H
12 R_1=O; R_2=β-OCOCH=C$(CH_3)_2$; R_3=CH_2OH; R_4=CH_3
13 R_1=H, H; R_2=β-OCOC(CH_3)=CHCH_3; R_3=CH_3; R_4=H
14 R_1=H,H; R_2=β-OCOCH=CHCH_3; R_3=CH_3; R_4=H

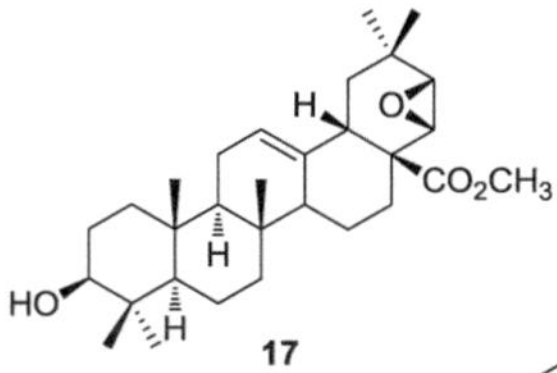

17

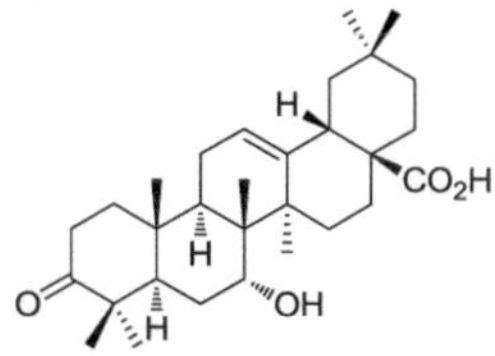

18

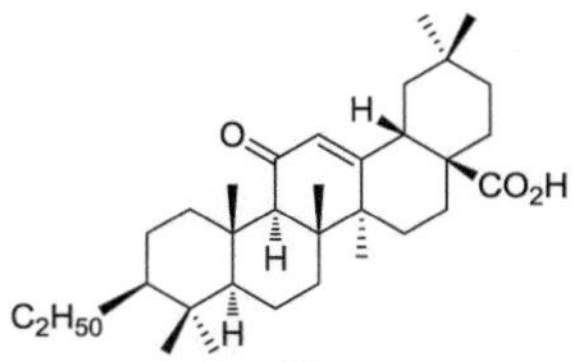

19

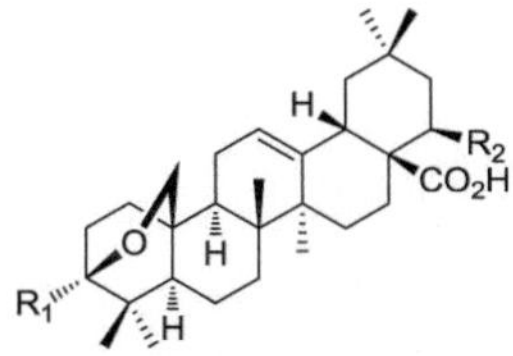

skeleton II

20 R_1=α-OH; R_2=β-OCOC(CH_3)=CHCH_3
21 R_1=α-OCH_3; R_2=β-OCOC(CH_3)=CHCH_3
22 R_1=α-OH; R_2=β-OCOCH=C$(CH_3)_2$
23 R_1=α-OH; R_2=H
24 R_1=α-OH; R_2=β-OCOC(CH_3)=CHCH_3
25 R_1=α-OH; R_2=β-OCOC(=CH_2)CHOHCH_3
26 R_1=α-OH; R_2=β-OCOCH=C$(CH_3)_2$
27 R_1=β-OH; R_2=β-OCOCH=C$(CH_3)_2$
28 R_1=α-OCH_2CH_3; R_2=β-OCOC(CH_3)=CHCH_3

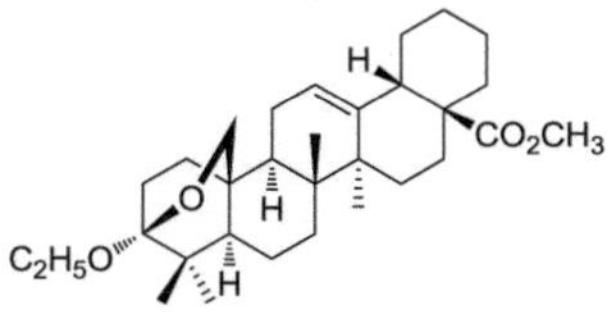

29

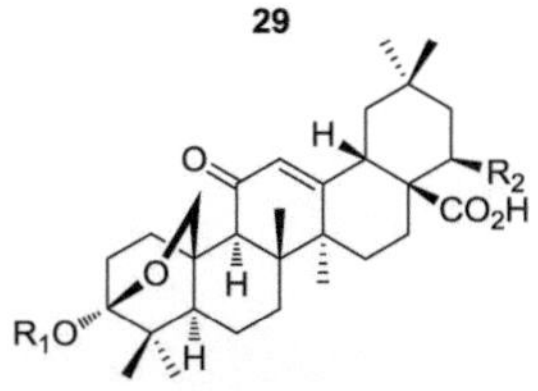

skeleton III

30 R_1=CH_3; R_2=OCOC(CH_3)=CHCH_3
31 R_1=H; R_2=OCOCH$(CH_3)_2$

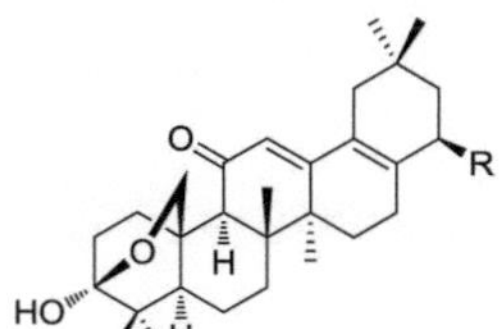

skeleton IV

32 R=β-OCOCH(CH3)2
33 R=H

Fig. 1 Structures of constituents isolated from *L. camara* and *L. montevidensis*

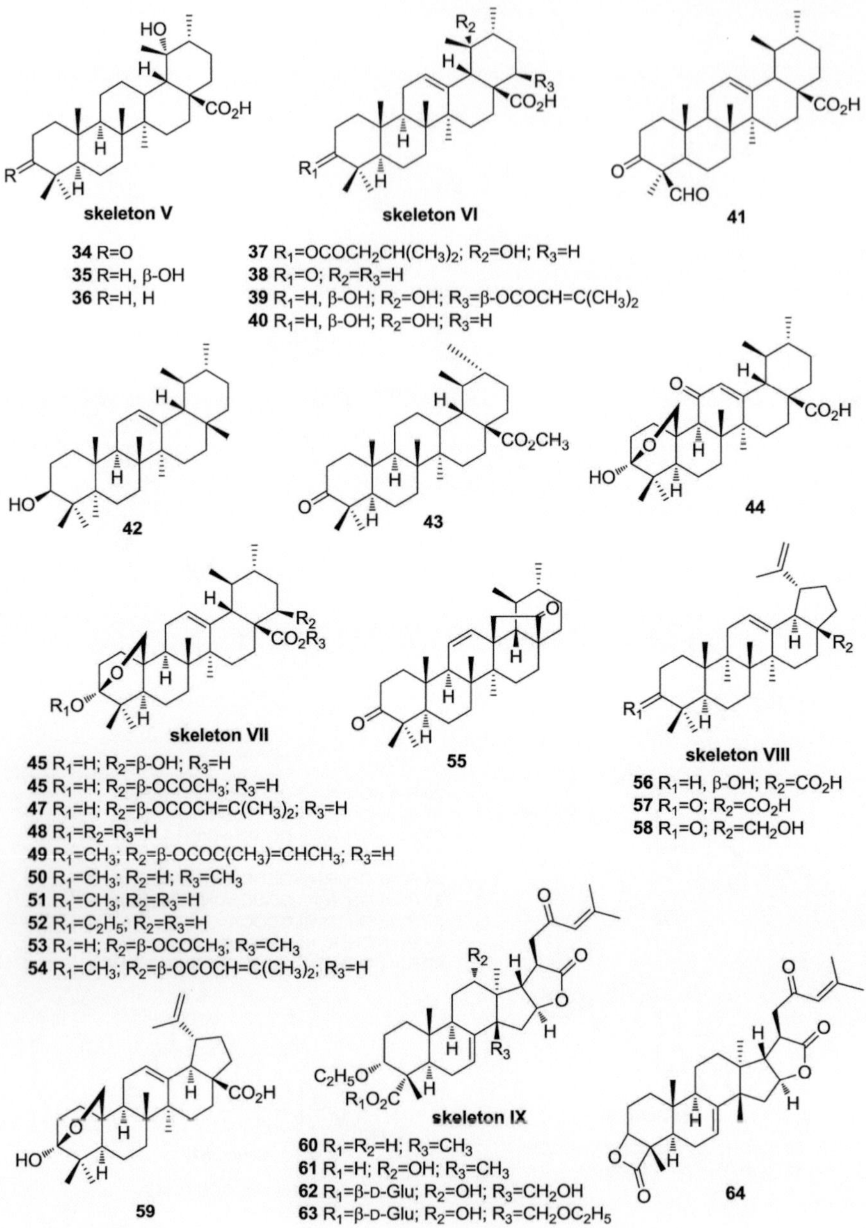

Fig. 1 (continued)

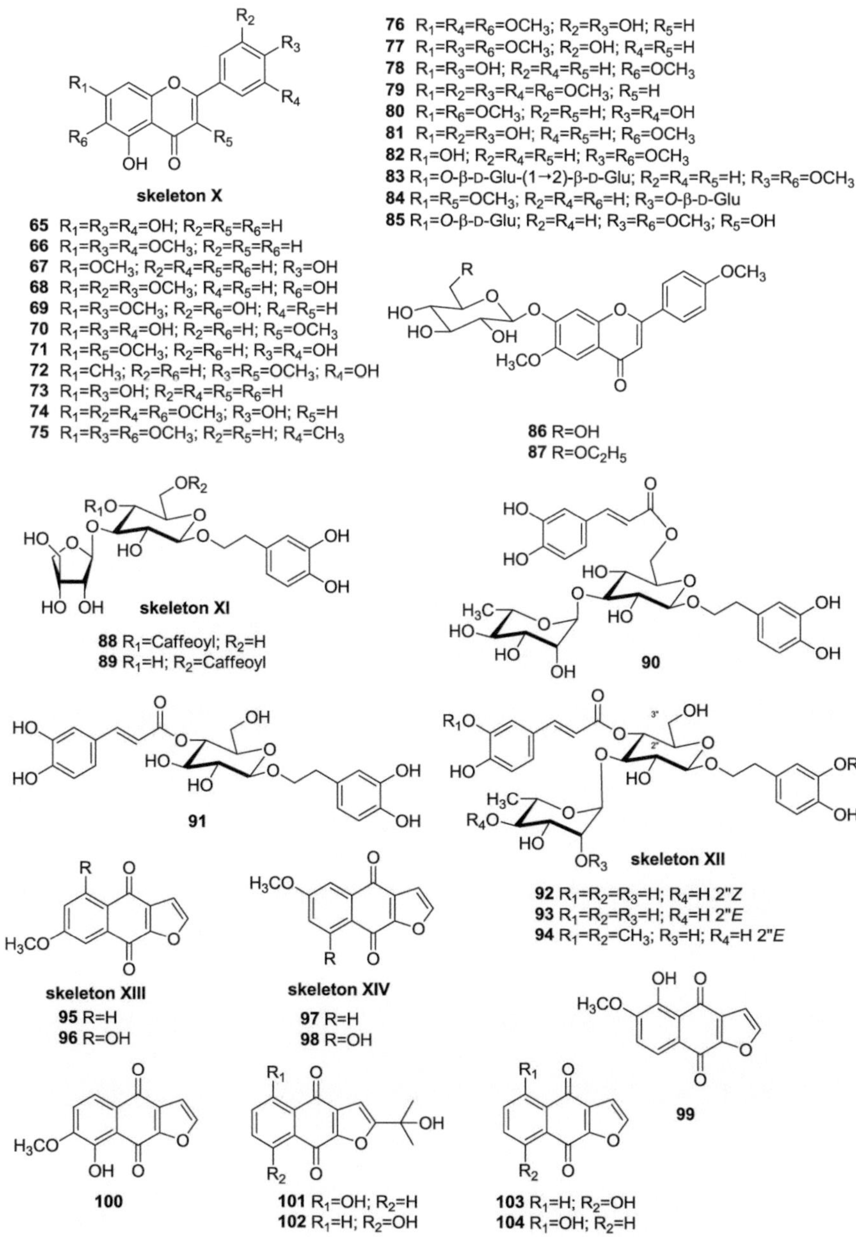

Fig. 1 (continued)

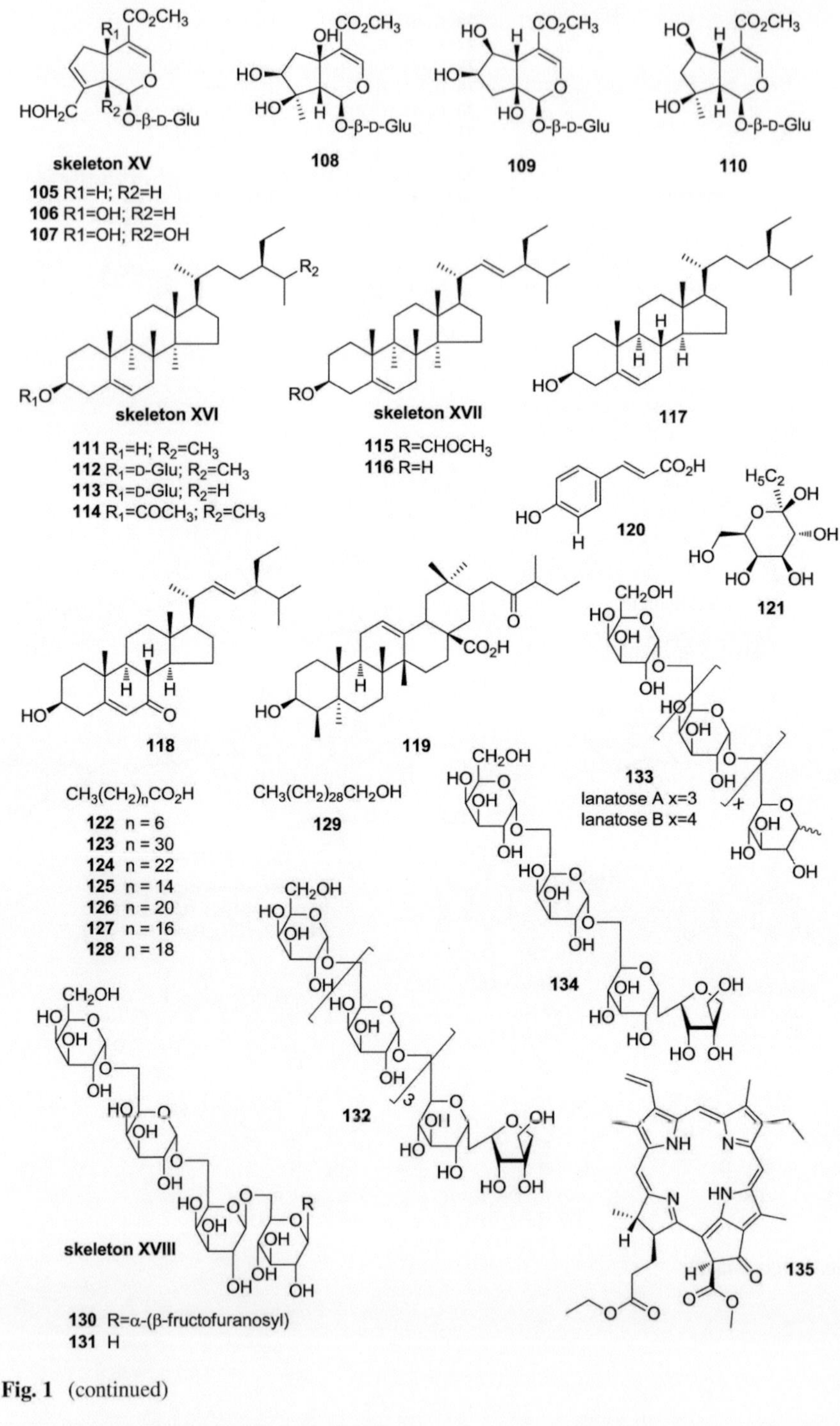

Fig. 1 (continued)

The isomers α and ß-caryophyllene were present among the main constituents of *L. camara's* essential oil from Northeast Brazil at different times of day (Sousa et al. 2010). In the seasonal evaluation of the same essential oil from Madagascar, the concentration of ß-caryophyllene was found to be consistently high throughout the year, independent of sampling seasons (Randrianalijaona et al. 2005).

3 Morphological Description

The genus *Lantana* includes herbaceous and shrubby plants, which can reach a height of over 2 m. They are very often planted for ornamental purposes which is due to the beauty of their flowers (Joly 1993). The species *L. camara* is an erect shrub; its quadrangular branches are armed with small curved spines, sometimes defenseless; opposite leaves, also short-petiolate, ovate-oblong, rounded at base, acuminate, crenate-sawn, rough-crosslinked, aromatic, very rough on the top page and pale or whitish on the bottom page; hard pubescent stems or rough hirsute or subinermes; flowers are white when bloom; fruits are purple-black and small berries. *L. montevidensis* is a hair-covered bush; strong root system; quadrangular branches, defenseless or aculeate; aculeate petioles; leaves are ovate-cordate, opposite, sawed-crenate, hairy or rough-hirsute and hispid on the top page and pale and hairy-stiff-hirsute on the bottom page; flowers are primarily gold yellow, then orange, pink or red and finally, vermilion, blooming from the center to the circumference, arranged in long-stalked chapters (Corrêa 1978).

4 Geographical Distribution

The genus *Lantana* as described by Linnaeus in 1753 contained 7 species, 6 from South America and 1 from Ethiopia; currently, they occur in approximately 50 countries with a very large number of species and subspecies. The recorded number of *Lantana* species varies from 50 to 270 specific and subspecific entities, but it appears that a better estimate is 150 species (Ghisalberti 2000).

L. camara is a shrub native from America and Africa and was introduced to many countries as an ornamental plant. Dutch explorers introduced it into the Netherlands from Brazil in the late 1600s and later explorers from other countries brought seeds to Europe, Great Britain and North America. *L. montevidensis* is a shrub native to Brazil and Uruguay and also was introduced to many countries as an ornamental plant (Ghisalberti 2000; Nagão et al. 2002).

5 Ecological Requirements

L. camara and *L. montevidensis* are shrubs that prefer full sun. They are quite resistant to pruning, undemanding in the soil, and bloom flowers virtually all year, which led floriculturists to consider them as ornamental species, thus spreading them everywhere while obtaining numerous varieties through plant breeding (Joly 1993). The plants grow luxuriantly at elevations up to 2000 m in tropical, sub-tropical and temperate regions.

6 Traditional Use (Part(s) Used) and Common Knowledge

In many parts of the world species of the genus *Lantana* are used to treat a wide variety of disorders, in the folk medicine, especially for tumours and cancer. A tea prepared from the leaves and flowers of *L. camara* was effective against fever, influenza and stomach-ache. In Central and South America, the leaves were made into a poultice to treat soreness, chicken pox and measles. Infusion of the whole plant was used, in Ghana, for bronchitis and the powdered root was added in milk then given to children for stomach-ache. In Asian countries, leaves are boiled for tea and the decoction is a remedy against coughing. The decoction of the whole plant is given as treatment against tetanus, rheumatism, malaria and ataxia of abdominal viscera. It is used as a lotion for wounds, too. Pounded leaves are applied to cuts, ulcers and swellings (Nagão et al. 2002). Their roots are used in the treatment of malaria, rheumatism and rash (Chharba et al. 1993). The leaves' infusions of *L. montevidensis* have been used in the treatment of scratching, stomachache, rheumatism, wound healing, biliary fever, toothache, bronchitis and antiseptic (Ghisalberti 2000).

7 Modern Medicine Based on Its Traditional Medicine Uses

In recent decades several studies have been directed to study the biological activities of species of the genus *Lantana*. In this sense, the following sequence of major biological activities of isolated constituents, extracts, fractions and essential oils can be established: oleanonic acid (**7**), oleanolic acid (**8**), camarin (**18**), lantanolic acid (**23**), camarinin (**31**), ursonic acid (**38**), lantacin (**39**), pomolic acid (**40**) and lantoic acid (**45**) isolated from extracts and fractions of aerial parts of *L. camara* showed promising anthelminthic activity (Begum et al. 2000; Misra et al. 2007). The dichloromethane and aqueous extracts of *L. camara's* leaves demonstrated anti-protozoal activity against cultures of chloroquine-sensitive and resistant strains of *Plasmodium falciparum* (Weenen et al. 1990).

Essential oils from the leaves of *L. camara* and *L. montevidensis* and extracts of leaves, twigs, stems and roots of *L. camara* showed toxic activity using *Artemia salina* larvae (Weenen et al. 1990; Fatore et al. 2002). A larvicidal potential of the essential oils was showed against *Aedes aegypti* larvae (Costa et al. 2010). The essential oil of *L. camara* leaves showed also insecticidal activity against adults of *Sitophilus oryzae* and *Tribolium castaneum* (Mohamed and Abdelgaleil 2008).

A study by Sousa et al. (2011a, b) showed the inhibitory activity of an ethanolic extract of *L. montevidensis* leaves against multiresistant strains of *Escherichia coli* and *Staphylococcus aureus*. The essential oil of the leaves of *L. camara* has been examined for antibacterial activity and also showed an inhibitory activity against these multiresistant strains of bacteria (Sousa et al. 2011a, b).

Two compounds isolated from *L. camara* leaves were found to possess strong antibacterial activity, the lactic acid (**48**) against *Escherichia coli* and *Bacillus cereus* and the carminic acid (**46**) against *Staphylococcus aureus* and *Salmonella typhi* (Saleh et al. 1999). The synergistic effect of gentamicin and amikacin against *Staphylococcus aureus* and *Pseudomonas aeruginosa* was observed in the presence of the essential oils and ethanolic extracts of leaves and roots of *L. camara* and *L. montevidensis* (Sousa et al. 2011a, b).

Both essential oils and ethanolic extracts from the leaves of *L. camara* and *L. montevidensis* presented a strong inhibition on DPPH free radical scavenging (Sousa et al. 2013). A study showed an antiproliferative activity of the flavonoid fraction of *L. montevidensis's* leaves against human gastric adenocarcinoma (MK-1), human uterine carcinoma (HeLa), and murine melanoma (B16F10) cells *in vitro*. In addition, the methanolic extracts of *L. camara* and *L. montevidensis's* leaves were very effective in inhibiting tumor cell growth (Nagão et al. 2002).

The compounds icterogenin (**11**) and 22β-dimethylacryloyloxy-24-hydroxy-3-oxo-olean-12-en-28-oic acid (**12**) isolated from leaves of *L. camara* were evaluated for their interaction with the antiapoptotic protein Bcl-xL/Bak association (Litaudon et al. 2009). The verbascoside (**113**) isolated from *L. camara* was shown to be an inhibitor of protein kinase C (PKC) from rat brain (Herbert et al. 1991). Lantadenes A (**1**), B (**2**) and C (**3**) isolated from leaves of *L. camara* displayed cytotoxic activity against four cancer cell lines: human oral epidermoid carcinoma (KB), human colon cancer (HCT-116), human breast cancer (MCF-7) and mouse lymphocytic leukemia (L1210) (Litaudon et al. 2009).

8 Conclusions

In this chapter a brief review of the ethnopharmacological, phytochemical and biological information of *L. camara* and *L. montevidensis* is given. Based on the above stipulations, the presence of terpenoids, flavonoids, phenylethanoid glycosides, furanonaphthoquinones, iridoid glycosides and steroids has been demonstrated. These species are a rich source of a variety of organic compounds with varying chemical structural patterns.

References

Abeygunawardena C, Kumar V, Marshall DS, Thomson RH, Wickramaratne DBM (1991) Furanonaphthoquinones from two *Lantana* species. Phytochemistry 30(3):941–945

Ahmed ZF, Shoaib AM, Wassel GM, El-Sayyad SM (1972) Phytochemical study of *Lantana camara*. terpenes and lactones II. Planta Med 22(1):34–37

Barre JT, Bowden BF, Coll JC, De Jesus J, De La Fuente V, Janairo GC et al (1997) Bioactive triterpene from *Lantana camara*. Phytochemistry 45(2):321–324

Barua AK, Chakrabarti P, Sanyal PK, Das B (1969) Triterpenoids XXXXII. Structure of lantic acid: a new triterpene from *Lantana camara*. J Indian Chem Soc 46:100–101

Begum S, Raza SM, Siddiqui BS, Siddiqui S (1995) Triterpenoids from the aerial parts of *Lantana camara*. J Nat Prod 58(10):1570–1574

Begum S, Wahab A, Siddiqui BS, Qamar F (2000) Nematicidal constituents of the aerial parts of *Lantana camara*. J Nat Prod 63(3):765–767

Begum S, Wahab A, Siddiqui BS (2003) Pentacyclic triterpenoids from the aerial parts of *Lantana camara*. Chem Pharm Bull 51(2):134–137

Begum S, Zehra SQ, Hassan S, Siddiqui BS (2008a) Noroleanane triterpenoids from the aerial parts of *Lantana camara*. Helv Chim Acta 91(3):460–467

Begum S, Zehra SQ, Siddiqui BS, Fayyaz S, Ramzan M (2008b) Pentacyclic triterpenoids from the aerial parts of *Lantana camara* and their nematicidal activity. Chem Biodivers 5(9):1856–1866

Chharba SC, Mahunnah RLA, Mshiu IN (1993) Plants used in traditional medicine in eastern Tanzania. J Ethnopharmacol 39(1–2):83–103

Corrêa MP (1978) Dicionário das plantas úteis do Brasil e das exóticas cultivadas. Imprensa Nacional, Rio de Janeiro

Costa JGM, Rodrigues FFG, Sousa EO, Junior DMS, Campos AR, Coutinho HDM et al (2010) Composition and larvicidal activity of the essential oils of *Lantana camara* and *Lantana montevidensis*. Chem Nat Compd 46(2):313–315

Dambolena JS, Zunino MP, Lucini EI, Zygadlo JA, Banchio E, Biurrun F, Rotman A, Ahumada O (2010) Aromatic plants of northwest Argentina. Constituents of the essential oils of aerial parts of seven Verbenaceae: *Lantana* and *Aloysia*. J Essent Oil Res 4(4):289–293

Fatore MO, Salihu L, Asante SK, Takeda T (2002) Larvicidal activity of extracts and triterpenoids from *Lantana camara*. Pharm Biol 40(8):564–567

Ford CW, Bendal L (1980) Identification of the iridoid glucoside theveside in *Lantana camara* (Verbenaceae), and determination of its structure and stereochemistry by means of N.M.R. Aust J Chem 33(3):509–518

Ghisalberti EL (2000) *Lantana camara* L. (Verbenaceae). Fitoterapia 71(5):467–486

Hart N, Lamberton JA, Sioumis AA, Suares H (1976) New triterpenes of *Lantana camara*. A comparative study of the constituents of several taxa. Aust J Chem 29(3):655–671

Herbert JM, Maffrand JP, Taoubi K, Augereau JM, Fouraste I, Gleye J (1991) Verbascoside isolated from *Lantana camara*, an inhibitor of protein kinase C. J Nat Prod 54(6):1595–1600

Jain R, Singh M, Dezman DJ (1989) Qualitative and quantitative characterization of phenolic compounds from *Lantana camara* leaves. Weed Sci 37(3):302–307

Joly AB (1993) Introdução à Taxonomia Vegetal. São Paulo, Companhia Editora Nacional

Juang FC, Chen YF, Lin FM, Huang KF (2005) Constituents from the leaves of *Lantana camara* (IV). J Chin Med 16(2–3):149–155

Lai J-S, Chan Y-F, Huang K-F (1998) Constituents from the stems of *Lantana camara* (II). Chin Pharm J 50:385–392

Litaudon M, Jolly C, Le Callonec C, Cuong DD, Retailleau P, Nosjean O et al (2009) Cytotoxic pentacyclic triterpenoids from *Combretum sundaicum* and *Lantana camara* as inhibitors of Bcl-xL/BakBH3 domain peptide interaction. J Nat Prod 72(7):1314–1320

Mahato SB, Kundu AP (1994) ^{13}C NMR spectra of pentacyclic triterpenoids D a compilation and some salient features. Phytochemistry 37(6):1517–1575

Misra L, Laatsch H (2000) Triterpenoids, essential oil and photo-oxidative 28413- lactonization of oleanolic acid from *Lantana camara*. Phytochemistry 4:969–974

Misra LN, Dixit AK, Sharma RP (1997) High concentration of hepatoprotective oleanolic acid from *Lantana camara* roots. Planta Med 63(6):582

Misra N, Sharma M, Raj K, Dangi A, Srivastava S, Misra-Bhattacharya S (2007) Chemical constituents and antifilarial activity of *Lantana camara* against human lymphatic filariid *Brugia malayi* and rodent filariid *Acanthocheilonema viteae* maintained in rodent hosts. Parasitol Res 100(3):439–448

Mohamed MIE, Abdelgaleil SAM (2008) Chemical composition and insecticidal potential of essential oils from Egyptian plants against *Sitophilus oryzae* (L.) (Coleoptera: Curculionidae) and *Tribolium castaneum* (Herbst) (Coleoptera: Tenebrionidae). Appl Entomol Zool 43(4):599–607

Nagão T, Abe F, Kinjo J, Okabe H (2002) Antiproliferative constituents in plants 10. Flavones from the leaves of *Lantana montevidensis* Briq. and consideration of structure-activity relationship. Biol Pharm Bull 25(7):875–879

O'Neill MJ, Lewis JA, Noble HM, Holland S, Mansat C, Farthing JE et al (1998) Isolation of translactone-containing triterpenes with thrombin inhibitory activities from the leaves of *Lantana camara*. J Nat Prod 61(11):1328–1331

Pan WD, Li Y, Mai LT, Ohtani K, Kasai R, Tanaka O (1992) Studies on chemical constituents of the roots of *Lantana camara*. Yao Xue Xue Bao 27(7):515–521

Pan WD, Li YJ, Mai LT, Ohtani K, Kasai R, Tanaka O et al (1993) Studies on triterpenoid constituents of the roots of *Lantana camara*. Yao Xue Xue Bao 28(1):40–44

Randrianalijaona JA, Ramanoelina PAR, Rasoarahona JRE, Gaydou EM (2005) Seasonal and chemotype influences on the chemical composition of *Lantana camara* L. Essential oils from Madagascar. Anal Chim Acta 545(1):46–52

Saleh M, Kamel A, Li X, Swaray J (1999) Antibacterial triterpenoids isolated from *Lantana camara*. Pharm Biol 37(1):63–66

Sena Filho JG, Xavier HS, Barbosa Filho JM, Duringer JM (2010) A chemical marker proposal for the *lantana* genus: composition of the essential oils from the leaves of *Lantana radula* and *L. canescens*. Nat Prod Commun 5(4):635–640

Sharma OP, Dawra RK (1991) Thin layer chromatographic separations of lantadenes the pentacyclic triterpenoids from *Lantana* (*Lantana camara*) plant. J Chromatogr 587(2):351–354

Sharma OP, Singh A, Sharma S (2000) Levels of lantadenes, bioactive pentacyclic triterpenoids, in young and mature leaves of *Lantana camara* var. aculeate. Fitoterapia 71(5):487–481

Siddiqui BS, Raza SM, Begum S, Siddiqui S, Firdous S (1995) Pentacyclic triterpenoids from *Lantana camara*. Phytochemistry 38(3):681–685

Singh SK, Singh A, Tripathi VJ, Finzi PV (1996) Minor constituents of *Lantana camara*. J Chem Soc 73:547–547

Sousa EO (2014) Perfil químico e atividade antioxidante e carrapaticida de *Lantana camara* L. e *Lantana montevidensis* Briq. [thesis]. Universidade Estadual do Ceará, Fortaleza

Sousa EO, Colares AV, Rodrigues FFG, Campos AR, Lima SG, Costa JG (2010) Effect of collection time on essential oil composition of *Lantana camara* Linn (Verbenaceae) Growing in Brazil Northeastern. Rec Nat Prod 4(1):31–37

Sousa EO, Almeida TS, Rodrigues FFG, Campos AR, Lima SG, Costa JGM (2011a) *Lantana montevidensis* Briq improves the aminoglycoside activity against multiresistant *Escherichia coli* and *Staphylococcus aureus*. Indian J Pharmacol 43(2):180–182

Sousa EO, Rodrigues FFG, Coutinho HDM, Campos AR, Lima SG, Costa JGM (2011b) Chemical composition and aminoglycosides synergistic effect of *Lantana montevidensis* Briq. (Verbenaceae) essential oil. Rec Nat Prod 5(1):60–64

Sousa EO, Rocha JBT, Barro LM, Barros AR, Costa JGM (2013) Phytochemical characterization and *in vitro* antioxidant properties of *Lantana camara* L. and *Lantana montevidensis* Briq. Ind Crop Prod 43:517–522

Syah YM, Pennacchio M, Ghisalberti EL (1998) Cardioactive phenylethanoid glycosides from *Lantana camara*. Fitoterapia 69(3):285–286

Taoubi K, Fauvel MT, Gleye J, Moulis C, Fouraste I (1997) Phenylpropanoid glycosides from *Lantana camara* and *Lippia multiflora*. Planta Med 63(2):192–193

Weenen H, Nkunya MHH, Bray DH, Mwasumbi LB, Kinabo LS, Kilimali VAEB (1990) Antimalarial activity of Tanzanian medicinal plants. Planta Med 56(1):368–370

Wollenweber E, Dorr M, Muniappan R, Siems K (1997) Flavonoid aglycones and triterpenoids from the leaf exudate of *Lantana camara* and *Lantana montevidensis*. Biochem Syst Ecol 25(3):269–270

Yadav SB, Tripath V (2003) A new triterpenoid from *Lantana camara*. Fitoterapia 74(3):320–321

Lippia alba (Mill.) N.E.Br. ex Britton & P. Wilson

Renata Evaristo Rodrigues da Silva, Isabel Cristina Santiago, Vanessa de Carvalho Nilo Bitu, Marta Regina Kerntopf, Irwin Rose Alencar de Menezes, and Roseli Barbosa

Abstract The species *Lippia alba* (Mill.) N.E.Br. ex Britton & P. Wilson is a subshrub belonging to the family Verbenaceae. It is widely distributed in Latin America. In Brazil, it occurs in almost all regions and is therefore known by various names, where the most common are "erva-cidreira," "falsa-melissa," "chá-de-tabuleiro," "salva-do-Brasil" and "erva-cidreira-brasileira," among others. It is an aromatic plant that contains a variety of volatile constituents including, citral, limonene, carvone, linalool, caryophyllene, myrcene, terpinene, 1,8-cineole and estragole. This variability of constituents results in a number of different chemotypes. *L. alba* is highly capable of adaptation to various environments as well as rapid spread and colonization, that enhance its industrial potential. Another advantage is that it grows and blooms year-round. *L. alba* is considered as one of the medicinal plants that is mostly used in traditional practices, in Brazil. Its pharmacological properties include analgesic, anti-inflammatory, antipyretic, sedative, digestive, anti-asthmatic, antihypertensive, antispasmodic, emmenagogue and diaphoretic, and it is used in the treatment of syphilis and gonorrhea. The leaves and roots are most frequently used in the form of infusions, alcoholic extracts, compresses, baths and syrups. Several preclinical studies have observed a variety of pharmacological activities related to its empirical use, especially antimicrobial, anti-ulcer, anti-nociceptive, muscle relaxant and antioxidant. In Brazil, *L. alba* is among the 66 regulated species with medicinal purposes. Clinical trials are needed, since this species has an not yet fully explored great potential for the future production of medicines.

Keywords *Lippia alba* · Pharmacological activity · Verbenaceae

R. E. R. da Silva (✉) · I. C. Santiago · V. de Carvalho Nilo Bitu · M. R. Kerntopf
I. R. A. de Menezes · R. Barbosa
Química Biológica, Universidade Regional do Cariri-URCA, Crato, Brazil

U. P. Albuquerque et al. (eds.), *Medicinal and Aromatic Plants of South America*, Medicinal and Aromatic Plants of the World 5,
https://doi.org/10.1007/978-94-024-1552-0_25

1 Taxonomic Characteristics

The genus *Lippia* is the second largest in the family Verbenaceae. Approximately 200 species of this genus have been found among herbs, shrubs and small trees. *Lippia alba* (Mill.) N.E.Br. ex Britton & P. Wilson is distributed in several tropical and subtropical regions, for example, Latin America, which has resulted that it has several common names, generally related to its characteristic aromatic odor and medicinal properties (Hennebelle et al. 2008b). In Colombia, for example, it is popularly known as "pronto alívio (ready relief)" and depending on the region it can also be called "curatodo (cure-all)" (Stashenko et al. 2003). In Brazil, the most common names are: erva-cidreira, falsa melissa, chá-de-tabuleiro, erva-cidreira-do-campo, salva-do-Brasil, salva-limão, erva-cidreira-brava, chá-de-febre, erva-cidreira-brasileira, alecrim-do-mato, alecrim-do-campo and alvia sija (Matos 2000; Holetz et al. 2002; Pascual et al. 2001).

Synonyms *L. alba* also has several botanical synonyms belonging to the genera *Lippia, Lantana, Filos, Verbena* and *Zapania*, such as: *Lippia asperifolia* A. Rich, *Lippia crenata* Sessé & Moc, *Lippia geminata microphylla* Griseb, *Lippia germinata* H.B. K, *L. glabriflora* Kuntze, *Lippia haanensis* Turcz, *Lippia lantanoides* Coult, *Lippia trifólia* Sessé & Moc, *Lantana alba* Mill, *Lantana canescens* Hort., *Lantana geminata* (H.B.K.) Spreng., *Lantana geminata* Spreng, *Lantana lippioides* Hook. & Arn., *Phyla geminata* H.B.K. and *Verbena lantanoides* Willd.

Taxonomy appears to confuse this huge diversity of synonyms in such a way that it makes botanical classification of the genus *Lippia* difficult. The taxonomic controversies are caused by morphological, anatomical and physiological versatility (differences) that could be attributed to different degrees of ploidy and wide phenotypic plasticity (Pascual et al. 2001).

Pierre et al. (2011) studied the karyology of three chemotypes of *L. alba* (citral, carvone and linalool) and found differences between them in the number and morphology of chromosomes, thus revealing that the citral chemotype had 2n = 30 chromosomes, while the carvone chemotype had a chromosome number of 2n = 60, which could be an autopolyploid of the citral chemotypes. On the other hand, the study showed numerical variation within linalool chemotypes, identifying a mixoploid that showed 2n = 12 to 2n = 60. Utilizing the FISH technique, the same study demonstrated that *L. alba* has an allopolyploid origin, where plants were found with differences in chromosome number. This has made it possible to distinguish two chemotypes: 1-citral and 2-linalool chemotypes. Studies thus revealed some of the existing variations in *L. alba* that thereby lead to karyotypic variations (Sousa et al. 2012).

2 Major Chemical Constituents and Bioactive Compounds

L. alba contains several volatile constituents. Substances, such as flavonoids, iridoids, naphthoquinones, tannins, resins and mucilages, are also frequently described. Even with great variability in its chemical composition, this species usually shows consistent profiles in its constitution. Frequently reported among the main aromatic constituents of the essential oils of *L. alba* are the monoterpenoids borneol, camphor, 1,8-cineole, citronellol, geranial, linalool, myrcene, neral, limonene, piperitone 2-undecanone, sabinine and the sesquiterpenoids, caryophyllene, murolene, cubebene, b-elemene, g-cadinene, allo-aromadendrene and caryophyllene oxide. These can vary both quantitatively and qualitatively, depending on various factors such as the season, flowering time, plant age, amount of circulating water and climatic and geographical factors. The range of essential oil content changes according to its physiological cycle. It has been found that *L. alba* produced the highest amount of essential oil outside the flowering period (Tavares et al. 2005).

Due to the high variability in the chemical composition of the essential oil of *L. alba*, it has been recommended to group this species into separate chemotypes differentiated by its major components (Julião et al. 2003; Matos et al. 1996). On analysing both the major chemical constituents of the essential oil and the plant's metabolic pathways, seven chemical types (chemotypes) were distinguished:

Chemotype 1 – citral, linalool and caryophyllene;
Chemotype 2 – tagetenone;
Chemotype 3 – limonene with varying amounts of carvone;
Chemotype 4 – myrcene;
Chemotype 5 – γ-terpinene;
Chemotype 6 – camphor-1,8-cineole;
Chemotype 7 – estragole,

One chemotype, in which citral was the major component accompanied with a small concentration of linalool, was classified as a subtype of chemotype 1 (Julião et al. 2003).

3 Morphological Description

L. alba is a shrub with variable morphological traits; it has whitish thin, brittle, curved branches, and the leaves have an elliptical shape, vary in width but with an acute apex, and are arranged oppositely (Matos 2000). It can grow up to 1 m in height, and bloom all around the year. It has inflorescences that vary in color and may be white, pink or violet, which form fruit calyx with seeds, which are dispersed by the wind (Salimena 2002).

Fig. 1 Photograph of stem and flower of *L. alba*. (Silva)

It is considered a hardy shrub with perennial cycles, as well as rapid growth and development. Thus, it easily colonizes through natural rooting from its branches in contact with the ground. It usually has a decumbent habit, and because of its colonizing potential, it is frequently found in sandy soils, as well as along the banks of rivers, lakes and reservoirs (Stefanini et al. 2001; Biasi and Costa 2003; Ehlert et al. 2003).

One of the ways of identifying *L. alba* is on the basis of its leaves, that are simple, whole, serrated, not round or square but oblong and acute, and arranged oppositely, with two per node. They are membranous, petiolate and pubescent and have a characteristic lemon-like scent (Castro et al. 2002), (Fig. 1).

4 Geographical Distribution

L. alba is found in all tropical and subtropical areas of South America, Central America, Caribbean Islands and the southern region of the United States. It also occurs in India and Australia. However, it has a wide distribution and it is traditionally used extensively in Latin America, from Mexico to Cuba, Uruguay, Paraguay and Brazil, where its great phenotypic variability is shown by its adaptation to different climatic conditions. In Brazil, it is found in almost all regions: North (Amapá, Pará, Amazonas, Acre), Northeast (Maranhão, Ceará, Rio Grande do Norte, Paraíba, Bahia), Central-West (Mato Grosso, Goiás, Mato Grosso do Sul), Southeast (Minas Gerais, São Paulo, Rio de Janeiro) and South (Paraná and Rio Grande do Sul) (Martins et al. 1995; Pascual et al. 2001; Hennebelle et al. 2008b).

5 Ecological Requirements

In the cultivation of medicinal plants, especially in the case of aromatic plants, several factors must be considered to produce good quality: in general, it is necessary to provide ideal conditions for germination, dissemination and rooting (Farias et al. 2003).

L. alba can withstand droughts of around 4–6 months without rainfall due to their morphological alterations. At a temperature of about 24 °C and relative humidity of 75%, in areas with defined rainy and dry seasons with an average annual rainfall of 1056 mm, this species shows good growth and development. The flexibility of *L. alba* to adapt to various environments increases its commercial potential. Due to its rapid colonization ability and spread, as well as vigor, it grows and blooms all year (Arambarri et al. 2006; Barbosa et al. 2006).

6 Traditional Use (Part(s) Used) and Common Knowledge

L. alba is one of the medicinal plants that is mostly used in traditional healing practices by the Brazilian population, as pointed out by the Central de Medicamentos (Center for Medications (CEME) (Ming 1994). Moreover, thanks to its wide dissemination and use by the people of Northeast Brazil, where it is popularly known as "erva cidreira", it was also included in the "Living Pharmacy" project of the Federal University of Ceará, the project "Herbal Medicine in Health Care" implemented by State Secretary of Health of Paraná and even projects promoted by the City Hall of Campinas (SP), providing herbal medicine assistance to the poor. In Brazil, it is currently among the 66 species that are regulated for medicinal purposes (Ming 1996; Matos 2000; Castro et al. 2002).

Various ethnopharmacological studies deal with a wide range of traditional uses for *L. alba*, with the main purposes being analgesic, anti-inflammatory, antipyretic, sedative, antispasmodic and cooking spice, and the treatment of dysentery, diarrhea, skin diseases, liver diseases, menstrual cramps, syphilis and gonorrhea. There are also investigations into its main uses for respiratory, digestive, cardiovascular conditions, as well as hypertensive ailments and as sedatives (Mattos et al. 2007; Hennebelle et al. 2008a).

L. alba is widely popular due to its use as a tranquilizer, analgesic, sedative, anxiolytic, antispasmodic and expectorant (Mattos et al. 2007). For these therapeutic purposes, there are various ways of preparing the herbals, i.e. using the leaves or roots, such as teas, infusions, baths, alcoholic extracts, compresses, and syrups, mainly because of its chemical constituents, especially in the essential oil (Julião et al. 2003).

Studies have reported the use of *L. alba* infusions and decoctions in treating gastrointestinal problems, particularly in South and Central America and tropical Africa (Agra and Barbosa Filho 1990; Vale et al. 1999; Pascual et al. 2001). Externally, it is commonly used in Brazil and Guatemala for skin problems such as burns, ulcers and wounds (Giron et al. 1991). Ethnobotanical studies have reported the use of this plant for the treatment of syphilis (Zamora and Nieto 1992).

Various ethnopharmacological studies have shown the extensive use of *L. alba* in traditional medicine. In a study conducted in three cities in the state of São Paulo, *L. alba* was the seventh most cited plant, used as an infusion for hypertension, digestive problems, nausea and colds, as a topical medication to heal wounds and as a syrup for cough and bronchitis (Di Stasi et al. 2002). In the state of Bahia, two more studies reported that *L. alba* was also the most cited, used as a sedative and for hypertension, pain and flatulence (Rodrigues and Guedes 2006). In Mexico, it is used by traditional healers for gastrointestinal problems (Heinrich et al. 1992). In Pernambuco, in the municipality of Igarassu, *L. alba* was cited for treating anemia and digestive problems (Gazzaneo et al. 2005).

7 Modern Medicine Based on Its Traditional Medicine Uses

Despite the widespread popular use of *L. alba*, there are only very few pharmacological studies aimed at elucidating its biological activities (Pascual et al. 2001). Considering the biological activities of extracts and/or polar fractions of *L. alba*, we found in vitro studies that showed antioxidant activity, which protects DNA from possible oxidative stress (Ramos et al. 2003), and that demonstrated antimicrobial activity against Gram-positive bacteria (*Staphylococcus aureus, Streptococcus pyogenes* and *Streptococcus pneumoniae*), causative agents of respiratory infections (Cáceres et al. 1991; Aquino et al. 2010). In another study, a hydroalcoholic extract using 90% alcohol showed no antimicrobial activity but did have a moderate antifungal effect against *Candida krusei* (Holetz et al. 2002). An ethanol extract of *L. alba* root showed antimicrobial activity against *Staphylococcus aureus* and *Klebsiella pneumonia* (Sena-Filho et al. 2006). Corroborating this study, Aguiar et al. (2008) also evaluated *L. alba* root, ethanolic as well as acetone and chloroform extracts, and found activity against *Staphylococcus aureus*, *Micrococcus luteus, Bacillus subtilis*, *Mycobacterium smegmatis*, *Monilia sitophila* and *Candida albicans*. In addition, they also studied hexane, ethanolic and methanolic extracts of *L. alba* leaves and observed growth inhibition of *Staphylococcus aureus*, *Micrococcus luteus*, *Bacillus subtilis*, *Mycobacterium smegmatis* and *Chrysonilia sitophila*. Another work reported the action of *L. alba* against other bacteria, namely *Bacillis subtilis*, *Sacrina lutea*, *Xanthomonas campestris* and *Escherichia coli* (Mamun-or-Rashid et al. 2012).

Antimicrobial activity was also observed in experiments with other species of microorganisms, using crude extracts, essential oil and honey from the nectar of *L. alba* flowers, such as against the fungus *Candida albicans* (Holetz et al. 2002) and against the replication of herpes simplex virus type I and poliovirus type 2 (Andreghetti-Frohner et al. 2005), influenza virus type A (H3N2) (Ruffa et al. 2004), yellow fever virus (Gomez et al. 2013). *L. alba* essential oil, rich in linalool, was also found to be effective against dermatophytic fungi (Costa et al. 2014).

In vivo tests have shown that *L. alba* infusion protects against the development of gastric ulcers induced by indomethacin, thereby supporting its purported antiulcer activity (Pascual et al. 2001), and there have also been reports of the same specified activity by its major constituents such as 1,8-cineole (Santos and Rao 2001), linalool (Barocelli et al. 2004), limonene (Moraes et al. 2009, Rozza et al. 2010) and citral (Ortiz et al. 2010). In studies conducted in vivo in catfish juveniles, it was observed that the essential oil of *L. alba* was effective in inducing sedation and anesthesia, as well as having antimicrobial activity (Cunha et al. 2010).

Studies to assess the sedative properties of *L. alba* found weak or moderate action on benzodiazepine receptors with the citral chemotype (Hennebelle et al. 2008a). Hatano et al. (2012) corroborated these types of findings when studying the activity of a carvone chemotype and showed that the essential oil had significant anxiolytic activity.

The essential oil of the citral, limonene, carvone and limonene chemotypes of *L. alba* showed significant antinociceptive and anti-edematogenic activity in the hot plate and writhing tests (Viana et al. 1998). In evaluating one of the main traditional uses, an extract of *L. alba* was tested in an experimental model of hypertension, and it was found that it reduced heart rate in the isolated rat heart but without changing its contractile force (Gazola et al. 2004). An evaluation of the effect of the essential oil of *L. alba* on isolated mesenteric artery of rats demonstrated vasorelaxation independent of the endothelium (Maynard et al. 2011), while the major component of a chemotype of *L. alba*, citronellol, lowered blood pressure in rats (Bastos et al. 2009).

The antioxidant activity of *L. alba* has also been investigated, where a study of essential oil from its leaves, obtained by hydrodistillation, showed significant results, comparable to vitamin E, the positive control (Stashenko et al. 2004). Corroborating this study, methanol extracts of the leaves of *L. alba* also demonstrated antioxidant properties, attributed to flavonoids and coumarins (Hennebelle et al. 2008a).

Recently, an uncontrolled prospective phase II clinical study reported the effects of hydroalcoholic extracts of the leaves of *L. alba* in patients with migraine headaches, observing that the chemotype that had carvone and geraniol as major compounds significantly reduced the frequency and intensity of pain (Conde et al. 2011).

8 Conclusions

L. alba is a promising plant for the pharmaceutical industry because it has great potential for use in drug development. This is due to its easy cultivation and the recent results on its popular use, as an analgesic, anti-inflammatory, antipyretic, sedative, and other purposes. Ethnopharmacological and pharmacological studies of *L. alba* have revealed various biological activities, for example, antioxidant, antimicrobial, anesthetic and protection against gastric ulcers, among others.

References

Agra MF, Barbosa Filho JM (1990) Survey of medicinal flora of Paraíba and phytochemical screening. Braz J Pharm., Rio de Janeiro 71(3):72–76

Aguiar JS, Costa MCCD, Nascimento SC et al (2008) Antimicrobial activity of *Lippia alba* (Mill.) N. E. Brown (Verbenaceae). Braz J Pharmacogn 18(3):436–440

Andreghetti-Fröhner CR, Sincero TCM, Silva AC et al (2005) Antiviral evaluation of plants from Brazilian Atlantic Tropical Forest. Phytotherapy 76:374–378

Aquino LCL, Santos GG, Trindade RC, Alves JAB, Santos PO, Alves PB, Balnk AF, Carvalho LM (2010) Antimicrobial activity of essential oils of cidreira-herb and basil against bacteria from bovine meat. Alim Nutr Araquara 21(4):529–535

Arambarri A, Freire S, Colares M, Bayon N, Novoa M, Monti C, Stenglein S (2006) Leaf anatomy of medicinal shrubs and tree from gallery forest of the paranaense province (Argentina). Part 1. Bull Argent Bot 41(3–4):233–268

Barbosa FF, Barbosa LCA, Melho EC et al (2006) Effect of drying air temperature upon the content and chemical composition of the essential oil from *Lippia alba* (Mill) N. E. Brown. New Chem 29:1221–1225

Barocelli E, Calcina F, Chiavarini M et al (2004) Antinociceptive and gastroprotective effects of inhaled and orally administered *Lavandula hybrida* Reverchon "Grosso" essential oil. Life Sci 76(26):213–223

Bastos JF, Moreira IJ, Ribeiro TP et al (2009) Hypotensive and vasorelaxant effects of citronellol, a monoterpene alcohol, in rats. Basic Clin Pharmacol Toxicol 106(4):331–337

Biasi LA, Costa G (2003) Vegetative propagation of *Lippia alba*. Ciênc Rural Santa Maria 33(3):455–459

Cáceres A, Alvarez AV, Ovando AEO, Samayoa BE (1991) Plants used in Guatemala for the treatment of respiratory diseases. 1. Screening of 68 plants against gram-positive bacteria. J Ethnopharmacol 31:193–208

Castro DM, Ming LC, Marques MOM (2002.; (ISHS)) Biomass production and chemical composition of *Lippia alba* (Mill.) N. E. Brown Britt & Wilson in leaves on different plant parts in different seasons. Acta Hort 569:111–115

Conde J, Correa VS, Carmona F et al (2011) Chemical composition and therapeutic effects of *Lippia alba* (Mill.) N. E. Brown leaves hydro-alcoholic extract in patients with migraine. Phytomedicine 15(18(14)):1197–1201

Costa DCM et al (2014) Inhibitory effect of linalool-rich essential oil from Lippia alba on the peptidase and keratinase activities of dermatophytes. J Enzyme Inhib Med Chem 29(1):12–17

Cunha MA, Barros FMC, Garcia LO, Veeck APL, Heinzmann BM, Loro VL, Emanuelli T, Baldisserotto B (2010) Essential oil of Lippia alba: a new anesthetic for silver catfish, Rhamdia quelen. Aquaculture 306:403–406

Di Stasi LC et al (2002) Medicinal plants popularly used in the Brazilian tropical Atlantic forest. Fitoterapia 73(1):69–91

Ehlert PAD et al (2003) Effect of harvest time on carvone and limonene of "erva cidreira brasileira" essential oil. Braz Hortic:21–28

Farias MR et al (2003) Assessment of quality of raw vegetables. Pharmacogn Plant Drug 5:262–288

Gazola R, Machado D, Ruggiero C et al (2004) *Lippia alba*, *Melissa officinalis* and *Cymbopogon citratus*: effects of the aqueous-extracts on the isolates hearts of rats. Pharmacol Res 50:477–480

Gazzaneo LRS, de Lucena RFP, de Albuquerque UP (2005) Knowledge and use of medicinal plants by local specialists in a region of Atlantic Forest in the state of Pernambuco (Northeastern Brazil). J Ethnobiol Ethnomed 1:9

GIRÓN, Lidia M. et al. Ethnobotanical survey of the medicinal flora used by the Caribs of Guatemala. Journal ofEthnopharmacology, v. 34, n. 2-3, p. 173-187, 1991.

Gomez LA, Stashenko E, Ocazionez RE (2013) Comparative study on in vitro activities of citral, limonene and essential oils from *Lippia citriodora* and *L. alba* on yellow fever virus. Nat Prod Commun 8(2):249–252

Hatano VY, Torricelli AS, Giassi AC et al (2012) Anxiolytic effects of repeated treatment with an essential oil from *Lippia alba* end (R)-(-)-carvone in the elevated T-maze. Braz J Med Biol Res 45(3):238–243

Heinrich M, Rimpler H, Barrera NA (1992) Indigenous phytotherapy of gastrointestinal disorders in a lowland Mixe community (Oaxaca, Mexico): ethnopharmacologic evaluation. J Ethnopharmacol 36:63–80

Hennebelle T, Sahpaz S, Gressier B et al (2008a) Antioxidant and neurosedative properties of polyphenols and iridoids from *Lippia alba*. Phytother Res 2(22):256–258

Hennebelle T, Sahpaz S, Joseph H et al (2008b) Ethnopharmacology of *Lippia alba*. J Ethnopharmacol 116:211–222

Holetz FB, Pessini GL, Sanches NR et al (2002) Screening of some plants used in the Brazilian folk medicine for the treatment of infectious diseases. Mem Inst Oswaldo Cruz 97:1027–1031

Julião LS, Tavares ES, Lage CLS et al (2003) Thin layer chromatography three chemotypes of *Lippia alba* extracts (Mill) NE Br. Braz J Pharmacogn 13:36–38

Mamun-or-Rashid ANM, Islam MR, Dash BK (2012) In vitro antibacterial effect of bushy mat-grass (*Lippia alba* Mill.) extracts. Res J Med Plant 6:334–340

Martins ER, Castro DM, Castellani DC et al (1995) Medicinal plants. University Press. UFV-Viçosa, Minas Gerais

Matos FJA (2000) Medicinal plants: selection chart and employment plants used in phytotherapy in Northeast Brazil, 2nd edn. University Press, Fortaleza

Matos FJA, Machado MIL, Craveiro AA, Alencar JW (1996) The essential oil composition of two chemotypes of *Lippia alba* grown in Northeast Brazil. J Essent Oil Res 8:695–698

Mattos SH, Innecco R, Marco CA, Araújo AV (2007) Medicinal and aromatic plants grown in Ceará: production technology and essential oils. Bank of Northeast Brazil, Fortaleza

Maynard LG, Santos KC, Cunha PS et al (2011) Chemical composition and vasorelaxant effect induced by the essential oil of *Lippia alba* (Mill.) N.E. Brown. (Verbenaceae) in rat mesenteric artery. Indian J Pharmacol 43(6):694–698

MING, Lin Chau. Coleta de plantas medicinais. PLANTAS medicinais: Arte e ciência. São Paulo: UNESP, p.69-86, 1996.

Ming LC (1994) Influence of organic fertilization in biomass production and essential oil content of *Lippia alba*. Braz Hortic 12:49–52

Moraes TM, Kushima H, Moleiro FC et al (2009) Effects of limonene and essential oil from *Citrus aurantium* on gastric mucosa: role of prostaglandins and gastric mucus secretion. Chem Biol Interact 14(180(3)):499–505

Ortiz MI, Ramirez-Montiel ML, González-Garcia MP et al (2010) The combination of naproxen and citral reduces nociception and gastric damage in rats. Arch Pharm Res 33(10):1691–1697

Pascual ME, Slowing K, Carretero E et al (2001) *Lippia*: traditional uses, chemistry and pharmacology: a review. J Ethnopharmacol 76:201–214

Pierre PMO, Sousa SM, Davide LC, Machado MA, Viccini LF (2011) Karyotype analysis, DNA content and molecular screening in *Lippia alba* (Verbenaceae). An Acad Bras Ciênc 83:993–1005

Ramos A, Visozo A, Piloto J, Garcia CA, Rodriguez A, Rivero R (2003) Screening of antimutagenicity via antioxidant activity in Cuban medicinal plants. J Ethnopharmacol 87:241–246

Rodrigues ACC, Guedes MLS (2006) Use of medicinal plants in town Sapucaia, Cruz das Almas – Bahia. Braz J Med Plants 8:1–7

Rozza AL, Moraes TDEM, Kushima H et al (2010) Gastroprotective mechanisms of Citrus lemon (Rutaceae) essential oil and its majority compounds limonene and β-pinene: involvement of heat-snock protein-70, vasoactive intestinal peptide, glutathione, sulfhydryl compounds, nitric oxide and prostaglandin E2. Chem Biol Interact 15(189(1–2)):82–89

Ruffa MJ, Wagner ML, Suriano M et al (2004) Inhibitory effect of medicinal herbs against RNA and DNA viruses. Antivir Chem Chemother 15(3):153–159

Salimena FRG (2002) News synonyms and typifications in Lippia sect Rhodolippia (Verbenaceae). Darwin 1–4:121–125

Santos FA, Rao VS (2001) 1,8-cineol, a food flavoring agente, prevents etanol-induced gastric injury in rats. Dig Dis Sci 46(2):331–337

Sena Filho JG, Melo JGS, Saraiva AM et al (2006) Antimicrobial activity and phytochemical profile from the roots of *Lippia alba* (Mill.) N. E. Brown. Braz J Pharmacogn 16:506–509

Sousa SM, Torres GA, Viccini LF (2012) Karyological studies in Brazilian species of *Lippia* L. (Verbenaceae). An Acad Bras Ciênc 84(4):1029–1037

Stashenko EE, Jaramillo BE, Martinez JR (2003) Comparación de la composición química y de la actividad antioxidante in vitro de los metabolitos secundários volátiles de plantas de la família Verbenaceae. Rev Acad Colomb Cienc Exactas Físicas Naturales 27:579–597

Stashenko EE, Jaramillo BE, Martinez JR (2004) Comparison of different extraction methods for the analysis of volatile secondary metabolites of *Lippia alba* (Mill.) N.E. Brown, grown in Colombia, and evaluation of its in vitro antioxidant activity. J Chromatogr A 1025:93–103

Stefanini MB, Rodrigues SD, Ming LC (2001) Phytoregulators action in the growth of the erva-cidreira-brasileira. Braz Hortic 20:18–23

Tavares ES, Julião LS, Lopes D et al (2005) Analysis of the essential oil from leaves of three *Lippia alba* (Mill.) N. E. Br. (Verbenaceae) chemotypes cultivated on the same conditions. Braz J Pharmacogn 1(15):1–5

Vale TG, Matos FJ, De Lima TC et al (1999) Behavioral effects of essential oils from *Lippia alba* (Mill.) N.E. Brown chemotypes. J Ethnopharmacol 1(67(2)):127–133

Viana GSB, Do Vale TG, Rao VSN, Matos FJA (1998) Analgesic and antiinflammatory effects of two chemotypes of *Lippia alba*: a comparative study. Pharm Biol 36:347–351

Zamora-Martinez MC, Nieto De Pascual C (1992) Medicinal plants used in some rural populations of Oaxaca, Puebla and Veracruz, México. J Ethnopharmacol 35:229–257

Lonchocarpus araripensis Benth. (Fabaceae)

Jackson Roberto Guedes da Silva Almeida, Ana Carolina Murta Ramalho, and Fernanda Guerra da Silveira

Abstract *Lonchocarpus* Kunth. is the most diverse genus of the Millettieae tribe, in the Neotropics. It is known for its problematic taxonomy due to their historical links with the genera *Deguelia*, *Derris*, *Muellera* and *Philenoptera*. Approximately 23 species of *Lonchocarpus* are recorded for Brazil. *Lonchocarpus araripensis* Benth. was previously classified as *Derris araripensis* Benth Ducke. This species is found in Northeast Brazil, where it is used in the folk medicine to treat pain and inflammation. Phytochemical investigations have proved that *Lonchocarpus* is a rich source of phenol compounds, including flavones, chalcones, flavonols, flavans, flavanones, and aurones. Flavonoids and one triterpenoid compound have been isolated and identified from *L. araripensis* which showed important biological activities, such as antinociceptive, anti-inflammatory and gastroprotective. *L. araripensis* could be considered a rich source of flavonoids, confirming previous investigations into this species. Chemical constituents isolated from *L. araripensis* possess promising biological activities. Structure-activity relationship studies are necessary to determine the true pharmacological potential of these metabolites.

Keywords *Lonchocarpus araripensis* · Fabaceae · Flavonoids · Biological activity

1 Taxonomic Characteristics

Lonchocarpus Kunth. is the most diverse genus of Millettieae tribe in the Neotropics. It is known for its problematic taxonomy taking their origin to their historical links with the genera *Deguelia* Aubl., *Derris* Lour., *Muellera* L.f. and *Philenoptera* Hochst. *ex* A. Rich. (Silva and Tozzi 2012). Tozzi (1989) has recorded 23 species of *Lonchocarpus* (s. lat.) for Brazil, but new occurrences, new taxa and new synonyms were appointed to the group from then (Tozzi 1995; Neubert and Miotto 1996; Tozzi

J. R. G. d. S. Almeida (✉) · A. C. M. Ramalho · F. G. da Silveira
Center for Studies and Research of Medicinal Plants (NEPLAME), Federal University of Vale do São Francisco (UNIVASF), Petrolina, Pernambuco, Brazil
e-mail: jackson.guedes@univasf.edu.br

U. P. Albuquerque et al. (eds.), *Medicinal and Aromatic Plants of South America*, Medicinal and Aromatic Plants of the World 5,
https://doi.org/10.1007/978-94-024-1552-0_26

and Silva 2007; Silva and Tozzi 2008; Silva and Tozzi 2010). As 16 species do not belong to *Lonchocarpus* s. str. so they should be transferred to genera *Muellera* or *Dahlstedtia* Malme.

The taxonomic revision (Silva and Tozzi 2012) of *Lonchocarpus* in Brazil has allowed recognize 23 species divided into two subgenera: *Lonchocarpus* subgenus *Lonchocarpus* with 15 species and *Lonchocarpus* subgenus *Punctati* (Benth) with 8 species, including *L. subglaucescens* Benth and *L. araripensis* Benth Ducke. The latter was previously classified as *Derris araripensis* Benth Ducke. *L. subglaucescens* is found in Southeast and *L. araripensis* in Northeast Brazil (Magalhães et al. 1996).

Some specialists include *Derris* and *Lonchocarpus* in the same genus. *Lonchocarpus* also shows a great vegetative and floral affinity with *Millettia*, *Pongamia* and *Piscidia*. Morphological complexity has resulted in the adoption of controversial taxonomical systems by different botanists (Magalhães et al. 1996).

Synonyms *Derris araripensis* (Benth.) N. F. Mattos, *Dahlstedtia araripensis* (Benth.) M. J. Silva & A. M. G. Azevedo (Silva and Tozzi 2015; IPNI 2015; Tozzi 1989).

2 Crude Drug Used

Plants belonging to the Family Fabaceae are among the most used plants in popular medicine. Their main use is in traditional medicine is to treat symptoms of rheumatism, arthritis, diabetes, intestinal cramps, chronic diarrhea as well as respiratory complaints (Corrêa 1984).

Derris (*Lonchocarpus*) *araripensis* Ducke is a large tree known as "angelim" (Nascimento and Mors 1981), "coção" or "sucupira branca". In some Brazilian regions the plants from the genus *Lonchocarpus* are traditionally used for the treatment of tumors, AIDS, headache, and skin diseases (Santos et al. 2009) as well as to relieve rheumatism, arthritis, diabetes, inflammations, gastritis, peptic ulcer and general wounds. Traditionally, the stem barks of the tree are used.

3 Major Chemical Constituents and Bioactive Compounds

Extensive phytochemical studies on the *Lonchocarpus* genus have led to the identification of numerous flavonoids, in addition to other metabolites such as alkaloids, amino acids, peptides, triterpenes, sterols, stilbenes and dibenzoylmethane derivatives (Hegnauer and Hegnauer 2001).

The *Lonchocarpus* genus is well known for its insecticidal properties which is due to the presence of rotenone derivatives (Ioset et al. 2001). The genus is also known for its pesticidal properties. Most of the species studied have been shown to contain flavonoids of a wide range of structural types (Bisby et al. 1994).

Previous phytochemical investigations have proved that *Lonchocarpus* is a rich source of phenol compounds, including flavones, chalcones, flavonols, flavans, flavanones, and aurones (Lima et al. 2009, 2014a). Furan and pyran moieties located at ring A in a linear or angular position, i.e. linked to either C-6/C-7 or C-7/C-8, respectively, are a common characteristic of the flavonoids exhibited by plants of this genus (Magalhães et al. 1996). This substitution pattern has also been observed for the flavonoids of *L. araripensis* (syn. *Derris araripensis*), as demonstrated in a study by Nascimento and Mors (1981).

A series of activities, such as antimicrobial, gastroprotective, cytotoxic, antiplatelet and antimalarial were related for flavonoids isolated from species of the *Lonchocarpus* genus (Pires et al. 2011).

Flavonoids were isolated from *L. araripensis* (Leguminoseae) and identified as 3-methoxy-6-*O*-prenyl-6″,6″-dimethylchromene-[2″,3″:7,8]-flavone (**1**), 3,6-dimethoxy-6″,6″-dimethylchromene-[2″,3″:7,8]-flavone (**2**) and 3,5,8-trimethoxy-[2″,3″:6,7]-furanoflavone (**3**). This was the first time that the compound **3** was described. Compound **2** has been previously isolated from roots while the compound **1** is reported in this species for the first time (Lima et al. 2014a).

The NMR study of the flavonoids 6a,11a-dihydro-9-methoxy-6H-benzofuran [3,2-C] benzopiran-3-ol (**4**) and (2,3-*cis*-3,4-*cis*-3,4,5,8-tetramethoxy-[1″,2″:6,7]-furanoflavan (**5**) was described. The relative stereochemistry at the asymmetric centers was established by NOE difference experiments. The compounds **4** and **5** are novel to *L. araripensis* (Lima et al. 2014b).

Two new polymethoxylated flavonoids, 2′,5′,6′-trimethoxy-[2″,3″:3′,4′]-furano dihydrochalcone and 2,4′,4,5-tetramethoxy-[2″,3″:6,7]-furanodihydroaurone, were isolated from the root barks of *L. araripensis*, along with the known compounds 3,4,5,6-tetramethoxy-[2″,3″:7,8]-furanoflavan, 3,6-dimethoxy-1″,1″-dimethylcromene-[2″,3″:7,8]-flavone, 3′,4′-methylenodioxy-5,6-dimethoxy-[2″,3″:7,8]-furanoflavone, 3,5,6-trimethoxy-[2″,3″:7,8]-furanoflavanone, 3,5,6-trimethoxy-[2″,3″:7,8]-furanoflavone, and 6α-hydroxy-medicarpin (Lima et al. 2009).

In another study, nine flavonoids, namely one dihydrochalcone, one flavone, four 3-methylflavonols, one flavanone, one 3-methylflavanonol and one flavan were isolated from the roots of *Derris araripensis* (*L. araripensis*). Eight of these compounds have been reported for the first time. Structures were established by spectral analysis and chemical degradation (Nascimento and Mors 1981). For a more complete list of compounds see Table 1.

4 Morphological Description

L. araripensis is a small to medium-sized tree, usually 3–5 m high, found in several plant formations, restricted to the Caatinga vegetation. It is a deciduous species, with woody branches, striated, hairless. The tree has seven or nine leaves, small

Table 1 Presence of compounds in *Lonchocarpus araripensis*

Compound type	Chemical structure	References
Flavonoids		
3-methoxy-6-*O*-prenyl-6″,6″-dimethylchromene-[2″,3″:7,8]-flavone	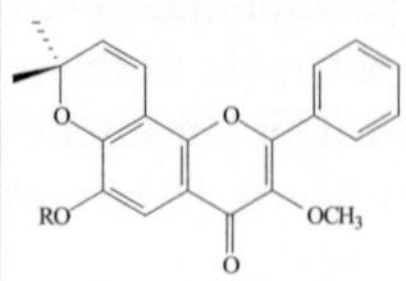 R= Prenyl	Lima et al. (2014a)
3,6-dimethoxy-6″,6″-dimethylchromene-[2″,3″:7,8]-flavone	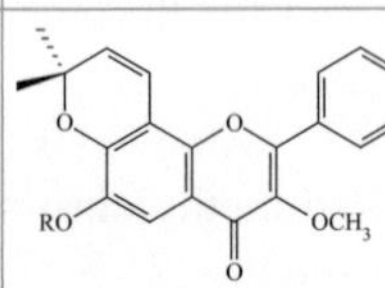 R= Methyl	Lima et al. (2014a) and Nascimento and Mors (1981)
3,5,8-trimethoxy-[2″,3″:6,7]-furanoflavone	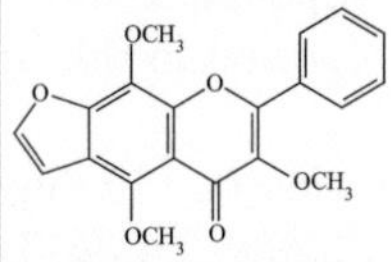	Lima et al. (2014a)
6a,11a-dihydro-9-methoxy-6H-benzofuran [3,2-C]-benzopiran-3-ol	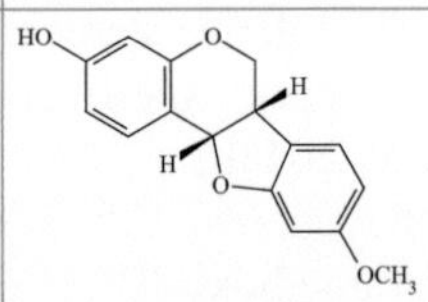	Lima et al. (2014b)
2,3-*cis*-3,4-*cis*-3,4,5,8-tetramethoxy-[1″,2″:6,7]-furanoflavan	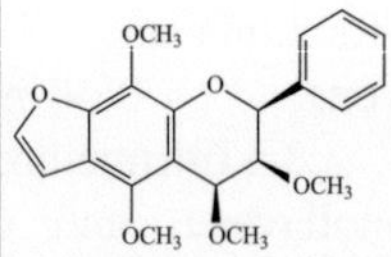	Lima et al. (2014b)
2′,5′,6′-trimethoxy-[2″,3″:3′,4′]-furanodihydrochalcone	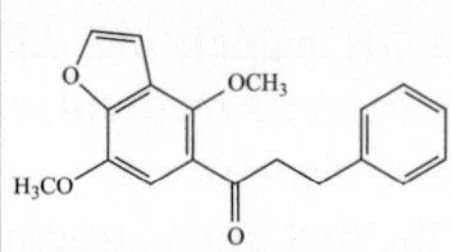	Lima et al. (2009)
2,4′,4,5-tetramethoxy-[2″,3″:6,7]-furanodihydroaurone	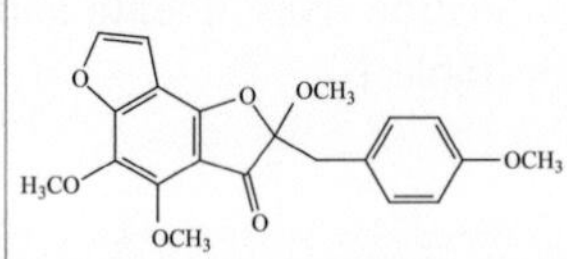	Lima et al. (2009)
3,4,5,6-tetramethoxy-[2″,3″:7,8]-furanoflavan	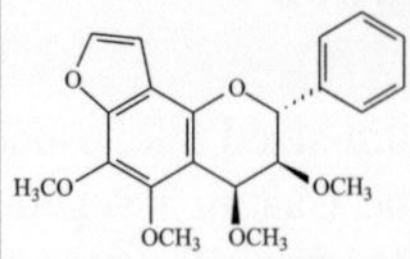	Lima et al. (2009)

(continued)

Table 1 (continued)

Compound type	Chemical structure	References
3,4,5,6-tetramethoxy-[2″,3″:7,8]-furanoflavan	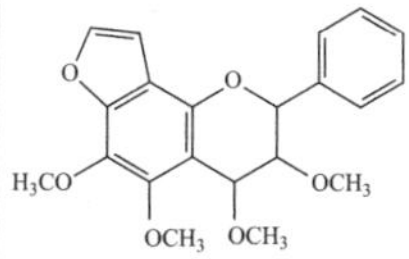	Nascimento and Mors (1981)
3′,4′-methylenodioxy-5,6-dimethoxy-[2″,3″:7,8]-furanoflavone	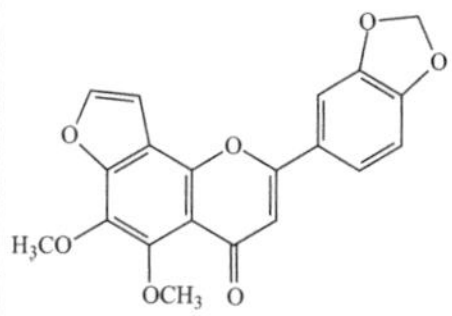	Lima et al. (2009) and Nascimento and Mors (1981)
3,5,6-trimethoxy-[2″,3″:7,8]-furanoflavanone	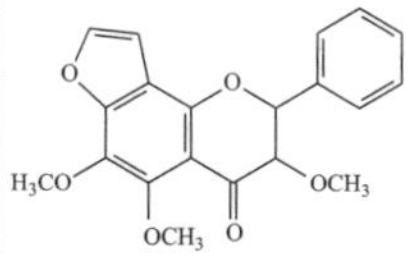	Lima et al. (2009)
3,5,6-trimethoxy-[2″,3″:7,8]-furanoflavone	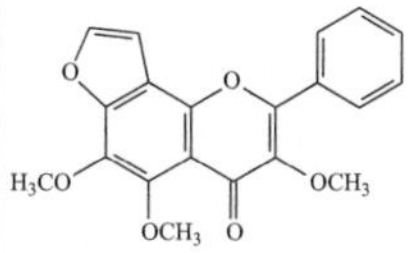	Lima et al. (2009) and Nascimento and Mors (1981)
6α-hydroxy-medicarpin	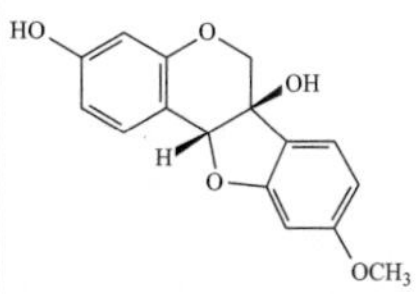	Lima et al. (2009)
Methylenedioxy-(3,4)-5′-hydroxy-2′,3′-methoxyfurano-(3′,4′,2″,3″)-dihydrochalcone	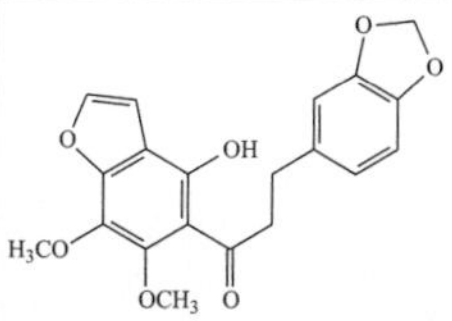	Nascimento and Mors (1981)
3′,4′-Methylenedioxy-3,6-dimethoxy-6″,6″-dimethylchromeno-(2″,3″:7,8)-flavone	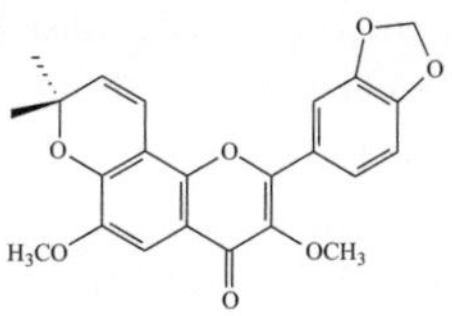	Nascimento and Mors (1981)
3′,4′-Methylenedioxy-3,5,6-trimethoxyfurano-(2″,3″:7,8)-flavone	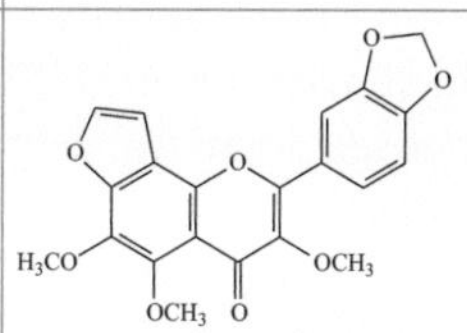	Nascimento and Mors (1981)

(continued)

Table 1 (continued)

Compound type	Chemical structure	References
3′,4′-Methylenedioxy-5-hydroxy-6-methoxyfurano-(2″,3″:7,8)-flavanone	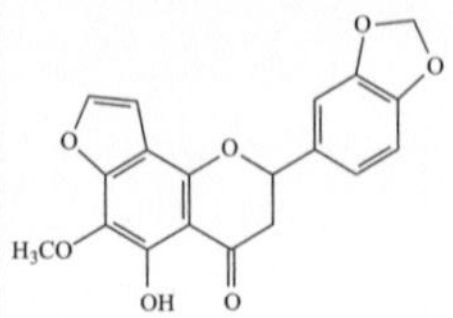	Nascimento and Mors (1981)
3′,4′-Methylenedioxy-3,5,6-trimethoxyfurano-(2″,3″:7,8)-flavanonol	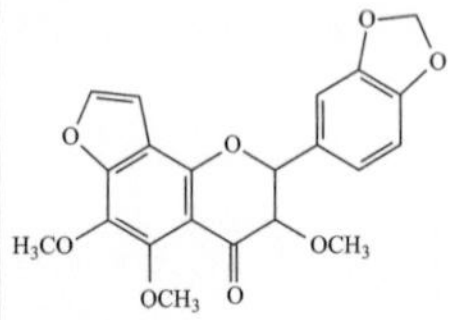	Nascimento and Mors (1981)
Triterpene		
Lupeol	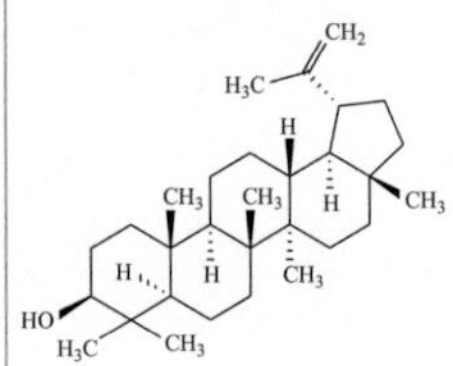	Lima et al. (2013)

petiole, 3–4 cm in length. Paniculate inflorescences. Fruits can reach 5 cm in length, usually with one seed. Reddish-brown seed with some black spots, about 2 mm thick and up to 1.5 cm (Fernandes 1964).

5 Geographical Distribution

Leguminosae (Fabaceae) is the third largest botanical family, with approximately 18,000 species in 619 genera, most of them belonging originally to the Brazilian flora (Joly 1993). The genus *Lonchocarpus* belongs to the subfamily Papilionoideae of the Leguminosae. The genus is represented by approximately 100 species distributed in the tropical America, Africa, and the Caribbean Islands (Magalhães et al. 1996), Madagascar and Australia (Allen and Allen 1981). *L. araripensis* is restricted to the Caatinga vegetation, a kind of seasonally dry tropical forest of Northeast Brazil growing under semi-arid climate (Queiroz 2006). Other authors mention that the genus *Lonchocarpus* comprises approximately 135 species, 24 of which are native to Brazil (Patel et al. 2010).

6 Ecological Requirements

Tree can be found dry forests and savannah, favor secondary formations, growing from the coast to moderate elevations and found in a wide range of soils and conditions including dry, rocky soil and moist, clayey, lowland soils (Lorenzi 2002). The main period of floration and frutification of *L. araripensis* collected in the Caatinga of Pernambuco is during the dry season (Lima et al. 2008).

7 Traditional Use (Part(s) Used) and Common Knowledge

The plant is popularly known in Northeastern Brazil as "sucupira-branca", "angelim", "coção", "rabo de cavalo", "pau de formiga" and "sucupira de concha", where it is used in folk medicine to treat symptoms of rheumatism, arthritis and diabetes. The tree is widely distributed in hot and dry areas of the states of Bahia, Ceará and Rio Grande do Norte, Paraíba, Pernambuco, Piauí and Maranhão, Brazil (Lima et al. 2011; Fernandes 1964).

8 Modern Medicine Based on Its Traditional Medicine Uses

Lupeol is other important compound isolated from this species. The antinociceptive properties of lupeol in models of inflammatory and post-operative pain, as well as its mechanisms of action were investigated. The effects of lupeol were tested against acetic acid-induced writhing, formalin test, carrageenan-induced hyperalgesia, and post-operative pain model. Pre-treatment with lupeol (50 and 100 mg/kg) inhibited the hyperalgesia and the local increase in tumor necrosis factor-α (TNF-α) and interleukin-1β (IL-1β) levels induced by carrageenan. In contrast, lupeol did not inhibit the post-operative pain. Lupeol-treated mice did not show any motor performance alterations or apparent systemic toxicity. The results demonstrated that lupeol has consistent antinociceptive properties during inflammatory pain, but not post-operative pain, acting through the inhibition of IL-1β and TNF-α production, constituting an attractive possibility to pharmacological development. A more indepth evaluation of the mechanisms involved will need to be performed (Lima et al. 2013).

The efficacy of lupeol isolated from *L. araripensis* in the treatment of bronchial asthma in BALB/c mice immunized with ovalbumin was evaluated. Administration of lupeol caused the reduction of cellularity and eosinophils in the bronchoalveolar lavage fluid. Treatment with lupeol also reduced the production of mucus and overall inflammation in the lung. Levels of Type II cytokines IL-4, IL-5 and IL-13 were significantly reduced in mice treated with lupeol, an effect that was similar to that observed in dexamethasone-treated mice. In contrast, IgE production was not significantly altered after treatment with lupeol. The results demonstrated that lupeol attenuates the alterations characteristic of allergic airway inflammation. The investigation of the mechanisms of action of this molecule may contribute for the development of new drugs for the treatment of asthma and other allergic diseases (Vasconcelos et al. 2008).

The antinociceptive activity of the flavonoid 3,6-dimethoxy-6″,6″-dimethyl-[2″,3″:7,8]-chromeneflavone (DDF) from *L. araripensis* was evaluated by measuring nociception by acetic acid, formalin and hot plate tests. The Rotarod test was used to evaluate motor coordination. The results demonstrated that DDF was able to prevent acetic-acid-writhing-induced nociception ($p < 0.001$) in mice. Furthermore, DDF produced a significant reduction of the nociceptive behavior at the early and late phases of paw licking in the formalin test. Also, DDF produced an inhibition of the nociceptive behavior during a hot-plate test. No alteration in motor coordination was observed. These results confirm the hypothesis that DDF reduces the nociceptive behavior in mice, probably through central mechanisms, but without compromising the motor coordination of animals (Almeida et al. 2015). The gastroprotective effect of DDF on gastric damage induced by absolute ethanol (96%, 0.2 ml/mouse) and indomethacin (30 mg/kg, p.o.) in mice was investigated. The intraperitoneal administration of DDF at dose levels of 50, 100 and 200 mg/kg markedly reduced the gastric lesions in the ethanol model by 62%, 72% and 96%, and in the indomethacin model by 34%, 70% and 75%, respectively, as compared with misoprostol (50 μg/kg, p.o.), the reference compound that caused lesion suppression by 67% in ethanol model and by 72% against indomethacin-induced ulceration. The ED_{50} of DDF in reducing gastric lesions induced by ethanol and indomethacin (dose of the DDF that reduced the gastric lesion area by 50% in relation to the control value) was 50.87 and 61.56 mg/kg, respectively. The results show that DDF provides gastroprotection against gastric damage induced by ethanol and indomethacin by different and complementary mechanisms, which include involvement of endogenous prostaglandins, nitric oxide release, activation of TRPV1 receptor or K^+-ATP channels, besides a sparing effect on NP-SH reserve (Campos et al. 2008).

9 Conclusions

Chemical constituents isolated from *L. araripensis* possess promising biological activities. *L. araripensis* could be considered a rich source of flavonoids, confirming previous investigations with this species. Farther structure-activity relationship studies are necessary to determine the true pharmacological potential of these metabolites.

References

Allen ON, Allen E (1981) The Leguminosae. Macmillan, New York

Almeida JRGS, Silva JC, Guimarães AL, Oliveira AP, Souza GR, Oliveira-Júnior RG, Lima-Saraiva SRG, Barbosa-Filho JM, Braz-Filho R, Queiroz DB, Botelho MA (2015) 3,6-dimethoxy-6″,6″-dimethyl-(7,8,2″,3″)-chromeneflavone, a flavonoid isolated from *Lonchocarpus araripensis* Benth. (Fabaceae), reduces nociceptive behaviour in mice. Phytother Res. https://doi.org/10.1002/ptr.5418

Bisby FA, Buckingham J, Harborne JB (1994) Phytochemical dictionary of the Leguminosae, vol 1. Chapman & Hall, London

Campos DA, de Lima AF, Ribeiro SR, Silveira ER, Pessoa OD, Rao VS, Santos FA (2008) Gastroprotective effect of a flavone from *Lonchocarpus araripensis* Benth. (Leguminosae) and the possible mechanism. J Pharm Pharmacol 60(3):391–397

Corrêa MP (1984) Dicionário das plantas úteis do Brasil e das exóticas cultivadas, vol 149. IBDF, Ministério da Agricultura, Rio de Janeiro

Fernandes AG (1964) *Lonchocarpus araripensis* Bentham. Bol Soc Cearense Agron 53(5):184–189

Hegnauer R, Hegnauer M (2001) Chemotaxonomie der Pflanzen, vol XIb-2. Birkhauser Verlag, Basle, pp 194–203

Ioset JR, Marston A, Gupta MP, Hostettmann K (2001) Five new prenylated stilbenes from the root bark of *Lonchocarpus chiricanus*. J Nat Prod 64(6):710–715

IPNI. The International Plant Names Index (2015) http://www.ipni.org/ipni/idPlantNameSearch. Access in 3/3/2015

Joly AB (1993) Botânica: Introdução a Taxonomia Vegetal. Ed. Nacional, São Paulo

Lima LCM, Barbosa DCA, Barbosa MCA (2008) Floração e frutificação das espécies lenhosas de Leguminosae e Euphorbiaceae na Caatinga em Pernambuco. Sitientibus Sér Ciênc Biológicas 8(2):235–246

Lima AF, Mileo PGM, Andrade-Neto M, Braz-Filho R, Silveira ER, Pessoa ODL (2009) ^{1}H and ^{13}C NMR assignments of new methoxylated furanoflavonoids from *Lonchocarpus araripensis*. Magn Reson Chem 47(2):165–168

Lima JT, Almeida JRGS, Mota KSL, Lúcio ASSC, Câmara CA, Barbosa-Filho JM, Silva BA (2011) Selective spasmolytic effect of a new furanoflavoquinone derivative from diplotropin on guinea-pig trachea. J Chem Pharm Res 3(1):249–258

Lima FO, Alves V, Barbosa-Filho JM, Almeida JRGS, Rodrigues LC, Soares MBP, Villarreal CF (2013) Antinociceptive effect of lupeol: evidence for a role of cytokines inhibition. Phytother Res 27(10):1557–1563

Lima AF, Ferreira DA, Monte FJQ, Braz-Filho R (2014a) Flavonoids from *Lonchocarpus araripensis* (Leguminoseae) – isolation, unequivocal assignment of NMR signals ^{1}H and ^{13}C and conformational analysis. Quim Nova 37(4):672–676

Lima AF, Ferreira DA, Monte FJQ, Braz-Filho R (2014b) Flavonoids from *Lonchocarpus araripensis* (Leguminosae): identification and total ^{1}H and ^{13}C resonance assignment. Am Int J Contemp Res 4(2):18–23

Lorenzi H (2002) Brazilian trees, vol 2, 4th edn. Instituto Plantarum De Estudos Da Flora, Brazil. ISBN:85-86714-15-1

Magalhães AF, Tozzi AMGA, Sales BHLN, Magalhães EG (1996) Twenty-three flavonoids fom *Lonchocarpus subglaucescens*. Phytochemistry 42(5):1459–1471

Nascimento MC, Mors WB (1981) Flavonoids from *Derris araripensis*. Phytochemistry 20(1):147–152

Neubert EE, Miotto STS (1996) O gênero *Lonchocarpus* Kunth (Leguminosae-Faboideae) no Rio Grande do Sul. Iheringia Sér Bot 47:73–102

Patel B, Das S, Prakash R, Yasir M (2010) Natural bioactive compound with anticancer potential. Int J Adv Pharm Sci 1:32–41

Pires AML, Silveira ER, Pessoa ODL (2011) Flavonoids from *Lonchocarpus campestris* (Leguminosae). Quim Nova 34(2):268–271

Queiroz LP (2006) The Brazilian Caatinga: phytogeographical patterns inferred from distribution data of the Leguminosae. In: Pennington RT, Lewis GP, Ratter JA (eds) Neotropical savannas and dry forests: diversity, biogeography, and conservation. Taylor & Francis CRC Press, Boca Raton, pp 113–149

Santos EL, Costa EV, Marques FA, Vaz NP, Maia BHLNS, Magalhães EG, Tozzi AMA (2009) Toxicity and antioxidant activity of flavonoids from *Lonchocarpus filipes* root bark. Quim Nova 32(9):2255–2258

Silva RR, Tozzi AMGA (2008) A new species of *Lonchocarpus* (Leguminosae, Papilionoideae) from Mato Grosso do Sul, Brazil. Brittonia 60:34–37

Silva MJ, Tozzi AMGA (2010) *Lonchocarpus*. In: Lista de espécies da Flora do Brasil. Jardim Botânico do Rio de Janeiro, Rio de Janeiro. http://floradobrasil.jbrj.gov.br/2010/FB022921

Silva MJ, Tozzi AMG (2012) Revisão taxonômica de *Lonchocarpus* s. str. (Leguminosae, Papilionoideae) do Brasil. Acta Bot Brassica 26(2):357–377

Silva MJ, Tozzi AMGA (2015) *Lonchocarpus* in Lista de Espécies da Flora do Brasil. Jardim Botânico do Rio de Janeiro. Disponível em: http://floradobrasil.jbrj.gov.br/jabot/floradobrasil/FB29731. Access in: 03/03/2015

Tozzi AMGA (1989) Estudos taxonômicos dos gêneros *Lonchocarpus* Kunth e *Deguelia* Aubl. no Brasil. Tese de Doutorado. Instituto de Biologia, Universidade Estadual de Campinas, Campinas. 341 pp.

Tozzi AMGA (1995) New species of *Lonchocarpus* Kunth (Leguminosae – Papilionoideae – Millettieae) from Brazil. Kew Bull 50:173–177

Tozzi AMGA, Silva MJ (2007) Sinonimizações em *Lonchocarpus* Kunth (Leguminosae – Papilionoideae – Millettieae). Rodriguésia 58:275–282

Vasconcelos JF, Teixeira MM, Barbosa-Filho JM, Lúcio ASSC, Almeida JRGS, Queiroz LP, Ribeiro-dos-Santos R, Soares MBP (2008) The triterpenoid lupeol attenuates allergic airway inflammation in a murine model. Int Immunopharmacol 8:1216–1221

Lychnophora pinaster Mart.

Paulo Sérgio Siberti da Silva, Maria Aparecida Ribeiro Vieira, and Marcia Ortiz Mayo Marques

Lychnophora pinaster Mart.
Photo: Maria A. R. Vieira

P. S. S. da Silva · M. A. R. Vieira · M. O. M. Marques (✉)
Instituto Agronômico, Campinas, São Paulo, Brazil
e-mail: mortiz@iac.sp.gov.br

U. P. Albuquerque et al. (eds.), *Medicinal and Aromatic Plants of South America*, Medicinal and Aromatic Plants of the World 5,
https://doi.org/10.1007/978-94-024-1552-0_27

Abstract *Lychnophora pinaster* Mart. (arnica mineira; arnica-da-serra) belongs to the family Asteraceae and is one of the main species of the genus *Lychnophora*. It is vulnerable to extinction and found exclusively in the Minas Gerais State-Brazil, with native populations showing disjunct distribution along the Espinhaço Range of the Minas Gerais State. The plant has high cultural value because of its intensive use in folk medicine, where its preparations are used for epidermal or oral administration. Its leaves, branches and flowers are obtained from predatory and indiscriminate harvesting in natural populations. Studies with the plant are promising. Its extracts contain active compounds against trypomastigote forms of *Trypanosoma cruzi*, and present anti-inflammatory and bactericidal activities. The plant is considered to be a potential source of chemoprophylactic agents.

Keywords Cerrado · Rupestrian fields · Arnica-mineira · Medicinal plant · Biological activity

1 Taxonomic Characteristics

Lychnophora pinaster Mart., synonym of *Lychnophora affins* Gardner, popularly known as "arnica", "arnica-da-serra" or "arnica-mineira", is a medicinal species belonging to the Class Equisetopsida, Subclass Magnoliidae, Order Asterales, Family Asteraceae and genus *Lychnophora* Mart. (Tropicos® 2013), the latter consisted of 64 species (Semir et al. 2011).

2 Crude Drug Used

Alcoholic preparations from branches, leaves and flowers of *L. pinaster* are traditionally indicated for the treatment of bruises, bumps, sprains, hematomas, insect bite disinfection (Rodrigues and Carvalho 2001), to soften the skin (Almeida et al. 1998), against earaches and as healing, anti-inflammatory, antirheumatic and analgesic.

3 Major Chemical Constituents and Bioactive Compounds

Among the major components identified in the essential oil and extracts of *L. pinaster* – including bioactive compounds – are found in the oil essential: *E*-methyl cinnamate, *E*-caryophyllene and α-humulene (Reis et al. 2010) and in the extracts: α-amyrin, lupeol, 3-*O* -acetyl-lupeol, 3-*O*-acetyl-pseudotaraxasterol, stigmasterol, sitosterol, quercetin (Ferreira et al. 2005; Abreu et al. 2011, 2013), 3-*O*-acetyl-α-amyrin, 4,4-dimethyl-cholesta-22,24-dien-5-ol, Δ^7-bauerenyl acetate (Abreu et al. 2011), isochlorogenic acid, vitexin, isovitexin, caffeic acid (Silveira et al. 2005b), *E*-lychnophoric acid or lychnophoic acid (Oliveira et al. 1996; Alcântara et al. 2005; Ferreira et al. 2005; Silveira et al. 2005a), goyazensolide, eremantholid, lychnopholide (Oliveira et al. 1996) and 15-deoxygoyazensolide (Duarte et al. 1993).

4 Morphological Description

According to Semir (1991) and Semir et al. (2011), *L. pinaster* varies from erect shrub with many branches to small ericoid shrubs and more rarely taller candelabriform shrubs with 0.4–2.4 m, rarely up to 3.6 m; branches alternate to subverticilated flexuous, delicate to more robust, densely tomentous to velutinous or shortly subvillosus, with branches 0.5–2.0 cm in diameter, the stem reaches 2.5–5.0 cm in diameter in older regions of larger shrubs; leaves very imbricated or ascending at the top of the branches and more patent even little reflex below, generally linear, linear-oblong, base rounded to auriculate sometimes slightly attenuated, apex obtuse to slightly rounded, rarely slightly acute, margin resolute, venation brochidodromous; main vein extended, tapering from base to apex; inflorescence in simple leafy glomerules, usually congested and hemispheric; color of flowers ranging from lilac to purple, measuring 8.0–10.0 mm in length; achene obconic to oval cylindrical, glabrous, olive glandule to brown, with 1.5–3.0 mm in length and 0.8–1.5 mm in diameter (Fig. 1).

Fig. 1 Individual of *Lychnophora pinaster* (**a**) and flowers (**b**). (Photos: Maria A.R. Vieira)

5 Geographical Distribution

L. pinaster is found exclusively in the Cerrado phytogeographic domain of Minas Gerais State, southeastern Brazil (Semir 1991). It is distributed in regions of high altitudes, above 900 m, as in the Espinhaço Range of the Minas Gerais State (Semir 1991; Semir et al. 2011; Carvalho 1992; Andrade 2013), a set of highlands boomerang-shaped with north-south direction centered on the meridian 43°W and west oriented-convexity (Saadi 1995).

6 Ecological Requirements

The phytophysiognomy of the rupestrian fields of the Espinhaço Range contains a vegetation typically xeric. Plants grow on oligotrophic and acidic soils that are subject to daily variations of temperature, exposure to wind and water restrictions (Giulietti et al. 1987).

The rupestrian fields arise from 900 m (Rapini et al. 2008), and therefore *L. pinaster* is found only at high altitudes. Its origin ecosystems are extremely rustic, with dry climate and soil, irregular topography and intense insolation; due to its endemism, *L. pinaster* may present edaphic limitation to specific substrates and even to different rainfall regimes (Coyle and Jones 1981; Semir 1991; Mansanares et al. 2002).

The plant can be associated with both the rupestrian fields linked to rock outcrops predominantly of quartz and the rupestrian field linked to hematite outcrops. The latter is common in the Ferriferous Quadrangle region of Minas Gerais State and is also known as ferruginous rupestrian field or canga vegetation (Viana and Lombardi 2007).

In the rupestrian field, populations of *L. pinaster* grow on lithologic and deeper soils. In the first, the soil has a higher proportion of fine particles and higher levels of organic matter and, in the second, that is poor in nutrients, the drainage is lower because of the sandy (Rapini et al. 2008). In the ferruginous rupestrian field, there are areas associated with huge iron ore deposits (Jacobi and Carmo 2008) and *L. pinaster* populations therein are thus metallophytes, having developed adaptations over evolutionary time that made them able to grow in the presence of heavy and toxic metals.

Rupicolous plant populations growing in these regions shows a disjunct distribution due to the discontinuity of the mountain ranges and rocky outcrops that make up the rupestrian fields and therefore have high endemism, considered one of the largest in Brazil (Santos et al. 2009).

7 Collection Practice

The method for gathering the plant material used popularly is strictly related to the disorderly extraction, done by local populations for their own use and for sale as phytotherapeutic agent, which is common in regions with the presence of the species; factor contributing for its classification into the category vulnerable to extinction by the State Council for Environmental Policy of the Minas Gerais State (COPAM), in 1997.

8 Traditional Use (Part(s) Used) and Common Knowledge

In traditional medicine, preparations of the branches, leaves and flowers – fresh or dried – of *L. pinaster* are used for epidermal administration in the form of compresses, alcoholic preparations (Rodrigues and Carvalho 2001), ointment and soap (Almeida et al. 1998) or for oral administration, macerated in "cachaça" (sugar cane spirit) or ethanol (Silveira et al. 2005b).

9 Modern Medicine Based on Its Traditional Medicine Uses

The search for natural products, as a source of effective drugs to combat the various diseases that affect humans, leads to increasing exploitation of plant resources in preclinical and toxicological studies. The search for compounds active against the flagellate protozoan *Trypanosoma cruzi* – etiologic agent of Chagas disease, that infects between seven and eight million people worldwide, mostly in endemic areas of 21 Latin American countries (Who 2014) – and to control inflammation and virulent microorganisms, makes *L. pinaster* a potential source of alternative chemoprophylactic agents.

Bioassays conducted on the species proved its trypanocidal effectiveness. The ethanol extract from the shoot, eliminated 100% Y strains of *T. cruzi* (Chiari et al. 1996). A trypanocidal component previously identified in the ethanol extract of *L. pinaster* was the sesquiterpene lactone 15-deoxygoyazensolide (Duarte et al. 1993), whose effectiveness has already been proven previously (Chiari et al. 1991). Elimination of 100% of *T. cruzi* Y strain was also observed for the lyophilized aqueous extract of the shoot of the plant (113.62 μg/mL), where the following compounds were identified: caffeic acid, vitexin, isovitexin, quercetin and isochlorogenic acid (Silveira et al. 2005b).

A study identified for the first time in the hexane extract of shoot of *L. pinaster* a compound related to caryophyllene, called lychnophoic acid (Oliveira et al. 1996) – later classified by Silveira et al. (2005b) as *E*-lychnophoric acid- capable of inhibiting by 50% the growth of Y and CL strains of *T. cruzi*. At concentrations of 5.68, 6.48 and 13.86 μg/mL, the effectiveness of *E*-lychnophoric acid and its ester and alcohol derivatives in controlling *T. cruzi* trypomastigote reached 100% (Alcântara et al. 2005).

From the hexane/dichloromethane extract of *L. pinaster* leaves, also the pentacyclic triterpene α-amyrin was isolated, which together with the non-polar extracts of the stem and leaves showed antibacterial activity against *Staphylococcus aureus* (Abreu et al. 2011), a virulent bacteria that can be fatal when infections are not treated (Shorr 2007).

There is evidence of anti-inflammatory and antinociceptive activities of the ethanol extract from the aerial part of the species (Guzzo et al. 2008). Anti-inflammatory activity of extracts and compounds isolated from the extracts was investigated by transdermal application via phonophoresis in rat paws with significant degeneration of muscle fibers (Abreu et al. 2013), in which it was observed that after injury, the hexane extract exerted moderate anti-inflammatory activity 72 h after application, while the aqueous extract drastically reduced the inflammatory process at the same time compared to the treatment with dexamethasone, a powerful anti-inflammatory drug (Guzzo et al. 1996; Cupolilo et al. 2007). The same was found for the flavonoid quercetin, the triterpene lupeol, a mixture of triterpenes α-amyrin and lupeol and a mixture of steroids stigmasterol and sitosterol, all isolated from the hexane extract of the plant; justifying the traditional use of the species.

Lethality assay with the *Artemia salina* (small crustacean) proved that the ethanol extract of the leaves of *L. pinaster* has a low mortality (LC_{50} = 678.73 μg mL^{-1}) (Ferraz-Filha et al. 2012) – toxicity tests were designed to evaluate or predict the toxic effects on biological systems and measure the relative toxicity of substances (Forbes and Forbes 1994) -, pointing to a possible selective toxic action of compounds with potential pharmacological activity.

10 Conclusions

To-date, researches on the biological activities of *L. pinaster* have accumulated a considerable body of data about its medicinal properties, justifying its popular use and, to some extent, the exploitation of native populations. Studies encompassing the management and conservation of this species in natural formations are lacking from the scientific literature, even though they would be extremely important for the better understanding and sustainable use of this species.

References

Abreu VGC, Takahashi JA, Duarte LP, Piló-Veloso D, Junior PAS, Alves RO, Romanha AJ, Alcântara AFC (2011) Evaluation of the bactericidal and trypanocidal activities of triterpenes isolated from the leaves, stems, and flowers of *Lychnophora pinaster*. Braz J Pharmacog 21(4):615–621

Abreu VGC, Correa GM, Silva TM, Fontoura HS, Cara DC, Piló-Veloso D, Alcântara AFC (2013) Anti-inflammatory effects in muscle injury by transdermal application of gel with *Lychnophora pinaster* aerial parts using phonophoresis in rats. BMC Complement Altern M 13(270):2–8

Alcântara AFC, Silveira D, Chiari E, Oliveira AB, Guimarães JE, Raslan DS (2005) Comparative analysis of the trypanocidal activity and chemical properties of *E*-lychnophoric acid and its derivatives using theoretical calculations. Eclét Quím 30(3):37–45

Almeida SP, Proença CEB, Sano SM, Ribeiro JF (1998) Cerrado: espécies vegetais úteis. EMBRAPA-CPAC, Planaltina

Andrade EA (2013) Composição florística e estrutura da vegetação de campo rupestre sobre quartzito no Complexo Serra da Bocaina-MG [Tese]. Universidade Federal de Lavras, Lavras. Lavras

Carvalho DA (1992) Flora fanerogâmica de campos rupestres da Serra da Bocaina, Minas Gerais: caracterização e lista de espécies. Ciên Prát 16:97–122

Chiari E, Oliveira AB, Raslan DS, Mesquita AAL, Tavares KG (1991) Screening in vitro of natural-products against blood forms of *Trypanosoma cruzi*. Trans Roy Soc Trop Med H 85(3):372–374

Chiari E, Duarte DS, Raslan DS, Saúde DA, Perry KSP, Boaventura MAD, Grandi TSM, Stehmann JR, Anjos AMG, Oliveira AB (1996) In vitro Screening of Asteraceae Plant Species Against *Trypanosoma cruzi*. Phytother Res 10(7):636–638

Conselho Estadual de Política Ambiental do Estado de Minas Gerais – Copam [Internet]. Lista das espécies ameaçadas de extinção da flora do Estado de Minas Gerais. Deliberação COPAM 85/97 [Cited 2014 Jul 15]. Available from: http://www.biodiversitas.org.br/florabr/mg-especies-ameacadas.pdf

Coyle NC, Jones SB (1981) *Lychnophora* (Compositae: Vernonieae), a genus endemic to the brazilian planalto. Brittonia 33(4):528–542

Cupolilo SMN, Gollner AM, Souza RLP, Tiago DAG, Lima VBR, Lima TS (2007) Investigation of damage effects by antiinflammatory and immunosuppressive doses of dexamethasone on mice's gastric mucosal. HU Rev 33:17–22

Duarte DS, Raslan DS, Chiari E, Oliveira AB (1993) Trypanocidal activity of *Lychnophora pinaster* Mart. Mem Inst Oswaldo Cruz 88:240

Ferraz Filha ZS, Lombardi JA, Guzzo LS, Saúde-Guimarães DA (2012) Brine shrimp (*Artemia salina* Leach) bioassay of extracts from *Lychnophoriopsis candelabrum* and different *Lychnophora* species. Rev Bras Plant Med 14(2):358–361

Ferreira AA, Azevedo AO, Silveira D, Oliveira PM, Castro MSA, Raslan DS (2005) Constituents of *Lychnophora pinaster* hydroalcoholic extract. Chem Nat Compd 41(4):466

Forbes VE, Forbes TL (1994) Ecotoxicology in theory and practice. Chapman and Hall, London

Giulietti AM, Menezes NL, Pirani JR, Meguro M, Wanderley MGL (1987) Flora da Serra do cipó, Minas Gerais: caracterização e lista de espécies. Bol Bot Univ São Paulo 9:1–151

Guzzo CA, Lazarus GS, Werth V (1996) Dermatological pharmacology. In: Hardman JG, Limbird LE, Molinoff PB, Ruddon RW, Gilman AG (eds) Goodman & Gilman's the pharmacological basis of therapeutics, 9th edn. McGraw-Hill, New York, pp 1604–1606

Guzzo LS, Saúde-Guimarães DA, Silva ACA, Lombardi JA, Guimarães HN, Grabe-Guimarães A (2008) Antinociceptive and anti-inflammatory activities of ethanolic extracts of *Lychnophora* species. J Ethnopharmacol 116:120–124

Jacobi CM, Carmo FF (2008) Diversidade dos campos rupestres ferruginosos no Quadrilátero Ferrífero, MG. Megadiversidade 4(1–2):26–33

Mansanares ME, Forni-Martins ER, Semir J (2002) Chromosome numbers in the genus *Lychnophora* Mart. (Lychnophorinae, Vernonieae, Asteraceae). Caryologia 55(4):367–374

Oliveira AB, Saúde DA, Perry KSP, Duarte DS, Raslan DS, Boaventura MAD, Chiari E (1996) Trypanocidal sesquiterpenes from *Lychnophora* species. Phytother Res 10(4):292–295

Rapini A, Ribeiro PL, Lambert S, Spirani JR (2008) A flora dos campos rupestres da Cadeia do Espinhaço. Megadiversidade 4(1–2):17–24

Reis ES, Pinto JEBP, Bertolucci SKV, Corrêa RM, Paula JR, Andrade ST, Ferri PH (2010) Seasonal variation in essential oils of *Lychnophora pinaster* Mart. J Essent Oil Res 22(2):147–149

Rodrigues VEG, Carvalho DA (2001) Levantamento etnobotânico de plantas medicinais no domínio dos cerrados na região do Alto Rio Grande – Minas Gerais. Ciênc Agrotec 25:101–123

Saadi A (1995) A geomorfologia da Serra do Espinhaço em Minas Gerais e de suas margens. Geonomos 3:41–63

Santos FR, Lacerda DR, Redondo RA, Nascimento AMA, Chartone-Souza E, Borba EL, Ribeiro RA, Lovato MB (2009) Diversidade genética. In: Drummond GM, Martins CS, Vieira F (eds) Biota Minas: Diagnóstico do conhecimento da diversidade genética. Universidade Federal de Minas Gerais, Minas Gerais, pp 390–410

Semir J (1991) Revisão taxonômica de *Lychnophora* Mart. (Vernoniaceae: Compositae) [Tese]. Universidade de Campinas, Campinas

Semir J, Rezende AR, Monge M, Lopes NP (2011) As Arnicas Endêmicas das Serras do Brasil. Editora UFOP, Ouro Preto

Shorr AF (2007) Epidemiology and economic impact of meticillin-resistant *Staphylococcus aureus*. Pharmacoeconomics 25:751–768

Silveira D, Souza Filho JD, Oliveira AB, Raslan DS (2005a) Lychnophoric acid from *Lychnophora pinaster*: a complete and unequivocal assignment by NMR spectroscopy. Eclet Quim 30:37–41

Silveira D, Wagner H, Chiari E, Lombardi JA, Assunção AC, Oliveira AB, Raslan DS (2005b) Biological activity of the aqueous extract of *Lychnophora pinaster Mart*. Braz J Pharmacog 15(4):294–297

Tropicos.org (2013) Missouri Botanical Garden [Internet]. *Lychnophora pinaster* Mart [cited 2014 Sep 13]. Available from: http://www.tropicos.org/Name/2738593

Viana PL, Lombardi JA (2007) Florística e caracterização dos campos rupestres sobre canga na Serra da Calçada, Minas Gerais, Brasil. Rodriguésia 58:159–177

Who (World Health Organization) (2014) [Internet]. Chagas disease (American trypanosomiasis) [cited 2014 Sep 14]. Available from: http://www.who.int/mediacentre/factsheets/fs340/en/

Marrubium vulgare L.

Valdir Cechinel Filho

Marrubium vulgare L.
Photo: Keir Morse
Available in: http://www.keiriosity.com/

V. Cechinel Filho (✉)
Programa de Pós-Graduação em Ciências Farmacêuticas e Núcleo de Investigações, Químico-Farmacêuticas (NIQFAR), Universidade do Vale do Itajaí (UNIVALI), Itajaí, SC, Brazil
e-mail: cechinel@univali.br

U. P. Albuquerque et al. (eds.), *Medicinal and Aromatic Plants of South America*, Medicinal and Aromatic Plants of the World 5,
https://doi.org/10.1007/978-94-024-1552-0_28

Abstract *Marrubium vulgare* L. (Lamiaceae) is a medicinal plant used as stimulant and antispasmodic, and to treat diabetes, headache, bladder or uteral pain, among others. It is not a native Brazilian plant, but is well adapted especially in the plateau, being known as marroio or marroio-branco. Several experimental studies have confirmed its therapeutic potential as anti-inflammatory, analgesic, antidiabetic, among others. The main bioactive principles are labdane diterpenes and flavonoids. Marrubiin, the major component, is produced from premarrubiin in conditions that use heating. Farther clinical studies are necessary to confirm the results evidenced in preclinical experiments.

Keywords *Marrubiun vulgare* · Folk medicine · Marrubiin · Therapeutic potential

1 Taxonomic Characteristics

Marrubium vulgare L. (Lamiaceae), known as horehound or common horehound, white horehound, marrube, houndsbane, marvel. In Brazil, it is called marroio or marroio- branco.

2 Crude Drug Used

The plant is used as a tea or infusion by the population. They also use its essential oil (Meyre Silva and Cechinel Filho 2010).

3 Major Chemical Constituents and Bioactive Compounds

Marrubium vulgare contains several biologically active, useful compounds. Its best known component is marrubiin (which is an artefact from pre-marrubiin); apigenin, apigenin 7-*O*-glucosideo, apigenin 7-lactate, apigenin 7-(6″-p-coumaroyl)-glucoside, luteolin, luteolin 7-*O*-β-D-glucosideo, luteolin 7-lactate, chrysoeriol, crysoeriol *O*-glucuronide, quercetin 3-*O*-α-L-ramnosil-glucoside, isoquercitrin, ursolic acid, gallic acid, caffeic acid, maleic cafeoil, vulgarol, vulgarin, β-sitosterol, stigmasterol, vitexin, acteoside, forsythoside B, arenarioside, ballotetroside, marruboside, acethyl marruboside, marrubenol, 6-octadecynoic acid, 5-*O*-caffeoylquinic (chlorogenic) acid; ladanein, 11-oxomarrubiin, vulgarcoside A, 3-hydroxyapigenin-4′-O-(6″-O-p-coumaroyl)-beta-D-glucopyranoside, phenylpropanoid esters and monoterpens (Sahpaz et al. 2002; Meyre Silva and Cechinel Filho 2010; Boudjelal et al. 2012; Ohtera et al. 2013; Shaheen et al. 2014).

4 Morphological Description

This plant is a perennial herb that can reach 30–50 cm in height, with stems covered in woolly hairs. Reproduced by seeds. Leaves arranged opposite along stem, decussate. Stems branch from the base of the plant, and along stems. The surfaces of this plant and reproductive organs are densely clothed with glandular and non-glandular trichomes, being the glandular trichomes of two main types: peltate and capitate. The non-glandular trichomes also present two main types, multicellular uniserrate and multicellular branched (Dmitruk and Haratym 2014). The seeds are elliptic in color dark-brown while the pollen morphology is psilate-perforate with aperture type tricolpate (Akgül et al. 2008).

5 Geographical Distribution

Europe, Asia, northern Africa and Americas (Sahpaz et al. 2002; Meyre Silva and Cechinel Filho 2010). And what about South America or Brazil?

6 Ecological Requirements

M. vulgare grows in temperate areas, on alkaline, calcareous soils (Sagliocco 2000). It grows in the area where temperatures are between 45 and 75 °F (7–24 °C) (Simon et al. 1984). It grows on poor, dry calcareous soils that have good drainage (Simon et al. 1984). *M. vulgare* is found on soils with a pH between 4.5 and 8.3 (Simon et al. 1984).

7 Collection Practice

In general, this plant was introduced in the countries as a medicinal herb, but it is also considered a weed, growing widely in the cattle pasture. This is not enough!

8 Traditional Use (Part(s) Used) and Common Knowledge

The whole plant is used as a tea or infusion by the population, for its stimulant and antispasmodic properties, and to treat diabetes, headache, bladder or uteral pain. It is also used as a diuretic, expectorant, digestive stimulant, anti-inflammatory for liver problems, and to treat flu and asthma. The essential oil is used to cure haemorrhoids (Meyre Silva and Cechinel Filho 2010; Popovic et al. 2014).

9 Modern Medicine Based on Its Traditional Medicine Uses

An investigation conducted with 21 patients using the extract of the leaves indicated that the extract reduced the following biochemical parameters: glucose 0.64%, cholesterol 4.16% and 5.78% triglycerides. This plant has demonstrated antimicrobial activity against gram-positive bacteria, especially *Staphylococcus aureus* and pronounced effect against methicillin-resistant *Staphylococcus aureus,* although was only moderately active against other microorganisms. Other therapeutic properties, as antioxidant, hypolipidemic, anti-inflammatory, cardioprotective and antiparasitic activities have been confirmed for this species. Previous biological studies conducted at our laboratory with marrubiin, the main compound of this plant, have revealed pronounced analgesic properties in different models of pain in mice, including antidiabetic, anti-hypertensive and antioedematogenic properties (Meyre Silva and Cechinel Filho 2010; Yousefi et al. 2013; Molina-Garza et al. 2014). Recent studies have indicated that *M. vulgare* exert inhibitory effects on mushroom tyrosinase activity and it could be considered as good food additives to prevent food browning and growth of microbes (Namjoyan et al. 2015).

10 Conclusions

M. vulgare is used by traditional medicine for the treatment of different kinds of human pathologies, particularly those related to respiratory, inflammatory and dolorous processes. And several experimental studies have confirmed the therapeutic potential of this plant, as well as the isolation and identification of many different active principles making this plant an important source of potential phytotherapeutic agents. Its best known chemical component, marrubiin is an artifact, produced from pre-marrubiin in conditions that use heating, still it seems to be the main marker of this plant.

References

Akgul G, Ketenoglu O, Pinar NM, Kurt L (2008) Pollen and seed morphology of the genus Marrubiun (Lamiaceae) in Turkey. Ann Bot Fenn 45:1–10

Boudjelal A, Henchiri C, Siracusa L, Sari M, Ruberto G (2012) Compositional analysis and in vivo anti-diabetic activity of wild Algerian *Marrubium vulgare* L. infusion. Fitoterapia 83(2):286–292

Dmitruk M, Haratym W (2014) Morphological differentiation of non-glandular and glandular trichomes on *Marrubium vulgare* L. Modern. Phytomorphology 6:85

Meyre Silva C, Cechinel Filho V (2010) A review of the chemical and pharmacological aspects of the genus Marrubium. Curr Pharm Des 16(31):3503–3518

Molina-Garza ZJ, Bazaldúa-Rodriguez AF, Quintanilha-Licea R, Galaviz-Silva L (2014) Anti-Trypanosoma cruzi activity of 10 medicinal plants used in northeast Mexico. Acta Trop 136:14–18

Namjoyan F, Jahangiri A, Azemi ME, Arkian E, Mousavi H (2015) Inhibitory effects of Physalis alkekengi L., Alcea rosea L., Bunium persicum B. Fedtsch. and *Marrubium vulgare* L. on mushroom tyrosinase. Jundishapur J Nat Pharm Prod 10(1):e23356

Ohtera A, Miyamae Y, Nakai N, Kawachi A, Han J, Isoda H, Neffati M, Akita T, Maejima K, Masuda S, Kambe T, Mori N, Irie K, Nagao M (2013) Identification of 6-octadecynoic acid from a methanol extract of *Marrubium vulgare* L. as a peroxisome proliferator-activated receptor γ agonist. Biochem Biophys Res Commun 440(2):204–209

Popovic Z, Smiljanic M, Kostic M, Nikic P, Jankovic S (2014) Wild flora and its usage in traditional phythoterapy (Deliblato Sands, Serbia, South East Europe). Indian J Tradit Knowl 13(1):9–35

Sagliocco JL (2000) The insect fauna associated with horehound (*Marrubium vulgare* L.) in western Mediterranean Europe and Morocco: potential for biological control in Australia. In: Horehound workshop; proceedings of a workshop held at the Victorian Institute for Dryland Agriculture in Horsham, April 19–20, 1999. Sponsored by the Co-operative Research Centre for Weed Management Systems. Plant Protect Quart 15(1):21–25

Sahpaz S, Garbacki N, Tits M, Bailleul F (2002) Isolation and pharmacological activity of phenylpropanoid esters from *Marrubium vulgare*. J Ethnopharmacol 79(3):389–392

Shaheen F, Rasoola S, Shah ZA, Soomro S, Jabeen A, Mesaik MA, Choudhry MI (2014) Chemical constituents of *Marrubium vulgare* as potential inhibitors of nitric oxide and respiratory burst. Nat Prod Commun 9(7):903–906

Simon JE, Chadwick AF, Craker LE (1984) Herbs; an indexed bibliography. 1971–1980. The scientific literature on selected herbs, and aromative and medicinal plants of the temperate zone. Archon Books, Hamden. 770 pp

Yousefi K, Soraya H, Fathiazad F, Khorrami A, Hamedeyazdan S, Maleki-Disaji N, Garjani A (2013) Cardioprotective effect of methanolic extract of *Marrubium vulgare* L. on isoproterenol-induced acute myocardial infarction in rats. Indian J Exp Biol 51(8):653–660

Maytenus ilicifolia Mart. ex Reissek

Larissa Lucena Périco, Vinícius Peixoto Rodrigues, Luiz Fernando Rolim de Almeida, Ana Paula Fortuna-Perez, Wagner Vilegas, and Clélia Akiko Hiruma-Lima

Maytenus ilicifolia Mart. ex Teissek
Photo: Julio Antonio Lombardi
Available in: https://www.kew.org/science/tropamerica/neotropikey/families/Celastraceae.htm

L. L. Périco · V. P. Rodrigues · C. A. Hiruma-Lima (✉)
Department of Physiology, São Paulo State University (UNESP), Institute of Biosciences, Botucatu, SP, Brazil
e-mail: clelia.hiruma@unesp.br

L. F. R. de Almeida · A. P. Fortuna-Perez
Departament of Botany, São Paulo State University (UNESP), Institute of Biosciences, Botucatu, SP, Brazil
e-mail: luizfernando@ibb.unesp.br; ana.fortuna@unesp.br

W. Vilegas
Coastal Campus of São Vicente, São Paulo State University (UNESP), Institute of Biosciences, São Vicente, SP, Brazil
e-mail: wagner.vilegas@unesp.br

U. P. Albuquerque et al. (eds.), *Medicinal and Aromatic Plants of South America*, Medicinal and Aromatic Plants of the World 5,
https://doi.org/10.1007/978-94-024-1552-0_29

Abstract Herbal medicines are widely used in Brazil and currently constitute an expanding market. Among the species with the highest number of registration entries is *Maytenus ilicifolia* Mart. ex Reissek, a plant species native to Brazil that has a high medicinal value.

Pharmacological pre-clinical studies have demonstrated the anti-ulcerogenic, anti-secretory, anti-inflammatory, anti-diarrhea, anti-oxidant, anti-microbial, anti-protozoal, anti-cancer and hypotensive properties of this medicinal plant. It has also been established that some of its pharmacological activities are due the presence of terpenoids, flavonoids, tannins, alkaloids and polysaccharides. The species *M. ilicifolia* has been used in traditional medicine since the mid-1920s. Presently, it is endangered due to the strong anthropic action in natural populations. As a medicinal plant, in Brazil, its leaves are used in homemade and industrial medicines to effectively treat stomach ulcers. Therefore, studies that validate the use of this important Brazilian native plant are warranted.

Keywords Espinheira-santa · Quebrachillo · Cancerosa · *Maytenus ilicifolia* · Celastraceae

1 Taxonomic Characteristics

The genus *Maytenus* is a large genus of approximately 300 species that is widely distributed in the tropics and subtropics of both the Old and New Worlds. Approximately 160 species of *Maytenus* grow in the New World, and nearly 50 species are known to be distributed in many regions of Brazil, including Amazonian forests, the Atlantic Rainforest, "caatinga" and "cerrado", including "campos rupestres" (Groppo et al. 2014). The name *Maytenus* is derived from the word, "Maytén," a name first used by the "Mapuche" people ("men of the land") of Chile (Niero et al. 2011).

The species *Maytenus ilicifolia* Mart. ex Reissek is a native Brazilian medicinal plant described in the 4th Brazilian Official Pharmacopoeia, 1988–1966 (Brandão et al. 2006). Its use was described in 1922, as a traditional medicine, for the treatment of gastric ulcer (Carlini and Frochtengarten 1988). It is characterized as a tree or shrub that is branched from the base, up to 5 m tall, features young twigs, and is angular, 4 or multi-carinate (Carvalho-Okano 1992). This medicinal plant has several common names; however, "espinheira-santa" seems to be the most common vernacular name in Brazil for this and other species, e.g., *Maytenus aquifolium, M. robusta* and *M. truncata* (Niero et al. 2011).

Synonyms *Celastrus spinifolius* Larrañaga; *M.* fo. *angustior* Briq.; *M. hassleri* Briq.; *M. muelleri* Schwacke; *M. officinalis* Mabb.; *M. pilcomayensis* Briq.; *Maytenus aquifolium* Mart

2 Crude Drug Used

The growing interest in *M. ilicifolia*, coupled with the increasing utilization of these species by the pharmaceutical industry, has accentuated the importance of developing analytical methods for use in the production of standardized preparations of *Maytenus*-derived phytomedicines (Tiberti et al. 2007). Many compounds or secondary metabolites are used to control the quality of medicinal herbs. Friedelin and β-friedelanol isolated from *Maytenus* species are specifically used to control the quality of *M. ilicifolia* (Valladão et al. 2009). According Souza et al. (2008), the governmental Brazilian drug agency (ANVISA) has approved the use and commercialization of phytotherapies derived from the leaves of *M. ilicifolia.* This species was approved as an herbal medicine in Brazil, and extracts obtained from the maceration of its leaves in alcohol, which are standardized by their tannin contents, have been commercialized (Cipriani et al. 2009). The Brazilian Ministry of Health (2009) published a list of medicinal plants called RENISUS that are of interest to Unified Health System (SUS), which contained more than 70 medicinal species. *M. ilicifolia* was cited in this list, making it a candidate for use in Brazilian health programs to benefit the population.

3 Major Chemical Constituents and Bioactive Compounds

The leaves of this *M. ilicifolia* contain flavonoids (mauritianin, trifolin, hyperin, afzelechin, epiafzelechin, quercetrin, quercitrin, rutin, kaempferol, gallocatechin, epicatechin, epigallocatechin and catechins), glycosylated flavonoids (monoglycosylated quercetin derivatives, quercetin-di-rhamno-hexoside, diglycosylated quercetin derivatives, kaempferol-di-rhamno-pentoside, tetra-glycoside kaempferol derivatives, diglycosylated kaempferol derivatives, condensed tannins (di-, tri-, tetra-, and penta-, hexa, and heptamers), terpenes (maytenin, tingenon, isotingenon II congorosin A and B maytenoic acid), triterpenes (friedelan-3-ol, friedelin), quinonemethide triterpenoid (pristimerin), glycolipids (monogalactosyldiacylglycerol, digalactosyldiacylglycerol, trigalactosildiacylglicerol, tetragalactosildiacylglicerol and sulfoquinovosyldiacylglycerol), glucosides (ilicifolinoside A–C) and alkaloids (mayteine, maitanprin and maitensin). These components are likely to be the active compounds (Alonso 1998; Carlini and Frochtengarten 1988; Costa et al. 2008; Mendes et al. 2006; Pereira et al. 2005; Souza-Formigoni et al. 1991; Gonzalez et al. 2001; Tiberti et al. 2007; Souza et al. 2008; Zhu et al. 1998; Leite et al. 2010). The leaves of *M. ilicifolia* also contain polysaccharides such as arabinogalactan, acidic heteroxylans and polygalacturonic acid (Cipriani et al. 2006, 2008, 2009). Mossi et al. (2004, 2010) extracted volatile and semi-volatile organic compounds from the leaves of native populations of *M. ilicifolia* such as phytol, squalene, vitamin E, limonene, stigmasterol, friedelan-3-ol, friedelin, fridelan-3-one, palmitic acid, dodecanoic acid and geranyl acetate.

The root bark of *M. ilicifolia* from Asuncion (Paraguay) contains triterpenes such as cangoronine, ilicifoline and friedelane-, pristimerin- and tingenone-type triterpenes (Itokawa et al. 1991). Sesquiterpene pyridine alkaloids (ilicifoliunines A and B), aquifoliunine E-I and mayteine (Santos et al. 2012) and terpenoids milicifolines A-D (Gutierrez et al. 2007) have also been extracted from the root bark of this species. Quinonemethide triterpenoids (maytenin and pristimerin) have also been isolated from the bark of the roots of mature *M. ilicifolia* from Brazil (Santos et al. 2010).

4 Morphological Description

The genus *Maytenus* consists of woody and shrubby species (Duarte and Debur 2005). *Maytenus* is characterized by its flattened or carinate young twigs, alternating leaves with crenate, spinose, or entire margins, and flowers with a conspicuous intrastaminal disc(Groppo et al. 2014). The fruit is characteristic of the genus; it is a capsule with two (or three) reflexing valves and one or two (up to four) arillate, erect seeds (Groppo et al. 2014). As *Maytenus* species feature rather uniform floral and inflorescence structures, vegetative characters are heavily (Groppo et al. 2014). The most recent comprehensive taxonomic treatments of Brazilian *Maytenus* species were conducted by Carvalho-Okano (1992) and based on examining a list of materials. *M. ilicifolia* has simple and entire leaves with an alternate phyllotaxy, lanceolate shape, acute apex and round base, measuring approximately 5 cm long and 2 cm wide. The margin features sparse spiny teeth, and the petiole is short. The foliar surface is coriaceous and glabrous, and the midrib is more prominent on the abaxial side (Duarte and Debur 2005). The leaves are dense, coriaceous and glabrous, with minute stipules and leaf blades 2.2–8.9 cm long and 1.1–3.0 cm wide. The leaves feature prominent veins on the abaxial surface are elliptical with an entire or spinose margin. They can feature one to several thorns that are regularly or irregularly distributed along the board and usually concentrated in the apical half of one or both semi-leaf blades. According to morpho-anatomical studies reported by Duarte and Debur (2005), the stem organization, in the secondary growth, shows a periderm beneath the remaining epidermis, conspicuous sclerenchymatic ring in the cortex and cambium that forms a phloem outside and a xylem inside. The leaf is simple, alternate and lanceolate and has sparse spiny teeth along the margin. Epidermal cells that contain calcium oxalate crystals, a thick cuticle that forms cuticular flanges, dorsiventral mesophyll and an amphicrival bundle in the midrib and petiole are observed. The inflorescences occur in multiflorous fascicles. The flowers feature sepals semi-circular and ciliated, ovate petals, and entire morphology. The stamens feature filaments flattened at the base, capitate stigma that is sessile or of different styles, and ovaries that protrude or are fused to the disc. The flowers are hermaphrodites, but strong evidence indicates that many of its flowers are functionally diclinous. The fruit is a bivalvar, orbicular capsule with a mature red-orange pericarp. The seeds are erect, suborbicular, ellipsoids or obovate and sometimes angular, and vary in number from 1 to 4 per fruit; they are entirely surrounded by aryl. The testa is hard, smooth and shiny and usually brown or black in color. The endosperm is abundant; the embryo is axial and membranaceous with a straight hypocotyl radicle, flat cotyledons and a short axis. The aryl is fleshy, white and covers the entire seed.

5 Geographical Distribution

According to Carvalho-Okano and Leitão Filho (2004), *M. ilicifolia* grows in several Brazilian states, such as Mato Grosso do Sul, Paraná, Santa Catarina and Rio Grande do Sul. It is referred to by several common names, such as cancerosa, cancorosa, cancorosa-de-sete-espinho, cancrosa, congorça, coromilho-do campo, espinheira-divina, espinho-de-deus, maiteno, salva-vidas, sombra-de-touros, erva-cancrosa and erva-santa (Lorenzi and Matos 2002). The species *M. ilicifolia* is also distributed in Argentina (Buenos Aires, Chaco, Corrientes, Entre Rios, Formosa, Misiones, Salta and Santa Fé), Paraguay, Bolivia (Cochabamba, La Paz and Santa Cruz) and Uruguay and is known by several common names, including quebrachillo, sombra-de-toro, and concorosa (Niero et al. 2011; Santos-Oliveira et al. 2009).

6 Ecological Requirements vs Cultural Practices

M. ilicifolia is sown in Rio Grande do Sul (Brazil) between the months of December and February, during which the seeds are in physiologically mature state and have brown in color. The germination rate is high during this phase, approximately 98% (Negrelle et al. 1999). Black, deep and bulky polyethylene bags should be used containers to ensure good root development, especially the taproot. Among the various recommended substrates for the production of *M. ilicifolia* seedlings, the importance of organic matter is emphasized because this species requires this type of fertilization. Scheffer (2001) recommends a mixture of soil, vermicompost and vermiculite at a ratio of 3:1:1. Mariot (2005) reported that soil mix, medium sand and cattle manure at a ratio of 1:1:1 achieved good results in the development of the seedlings in wooded areas. Nicoloso et al. (2000) recommended a 1:1 mixture of soil and charred rice hulls. Montanari et al. (2004) recommend the used of vermiculite and sand at a ratio of 2:1, which allowed the adequate development of seedlings. The producer must prepare the substrate using materials that are readily available, but organic matter and sand or carbonized rice husk should be used to ensure good drainage (Mariot and Barbieri 2006). The seeds should be buried at depths between 10 and 15 mm to maintain a constant moisture content in the substrate (Montanari et al. 2004). Shading (50%) or bamboo can be used to protect the plant throughout the hot season (Mariot and Barbieri 2006). "Espinheira-santa" grows very slowly. The seedlings remain in polyethylene bags for a long period until they are transplanted to the site in September when they reach a height of approximately 20 cm. The rainy season begins at this time and temperature is rising, which helps the establishment of seedlings (Mariot and Barbieri 2006). Weed are controlled by weeding, and this process can be mitigated by intercropping with legumes, such as the peanut (*Arachis pintoi* Krapov. & W.C.Greg.). This species, whose root system is superficial, does not compete with *M. ilicifolia* for water and nutrients. In addition, it forms an excellent vegetation cover that reduces the incidence of invasive plants, conserves soil moisture and reduces the thermal oscillation on the soil surface. Because it is a legume, it has the advantage of returning nitrogen to the soil, which nutritionally benefits *M. ilicifolia* (Mariot and Barbieri 2006). Pests, such as scale insects, mites and aphids, have been observed, but these do not cause serious damage (Magalhães 2002). Large infestations of aphids cause leaf crinkle. However, an attack of leaf-cutting ants can seriously damage the crop during the crop installation phase, soon after transplantation. Many authors have reported the occurrence of two fungal diseases in this plant: the sooty mold, which is associated with the presence of scale insects on the leaves, and powdery mildew in early spring (Mariot and Barbieri 2006). "Espinheira-santa" is harvested by pruning branches and subsequently removing the leaves, which are the part of the plant that is used. Defoliation is not recommended because pruning encourages greater growth (Carvalho et al. 2003).

According to various authors, the plants should be harvested in the fall after the reproductive stage to ensure the production of seeds. The first harvest should be carried out after the 2nd or 3rd year, due to the slow growth of plants (Castro and Ramos 2003). Harvesting can be performed manually or by machine (Montanari et al. 2004). Manual harvesting, pruning shears are used and appropriate gloves due to the presence of spines on the leaves. Magalhães (2002) recommends that the plants should be pruned at the height of 50 cm during the first harvest, and subsequent harvests should consist of pruning just above the height of pruning during the previous year. Mechanized harvesting utilizes a side mower attached to a tractor (Montanari et al. 2004). The machine must remove a horizontal section at a height of 50 cm, which ensures that the lower leaves remain on the plant. The harvested branches should be placed on plies or clean containers to avoid contamination by microorganisms, which are more abundant on leaves near the ground (Montanari et al. 2004). Plants should be then sent to the processing facility.

The yield of the *M. ilicifolia* is highly variable and depends on the soil and climatic factors, the age of the plants, the cultivation system, the technologies employed and the genetic potential of plants (Mariot and Barbieri 2006).

It is estimated that 160 tons/year of plant matter is sold as *M. ilicifolia,* in Brazil. Remarkably, only 21% of this amount consists of *M. ilicifolia* and *M. aquifolium* (Mariot and Barbieri 2006). "Espinheira-santa" is easily found on the common market. However, the species offered often is not *M. ilicifolia* but *Sorocea bomplandii* Bailon (Moraceae), a common **adulteration** of "espinheira-santa".

Many researchers have carried out a phytochemical and pharmacological study of *S. bomplandii.* They have verified the presence of some flavonoids with analgesic and anti-ulcerogenic actions similar to that of *M. ilicifolia* (Calixto 1993; Gonzalez et al. 2001). The effectiveness of the two species has, however, not been compared and the possibility of the chronic toxicity of *S. bomplandii*, which can become a risk for people who inadvertently consume this species, believing it to be "espinheira-santa", has not yet been studied (Santos-Oliveira et al. 2009). The high medicinal value of *M. ilicifolia* has led to intense anthropic action in natural populations, resulting in the genetic erosion of the species by predatory extraction. It was therefore to be considered as a priority for conservation (Vieira 1999).

7 Traditional Use (Part(s) Used) and Common Knowledge

Maytenus iliciIifolia, which is popularly called "espinheira-santa" due to the appearance of its leaves and its therapeutic properties, is widely used in popular medicine to treat stomach conditions, including nausea, gastritis, and ulcers (Balbach 1980; Cruz 1982).

It is administered as a tea to treat gastric disorders (atonia, hyperacidity, gastric and duodenal ulcers and chronic gastritis), which is prepared by adding one tablespoon of chopped leaves to one cup of boiling water, at a dose of 1 cup of infusion before main meals (Panizza 1998). The leaves of this species have also been used to treat hangovers caused by drinking an excess of alcohol (Simões 1989), and prepare tonics, antiseptics, carminatives, diuretics, laxatives (Teske and Trentini 1995) and emmenagogic agents (Niero et al. 2011). Scheffer (2004) also mentions that this folk medicine can be used as a contraceptive, abortifacient, vulnerary, to treat liver diseases and hydropsy due to alcohol abuse and a drug to reduce milk production during breastfeeding. Additionaly, *M. ilicifolia* has been used in Brazilian folk medicine to treat diabetes, urinary tract infections, intestinal problems, nervous diseases, kidney and blood disorders (Mariot 2005; Mariot and Barbieri 2007a). Traditionally, only the leaves of this species have been used, but the use of the root has also been reported, particularly for the treatment of diabetes (Mariot and Barbieri 2007b). The leaves o this plant have also reportedly been used to prepare a paste for the topical treatment of skin cancer (Lorenzi and Matos 2002). It has been used in Argentinean folk medicine as a sialogogue, antihistamine, antiseptic and vulnerary. It has also been employed as an indigenous antitumor remedy in Brazil. This plant is also used by Indian tribes and rural populations in Paraguay to regulate fertility (Zhu et al. 1998).

8 Modern Medicine Based on Its Traditional Medicine Uses

The traditional medicinal use of *M. ilicifolia* as antiulcer has been extensively studied with different extracts. A pharmacological study in rodents confirmed that a simple extract of leaves prepared with hot water was an effective antiulcer agent because it increased the volume and pH of the gastric juices (Souza-Formigoni et al. 1991). The lyophilized aqueous extract of *M. ilicifolia* could reduce acid secretion *in vitro*, via the same mechanism as cimetidine (anti-secretory anti-ulcer drug): it antagonizes the histamine H2 receptor (Ferreira et al. 2004). The hexane and ethylacetate extracts obtained from the leaves of *M. ilicifolia* yield anti-edematogenic and anti-ulcer effects in mice (Jorge et al. 2004). Queiroga et al. (2000) isolated the triterpenes, friedelan-3b-ol and friedelin from leaves of *M. ilicifolia* and described that these compounds are not the active components responsible for the anti-ulcer effect of the leaves of *M. ilicifolia*. Leite et al. (2010) suggest that only the fraction that contains the tri- and tetra-flavonoids glycosides, mauritianin and hyperin exerted a significant gastroprotective effect by increasing the volume and pH of gastric juices. The flavonoid-rich fraction containing galactitol (25%), epicatechin (3.1%) and catechin (2%) obtained from the leaves of *M. ilicifolia* also exerts gastroprotective effect by inhibiting gastric acid via the inhibition of gastric H^+- and K^+- ATPase and the modulation of nitric oxide formation (Baggio et al. 2007). The anti-ulcer effect of *M. ilicifolia* does not appear to be restricted to phenolic compounds. Pre-clinical studies have shown that polysaccharides obtained from the leaves by aqueous extraction, such as polygalacturonic acid (Cipriani et al. 2009), acidic heteroxylans (Cipriani et al. 2008) and arabinogalactans (Cipriani et al. 2006), protect from gastric ulcers. However, medicines obtained via the maceration of leaves from *M. ilicifolia* in alcohol do not contain this polysaccharide. Thus, consuming this medicine as a tea (water extractable polysaccharides) improves its pharmacological effect in treating gastric ulcers (Cipriani et al. 2009).

In addition to the efficacy of *M. ilicifolia* as an anti-ulcer agent, this species is also effective in treating other gastric disturbances, such as diarrhea. Baggio et al. (2009) proved that flavonoid-rich extracts reduce the gastrointestinal motility of mice in vivo. This result indicates that this plant may have anti-diarrhea and/or spasmolytic properties.

The crude ethanolic extract from the root bark exerts *in vitro* antioxidant activity (Vellosa et al. 2006). This antioxidant effect is likely related to the quinonemethide triterpenes and/or phenolic substances present in this root (Santos et al. 2010). Vargas et al. (1991) showed that the aqueous extract of leaves from *M. ilicifolia* did not exert *in vitro* genotoxicity, as assessed with the Ames test. Camparoto et al. (2002) also proved that the infusion of leaves from *M. ilicifolia* was free of mutagenic and cytotoxic effects by analyzing the number of chromosome alterations and rates of cell division.

Horn and Vargas (2003) described the anti-mutagenic effect of the aqueous extract of leaves in *Salmonella*/microsome assays.

The anti-cancer effect of *M. ilicifolia* has also been studied. Pristimerin, a quinonemetride triterpenoid, is present in several plants, including *M. ilicifolia.* This compound is cytotoxic to several cancer cell lines. Costa et al. (2008) found that the anti-proliferative effect of pristimerin is due to its ability to inhibit DNA synthesis and trigger apoptosis in leukemic human cells. However, this anticancer effect is not restricted to the isolated triterpenenoid pristimerin. The spray-dried extract of the leaves of *M. ilicifolia* could protect normal cells and induce apoptosis in human carcinoma cells by down-regulating Bcl-2 and activating the caspase-2-dependent signaling pathway (Araújo-Júnior et al. 2013).

Leme et al. (2013) revealed that the purified fraction obtained from *M. ilicifolia* contains compounds responsible for diuretic and hypotensive activities, and this effect could involve the prostaglandin/cAMP pathway. Crestani et al. (2009) also proved the hypotensive effect of fractions from this plant *in vivo*, and they attributed this effect to the nitric oxide/guanylate cyclase pathway. Rattmann et al. (2006) showed that this same pathway was responsible for inducing vasorelaxation, which certainly contributed to the hypotensive action of *M. ilicifolia*. An extract of this species also has sedative activity, which synergistically enhances barbiturate-induced sleep in mice (Alonso 1998).

The traditional medicinal use of the leaves of this species as an antiseptic was also studied. Maytenin isolated from the bark of the roots of mature *M. ilicifolia* exhibits strong antimicrobial activity against Gram-positive (*Staphylococcus aureus* and *Streptococcus sp.*) and Gram-negative bacteria (Gonçalves de Lima et al. 1969). According to Singh and Dubey (2001), the friedelin and friedelanin-3-β-ol in *M. ilicifolia* also exerts *in vitro* antimicrobial activity against *Staphylococcus aureus*, *Escherichia coli*, and *Aspergillus niger*. The antifungal effects of maytenin and pristimerin were evaluated, but maytenin yield better results (Gullo et al. 2012). Both of these triterpenoids were also effective against the Trypanosomatidae *Leishmania amazonensis* and *Leishmania chagasi* and *Trypanosoma cruzi*, which are etiologic agents of leishmaniasis and Chagas disease, respectively (Santos et al. 2013). The *in vitro* antiprotozoal activity against *Leishmania chagasi* and *Trypanosoma cruzi* was also assessed with the alkaloid aquifoliunine E-I isolated from the root bark of *M. ilicifolia* suggesting that these compounds should be considered in the development of a new drug for the treatment of leishmaniasis and Chaga's disease (Santos et al. 2012).

The common use of *M. ilicifolia* in folk medicine as an abortifacient was also studied. Cunha-Laura et al. (2014) studied the hydro-acetonic extract of this species and showed that it was non-toxic to pregnant rats and did not interfere with embryo-fetal development or maternal reproductive parameters. However, Montanari and Bevilacqua (2002) showed that the hydro-alcoholic extract of the leaves of *M. ilicifolia* reduced the rate of embryo implantation during early pregnancy in mice at dose of 1 g/kg. These data indicate that this medicinal plant should be used with caution in pregnant woman. Montanari et al. (1998) also studied the ethanolic extract of *M. ilicifolia* in male rats and concluded that it did not induce changes in spermatogenesis. In addition to the non-toxicity of *M. ilicifolia* the accurate identification and collection of this medicinal herb is vital to enhance the drug's efficacy and avoid adulterants. For example, Gonzales et al. (2001) studied three species of native plants known as "espinheira santa" from Tropical Atlantic forests and showed that *Zolernia ilicifolia* exerted a significant toxic effect.

9 Conclusions

The ethnobotanical, ethnopharmacological, agronomic and toxicological studies of *M. ilicifolia* explain the growing interest in this species, as well as the importance of this plant as a phytomedicine for the treatment of inflammations, ulcers, microbial and protozoan infections and cancer. The growing interest in this species should, however, be accompanied by both new pharmacokinetic studies. The elaboration of new analytical methods are needed in order to generate standardized preparations of *M. ilicifolia* could be used to eliminate the common adulterations.

References

Alonso JR (1998) Phytomedicine Treaty clinical and pharmacological bases. Isis Ediciones SRL, Buenos Aires, pp 828–834 Spanish

Araújo-Júnior RF, Oliveira AL, Pessoa JB, Garcia VB, Guerra GC, Soares LA, Souza TP, Petrovick PR, Araújo AA (2013) *Maytenus ilicifolia* dry extract protects normal cells, induces apoptosis and regulates Bcl-2 in human cancer cells. Exp Biol Med 238:1251–1258

Baggio CH, Freitas CS, Otofuji GM, Cipriani TR, Souza LM, Sassaki GL, Iacomini M, Marques MCA, Mesia-Vela S (2007) Flavonoid-rich fraction of *Maytenus ilicifolia* Mart. ex. Reiss protects the gastric mucosa of rodents through inhibition of both H+,K+-ATPase activity and formation of nitric oxide. J Ethnopharmacol 113:433–440

Baggio CH, Freitas CS, Mayer B, Santos AC, Twardowschy A, Potrich FB, Cipriani TR, Souza LM, Sassaki GL, Iacomini M, Marques MCA, Mesia-Vela S (2009) Muscarinic-dependent inhibition of gastric emptying and intestinal motility by fractions of *Maytenus ilicifolia* Mart ex. Reissek. J Ethnopharmacol 123:385–391

Balbach A (1980) The national flora in domestic medicine. EDEL, São Paulo Portuguese

Brandão MGL, Cosenza GP, Moreira RA, Monte-Mor RLM (2006) Medicinal plants and other botanical products from the Brazilian Official Pharmacopeia. Braz J Pharmacogn 16(3):408–420

Brazilian Ministry of Health (2009) Relação Nacional de Plantas Medicinais de Interesse ao Sistema Único de Saúde (RENISUS), Ministério da Saúde. http://portalsaude.saude.gov.br/images/pdf/2014/maio/07/renisus.pdf. (Brazil)

Calixto JB (1993) Pharmacological analysis of the methanolic extract and sorocein A, a new Diels-Alder compound isolated from the roots of *Sorocea bonplandii* Bailon in the isolated rat uterus and guinea pig ileum. Gen Pharmacol 24:983–989

Camparoto ML, Teixeira RO, Mantovani MS, Vicentini VEP (2002) Effects of *Maytenus ilicifolia* Mart. and *Bauhinia candicans* Benth infusions on onion root-tip and rat bone-marrow cells. Genet Mol Biol 25(1):85–89

Carlini EA, Frochtengarten ML (1988) Clinical toxicology (Phase I) of the espinheira-santa (*Maytenus ilicifolia*). Brasília-Distrito Federal. 67–73. Portuguese

Carvalho RIN, Cardon LM, Jaremtchuk CC, Kanawate EM, Silva JEC (2003) Carqueja e Espinheira-Santa in the Metropolitan region of Curitiba: from production to trade. (Curitiba) Life. 44. Portuguese

Carvalho-Okano RM (1992) Taxonomic studies of the genus *Maytenus* Mol. Emend. Mol. (Celastraceae) of extra-Amazonian Brazil. Thesis (Doctorate in Sciences – Plant Biology), University of Campinas. 253. Portuguese

Carvalho-Okano R, Leitão Filho HF. Taxonomic studies of the genus *Maytenus* Mol. Emend. Mol. (Celastraceae) of extra-Amazonian Brazil. Reis, M.S.; Silva, S.R. Conservation and sustainable use of medicinal and aromatic plants: *Maytenus* spp., espinheira-santa. Brasília IBAMA. 2004; 11–51. Portuguese

Castro LO, Ramos RLD (2003) Botanical description, cultivation and use of *Maytenus ilicifolia* Mart. ex Reiss. Cancorosa ou espinheira-santa (Celastraceae). FEPAGRO, Porto Alegre, p 12 Portuguese

Cipriani TR, Mellinger CG, Souza LM, Baggio CH, Freitas CS, Marques MCA, Gorin PAJ, Sassaki GL, Iacomini MA (2006) Polysaccharide from a Tea (Infusion) of *Maytenus ilicifolia* Leaves with anti-ulcer protective effects. J Nat Prod 69:1018–1021

Cipriani TR, Mellinger CG, Souza LM, Baggio CH, Freitas CS, Marques MCA, Gorin PAJ, Sassaki GL, Iacomini M (2008) Acidic heteroxylans from medicinal plants and their anti-ulcer activity. Carbohydr Polym 74:274–278

Cipriani TR, Mellinger CG, Souza LM, Baggio CH, Freitas CS, Marques MCA, Gorin PAJ, Sassaki GL, Iacomini M (2009) Polygalacturonic acid: Another anti-ulcer polysaccharide from the medicinal plant *Maytenus ilicifolia*. Carbohydr Polym 78:361–363

Costa PM, Ferreira PMP, Bolzani VS, Furlan M, Santos VAFFM, Corsino J, Moraes MO, Costa-Lotufo LV, Montenegro RC, Pessoa C (2008) Antiproliferative activity of pristimerin isolated from Maytenus ilicifolia (Celastraceae) in human HL-60 cells. Toxicol In Vitro 22:854–863

Crestani S, Rattmann YD, Cipriani TR, Souza LM, Iacomini M, Kassuya CAL, Marques MCA, Silva-Santos JE (2009) A potent and nitric oxide-dependent hypotensive effect induced in rats by semi-purified fractions from *Maytenus ilicifolia*. Vasc Pharmacol 51:57–63

Cruz GL (1982) Dictionary of useful plants in Brazil, 2nd edn. Civilização Brasileira, São Paulo Portuguese

Cunha-Laura AL, Auharek SA, Oliveira RJ, Siqueira JM, Vieira MC, Leite VS, Portugal LC (2014) Effects of *Maytenus ilicifolia* on reproduction and embryo-fetal development in Wistar rats. Genet Mol Res 13(2):3711–3720

Duarte MR, Debur MC (2005) Stem and leaf morphoanatomy of *Maytenus ilificolia*. Fitoterapia 76:41–49

Ferreira PM, Oliveira CN, Oliveira AB, Lopes MJ, Alzamora F, Vieira MAR (2004) A lyophilized aqueous extract of *Maytenus ilicifolia* leaves inhibits histamine-mediated acid secretion in isolated frog gastric mucosa. Planta 219:319–324

Gonçalves de Lima O, Dálbuquerque IL, Coêlho JSB, Martins DG, Lacerda AL, Maciel GM (1969) Antimicrobial substance of higher plants. XXXI communication. Maitenin, antimicrobial with new antineoplastic isolated action Celastraceae of Pernambuco. Rev Inst Antibióticos 9(2):17–25 Portuguese

Gonzalez FG, Portela TY, Stipp EJ, Di Stasi LC (2001) Antiulcerogenic and analgesic effects of *Maytenus aquifolium*, *Sorocea bomplandii* and *Zolernia ilicifolia*. J Ethnopharmacol 77:41–47

Groppo M, Simmons MP, Cappa JJ, Biral L, Lombardi JA (2014) A new species of *Maytenus* (Celastraceae) with fleshy fruits from eastern Brazil, with notes on the delimitation of *Maytenus*. Syst Bot 39(2):478–484

Gullo FP, Sardi JCO, Santos VAFFM, Sangalli-Leite F, Pitangui NS, Rossi AS, Silva ACAP, Soares LA, Silva JF, Oliveira HC, Furlan M, Silva DHS, Bolzani VS, Mendes-Giannini MJS, Fusco-Almeida AM (2012) Antifungal activity of Maytenin and Pristimerin. Evid-Based Compl Alternat Med 2012:1–6

Gutiérrez F, Estévez-Braun A, Ravelo AG, Astudillo L, Zárate R (2007) Terpenoids from the medicinal plant *Maytenus ilicifolia*. J Nat Prod 70:1049–1052

Horn RC, Vargas VM (2003) Antimutagenic activity of extracts of natural substances in the *Salmonella*/microsome assay. Mutagenesis 18:113–118

Itokawa H, Shirota O, Ikuta H, Morita H, Takeya K, Iitaka Y (1991) Triterpenes from *Maytenus ilicifolia*. Phytochemistry 30(11):3713–3716

Jorge RM, Leite JPV, Oliveira AB, Tagliati CA (2004) Evaluation of antinociceptive, anti-inflammatory and antiulcerogenic activities of *Maytenus ilicifolia*. J Ethnopharmacol 94:93–100

Leite JPV, Braga FC, Romussi G, Persoli RM, Tabach R, Carlini EA, Oliveira AB (2010) Constituents from *Maytenus ilicifolia* leaves and bioguided fractionation for gastroprotective activity. J Braz Chem Soc 21:248–254

Leme TSV, Prando TBL, Gasparotto FM, Souza P, Crestan S, Souza LM, Cipriani TR, Lourenço ELB, Junior AG (2013) Role of prostaglandin/cAMP pathway in the diuretic and hypotensive effects of purified fraction of *Maytenus ilicifolia* Mart ex Reissek (Celastraceae). J Ethnopharmacol 150:154–161

Lorenzi H, Matos FJA (2002) Medicinal plants in Brazil: natives and exotics. Instituto Plantarum, Nova Odessa, pp 120–121 Portuguese

Magalhães PM (2002) Agrotechnology for espinheira-santa. Available in: http://www.cpqba.unicamp.br/plmed/artigos/agroespsant.htm. Portuguese

Mariot MP (2005) Genetics resources of espinheira-santa (*Maytenus ilicifolia* e *M. aquifolium*) in Rio Grande do Sul. Thesis (Doctorate in Plant Breeding) – Postgraduate Course in Agronomy, Federal University of Pelotas. 131

Mariot MP, Barbieri RL (2006) Espinheira-santa: an alternative of production in small properties. Embrapa Clima Temperado, Pelotas, p 30 Portuguese

Mariot MP, Barbieri RL (2007a) Secondary metabolites and medicinal properties of espinheira-santa (*Maytenus ilicifolia* Mart. Ex Reiss. e *M. aquifolium* Mart.). Braz J Med Plants 9(3):89–99 Portuguese

Mariot MP, Barbieri RL (2007b) The associated popular knowledge to use of Espinheira-santa (*Maytenus ilicifolia* e *M. aquifolium*). Braz J Biosci 5(1):666–668 Portuguese

Mendes BG, Machado MJ, Falkenberg M (2006) Glycolipids screening of medicinal plants. Braz J Pharmacogn 16:568–575 Portuguese

Montanari T, Bevilacqua E (2002) Effect of *Maytenus ilicifolia* Mart. on pregnant mice. Contraception 65:171–175

Montanari T, Carvalho JE, Dolder H (1998) Effect of *Maytenus ilicifolia* Mart. Ex Reiss on espermatogenesis. Contraception 57:335–339

Montanari JR, Scheffer MC, Radomski MI (2004) Cultivation of espinheira-santa. In Reis MS, Silva SR (org). Conservation and sustainable use of medicinal and aromatic plants: *Maytenus* spp., espinheira-santa. IBAMA, Brasília, Portuguese, pp 163–180

Mossi AJ, Cansian RL, Carvalho AZ, Dariva C, Oliveira JV, Mazutti M, Filho IN, Echeverrigaray S (2004) Extraction and characterization of volatile compounds in *Maytenus ilicifolia*, using high-pressure CO_2. Fitoterapia 75(2):168–178

Mossi AJ, Mazutti MA, Cansian RL, Oliveira D, Oliveira JV, Dallago R, Leontiev-Orlov O, Treichel H (2010) Chemical variability of volatile organic compounds and semi-volatile native populations *Maytenus ilicifolia*. Quim Nova 33(5):1067–1070. Portuguese

Negrelle RRB, Doni ME, Ohlson OC, Herr S (1999) [Seed production technology espinheira-santa] (*Maytenus ilicifolia* Mart. ex Reiss. – Celastraceae). Rev Bras Sementes 21(1):76–81. Portuguese

Nicoloso FT, Fortunato RP, Zanchetti F, Cassol LF, Eisinger SM (2000) Containers and substrates in the production of seedlings *Maytenus ilicifolia* e *Apuleia leocarpa*. Ciência Rural 30(6):987–992. Portuguese

Niero R, Andrade SF, Filho VC (2011) A Review of the Ethnopharmacology, Phytochemistry and Pharmacology of Plants of the *Maytenus* Genus. Curr Pharm Des 17:1851–1871

Panizza S (1998) Plants that heal, vol 280, 3rd edn. IBRASA, São Paulo. Portuguese

Pereira AMS, Januário AH, Queiroz MEE, Biondo R, França SC (2005) Evaluation of *Maytenus aquifolia* Mart. and *Maytenus ilicifolia* Mart. chemotypes for tannins, total phenols and triterpenes. Braz J Med Plants 8:13–17

Queiroga CL, Silva GF, Dias PC, Possenti A, Carvalho JE (2000) Evaluation of the antiulcerogenic activity of friedelan-3b-ol and friedelin isolated from *Maytenus ilicifolia* (Celastraceae). J Ethnopharmacol 72:465–468

Rattmann YD, Cipriani TR, Sassaki GL, Iacomini M, Rieck L, Marques MCA, Silva-Santos JE (2006) Nitric oxide-dependent vasorelaxation induced by extractive solutions and fractions of *Maytenus ilicifolia* Mart ex Reissek (Celastraceae) leaves. J Ethnopharmacol 104:328–335

Santos VAFFM, Santos DP, Castro-Gamboa I, Zanoni MVB, Furlan M (2010) Evaluation of antioxidant capacity and synergistic associations of Quinonemethide Triterpenes and Phenolic substances from *Maytenus ilicifolia* (Celastraceae). Molecules 15:6956–6973

Santos VAFFM, Regasini LO, Nogueira CR, Passerini GD, Martinez I, Bolzani VS, Graminha MAS, Cicarelli RMB, Furlan M (2012) Antiprotozoal Sesquiterpene Pyridine Alkaloids from *Maytenus ilicifolia*. J Nat Prod 75:991–995

Santos VAFFM, Leite KM, Siqueira MC, Regasini LO, Martinez I, Nogueira CT, Galuppo MK, Stolf BS, Pereira MAS, Cicarelli RMB, Furlan M, Graminha MAS (2013) Antiprotozoal activity of Quinonemethide Triterpenes from *Maytenus ilicifolia* (Celastraceae). Molecules 18:1053–1062

Santos-Oliveira R, Coulaud-Cunha S, Colaço W (2009) Review on *Maytenus ilicifolia* Mart. ex Reissek, Celastraceae. Contribution to the study of the pharmacological properties. Braz J Pharmacogn 19:650–659. Portuguese

Scheffer MC (2001) Mating system and genetic variation between populations and progeny "espinheira-santa" Thesis (Doctorate in Forest Engineering – Forestry). Federal University of Paraná, Curitiba. 104. Portuguese

Scheffer MC (2004) Traditional and current use of species *Maytenus*. In: Reis MS, Silva SR (eds) Conservation and sustainable use of medicinal and aromatic plants: *Maytenus* spp., espinheira-santa. IBAMA, Brasília, pp 53–66. Portuguese

Simões CMO (1989) Plants of folk medicine of Rio Grande do Sul. Ed. Da Universidade/UFRGS, Porto Alegre. Portuguese

Singh B, Dubey MM (2001) Estimation of triterpenoids from *Heliotropium maifolium* Kohen ex Retz *in vivo* and *in vitro*: antimicrobial screening. Phytother Res 15:231–234

Souza LM, Cipriani TR, Iacomini M, Gorin PAJ, Sassaki GLHPLC (2008) ESI-MS and NMR analysis of flavonoids and tannins in bioactive extract from leaves of *Maytenus ilicifolia*. J Pharm Biomed Anal 47:59–67

Souza-Formigoni MLO, Oliveira MGM, Monteiro MG, Silveira-Filho NG, Braz S, Carlini EA (1991) Antiulcerogenic effects of two *Maytenus* species in laboratory animals. J Ethnopharmacol 34:21–27

Teske M, Trentini AMM (1995) Compendium of herbal medicine. Herbarium. 2 ed. 128–129

Tiberti LA, Yariwake JH, Ndjoko K, Hostettmann K (2007) Identification of flavonols in leaves of *Maytenus ilicifolia* and *M. aquifolium* (Celastraceae) by LC/UV/MS analysis. J Chromatogr B 846:378–384

Valladão FN, Miranda RRS, Vale FH, Valladão SA, Silva GDF, Duarte LP, Carvalho-Okano RM, Messias MCTB, Filho SAV (2009) Four Brazilian *Maytenus salicifolia* Reissek (Celastraceae) groups studied by TLC and UV/Vis spectrophotometry. Brazilian. J Pharmacogn 19:733–739

Vargas VM, Guidobono RR, Henriques JA (1991) Genotoxicity of plant extracts. Mem Inst Oswaldo Cruz 86(2):67–70

Vellosa JCR, Khalil NM, Formenton VAF, Ximenes VF, Fonseca LM, Furlan M, Brunetti IL, Oliveira OMMF (2006) Antioxidant activity of *Maytenus ilicifolia* root bark. Fitoterapia 77:243–244

Vieira RF (1999) Conservation of medicinal and aromatic plants in Brazil. In: Janick J (ed) Perspectives on new crops and new uses. ASHS Press, Alexandria, pp 152–159

Zhu N, Sharapin N, Zhang J (1998) Three glucosides from *Maytenus ilicifolia*. Phytochemistry 47(2):265–268

Mikania glomerata Spreng. & Mikania laevigata Sch.Bip. ex Baker

Letícia M. Ricardo and Maria G. L. Brandão

Mikania laevigata Spreng
Photo: Sérgio Bordignon
Available in: http://www.ufrgs.br/fitoecologia/florars/open_sp.php?img=11914

L. M. Ricardo
CEPLAMT, Museu de História Natural e Jardim Botânico & Faculdade de Farmácia, Universidade Federal de Minas Gerais, Belo Horizonte, Brazil

Departamento de Assistência Farmacêutica e Insumos Estratégicos, Secretaria de Ciência, Tecnologia e Insumos Estratégicos, Ministério da Saúde, Brazil

M. G. L. Brandão (✉)
CEPLAMT, Museu de História Natural e Jardim Botânico & Faculdade de Farmácia, Universidade Federal de Minas Gerais, Belo Horizonte, Minas Gerais, Brazil
e-mail: mbrandao@ufmg.br

U. P. Albuquerque et al. (eds.), *Medicinal and Aromatic Plants of South America*, Medicinal and Aromatic Plants of the World 5,
https://doi.org/10.1007/978-94-024-1552-0_30

Abstract Guaco is the popular name of both *Mikania glomerata* Spreng. and *Mikania laevigata* Schultz Bip. ex Baker. Both species are lianas. Their leaves and stems have been used for centuries in Brazil to treat snake bites and respiratory troubles. Studies have validate their bronchodilator and expectorant activities, properties that are commonly associated with the presence of coumarins (1,2-benzopyrone). Other substances identified in the extracts also contribute to the pharmacological effects. Studies have demonstrated also anti-inflammatory, antimicrobial and anti-ulcerogenic activities. Morphological, anatomical and molecular studies recently performed in order to verify the differences between the two species, show a large degree of similarity. These results signalize that both species could be unified in terms of nomenclature.

Keywords Guaco · *Mikania glomerata* · *Mikania laevigata* · Compositae

1 Taxonomic Characteristics

The genus *Mikania* belongs to the Asteraceae Family and tribe Eupatoriae. It has about 430 species, mainly distributed in South America. In Brazil, the genus is represented by approximately 171 species. Two species, *Mikania glomerata* Spreng. and *Mikania laevigata* Sch. Bip. ex Baker are commonly called guaco and used in traditional medicine (Correa 1984).

Synonyms Only *M. glomerata* has synonymus: *Cacalia trilobata* Vell., *M. hederaefolia* DC., *Corynanthelium moronoa* Kunze, *Corynanthelium moronoa* Kunze, *Mikania glomerata* var. *glomerata*, *Mikania glomerata* var. *montana* Hassl., *Mikania scansoria* DC., *Morrenia odorata* Hort. ex D.G. Kuntze, *Willoughbya glomerata* (Spreng.) Kuntze.

2 Crude Drug Used

The National Formulary of Herbal Medicines of the Brazilian Pharmacopoeia, published in 2011 by the Brazilian Health Surveillance Agency, determines the use of dried leaves for *M. laevigata* and *M. glomerata* (Brasil 2011).

3 Major Chemical Constituents and Bioactive Compounds

The chemical composition of *M. glomerata* and *M. laevigata* is very similar (Bolina et al. 2009). They have coumarins, lupeol, volatile oils rich in sesquiterpene and diterpene of kaurane type, β-sitosterol, friedeline, stigmasterol, tanins, flavonoids and saponins (Bertolucci et al. 2013).

Santana et al. (2014) report that most of these compounds have a proven pharmacological activity: lupeol has anti-inflammatory activity; kaurenoic acid is a potential antimicrobial, hypotensive and anti-inflammatory; stigmasterol has antinociceptive, anti-inflammatory and hypocholesterolemic activity. Coumarin as the major chemical marker is responsible for the anti-inflammatory, immunosuppressive, anti-hypertensive and antioxidant effects (Gasparetto et al. 2010).

4 Morphological Description

M. glomerata and *M. laevigata* are lianas sub-woody, perennial, obtuse leaves at the base, almost deltoid shape, dark green, with three ribs highlighted. The morphological and anatomical leaf features indicate substantial similarity between the two species. Molecular data corroborate the morphological data in pointing to the total similarity between the two species observed in the loci used. Based on these results *M. glomerata* and *M. laevigata* could be unified in terms of nomenclature (Bastos et al. 2011).

5 Geographical Distribution

M. glomerata and *M. laevigata* are native to South Brasil, though currently they are cultivated in many other parts of the country. The species are not endemic of Brazil. The confirmed occurrences are Bahia, Espírito Santo, Minas Gerais, Rio de Janeiro, São Paulo, Paraná, Rio Grande do Sul and Santa Catarina. The phytogeographical domain are Cerrado and Mata Atlântica.

6 Ecological Requirements and Collection Practice

"Guaco" has its habitats along river banks, growing spontaneously in primary forests, secondary forests, coppices, edge of forests, alluvial land, wetlands that are frequently subject to flooding. The species has good adaptation ability to domestication and cultivation. The plant is frequented by honey bees during the time of flowering. It reproduces by seed or planting stem cuttings, preferably in sandy and wetland (Czelusniak et al. 2012).

7 Traditional Use (Part(s) Used) and Common Knowledge

Historical data indicate the use of guaco in Brazil for the treatment of snake bites, gout, rheumatism, influenza, as an antipyretic and tonic for decades (Correa 1984). According to recent ethnobotanical studies, the main uses are related to the treatment of disorders of the respiratory system, especially in view of the bronchodilator and expectorant properties (David and Pasa 2015; Messias et al. 2015; Ferrão et al.

2014; Soares et al. 2013; Pasa 2011; Agra et al. 2008; Calábria et al. 2008). These effects appear to be directly related to the coumarins content of the species. Other uses referred to in the special literature are the treatment of rheumatism and nevralgias (Agra et al. 2008; Boscolo and Senna-Valle 2008; Alice et al. 1991), liver detox, nausea and intestine cramps (Coelho-Ferreira 2009).

8 Modern Medicine Based on Its Traditional Medicine Uses

Several studies have confirmed the bronchodilator effect of both species of "guaco" in traditional medicinal use, mainly in the treatment of inflammatory conditions, ulcers, ophidic venom, ulcers (Napimoga and Yatsuda 2010), diarrhea (Salgado et al. 2005), antibacterial and antiparasitic activity, although the efficacy of the antibacterial activity is so far controversial (Napimoga and Yatsuda 2010).

Studies have reported, for example, that extracts of "guaco" act directly causing bronchodilation and smooth muscle relaxation of the respiratory system. This activity is related to the blocking of calcium channels, together with anti-inflammatory actions (Alves et al. 2009; Freitas et al. 2008; Graça et al. 2007). The coumarin seemed to be partially responsible for the bronchodilator activity of the plant through the relaxation of smooth muscle. In addition, aqueous and hydro-alcoholic extracts obtained from *M. glomerata* induced a significant inhibition of the histamine contractions on the isolated guinea-pig trachea (Soares de Moura et al. 2002). In studies evaluating the effect of aqueous and hydro-alcoholic extracts from *M. laevigata* show that the extract produced a dose-dependent relaxation in denuded and intact rat epithelium tracheal, pre-contracted with acetylcholine (Gasparetto et al. 2010). These data support the indication that both *M. glomerata* and *M. laevigata* are useful in treating bronchoconstrictive respiratory diseases.

Another important activity observed in some species is the anti-allergic activity. A fraction obtained from the ethanolic extract used as an anti-allergic and anti-inflammatory agent was evaluated for these properties on ovalbumin-induced allergic pleurisy and in models of local inflammation induced by biogenic amines, carrageenan and PAF. Plasma exudation, as well as neutrophil and eosinophil infiltration evoked by the intrapleural injection of the antigen, were significantly reduced by the plant (Fierro et al. 1999). Guaco extract administered subcutaneously reduces vascular permeability, leukocyte migration and adhesion to inflamed tissues. This anti-inflammatory effect of the herbal medicine may be due to inhibition of pro-inflammatory cytokines at the site of inflammation (Alves et al. 2009). The effects of hydroalcoholic extract of *M. glomerata* and solution of coumarin, undergoing tests in vivo (paw edema) were assessed. A different intensity on pharmacological effects indicates that coumarin has contributed to the pharmacological effect with other chemicals in the extract in a synergic action (Freitas et al. 2008). The analgesic and anti-inflammatory activities of "guaco" tea were also previously observed by evaluating the number of contortions in mice and diffusion of Evans blue dye in the peritoneum (Ruppelt et al. 1991). Napimoga and Yatsuda (2010) affirm that the studies on *M. glomerata M. laevigata* have provided scientific evidence that those plants have a considerable anti-inflammatory therapeutic potential.

The crude hydroalcoholic 70% extract of *M. laevigata* presents antiulcerogenic activity when applied in male Wistar rats decreasing the ulcerative index produced by indomethacin, ethanol, stress and reserpine (Bighetti et al. 2005). In this way, both species of "guaco" show activity in the digestive system.

The antiophidic effect of coumarin present in *M. glomerata* was confirmed with the venom of *Bothrops jararaca* snake and the animal survival rate was higher as compared to 0% in the control group (Pereira et al. 1994). *M. glomerata* root extracts also reduced the hemorrhage zone stimulated by the intradermal injection of *Bothrops* venom by 80% in rats (Maiorano et al. 2005). Mourão et al. (2014) showed that intradermal administration of *Botrhops* venom incubated with the hydroalcoholic extract in rats promoted a significant reduction in the number of inflammatory cells, a marked decrease in edema after the third hour and a significant antihemorrhagic activity.

A study has shown the potential of *M. glomerata* as anti-diarrheal (Salgado et al. 2005). Aqueous extract of leaves (1000 mg/mL) showed a decrease in the propulsive movements of the intestinal contents in mice, in comparison as loperamide, a reference antidiarrheal drug. These findings suggested that the aqueous extract of the leaves of *M. glomerata* might elicit an antidiarrheal effect by inhibiting intestinal motility.

Ushimaru et al. (2012), conducting a study with 14 *E. coli* strains isolated from human specimens, verified that *M. glomerata* shows antagonism with some antibiotics, like cephalotin, cefoxitin, ciprofloxacin, gentamicin, sulphamethoxazole and trimethoprim and tetracycline. Essential oil obtained from leaves of *M. glomerata* showed a strong activity against *Candida albicans* (Duarte et al. 2005). Extracts from *M. glomerata* and *M. laevigata* were also active against different microorganisms, among them *Staphylococcus aureus* (Amaral et al. 2003; Pessini et al. 2003; Holetz et al. 2002).

Dry extracts of guaco may interact synergistically with anticoagulants, like warfarin, as well as certain antibiotics such as tetracyclines, chloramphenicol, gentamycin, penicillin and vancomycin, however, the action mechanism is still unknown (Betoni et al. 2006).

M. glomerata and *M. laevigata* are included in the List of traditional herbal products simplified registration published by the Brazilian Health Regulatory Agency (Anvisa) like expectorant and bronchodilator; thus, the registration of these phytomedicines by industries is facilitated (Brasil 2014a). Currently, phytomedicine containing *M. glomerata* are prepared by different Brazilian pharmaceutical companies and available in the market. Among these, there are simple and compounds syrups and oral solutions without sugar, at different concentrations. In addition, *M. glomerata* and *M. laevigata* are in the National Formulary of Herbal Medicines of the Brazilian Pharmacopoeia, facilitating preparations in pharmacies (Brasil 2011). *M. glomerata*, due to its expectorant and bronchodilator actions, also integrates the National Relation of Essential Medicines in the Unified Health System of Brazil and since 2007 it can be purchased with governmental funds (Brasil 2014b).

9 Conclusions

The species *M. glomerata* and *M. laevigata* are native to Brazil. They are mainly used against respiratory diseases. There are many products made with these species registered in Anvisa and since 2007 they are funded by the Unified Health System of Brazil. In most cases, the species are presented as distinct species, although morphological, anatomical and molecular studies have recently revealed a large degree (total) similarity between the two species and suggest the use of unified terms of nomenclature.

References

Agra MF, Silva KN, Basílio IJLD, Freitas PF, Barbosa-Filho J (2008) Survey of medicinal plants used in the region Northeast of Brasil. Rev Bras Farmacogn 18(3):472–508

Alice CB, Vargas VMF, Silva GAAB, Siqueira NCS, Schapoval EES, Gleye J, Henriques JAP, Henriques AT (1991) Screening of plants used in south Brazilian folk medicine. J Ethnopharmacol 35(2):165–171

Alves CF, Alves VBF, Assis IP, Clemente-Napimoga JT, Uber-Bucek E, Dal-Secco D, Cunha FQ, Rehder VLG, Napimoga MH (2009) Anti-inflammatory activity and possible mechanism of extract from *Mikania laevigata* in carrageenan-induced peritonitis. J Pharm Pharmacol 61(8):1097–1104

Amaral RR, Arcenio-Neto F, Carvalho ES, Teixeira LA, Araújo GL, Sharapin N, Testa B, Gnerre C, Rocha L (2003) Avaliação da atividade IMAO e antibacteriana de extratos de *Mikania glomerata* Sprengel. Rev Bras Farmacogn 13(Suppl 1):24–27

Bastos CL, Mata CS, Maia VH, Borges RAX, Franco LO, Ferreira PCG, Tamaio N (2011) Anatomical and molecular identification of "guaco" Mikania glomerata and *Mikania laevigata* (Asteraceae), two important medicinal species from Brazil. J Med Plant Res 5(18):4579–4583

Bertolucci SK, Pereira AB, Pinto JE, Oliveira AB, Braga FC (2013) Isolation and HPLC quantitation of kaurane-type diterpenes and cinnamic acid derivatives of long-term stored leaves of Mikania laevigata and Mikania glomerata. An Acad Farma Cienc 85(2):473–485

Betoni JEC, Mantovani RP, Barbosa LN, Di Stasi LC, Fernandes Junior A (2006) Synergism between plant extract and antimicrobial drugs used on *Staphylococcus aureus* diseases. Mem Inst Oswaldo Cruz 101(4):387–390

Bighetti AE, Antônio MA, Kohn LK, Rehder VLG, Foglio MA, Possenti A (2005) Antiulcerogenic activity of a crude hydroalcoholic extract and coumarin isolated from *Mikania laevigata* Schultz Bip. Phytomedicine 12(1–2):72–77

Bolina RC, Garcia EF, Duarte MGR (2009) Estudo comparativo da composição química das espécies vegetais *Mikania glomerata* Sprengel e *Mikania laevigata* Schultz Bip. ex Baker. Rev Bras Farmacogn 19(Suppl 1B):294–298

Boscolo OH, Senna-Valle L (2008) Plantas de uso medicinal em Quissamã, Rio de Janeiro, Brasil. Iheringia 63(2):263–277

Brasil (2011) Agência Nacional de Vigilância Sanitária. Formulário de Fitoterápicos da Farmacopeia Brasileira. Anvisa, Brasília, p 126p

Brasil (2014a) Agência Nacional de Vigilância Sanitária. Instrução Normativa n. 2, de 13 de maio de 2014a. Publica a "Lista de medicamentos fitoterápicos de registro simplificado" e a "Lista de produtos tradicionais fitoterápicos de registro simplificado". http://bvsms.saude.gov.br/bvs/saudelegis/anvisa/2014/int0002_13_05_2014.pdf. Acessed in 24 Mar 2015.

Brasil (2014b) Ministério da Saúde. Secretaria de Ciência, Tecnologia e Insumos Estratégicos. Departamento de Assistência Farmacêutica e Insumos Estratégicos. Relação Nacional de Medicamentos Essenciais: Rename, 9th edn. Ministério da Saúde, Brasília 230p

Calábria L, Cuba GT, Hwang SM, Marra JCF, Mendonça MF, Nascimento RC, Oliveira MR, Porto JPM, Santos DF, Silva BL, Soares TF, Xavier EM, Damasceno AA, Milani JF, Rezende CHA, Barbosa AAA, Canabrava HAN (2008) Levantamento etnobotânico e etnofarmacológico de plantas medicinais em Indianópolis, Minas Gerais, Brasil. Rev Bras Pl Med 10(1):49–63

Coelho-Ferreira M (2009) Medicinal knowledge and plant utilization in an Amazonian coastal community of Marudá, Pará State (Brazil). J Ethnopharmacol 126(1):159–175

Correa MP (1984) Dicionário das plantas úteis do Brasil e das exóticas cultivadas, 2nd edn. Imprensa Nacional, Rio de Janeiro

Czelusniak KE, Brocco A, Pereira DF, Freitas GBL (2012) Farmacobotânica, fitoquímica e farmacologia do Guaco: revisão considerando Mikania glomerata Sprengel e Mikania *laevigata* Schulyz Bip. Ex Baker. Rev Bras Planta Med 14(2):400–409

David M, Pasa MC (2015) As plantas medicinais e a etnobotânica em Várzea Grande, MT Brasil. Interações 16(1):97–108

Duarte MCT, Figueira GM, Sartoratto A, Rehder VLG, Delarmelina C (2005) Anti-*Candida* activity of Brazilian medicinal plants. J Ethnopharmacol 97(2):305–311

Ferrão BH, Oliveira HB, Molinari RF, Teixeira MB, Fontes GG, Amaro MOF, Rosa B, Carvalho CA (2014) Importância do conhecimento tradicional no uso de plantas medicinais em Buritis, MG, Brasil. CeN 36:321–334

Fierro IO, Silva ACB, Lopes CS, Moura RS, Barja-Fidalgo C (1999) Studies on the anti-allergic activity of *Mikania glomerata*. J Ethnopharmacol 66(1):19–24

Freitas TP, Silveira PC, Rocha LG, Rezin GT, Rocha J, Citadini-Zanette V, Romão PT, Dal-Pizzol F, Pinho RA, Andrade VM, Streck EL (2008) Effects of *Mikania glomerata* Sprengel. and *Mikania laevigata* Schultz Bip. ex Baker (*Asteraceae*) extracts on pulmonary inflammation and oxidative stress caused by acute coal dust exposure. J Med Food 11(4):761–766

Gasparetto JC, Campos FR, Budel JM, Pontarolo R (2010) *Mikania glomerata* Spreng. e *M. laevigata* Sch. Bip. ex Baker, Asteraceae: estudos agronômicos, genéticos, morfoanatômicos, químicos, farmacológicos, toxicológicos e uso nos programas de fitoterapia do Brasil. Rev Bras Farmacogn 20(4):627–640

Graça C, Baggio CH, Freitas CS, Rattmann YD, Souza LM, Cipriani TR, Sassaki GL, Rieck L, Pontarolo R, Silva-Santos JE, Marques MCA (2007) In vivo assessment of safety and mechanisms underlying in vitro relaxation induced by *Mikania laevigata* Schultz Bip. Ex Baker in the rat trachea. J Ethnopharmacol 112(3):430–439

Holetz FB, Pessini GL, Sanches NR, Cortez DAG, Nakamura CV, Dias Filho BP (2002) Screening of some plants used in the Brazilian folk medicine for the treatment of infectious diseases. Mem Inst Oswaldo Cruz 97(7):1027–1031

Maiorano VA, Marcussi S, Daher MA, Oliveira CZ, Couto LB, Gomes OA, França SC, Soares AM, Pereira PS (2005) Antiophidian properties of the aqueous extract of *Mikania glomerata*. J Ethnopharmacol 102(3):364–370

Messias MCTB, Menegato MF, Prado ACC, Santos BR, Guimarães MFM (2015) Uso popular de plantas medicinais e perfil socioeconômico dos usuários: um estudo em área urbana em Ouro Preto, MG, Brasil. Rev Bras Pl Med 17(1):76–104

Mourão VB, Giraldi GM, Neves LM, Gaspi FO, Rodrigues RA, Alves AA, Esquisatto MA, Mazzi MV, Mendonça FA, Santos GM (2014) Anti-hemorrhagic effect of hydro-alcoholic extract of the leaves of Mikania glomerata in lesions induced by *Bothrops jararaca* venom in rats. Acta Cir Bras 29(1):30–37

Napimoga MH, Yatsuda R (2010) Scientific evidence for *Mikania laevigata* and *Mikania glomerata* as a pharmacological tool. J Pharm Pharmacol 62(7):809–820

Pasa MC (2011) Saber local e medicina popular: a etnobotânica em Cuiabá, Mato Grosso, Brasil. Bol Mus Para Emílio Goeldi Cienc Hum 6(1):179–196

Pereira NA, Pereira BM, Nascimento MC, Parente JP, Mors WB (1994) Pharmacological screening of plants recommended by folk medicine as snake venom antidotes; IV. Protection against jararaca venom by isolated constituents. Planta Med 60(2):99–100

Pessini GL, Holetz FB, Sanches NR, Cortez DAG, Dias Filho BP, Nakamura CV (2003) Avaliação da atividade antibacteriana e antifúngica de extratos de plantas utilizados na medicina popular. Rev Bras Farmacogn 13(Suppl 1):21–24

Ruppelt BM, Pereira EF, Gonçalves LC, Pereira NA (1991) Pharmacological screening of plants recommended by folk medicine as anti-snake venom- I. Analgesic and anti-inflammatory activities. Mem Inst Oswaldo Cruz 86(Suppl 2):203–205

Salgado HRN, Roncari AFF, Moreira RRD (2005) Antidiarrhoeal effects of *Mikania glomerata* Sprengel (Asteraceae) leaf extract in mice. Rev Bras Farmacogn 15(3):205–208

Santana LCLR, Brito MRM, Oliveira GLS, Citó AMGL, Alves CQ, David JP, David JM, Freitas RM (2014) Mikania glomerata: phytochemical, pharmacological, and neurochemical study. Evid Based Complement Alternat Med 2014:1–11

Soares de Moura R, Costa SS, Jansen JM, Silva CA, Lopes CS, Bernardo-Filho M, Nascimento da Silva V, Criddle DN, Portela BN, Rubenich LMS, Araújo RG, Carvalho LCRM (2002) Bronchodilator activity of *Mikania glomerata* Sprengel on human bronchi and guinea-pig trachea. J Pharm Pharmacol 54(2):249–256

Soares NP, Neres AC, Abreu T, Pfrimer GA, Nishijo H, Aversi-Ferreira TA (2013) Medicinal plants used by the population of Goianápolis, Goiás State, Brazil. Acta Sci Biol Sci 35(2):263–271

Ushimaru PI, Barbosa LN, Fernandes AAH, Di Stasi LC, Fernandes Junior A (2012) In vitro antobacterial activity of medicinal plant extracts against *Escherichia coli* from human clinical specimens and interactions with antimicrobial drugs. Nat Prod Res 26(16):1553–1557

Mimosa tenuiflora (Willd.) Poir.

Andrêsa Suana Argemiro Alves, Gilney Charll Santos, and Ulysses Paulino Albuquerque

Mimosa tenuiflora ([Willd.] Poir.)
Photo: G.P. Lewis
Available in: https://www.kew.org/science/tropamerica/neotropikey/families/Leguminosae_(Mimosoideae).htm

A. S. A. Alves · G. C. Santos
Laboratório de Ecologia e Evolução de Sistemas Socioecológicos, Departamento de Botânica, Universidade Federal de Pernambuco, Recife, PE, Brazil

U. P. Albuquerque (✉)
Departamento de Botânica, Centro de Biociências, Universidade Federal de Pernambuco, Recife, Brazil

U. P. Albuquerque et al. (eds.), *Medicinal and Aromatic Plants of South America*, Medicinal and Aromatic Plants of the World 5,
https://doi.org/10.1007/978-94-024-1552-0_31

Abstract The use of *Mimosa tenuiflora* ([Willd.] Poir.) dates back to pre-colonial American civilizations. This plant has been used for a variety of purposes, such as magic-religious rituals; as a medicinal resource with anti-inflammatory, antimicrobial and cicatrization properties; for fence construction and as a fuel. It contains high concentrations of tannins and flavonoids, which are mainly used to treat skin diseases. *M. tenuiflora* also contains N,N-dimethyltryptamine, a tryptamine alkaloid with psychoactive properties, which causes changes in humans' mental states. Therefore, most of the studies on the bioactive compounds of *M. tenuiflora* have focused on its psychoactive actions and on the effectiveness of the flavonoids and tannins for cicatrization.

Keywords Skin tree · Fabaceae · Jurema · Psychotropic plants · Tepezcohuite

1 Taxonomic Characteristics

Mimosa tenuiflora (Willd.) Poir. is popularly known in Latin America as "cabrera," "cabrero," "carbón," "carbonal" (Colombia, Honduras and Venezuela), "calumbi," "jurema," "jurema-preta" (Brazil), "tepescohuite," "tepesquehuite" and "tepezcohuite" (México). "Jurema" is also the common name of several other species of the genus *Mimosa,* family Fabaceae, and even of other plant families. In addition, some uses that are attributed to *M. tenuiflora* are common to the different species also named "jurema" in Brazil.

Synonyms *Acacia hostilis* Mart., *Acacia tenuiflora* Willd., *Mimosa cabrera* Karsten, *Mimosa hostilis* (C. Mart.) Benth., and *Mimosa limana* Rizzini.

2 Crude Drug Use

Different parts of *M. tenuiflora*, including the stem bark, branches and leaves, are used in the pharmacopoeias of different Latin American populations (Albuquerque et al. 2007).

Regarding its magic-religious use, in Northeast Brazil, the indigenous and afro-descendant communities use the roots and branch bark of *M. tenuiflora* to produce "jurema," a beverage used in rituals that has psychoactive properties due to the presence of N,N-dimethyltryptamine, a tryptamine alkaloid (Souza et al. 2008; Gaujac et al. 2013).

3 Major Chemical Constituents and Bioactive Compounds

Due to its wide use, *Mimosa tenuiflora* has been studied extensively. A large number of these studies have focused on the biological activity of tannins and flavonoids, which are responsible for its anti-inflammatory, antimicrobial and cicatrization activities (Bitencourt et al. 2014). In addition, *M. tenuiflora* also contains saponins, chalcones, steroids, terpenoids and indole alkaloids (Anton et al. 1993; Meckes-Lozoya et al. 1990a, b; Camargo-Ricalde 2000).

These compounds have often been studied in *M. tenuiflora* stem and root bark. The different organs of plants have been reported to contain the following steroids and terpenoids: steroid saponins (3-O-β-D-glucopyranosyl campesterol, 3-O-β-D-glucopyranosyl stigmasterol, and 3-O-β-D-glucopyranosyl β-sitosterol), triterpenoid saponins (mimonosides A, B and C), lupeol, campesterol, stigmasterol and β-sitosterol (Meckes-Lozoya et al. 1990b; Jiang et al. 1991; Anton et al. 1993). In addition, indole alkaloids (5-hydroxytryptamine and N,N-dimethyltryptamine) (Souza et al. 2008; Gaujac et al. 2013) and the chalcones Kukulkanin A (2′,4′-dihydroxy-3′,4′dimethoxychalcone) and Kukulkanin B (2′,4′,4′-trihydroxy-3′methoxychalcone) were also identified (Camargo-Ricalde 2000). Other compounds identified in the bark and bast of *M. tenuiflora* include anthocyanins, anthocyanidins, leucoanthocyanidins, catechins, flavones, flavonols, flavanones, flavononols, xanthones and lipids (Camargo-Ricalde 2000; Mucci et al. 2006; Bezerra et al. 2011). Anthocyanins, anthocyanidins, flavonoids, flavonols, flavanones, flavanonols, xanthones, steroids, triterpenoids and saponins have been identified in the leaves (Bezerra et al. 2011). The following flavonoids were identified, in both the leaves and flowers: 6-methoxy-4′-O-methylnaringenin, santin, 6-methoxynaringenin, 5,7,4′-trihydroxy-3,6-dimethoxyflavone, 6-demethoxy-4′-O-methylcapilarisine, 6-methoxykaempferol and tenuiflorin A and C (Bautista et al. 2011).

4 Morphological Description

Mimosa tenuiflora is a shrub that grows to approximately 2–2.5 m high. It has dark branches, aculei and deciduous stipules; the leaves are bipinnate with gland dots on the adaxial side of leaflets. *Mimosa tenuiflora* displays tetramerous, sessile, campanulate, whitish flowers that are 2–2.5 mm in length; spiciform, solitary, axillary, multifloral inflorescences (ca. 150 flowers); and craspedium fruits that are 2.5–4.0 mm length, with four to eight articles (Dourado et al. 2013).

5 Geographical Distribution

Mimosa is one of the most diverse genera within the family Fabaceae. It includes approximately 540 species, of which approximately 500 occur in neotropical regions (Simon et al. 2011). Argentina, Brazil, Paraguay and Uruguay in South America, and Mexico in North America are the centers of diversity of this genus (Barneby 1991; Lewis et al. 2005). *Mimosa* species grow in several environments, from humid to dry forests, open areas such as savannas, deserts and pastures (Simon et al. 2011).

Mimosa tenuiflora is widely distributed in Brazil, Colombia, El Salvador, Honduras, Mexico and Venezuela (Barneby 1991; Santos-Silva and Sales 2010), where it forms large populations in semi-deciduous forests (Rivera-Arce et al. 2007b).

6 Ecological Requirements

Mimosa species are quite diversified and can grow to large populations in semiarid environments with open vegetation. In addition, many *Mimosa* species are considered invasive, such as *M. pigra* L. and *M. tenuiflora*; therefore, this genus is characterized as one of the largest representative genera of invasive plants on the planet (Simon et al. 2011). *M. tenuiflora* may be considered a pioneer species, as it forms large populations in areas with good light availability and semiarid vegetation, including areas with a water deficit and high anthropic pressure (Figueirôa et al. 2006; Diesel et al. 2014). This species also exhibits a high regeneration capacity and fast growth (Carmargo-Ricalde and Grether 1998; Figueirôa et al. 2006; Mattos et al. 2015), indicating that it has significant potential for use in the recovery of degraded areas to avoid erosion, and facilitate the establishment of other plant species (Camargo-Ricalde and García-García 2001; Lucena et al. 2014). In some regions in Mexico, *M. tenuiflora* is considered to be an invasive species in corn (*Zea mays* L.) plantations (Cadena-Iñiguez et al. 2014) and abandoned agricultural areas, mainly due to its high seed production (Camargo-Ricalde and Grether 1998).

7 Collection Practice

The collection of *M. tenuiflora* for medicinal purposes consists of the extraction of the stem bark (Agra et al. 2007; Albuquerque et al. 2007), bast, leaves and flowers (Albuquerque et al. 2007). The stem is also cut to obtain wood to construct fences (Nascimento et al. 2009), and/or to be used as a fuel (Camargo-Ricalde and Grether 1998; Mattos et al. 2015). Despite its high demand for use, its conservation status does not seem to be threatened, as it is present in large populations in the areas where it grows.

8 Traditional Use and Common Knowledge

The use of *M. tenuiflora* dates back to pre-colonial American civilizations, prior to the influence of European colonization on the native American ethnic groups. In Latin America, the indigenous, afro-descendant and rural communities use *M. tenuiflora* for several purposes, such as medicinal and magic-religious purposes, to build fences, as a fuel, and to dye leather and fabrics (Camargo-Ricalde and Grether 1998; Camargo-Ricalde 2000; Albuquerque et al. 2007; Rivera-Arce et al. 2007b). Its medicinal uses include the treatment of bronchitis, cough (Agra et al. 2007; Albuquerque et al. 2007), bruises, inflammations, toothaches, menstrual pain, headaches, hypertension, fever (Almeida et al. 2005; Albuquerque et al. 2007; Cartaxo et al. 2010; Martel-Estrada et al. 2015), skin tumors (Vilarreal et al. 1992), skin diseases (Mucci et al. 2006; Cadena-Iñiguez et al. 2014), gastrointestinal problems (Camargo-Ricalde and Grether 1998), and varicose veins (Rivera-Arce et al. 2007a; Martel-Estrada et al. 2015); it is also used as an antiseptic (Cartaxo et al. 2010).

For some Brazilian populations, the magic-religious use of *M. tenuiflora* is as important as its medicinal uses. Its origins are related to the worship of "jurema preta," a common name of *M. tenuiflora*, and the religious cult associated with it. Within indigenous cultures, these cults originate in the "toré" and "pajelança," which are based on the indigenous structure of the sacred (Rodrigues and Campos 2013). Afro-descendant and indigenous groups, particularly those in the Northeast region of Brazil, use the *M. tenuiflora* roots and branch bark to make a beverage. The result of this preparation is the "vinho de jurema," "jurema," "ajucá" or "anjucá," which has psychoactive properties due to the presence of N,N-dimethyltryptamine, a bioactive alkaloid (Souza et al. 2008; Gaujac et al. 2013). According to the indigenous and afro-descendant groups that use *M. tenuiflora* in their rituals, the worship of "jurema" leads to passage into the spiritual world and to the invocation of spirits that assist in the cure or counseling processes (Mota and Albuquerque 2006). It should be noted that the inclusion of this plant in religious rituals with African origins, resulted probably from the contact between the native Indians from the Brazilian territory and the Africans, who were broght to Brazil with their descendants (Albuquerque and Andrade 2005). In addition, the non-indigenous rural populations have included the "jurema preta" into their pharmacopoeias, mainly due to its cicatrization and anti-inflammatory properties.

In Mexico, the use of *M. tenuiflora* (tepezcohuite) was popularized for its cicatrization properties following a series of catastrophes in the 1980s, namely the Chichonal volcano eruption (1982), a natural gas explosion in San Juan Ixhuatepec (1984), the Mexico City earthquake (1985), and an airplane crash in Toluca (1986), which caused a great number of burns and skin wounds in the region's inhabitants (Camargo-Ricalde 2000; Mucci et al. 2006). However, "tepezcohuite" had been used by the Mayans since pre-Hispanic times for several purposes, including the cure of skin afflictions, such as wounds, burns and ulcers (Mucci et al. 2006; Cadena-Iñiguez et al. 2014).

9 Modern Medicine Based on Its Traditional Medicine Uses

Several studies have focused on the biological activity of *M. tenuiflora*, specifically the flavonoids, on anti-inflammatory conditions. This is the case in the studies by Tellez and Dupoy de Guitard (1990), Zippel et al. (2009), and Shrivastava (2011), who observed the efficacy of *M. tenuiflora* for cicatrization. Mucci et al. (2006) evaluated the therapeutic efficacy of *M. tenuiflora* for the treatment of nipple rhagades and observed that more than 95% of the skin regenerated, with no indications of adverse effects on the nursing women or their infants. Several studies studied the effect of *M. tenuiflora* extracts for their ability to treat eczema and varicose veins, and observed satisfactory cicatrization performance (Tellez and Dupoy de Guitard 1990; Rivera-Arce et al. 2007a; Lammoglia-Ordiales et al. 2012). Martel-Estrada et al. (2015) investigated the osteogenic activity that has been attributed to the plant's cortex, and observed increased osteoblast proliferation and no cytotoxic effects. Some studies have demonstrated that the *M. tenuiflora* tannins have biological activity (Heinrich et al. 1992; Padilha et al. 2010; Bezerra et al. 2011; Siqueira et al. 2012) and bacteriostatic and bactericidal efficacy, indicating the potential use of this species as an antimicrobial agent. The activity of *M. tenuiflora* saponins has also been investigated. Jiang et al. (1992) and Anton et al. (1993), tested the cytotoxicity of the saponins on lymphocytes and lymphoma cells and observed a significant effect on lymphocyte growth and an inhibition of lymphoma cell growth, which resulted from a synergistic effect. The hemolytic action of the *M. tenuiflora* saponin extracts has also been tested (Banerji et al. 1981; Meckes-Loyoza et al. 1990a, b; Heinrich et al. 1992), and *M. tenuiflora* alkaloids were confirmed to inhibit intestinal peristalsis (Meckes-Loyoza et al. 1990a).

Other studies successfully tested the biological activity of *M. tenuiflora* as an antiprotozoal (Muelas-Serrano et al. 2000; Bautista et al. 2011), molluscicide and larvicide (Santos et al. 2012). The *M. tenuiflora* extracts were confirmed to exhibit antimutagenic activity but did not have genotoxic or mutagenic activities (Silva et al. 2013). A possible teratogenic effect of the *M. tenuiflora* extracts was also reported (Gardner et al. 2014). Positive effects in the treatment of chemical dependency (Brierley and Davidson 2012) and serum therapy (Bitencourt et al. 2014) have also been observed.

10 Conclusions

The studies on the biological activity of compounds isolated from *M. tenuiflora* indicate that this species is quite promising for obtaining new anti-inflammatory, antimicrobial and cicatrization drugs. However, it should be noted that most of the studies focused on the biological activity of the tannins and flavonoids present in the bark and bast, which are the plant parts used and known to be efficacious in the traditional medicine. As the extraction of these plant parts has a high regeneration cost for the plant, studies of the bioactive potential of the plant parts with lower regeneration costs are needed. Additionally, other bioactive components, such as steroids, terpenoids, alkaloids and chalcones, should be further investigated.

Acknowledgments We are especially grateful to the National Institute of Science and Technology in Ethnobiology, Bioprospecting and Nature Conservation, certified by CNPq, with financial support from FACEPE (Foundation for the Support of Science and Technology of the State of Pernambuco).

References

Agra MF, Baracho GS, Nurit K, Basílio IJLD, Coelho VPM (2007) Medicinal and poisonous diversity of the flora of "Cariri Paraibano", Brazil. J Pharmacogz 111:383–395

Albuquerque UP, Andrade LHC. As plantas na medicina e na magia dos cultos afro-brasileiros. Tópicos em Conservação, Etnobotânica e Etnofarmacologia de Plantas Medicinais (Albuquerque UP, Almeida CFCBR, Andrade JFM). NUPEEA, Recife, 51–76, 2005

Albuquerque UP, Medeiros PM, Almeida ALS, Monteiro JM, Lins Neto EMF, Melo JG, Santos JP (2007) Medicinal plants of the caatinga (semi-arid) vegetation of NE Brazil: a quantitative approach. J Ethnopharmacol 114:325–354

Almeida CFCBR, e Silva Lima TC, ELC A, MBS M, Albuquerque UP (2005) Life strategy and chemical composition as predictors of the selection of medicinal plants from the caatinga (Northeast Brazil). J Arid Environ 62:127–142

Anton R, Jiang Y, Weniger B, Beck JP, Rivier L (1993) Pharmacognosy of *Mimosa tenuiflora* (Willd.) Poiret. J Ethnopharmacol 38:153–157

Banerji R, Prakash D, Misra G, Nigam SK (1981) Cardiovascular and hemolytic activity of Saponins. Indians Drugs 52:121–124

Barneby CR (1991) Sensitivae censitae. The genus *Mimosa* L. (Mimosoideae) in the new world. N Y Bot Gard 65:1–835

Bautista E, Calzada F, Ortega A, Yépez-Mulia L (2011) Antiprotozoal activity of flavonoids isolated from *Mimosa tenuiflora* (Fabaceae-Mimosoideae). J Mex Chem Soc 55(4):251–253

Bezerra DAC, Rodrigues FFG, Costa JGM, Pereira AV, Sousa EO, Rodrigues OG (2011) Abordagem fitoquímica, composição bromatológica e atividade antibacteriana de *Mimosa tenuiflora* (Willd.) Poiret e *Piptadenia stipulacea* (Benth.) Ducke. Acta Sci Biol Sci. https://doi.org/10.4025/actascibiolsci.v33i1.5366

Bitencourt MAO, Lima MCJS, Torres-Rêgo M, Fernandes JM, Silva-Júnior AA, Tambourgi DV, Zucolotto SM, Fernandes-Pedrosa MF (2014) Neutralizing effects of *Mimosa tenuiflora* extracts against inflammation caused by *Titus serrulatus* scorpion venom. BioMed Res Int, ID 378235, 8 pages https://doi.org/10.1155/2014/378235

Brierley DI, Davidson C (2012) Developments in harmine pharmacology—implications for ayahuasca use and drug-dependence treatment. Prog Neuro-Psychopharmacol Biol Psychiatry 39:263–272

Cadena-Iñiguez P, Cruz-Morales FDC, Ballinas-Albores E (2014) Tepezcohuite (*Mimosa tenuiflora* (L.) Willd) el árbol de la piel. Agroproductividad 7:10–16

Camargo-Ricalde SL (2000) Descripción, distribuición, anatomía, composición química y usos de *Mimosa tenuiflora* (Fabaceae-Mimosoideae) en México. Rev Biol Trop 48(4):939–954

Camargo-Ricalde SL, García-García V (2001) El género *Mimosa* L. (Fabaceae) y la restauración ecológica. Contactos 39:34–42

Camargo-Ricalde SL, Grether R (1998) Germinación, dispersión y estabelecimiento de plântulas de *Mimosa tenuiflora* (Leguminosae) em México. Rev Biol Trop 46(3):543–554

Cartaxo SL, Souza MMA, Albuquerque UP (2010) Medicinal plants with bioprospecting potential used in semi-arid northestearn Brazil. J Ethnopharmacol 131:326–342

Diesel KMF, Costa FSL, Pimenta AS, Lima KMG (2014) Near-infrared spectroscopy and wavelength selection for estimating basic density in *Mimosa tenuiflora* [Willd.] Poiret wood. Wood Sci Technol 48:949–959

Dourado DAO, Conceição AS, Santos-Silva J (2013) O gênero *Mimosa* L. (Leguminosae: Mimosoideae) na APA Serra Branca/Raso da Catarina, Bahia, Brasil. Biota Neotrop 13(4):225–240

Figueirôa JM, Pareyn FGC, Araújo EL, Silva CE, Santos VF, Cutler DF, Baracat A, Gasson P (2006) Effects of cutting regimes in the dry and wet season on survival and sprouting of woody species from the semi-arid caatinga of northeast Brazil. Forest Ecol Manag 229:294–303

Gardner D, Riet-Correa F, Lemos D, Welch K, Pfister J, Panter K (2014) Teratogenic effects of *Mimosa tenuiflora* in a rat model and possible role of N-Methyl- and N,N-Dimethyltryptamine. J Agric Food Chem 62:7398–7401

Gaujac A, Martinez ST, Gomes AA, Andrade SJ, Pinto AC, David JM, Navickiene S, Andrade JB (2013) Application of analytical methods for the structural characterization and purity assessment of N,N-dimethyltrhryptamine, a potent psychedelic agent isolated from *Mimosa tenuiflora* inner barks. Microchem J 109:78–83

Heinrich M, Kuhnt M, Wright CW, Rimpler H, Phillipson JD, Schandelmaier A, Parasitological WDC (1992) Microbiological evaluation of Mixe Indian medicinal plants (Mexico). J Ethnopharmacol 36:81–85

Jiang Y, Massiot G, Lavaud C, Teulon J, Guéchot C, Haag-Berrurier M, Anton R (1991) Triterpenoid glycosides from the bark of *Mimosa tenuiflora*. Phytochemistry 30:2357–2360

Jiang YL, Weniger B, Haag-Berrurier M, Anton R, Beck JP, Italiano L (1992) Effects of saponins from *Mimosa tenuiflora* on lymphoma cells and lymphocytes. Phytoter Res 6:310–313

Lammoglia-Ordiales L, Vega-Memije ME, Herrera-Arellano A, Rivera-Arce E, Agüero J, Vargas-Martinez F, Contreras-Ruiz J (2012) A randomised comparative trial on the use of a hydrogel with tepescohuite extract (*Mimosa tenuiflora* cortex extract-2G) in the treatment of venous leg ulcers. Int Wound J 9(4):412–441

Lewis GP, Schrire B, MacKinder B, Lock M (2005) Legumes of the world. Royal Botanical Gardens, Kew

Lucena RFP, Abreu DBO, Leal JLM, Guerra NM, Leite AP, Ribeiro JES, Ribeiro JPO, Nunes EN, Anselmo MGV, Alves CAB, Sousa Júnior SP, Nunes AT, Souto JS, Carvalho TKN, Sousa RF (2014) Traditional knowledge and use of *Mimosa tenuiflora* (Wild.) Poir. (jurema-preta) in the semi-arid region from Northeastern Brazil. Gaia Scientia 8(1):34–50

Martel-Estrada SA, Rodríguez-Espinoza BR, Santos-Rodrígues E, Jiménez-Vega F, García-Cassillas PE, Martinéz-Pérez CA, Armendáriz IO (2015) Biocompatibility of chitosan/*Mimosa tenuiflora* scaffolds for tissue engineering. J Alloy Comp 643(1):S119–S123

Mattos PP, Braz EM, Domene VD, Sampaio EVSB, Gasson P, Payren FGC, Alvarez IA, Baracat A, Araújo EL (2015) Climate-tree growth relationships of *Mimosa tenuiflora* in seasonally dry tropical forest, Brazil. Cerne 21:141–149. https://doi.org/10.1590/01047760201521011460

Meckes-Lozoya M, Lozoya X, Marles R, Soucy-Breau C, Avalokitesvarasen AJ (1990a) N,N-dimethyltryptamine alkaloid in *Mimosa tenuiflora* bark (Tepescohuite). Arch Invest Med 21:175–177

Meckes-Lozoya M, Lozoya X, Gonzalez J (1990b) Propiedades farmacológicas in vitro de algunos extractos de *Mimosa tenuiflora* (tepescohuite). Arch Invest Med 21:163–169

Mota CN, Albuquerque UP (2006) As muitas faces da jurema: de espécies botânica à divindade afro-indígena, 2ª edn. NUPEEA, Recife

Mucci M, Sciocchetti M, Benvenuti C (2006) Clinical efficacy and safety of *Mimosa tenuiflora* bark extract in the rhagades of the nipple. G Ital Ostetricia Ginecol 23(3):106–114

Muelas-Serrano S, Nogal JJ, Martinéz-Díaz RA, Escario JA, Martínez-Fernández AR, Gómez-Barrio A (2000) In vitro screening of American plant extracts on *Trypanosoma cruzi* and *Trichomonas vaginalis*. J Ethnopharmacol 71:101–107

Nascimento VT, Sousa LG, Alves AGC, Araújo EL, Albuquerque UP (2009) Rural fences in agricultural landscapes and their conservation role in an area of *caatinga* (dryland vegetation) in Northeast Brazil. Environ Dev Sustain 11:1005–1029

Padilha IQM, Pereira AV, Rodrigues OG, Siqueira-Júnior JP, Pereira MS (2010) Antimicrobial activity of *Mimosa tenuiflora* (Willd.) Poir. From northeast Brazil against clinical isolates of *Staphylococcus aureus*. Rev Bras Farmacog 20(1):45–47

Rivera-Arce E, Chávez-Soto MA, Herrera-Arellano A, Arzate S, Agüero J, Feria-Romero IA, Cruz-Guzmán A, Lozoya X (2007a) Therapeutic effectiveness of a *Mimosa tenuiflora* cortex extract in venous leg ulceration treatment. J Ethnopharmacol 109:523–528

Rivera-Arce E, Gattuso M, Alvarado R, Zárate E, Agüero J, Feria I, Loyoza X (2007b) Pharmacognostical studies of the plant drug *Mimosa tenuiflorae* cortex. J Ethnopharmacol 113:400–408

Rodrigues MG, Campos RBC (2013) Caminhos da visibilidade: a ascensão do culto a jurema no campo religioso do Recife. Afro-Ásia 47:269–291

Santos EA, Carvalho CM, Costa ALS, Conceição AS, Moura FBP, Santana AEG (2012) Bioactivity evaluation of plant extracts used in indigenous medicine against the snail, *Biomphalaria glabrata*, and the larvae of *Aedes aegypti*. Evid Based Complement Altern Med. https://doi.org/10.1155/2012/846583

Santos-Silva J, Sales MF (2010) Diversidade e potencial econômico de *Mimosa* L. (Leguminosae-Mimosoideae) em Pernambuco, Brasil. In: Albuquerque UP, Moura AN, Araújo EL (eds) Biodiversidade, potencial econômico e processos ecofisiológicos em ecossistemas nordestinos. NUPEEA, Recife, pp 283–313

Shrivastava R (2011) Clinical evidence to demonstrate that simultaneous growth of epithelial and fibroblast cells is essencial to deep wound healinhg. Diabetes Res Clin Pr 92:92–99

Silva VA, Gonçalves GF, Pereira MSV, Gomes IF, Freitas AFR, Diniz MFFM, Pêssoa HLF (2013) Assessment of mutagenic, antimutagenic and genotoxicity effects of *Mimosa tenuiflora*. Rev Bras Farmacog 23(2):329–334

Simon MF, Grether R, Queiroz LP, Särkinen TE, Dutra VF, Hughes CE (2011) The evolutionary history of *Mimosa* (Leguminosae): toward a phylogeny of the sensitive plants. Am J Bot 98(7):1201–1221

Siqueira CFQ, Cabral DLV, Sobrinho TJSP, Amorim ELC, Melo JG, Araújo TAS, Albuquerque UP (2012) Levels of tannins and flavonoids in medicinal plants: evaluating bioprospecting strategies. Evid Based Complement Altern Med. https://doi.org/10.1155/2012/434782

Souza RSO, Albuquerque UP, Monteiro JM, Amorim ELC (2008) Jurema-preta (*Mimosa tenuiflora* [Willd.] Poir.): a review of its tradicional use, phytochemistry and pharmacology. Braz Arch Biol Technol 51(5):937–947

Tellez PJ, Dupoy de Guitard J (1990) Pharmaceutical preparation containing *Mimosa tenuiflora* extract with skin-regenerating properties. Patent Eur Pat Appl 349:469

Vilarreal ML, Alonso D, Melesio G (1992) Cytotoxic activity of Mexican plants used in traditional medicine. Fitoterapia 43:518–521

Zippel J, Deters A, Hensel A (2009) Arabinogalactans from *Mimosa tenuiflora* (Willd.) Poiret bark as active principles for wound-healing properties: specific enhancement of dermal fibroblast activity and minor influence on HaCaT keratinocytes. J Ethnopharmacol 124:391–396

Oxalis adenophylla Gillies ex Hook. & Arn.

Juan J. Ochoa and Ana Haydeé Ladio

Oxalis adenophylla Gillies ex Hook. & Arn.
Photo: David Stang
Available in: http://www.tropicos.org/Image/100117061

J. J. Ochoa
Instituto de Investigaciones en Diversidad Cultural y Procesos de Cambio – CONICET
Universidad Nacional de Río Negro, San Carlos de Bariloche, Río Negro, Argentina

A. H. Ladio (✉)
Instituto de Investigaciones en Biodiversidad y Medioambiente – CONICET – Universidad Nacional del Comahue, San Carlos de Bariloche, Río Negro, Argentina

U. P. Albuquerque et al. (eds.), *Medicinal and Aromatic Plants of South America*, Medicinal and Aromatic Plants of the World 5,
https://doi.org/10.1007/978-94-024-1552-0_32

Abstract *Oxalis adenophylla* Gillies ex Hook. & Arn., is a native herb of the Subantartic Forest and Stepps of Patagonia. *O. adenophylla* has multiple uses and a high cultural value for many local populations of Patagonia. The leaves are used to treat fever, their roots are edible, and the plant is employed as ornamental. Despite their local cultural importance as medicine, there is little understanding of the phytochemistry and bioactivity of its property to treat fever, and the nutritional characteristics of its edible root. Similarly, its ornamental potential and growing marketing have not been investigated in the region. The ecological knowledge of local populations, that have historically used and currently use this species, seems to be essential to promote the sustainable management and conservation of *O. adenophylla* in Patagonia.

Keywords Multi-purpose native plant · Treatment for fever · Patagonia · Cuye

1 Taxonomic Characteristics

Oxalis L. (Oxalidaceae) is a cosmopolitan genus of about 500 species distributed in three centers of abundance. The largest of these centers is located in South America, with more than half of the species and the largest morphological variation, ranging from herbs to shrubs (Lourteig 1994). Based on the characteristics of the leaves, *Oxalis* can be divided into four subgenera (Lourteig 2000): *Oxalis*, *Monoxalis*, *Trifidus*, and *Thamnoxys* (Lourteig 1994). The *Oxalis* sub-genus is characterized by the presence of leaves with multiple sub-sessile leaflets, divided into 19 sections. *Oxalis adenophylla* Gillies ex Hook. & Arn. is found within the Palmatifoliae section of de Candolle (1824), which differs from other sections in that it includes stemless or short bare stemmed species, with stalked palmate leaves, with 5–13 leaflets and no glands.

From partial molecular phylogenies leaves, the species belonging to Palmatifoliae have been identified as monophyletic (Heibl and Renner 2012). There are five more species besides *O. adenophylla*: *O. enneaphylla*, *O. enneaphylla* subsp. *ibari* (Philippi 1879; Lourteig 1988), *O. laciniata*, *O. loricata*, *O. squamoso-radicosa* and *O. morronei* (López and Múlgura 2011).

O. adenophylla is clearly distinguishable from other species of this section by the presence of pseudo-bulbs, lack of nurturing scales and the presence of bi-floral tops.

Synonyms *Acetosella adenophylla* (Gillies ex Hook. & Arn.) Kuntze; *Acetosella bustillosii* (Phil.) Kuntze; *Oxalis bustillosii* Phil., *Oxalis bustillosii* Phil. var. *biflora*

2 Major Chemical Constituents and Bioactive Compounds

From the phytochemical and pharmacological point of view, studies on several species of the genus *Oxalis* highlight its potential as a source of antioxidants, antitumor and antidiabetic compounds (Kathiriya et al. 2010; Sircelj et al. 2010; Agila 2012);

benzoquinones and phenols with bactericidal properties (Feresin et al. 2003) and which inhibit pigmentation of the skin (Huh et al. 2010). Other studies have addressed the nutritional content of *O. tuberosa*, a species of importance in the economy and diets of American populations (Repo-Carrasco Valencia 2011). We can highlight the presence of coating, a reserve protein with antimicrobial properties (Flores et al. 2002), and a high concentration of digestible amino acids and sugars (Hodge 1957). For this reason, it is considered a nutraceutical food (Campos et al. 2006).

Chemical compounds and bioactivity of *O. adenophylla* have been poorly studied. The only study on its bioactivity was a test of the inhibitory effect of enzyme acetilcolinestarese (Rhee et al. 2003). In contrast, there has been greater effort spent in the study of other Patagonian species such as *O. rosea* (Schmeda-Hirschmann et al. 1992; Rodriguez et al. 1994; Inzunza and Aballay 1995) and *O. erythrorryza* (Feresin et al. 2003).

3 Morphological Description

The genus *Oxalis* shows high variability in their vegetative characters, and usually are annual or perennial herbs with underground structures like rhizomes, corms, tubers, tuberous roots and bulbs (Salter 1944). *O. adenophylla* is an herb that grows in the form of 4–15 cm pads (Fig. 1), with one or more ob-triangular roots (Fig. 2) and fibrous branches. It presents pseudo-bulbs (Fig. 3) consisting of a vertical, hollow rhizome of 20 × 5 mm, covered with linear protective scales of 1–3.5 × 10–35 mm: reddish brown, membranous, of sharp apex, with a densely ciliated margin, and undulating cilia of up to 5 mm. Stipules of 3–7 × 1–2 mm, fully adnate to the petiole, narrowing towards the apex, reddish, hyaline, glabrous to pubescent on both surfaces. Petioles of 4–15 cm, glabrous (or barely pubescent). Eight to 12 leaflets, up to 8 × 8 mm, incised 1/6–2/3, divergent lobes, unequal, oblong, hairless (or with

Fig. 1 General aspect of the aerial parts of *Oxalis adenophylla* Gillies ex Hook. & Arn

Fig. 2 Edible triangular tuberous root of *Oxalis adenophylla*

Fig. 3 Pseudo-bulb of *Oxalis adenophylla*

few fine and wavy trichomes), calluses sometimes present. Inflorescences bear 1 (−2) flowers, stalks up to 15 cm, glabrous; bracts 0.3–2.5 × 9.3 mm; 4–10 mm peduncles; bracteoles of up to 5 × 1 mm; sepals broadly ovate, 4–8 × 2.5–5 mm, moderately uneven, acuminate or acute, rarely obtuse, ciliated apex, undulating cilia, sometimes with calluses. Flowers up to 45 mm in diameter, petals obovate or spatula, pink to purple, white at the base, veins and throat purple, unguiculate base, margin finely ciliated at the apex. Its fruit is a globose capsule of 6–7 mm diam.; glabrous or with simple glandular trichomes; with pubescent carpels inside; with 1 or 2 seeds. Asymmetric ellipsoids seeds, ±2 × 1 mm, of ocher color (Lourteig 1994).

4 Geographical Distribution

In America, the genus *Oxalis* is distributed in mountainous areas of Patagonia and the Northeastern United States. *O. adenophylla* is endemic to the sub-Antarctic Forests and the Patagonian Steppes. In Argentina it is distributed from southern Mendoza to Santa Cruz, and in Chile from the IV Metropolitan Region. In addition to this distribution in the wild, we should consider that the species is also cultivated and sold for ornamental purposes in North America and Europe (Ochoa and Ladio 2014; Ochoa 2015).

5 Ecological Requirements

O. adenophylla can grow from sea level to 2600 m. It grows in arid environments (Fig. 4) with slopes of between 0 and 60°, in soils ranging from completely bare to soils with 60% of covering (Ochoa 2015). In the environment of the Patagonian steppe, it is usually found in herbaceous-xerophytic plant communities, or in herbaceous-shrub communities dominated by *Mulinum spinosum*, and grasses of the genus *Pappostipa*. In high mountain environments, it is part of stony plant communities along species such as *Oreopulus gracilis*, *Stipa, Poa, Sisyrinchium, Phacelia, Pozoa, Cerastium, Mulinum, Oreopulus, Leucheria,* and *Rhodophiala*, among others. In ecotone areas, it can be found between patches of cypress forests (*Austrocedrus chilensis*), and laura scrubs (*Schinus patagonicus*), radal (*Lomatia hirsuta*), maitén (*Maytenus boaria*); and associated with species of the genera *Mulinum*, *Geranium*, *Lathyrus*, *Euphorbia*, *Balbisia* and *Stipa.* and other shrubs typical of the Andean Patagonian forests.

Regarding its phenology, this geophyte usually emerges in November, blooms during the month of December, and disappear during March (Ochoa 2015). While there are no specific studies related to the biology of this species, the characteristics of its flowers, fruits and seeds indicate that pollination is entomophilous, and its dispersion is of the bacoría type (by force of gravity). Regarding the interaction with domestic and wild animals, Ochoa and Ladio (2014) documented, from the perspective of local people, that this species is not preferred by sheep and goats, which can occasionally consume its leaves, while its root is consumed by wild boar (*Sus scrofa*).

O. adenophylla has not been evaluated in the reports of the International Union for Conservation of Nature and Natural Resources (IUCN) and is not included in the CITES red list of endangered plants. The species is represented in different protected areas of Argentina (Lanin National Park, Nahuel Huapi NP,

Fig. 4 Environment where *Oxalis adenophylla* grows in Mountains of Patagonia

Lago Puelo NP, Los Alerces NP; and Los Glaciares Perito Moreno PN). In the case of Chile, we could not access data on their status in the system of protected areas of the country.

Surveys on the state of conservation of this species in the northwest of Patagonia Argentina (which consider variables such as the degree of endemicity, the area of distribution, ecological amplitude, coverage, the slope of the environments where it grows, livestock use, and the type and intensity of human use) have shown that populations of *O. adenophylla* have a medium risk index value (Ochoa and Ladio 2014). The steep slopes where this species grows seem to be the variable that most contributes to increasing its risk value. On the other hand, the low frequency and intensity of medicinal use by the residents, the existence of local rules governing its extraction, and local cultivation practices could encourage the conservation of this species in the areas under study (Ochoa and Ladio 2014).

6 Traditional Uses and Common Knowledge

Different species of *Oxalis* have cultural and economic value for their ornamental attributes (e.g.: *O. articulata, O. corymbosa, O. Boweiana, O.adenphylla*) (von Hentig 1995; Ochoa and Ladio 2014); its potential as a source of food coloring (*O. triangularis*) (Alexandra et al. 2001); its edible leaves (e.g.: *O. acetosella*; *O corniculata, O. stricta, O. adenophylla, O. valdiviensis*) (Zennie and Ogzewalla 1977; Rapoport et al. 2003; Sircelj et al. 2010; Jain et al. 2010); its edible tubers (*O. tuberousa, O. adenophylla*) (National Research Council 1989; Rapoport et al. 2003); for acting as an invader of native ecosystems (e.g.: *O. crassipes, O. valdiviensis, O. micrantha*) (Doust et al. 1985; Rottenberg and Parker 2004); and for being part of local pharmacopoeia (e.g.: *O. corniculata, O crassipes, O. triangularis, O. rosea, O.adenophylla, O. valdiviensis*) (Anonymous 1996; Leonard 2010; Molares and Ladio 2009).

In the Patagonia region several rural communities collect native species of *Oxalis* for food and/or medicinal purposes (Ladio 2001, 2011; Molares and Ladio 2008; Ochoa and Ladio 2014). Of the 18 species of the genus in Patagonia Argentina, at least seven have ethnobotanical history: *O. valdiviensis* are used as a substitute for lemon (Martinez-Crovetto 1980; Ladio 2006; Rapoport et al. 2003) and as analgesic (Conticello et al. 1997). *O. erythroriza* is used for heart and liver problems (Saúde-Guimarães and Farias 2007). *O. rosea* is used as an emmenagogue, abortive, to treat symptoms of fever (Montecino and Conejeros 1985), for cough, scurvy and dull sight (Houghton and Manby 1985), and also for its edible stems (Villagrán et al. 1983). *O. lobata* has carminative properties, and *O. perdicaria* has been used for its edible bulbs (Houghton and Manby 1985). *O. nahuelhuapensis* and *O. adenophylla* are used as antipyretic and for their edible roots (Ochoa et al. 2010).

The analysis of historical documents attesting to the use of *O. adenophylla* in the region shows its first recording in the twentieth century. However, the vulgar word "culle" is mentioned in previous documents. And, considering the fragmentary nature of historical sources, and the generic nature of the common name, it is likely that the species referred to by this term, and identified as *O. rosea* in documents of the sixteenth century (Ochoa and Ladio 2011), may also include *O. adenophylla*, which has similar morphological features.

In the Patagonia Argentina they have been documented its use in the provinces of Chubut (Molares and Ladio 2009; Ochoa and Ladio 2014), Río Negro (Ochoa et al. 2010) and Neuquén (Ladio 2001; Duzevich 2011; Ochoa and Ladio 2014); mainly in towns located in ecotonal or mountainous areas. The absence of ethnobotanical data in the provinces of Santa Cruz and Mendoza, where these species grow, must be due more to lack of ethnobotanical efforts in these regions than to the absence of local applications. In the Chilean case, no ethnobotanical records of this species have been found.

O. adenophylla is popularly known by the name of *culle*, *cuye*, *uyi*, *cuye colorado*, and *vinagrillo*, among others. From the phytonymic and ethno-taxonomic point of view the name "culle" comes from the indigenous term *kulle* or *kulli* (Febres 1846) and represents a similar class to genus, encompassing several species in the region (Villagrán 1998). Although it is often named simply as "cuye", in various populations, it is often distinguished from other species of the genus by the use of compound nouns. For example, in the rural population of Arroyo Las Minas, it is known as "red cuye" or "true cuye", while the simple name of cuye is usually applied to *O. nahuelhuapensis*, a species less preferred but used for the same purposes as *O. adenophylla* (Ochoa et al. 2010).

O. adenophylla is mainly reported to have analgesic and anti-inflammatory action, (Estomba et al. 2006). Among the published ethnobotanical reports, the medicinal and edible uses are noteworthy. On the one hand, it has reputed antipyretic properties associated with flu-like conditions (Ladio 2001; Molares and Ladio 2012). To this end, its leaves are collected before flowering between the months of October and December. The practice of its use consists of the selection of

Fig. 5 "Tortilla" of *Oxalis adenophylla* (leaves compacted on a wooden board or wood stove, forming an omelet that is stored in a dark and dry place) used to treat cases of fever

an exemplary of good size and the collection of leaves with a knife. The organs are then compacted on a wooden board or wood stove, forming an omelet that is stored in a dark, dry place for the winter (Fig. 5). In cases of fever, caused primarily by flu states, some of these tortillas are consumed. The dry parts (one teaspoon of leaves) are rehydrated in boiling water, or in some cases, directly from the fresh produce (Ladio 2001; Ochoa and Ladio 2014). Some families take the tea of *culle* with cachanlahue culle (*Centaurium cachanlahuen*) to enhance this anti-fever action (Ladio et al. 2007). Others also use the red *culle* with an aspirin and lemon juice to enhance its action (Igon et al. 2007). It has also been said to cleanse the kidneys, and to be effective against nosebleeds and menstrual problems (Ladio et al. 2007). Among the residents, its use in people with kidney problems and pregnant women is not recommended (Igon et al. 2007).

Additionally, the leaves of the plant are eaten raw for their acidic taste, like lemon, or sour, according to different informants. This organoleptic criterion appears to be key in its recognition and use (Molares and Ladio 2008). There is also recorded use of the juice of its leaves, mixed with sugar and water, as a refreshing drink similar to lemonade (Muñoz et al. 1981; Rapoport et al. 2003). Another recorded instance is the occasional use of the tuberous root that this plant develops (Fig. 2) (Ochoa et al. 2010; Ochoa and Ladio 2014). It is an activity carried out in specific contexts, during traditional activities such as searching for the sheep, goats and horses, gathering medicinal plants or fire woods, and recreational childhood activities (Ochoa and Ladio 2014). In these contexts, "large plants" are dug out, which develop this white colored root, and are consumed as a snack in the place of harvest.

Finally, it is used as an ornamental plant because of its beautiful flowers, easy reproduction and its non-invasive characteristics. It is used for landscape design (Seydouglu et al. 2009) and as a potted plant (Van Leeuwen 1991; Armitage et al. 1996). To this end, the plant is sold incipiently in Argentina (*pers. com.*), increasingly in Chile, and most commonly in horticultural circles in Europe and the United States, where you can buy its bulbs and seeds (e.g. www.bulbsdirect. com, www.rhs. org.uk, among others). In the rural town of Villa Llanquín, it has been documented that some people, inspired by the beauty of its flowers, transplant it from wild populations to home gardens or around houses (Fig. 6). They protect and take care of these specimens in the manner of domesticated plants (Ochoa and Ladio 2014).

Fig. 6 Individual plants of *Oxalis adenophylla* transplant from wild populations to home gardens or around houses

7 Conclusions

O. adenophylla is a wild species with multiple uses and a high cultural value for many local populations of Patagonia. It is most valued and extended, locally and regionally. Its value is in the use of its leaves in the preparation of a febrifuge remedy. On the other hand, the knowledge and use of its edible root, as well as the appreciation and use of this species for its ornamental qualities, seems to be more restricted in rural populations and there is little knowledge of it in urban populations of the region. Despite all this, it has a growing commercialization for ornamental purposes, in cities such as El Bolsón and Bariloche (Rio Negro, Argentina) (pers. com.), as well as increasing cultivation and commercial exploitation, in Europe and the United States.

Despite these multiple values (properties) of *O. adenophylla*, there is little understanding of the phytochemistry and bioactivity of this valuable species, its property to treat fever, and the nutritional characteristics of its edible root. Similarly, its ornamental potential and its growing marketing have not been investigated in the region, in order to account for its reproduction in nurseries. On the other hand, there are studies in other countries that document, for example, the influence of cold storage in its underground organs and its moisture regimes in flowering (Armitage et al. 1996).

The available ethnobotanical data suggest that in future the plant could be subject to greater use and related market pressures, so it would be essential to elaborate a research plan to deepen our understanding of the chemical, pharmacological and nutritional aspects of this species. Similarly, more knowledge is needed on the ex-situ cultivation practices that ultimately could favor its production for commercial purposes. Considering the ecological knowledge of local populations, that have historically used and currently use this species, it is also essential to promote the sustainable management and conservation of *O. adenophylla* in Patagonia.

References

Agila KN (2012) Antidiabetic, antihyperlipidaemic and antioxidant activity of *Oxalis corniculata* in alloxan induced diabetic mice. J Nat Sci Res 2(7):9–17

Alexandra PE, Monica GM, Wrolstad RE, Gloria MBA (2001) Anthocyanins from *Oxalis triangularis* as potential food colorants. Food Chem 75:211–216

Anonymous (1996) Pharmacopiea of India, 3rd edn. Govt. of India, New Delhi Ministry of Health and Hamily Welfare, New Delhi

Armitage AM, Copeland L, Gross P, Green M (1996) Cold storage and moisture regime influence flowering of *Oxalis adenophylla* and *Ipheionuniflorum*. Hort Sci 31(7):1154–1155

Campos D, Noratto G, Chirinos R, Arbizu C, Roca W, Cisneros-Zevallos L (2006) Antioxidant capacity and secondary metabolites in four species of Andean tuber crops: native potato (*Solanum* sp.), mashua (*Tropaeolum tuberosum* Ruiz & Pavon), Oca (*Oxalis tuberosa* Molina) and ulluco (*Ullucus tuberosus* Caldas). J Sci Food Agric 86(10):1481–1488

Conticello L, Gandullo R, Bustamante A, Tartaglia C (1997) El uso de plantas medicinales por la comunidad Mapuche de San Martín de los Andes, provincia de Neuquén (Argentina). Parodiana 10(1–2):165–180

de Candolle AP (1824) Oxalideae. In: de Candolle AP (ed) Prodromus systematis naturalis regni vegetabilis, vol 1. Treutell & Würst, Paris, pp 689–702. https://doi.org/10.5962/bhl.title.286

Doust LL, MacKinnon A, Dousp JL (1985) Biology of Canadian Weeds: 71. *Oxalis stricta* L., *O. corniculata* L., *O. dillenii* Jacq. ssp. *dillenii* and *O. dillenii* Jacq. ssp. *filipes* (Small) Eiten. Can J Plant Sci 65(3):691–709

Duzevich S (2011) Plantas patagónicas medicinales empleadas por poblaciones mapuches para afecciones de la piel. Hort Arg 30(73):481

Estomba D, Ladio AH, Lozada M (2006) Medicinal wild plant knowledge and gathering patterns in a Mapuche community of North-western Patagonia. J Ethnopharmacol 103:109–119

Febres A (1846) Diccionario Hispano Chileno. Imprenta del Progreso Santiago de Chile, Santiago

Feresin GE, Tapia A, Sortino M, Zacchino S, de Arias AR, Inchausti A, Schmeda-Hirschmann G (2003) Bioactive alkyl phenols and embelin from *Oxalis erythrorhiza*. J Ethnopharmacol 88(2):241–247

Flores T, Alape GA, Flores DM, Flores HE (2002) Ocatin. A novel tuber storage protein from the andean tuber crop oca with antibacterial and antifungal activities. Plant Physiol 128:1291–1302

Heibl C, Renner SS (2012) Distribution models and a dated phylogeny for Chilean *Oxalis* species reveal occupation of new habitats by different lineages, not rapid adaptive radiation. Syst Biol 61(5):823–834

Hodge WH (1957) Three native tubers of the high Andes. Econ Bot 5:185–201

Houghton PJ, Manby J (1985) Medicinal plants of the Mapuche. J Ethnopharmacol 13:89–103

Huh S, Kim YS, Jung E, Lim J, Jung KS, Kim MO, Park D (2010) Melanogenesis inhibitory effect of fatty acid alkyl esters isolated from *Oxalis triangularis*. Biol Pharm Bull 33(7):1242–1245

Igon P, Ladio A, Lozada M. (2007) Plantas Medicinales utilizadas en las Comunidades de Villa Traful y Cuyín Manzano. Ediciones Imaginaria. Bariloche

Insunza V, Aballay E (1995) Evaluation of 16 plants with nematicidal properties as hosts for *Xiphinema americanum* sensu lato in Chile. Investig Agr (Chile) 15(1–2):39–42

Jain AK, Tiwari P, Bashir M (2010) Nutritive aspects of *Oxalis corniculata* L. used by tribals of Central India during scarcity of food. Bot Res Int 3:35–37

Kathiriya A, Das K, Kumar EP, Mathai KB (2010) Evaluation of antitumor and antioxidant activity of *Oxalis Corniculata* Linn. against ehrlich ascites carcinoma on mice. Iran J Cancer Prev 3:157–165

Ladio AH (2001) The maintenace of wild edible plant gathering in a Mapuche community of Patagonia. Econ Bot 55(2):243–254

Ladio AH (2006) Uso y conservación de plantas silvestres con órganos subterráneos comestibles en comunidades Mapuche de la estepa patagónica argentina. In: Alburquerque UP, Andrade Maris JF, Almeida CBR (eds) Tópicos em conservação e etnobotánica de plantas comestíveis. Universidade Federal Rural de Pernambuco, Recife, pp 53–72

Ladio AH (2011) Underexploited wild plant foods of North-Western Patagonia. In: Filipi R (ed) Multidisciplinary approaches on food science and nutrition for the XXI century. Transworld Research Network, Kerala, pp 1–16

Ladio AH, Lozada M, Weigandt M (2007) Comparison of traditional wild plant knowledge between aboriginal communities inhabiting arid and forest environments in Patagonia. Argentina J Arid Environ 69:695–715

Leonard DB (2010) Medicine at your feet: healing plants of the Hawaiian kingdom, vol 1. Roast Duck Productions, Richmond

López A, Múlgura ME (2011) A new species of Oxalis section Palmatifoliae (Oxalidaceae) from southern Argentina. Phytotaxa 33:41–45

Lourteig A (1988) Oxalis. In: Correa MN (ed) Flora Patagónica, vol 8(5). Colección Científica del Instituto Nacional de Tecnología Agropecuaria, Buenos Aires, pp 1–29

Lourteig A (1994) Oxalis L. subgénero Thamnoxys (Endl.) Reiche emend. Lourt. Bradea 7:1–199

Lourteig A (2000) Oxalis L. subgéneros Monoxalis (Small) Lourt., Oxalis y Trifidus Lourt. Bradea 7:201–629

Martínez-Crovetto R (1980) Apuntes sobre la vegetación de los alrededores del Lago Cholila. Publicación Técnica:1–22

Molares S, Ladio AH (2008) Plantas medicinales en una comunidad Mapuche del NO de la Patagonia Argentina: clasificación y percepciones organolépticas relacionadas con su valoración. Bol Latinoam Caribe Plant Med Aromat 7(3):149–155

Molares S, Ladio AH (2009) Ethnobotanical review of the Medicinal Mapuche Flora: use patterns on a regional scale. J Ethnopharmacol 122:251–260

Molares S, Ladio AH (2012) Plantas aromáticas con órganos subterráneos de importancia cultural en la Patagonia Argentina: una aproximación a sus usos desde la etnobotánica, la percepción sensorial y la anatomía. Darwiniana 50:7–24

Montecino S, Conejeros A (1985) Mujeres Mapuches. El saber tradicional en la curación de enfermedades comunes. Centro de estudios de la mujer, Santiago de Chile

Muñoz MS, Barrera E, Meza I (1981) El uso medicinal y alimenticio de plantas nativas y naturalizadas en Chile. Publicación Ocasional del Museo de Historia Natural de Chile, 91 p

National Research Council (1989) Lost crops of the Incas: little known plants of the Andes with promise for worldwide cultivation. National Academy Press, Washington, DC

Ochoa JJ (2015) Uso de plantas silvestres con órganos de almacenamiento subterráneos comestibles en la Patagonia: perspectivas etno-ecológicas. Tesis doctoral. Universidad Nacional de Comahue, Argentina

Ochoa JJ, Ladio AH (2011) Pasado y presente del uso de plantas silvestres con órganos subterráneos de almacenamiento comestibles en Patagonia. Bonplandia 20(2):265–289

Ochoa JJ, Ladio AH (2014) Ethnoecology of *Oxalis adenophylla* Gillies ex Hook. & Ar. J Ethnopharmacol 155(1):533–542

Ochoa JJ, Ladio AH, Lozada M (2010) Uso de recursos herbolarios entre mapuches y criollos de la comunidad campesina de Arroyo Las Minas (Río Negro, Patagonia Argentina). Bol Latinoam Caribe Plant Med Aromat 9(4):269–276

Philippi RA (1879) Relación de los estudios hechos en el Estrecho de Magallanes i la Patagonia Austral. Anu Hidrográfico Mar Chile 5:24–25

Rapoport EH, Ladio AH, Sanz HA. (2003) Plantas nativas comestibles de la Patagonia andina: argentino/chilena. Parte 1. Ed. Imaginaria. San Carlos de Bariloche

Repo-Carrasco Valencia R (2011) Andean indigenous food crops: nutritional value and bioactive compounds. University of Turku, Turku

Rhee IK, Appels N, Luijendijk T, Irth H, Verpoorte R (2003) Determining acetylcholinesterase inhibitory activity in plant extract using a fluorimetric flow assay. Phytochem Anal 14:145–149

Rodriguez J, Loyola JI, Maulen G, Schmeda-Hirschmann G (1994) Hypoglycaemic activity of *Geranium core-core*, *Oxalis rosea* and *Plantago major* extract in rats. Phytother Res 8(6):372–374

Rottenberg A, Parker JS (2004) Asexual populations of the invasive weed *Oxalis pes–caprae* are genetically variable. Proc Royal Soc London B: Biol Sci 271(4):206–208

Salter TM (1944) The genus Oxalis in South Africa: a taxonomic revision. S Afr J Bot 31:574–598

Saúde-Guimarães DA, Faria AR (2007) Substâncias da natureza com atividade anti-*Trypanosoma cruzi*. Rev Bras Farmacog 17:455–465

Schmeda-Hirschmann G, Loyola J, Sierra J, Retamal R, Rodríguez J (1992) Hypotensive effect and enzyme inhibition activity of Mapuche medicinal plants extracts. Phytother Res 6:184–188

Seydouglu N, Zencirkiran M, Yahya A (2009) Position and application areas of geophytes within landscape design. Afr J Agr Res 4(12):1351–1357

Šircelj H, Mikulič-Petkovšek M, Batič F (2010) Antioxidants in spring leaves of *Oxalis acetosella* L. Food Chem 123(2):351–357

Van Leeuwen PJ (1991) *Oxalis adenophylla* goed to forcerenal spotplant. Bloem-bollencultuur 102:30–31

Villagrán C (1998) Etnobotánica indígena de los bosques de Chile. Sistema de clasificación de recursos de uso múltiples. Rev Chile Hist Nat 71:245–268

Villagrán C, Meza I, Silva E, Vera N (1983) Nombres folklóricos y usos de la flora de la Isla Quinchao, Chiloé. Publicación Ocasional 39:1–58

Von Hentig WU (1995) The development of “new ornamental plants” in Europe. Acta Hort 397:9–29

Zennie TM, Ogzewalla D (1977) Ascorbic acid and vitamin A content of edible wild plants of Ohio and Kentucky. Econ Bot 31(1):76–79

Phyllanthus niruri L.

Valdir Cechinel Filho

Phyllanthus niruri L.
Photo: G. A. Parada
Available in: http://www.tropicos.org/Image/100168185

V. Cechinel Filho (✉)
Programa de Pós-Graduação em Ciências, Farmacêuticas e Núcleo de Investigações Químico-Farmacêuticas (NIQFAR), Universidade do Vale do Itajaí (UNIVALI), Itajaí, SC, Brazil
e-mail: cechinel@univali.br

U. P. Albuquerque et al. (eds.), *Medicinal and Aromatic Plants of South America*, Medicinal and Aromatic Plants of the World 5,
https://doi.org/10.1007/978-94-024-1552-0_33

Abstract *Phyllanthus niruri* L. (Euphorbiaceae), known as "quebra-pedra" or "stone breaker" is widely employed in folk medicine to treat ailments including disturbances of kidney and urinary bladder, intestinal infections, diabetes, hepatitis B virus, pain disorders, dyspepsia, vaginitis, tumors, diarrhea, epilepsy, malaria, hypertension, fever, inflammatory and dolorous processes. It is the most studied species of the Genus as regards chemical and biological aspects. Several experimental models have confirmed its medicinal properties, which are in general, related to the presence of phenolic compounds (flavonoids, tannins) and lignans. Interestingly, this plant was one of the first clinically studied species in Brazil, demonstrating the significant increase in renal calculi elimination.

Keywords *Phyllanthus niruri* · Renal and urinary problems · Phenolic compounds

1 Taxonomic Characteristics

Phyllanthus niruri L. (Euphorbiaceae), is known as "quebra-pedra" in Brazil and "chanca piedra" in Latin America meaning "stone breaker".

2 Crude Drug Used

The whole plant is used as a tea or decoctions as a remedy against many ailments, particularly those related to the urinary tract and hepatitis (Calixto et al. 1998).

3 Major Chemical Constituents and Bioactive Compounds

This plant is rich in active principles, being isolated several classes with pharmacological potential. Rutin, quercetin, quercitrin, astragalin, nirurin, quercetol, niruflavone, limonene, p-cymene, lupeol, lupeol aceate, ellagic acid, gallic acid, methyl brevifolincarboxylate, brevifolin, phyllanthin, hypophyllanthin, niranthin, 2,3-desmethoxy seco-isolintetralin, 2,3-desmethoxy seco-isolintetralin diacetate, linnanthin, nirphyllin, phyllnirurin, seco-4-hydroxylintetralin, hydroxyniranthin, geraniin, repandusinic acid, corilagin, norsecurinine, securinine, allosecurinine, phyllochrysine, niruriside, β-glucogallin, phyllanthone, phyllanthenol, phyllanthenone, E-phythol, orthosiphol G, orthosiphol I, hinokinin, epigallocatechin, kaempferol 4'-O-a-l-rhamnopyranoside (Calixto et al. 1998; Bagalkotkar et al. 2006; Qi et al. 2014).

4 Morphological Description

P. niruri is an herb and grows to about 12–20 cm in height. It has lateral horizontal branches, very thin, 3–7 cm in length with 7–28 leaves. The leaves are small (4–12 mm), green and oval. The flower is either male or female, with both types appearing on one plant (monoicous). The fruits measure 2–2.5 mm in diameter. Seeds are small (about 1 mm), round and smooth (Ulysséa and Amaral 1997).

5 Geographical Distribution

According to Webster (1956, 1970) and other authors (Gupta 2008), this plant is native and well distributed from Mexico until Argentina. In Brazil, it grows in practically all over in the country, as a weed. According to certain descriptors it occurs in several foreign countries, including Malaysia, Indonesia, India, USA, etc. (Bagalkotar et al. 2006).

6 Ecological Requirements

P. niruri grows preferably in dark places and forest border as well as associated with gallery forest, rocky fields and forest Atlantic coastal or mountain (Ulysséa and Amaral 1997; Da Silva and Sales 2008).

7 Traditional Use (Part(s) Used) and Common Knowledge

Among the more than 600 species of the genus *Phyllanthus*, *P. niruri* is considered to be the most widely used in world folk medicine. In general, the whole plant is used as a remedy in the form of tea, infusion or decoction, to treat a great variety of ailments, including disturbances of kidney and urinary bladder, intestinal infections, diabetes, hepatitis B virus, pain disorders, dyspepsia, vaginitis, tumors, diarrhea, epilepsy, malaria, hypertension, fever, inflammatory and dolorous processes, etc. (Calixto et al. 1998; Bagakotar et al. 2006; Gupta 2008; Qi et al. 2014).

8 Modern Medicine Based on Its Traditional Medicine Uses

Substantial pharmacological preclinical (also clinical but in less extension) studies have used extracts, fractions or pure compounds from *P. niruri*. The extract obtained from the whole plant exhibited pronounced antispasmodic effects against several smooth muscles (Calixto et al. 1998). Some of the mechanisms underlying the analgesic effects of the hydroalcoholic extract from *P. niruri* against formalin-induced nociception was studied in mice with promising results. Marked and dose-related inhibition of capsaicin-induced pain was observed, as well as potent effects against the second phase of formalin-induced pain (Santos et al. 1995). Other pharmacological effects (such as a diuretic, protection against liver damage, anti-HIV, anti-hepatitis virus, anti-plasmodial, antimalarial, among other biological properties) have been confirmed in in vitro and in vivo studies (Bagalkotar et al. 2006). Recently it was demonstrated that the methanolic extract from this plant exhibits promising antibacterial efficiency against pathogenic bacteria responsible for common infections of the skin, and urinary and gastrointestinal tracts (Ibrahim et al. 2013). More recently, Mediani and co-workers (2015) demonstrated that *P. niruri* extracts present strong α-glucosidase inhibitory and antioxidant activities. Regarding the antinociceptive effects, the ellagitannins geraniin and corilagin, isolated from *P. niruri* and also present in several species of the genus Phyllanthus, are, at least in part, responsible for the antinociceptive actions reported previously for these plants (Miguel et al. 1996; Moreira et al. 2013).

Experimental data have demonstrated that the lignans niranthin and nirtretalin exhibits anti-hepatitis B virus activity both in vitro and in vivo (Liu et al. 2014a, b).

Recently de Melo and co-workers (2015) demonstrated that the spray-dried extract obtained from the aerial parts reduces mucosal damage in rats with intestinal inflammation, suggesting that such pharmacological effect is related to the antioxidant potential of this plant.

Clinical (human) studies have demonstrated diuretic, hypotensive and hypoglycaemic effects as well as reduction of blood glucose in diabetic patients. It was also verified that this plant species exerted a significant increase in renal calculi elimination, not associated with the diuretic action (Calixto et al. 1998; Bagalkotar et al. 2006).

9 Conclusions

P. niruri is the most used species of the genus *Phyllanthus* in popular medicine against a variety of diseases. Many experimental preclinical and clinical studies have confirmed important therapeutic properties of this plant (extracts, fractions and its main constituents), including renal and urinary problems, infections, diabetes, hypertension, dolorous processes, etc. The main active principles responsible for these pharmacological or biological actions were determined as phenolic compounds, particularly flavonoids and tannins, lignans, terpenes and alkaloids.

References

Bagalkotar G, Sagineedu SR, Saad MS, Stanslas J (2006) Phytochemical from *Phyllanhtus niruri* Linn. and their pharmacological properties: a review. J Pharm Pharmacol 58:1559–1570

Calixto JB, Santos ARSS, Cechinel Filho V, Yunes RA (1998) A review of the plants of the genus *Phyllanthus*: their chemistry, pharmacology and therapeutic potential. Med Res Rev 18(4):225 258

Da Silva MJ, Sales MF (2008) Sinopse do gênero *Phyllanthus* (Phyllanthaceae) no nordeste do Brasil. Rodriguésia 59(2):407–422

De Melo MN, Soares LA, Porto CR, De Araújo AA, Almeida MD, De Souza TP, Petrovick PR, De Araújo RF Jr, Guerra GC (2015) Spray-dried extract of *Phyllanthus niruri* L. reduces mucosal damage in rats with intestinal inflammation. J Pharm Pharmacol 67:1107–1118

Gupta MP (ed) (2008) Plantas medicinales iberoamericanas. Convenio Andrés Bello y CYTED, Bogotá, pp 415–425

Ibrahim D, Hong LS, Kuppan N (2013) Antimicrobial activity of crude methanolic extract from *Phyllanthus niruri*. Nat Prod Comm 8(4):493–496

Liu S, Wei W, Li Y, Lin X, Shi K, Cao X, Zhou M (2014a) In vitro and in vivo anti-hepatitis B virus activities of the lignan nirtetralin B isolated from *Phyllanthus niruri* L. J Ethnopharmacol 157:62–68

Liu S, Wei W, Shi K, Cao X, Zhou M, Liu Z (2014b) In vitro and in vivo anti-hepatitis B virus activities of the lignan niranthin isolated from *Phyllanthus niruri* L. J Ethnopharmacol 155:1061–1067

Mediani A, Abas F, Khatib A, Tan CP, Ismail IS, Shaari K, Ismail A, Lajis NH (2015) Relationship between metabolites composition and biological activities of *Phyllanthus niruri* extracts prepared by different drying methods and solvents extraction. Plant Foods Hum Nutr 70, 184–192

Miguel OG, Calixto JB, Santos AR, Messana I, Ferrari F, Cechinel Filho V, Pizzolatti MG, Yunes RA (1996) Chemical and preliminary analgesic evaluation of geraniin and furosin isolated from *Phyllanthus sellowianus*. Planta Med 62(2):146–149

Moreira J, Klein-Júnior LC, Cechinel Filho V, de Campos Buzzi F (2013) Anti-hyperalgesic activity of corilagin, a tannin isolated from *Phyllanthus niruri* L. (Euphorbiaceae). J Ethnopharmacol 146(1):318–323

Qi W, Hua L, Gao K (2014) Chemical constituents of the plants from the Genus *Phyllanthus*. Chem Biodivers 11:364–395

Santos ARS, Cechinel Filho V, Yunes RA, Calixto JB (1995) Analysis of the mechanisms underlying the antinociceptive effect of the extracts of plants from the genus *Phyllanthus*. Gen Pharmacol 26(7):1499–1506

Ulysséa M, Amaral LG (1997) Contribution to the study of the genus *Phyllanthus* (Euphorbiaceae) in Santa Catarina island. Brazil Insula 26:1–28

Webster GL (1956) Studies of the Euphorbiaeeae, Phyllanthoideae. 11. The american species of *Phyllanthus* described by Linnaeus. J Arnold Arb 37:1–14

Webster GL (1970) A revision of *Phyllanthu*s (Euphorbiaeeae) in the Continental United States. Brittonia 22:44–76

Pluchea carolinensis (Jacq.) G. Don

Carles Roersch

Pluchea carolinensis (Jacq.) G. Don.
Photo: Jessie Harris
Available in: http://www.tropicos.org/Image/100546070

C. Roersch (✉)
Herbario "Dr. Henri Alain Liogier", Universidad Nacional Pedro Henriquez Ureña (UNPHU), Santo Domingo, Dominican Republic
e-mail: croersch@unphu.edu.do; croersch@imd-medicina-dominicana.org

U. P. Albuquerque et al. (eds.), *Medicinal and Aromatic Plants of South America*, Medicinal and Aromatic Plants of the World 5,
https://doi.org/10.1007/978-94-024-1552-0_34

Abstract *Pluchea carolinensis*, is widely distributed in Central America, the Caribbean, the North of South America, and is naturalized in Florida, Hawaii, Islands of the Pacific, and Taiwan. The large number of common names given to *P. carolinensis* indicates the popularity of this medicinal plant. Also, the conformity in traditional uses between Spanish, French and English speaking cultures is remarkable. However, until now very few biological, pharmacological experiment are carried out to corroborate the traditional uses. Clinical experiments are completely absent. The anti-Leishmania activity of the extracts and pure compounds are promising.

Keywords *Pluchea carolinensis* · Traditional uses · Chemical compounds · Anti-Leishmania activity

1 Taxonomic Characteristics

Synonyms (Basionym), based on Tropicos: *Conyza carolinensis* Jacq.

The genus Pluchea consists of about 80 species distributed in tropical areas in North and South America, the Caribbean, Africa, Asia and Australia (Sharma and Goyal 2011). There has been some indistinctness about the nomenclature of *P. carolinensis*. The first confusion concerns the application of the name *Pluchea odorata* (L.) Cass (Godfrey 1952, in Villaseñor and Villareal 2006). In many publications, this name is used as a synonym for *P. carolinensis*. In 1977, William T. Gillis published a revision of this genus and concluded that *P. carolinensis* should be named as *Pluchea symphytifolia* (with *Conyza symphytifolia* as basionym) (Gillis 1977). Twelve years later Khan and Jarvis (1989) repeated the work of Gilles and concluded that the interpretation of the original material associated with the name *Conyza symphytifolia* was erroneous. They reestablished the former name, *P. carolinensis* (Jacq.) G.Don (with *Conyza carolinensis* as basionym) as the correct one. In this monograph, publications which use the name *P. symphytifolia* will be considered as *P. carolinensis* (Villaseñor and Villareal 2006; José Luis Villaseñor, pers. communication, April, 2011). The difference between *P. odorata* and *P. carolinensis* is quite obvious. The latter is a shrub, 1–2.5 m tall with big leaves, which are longer than wide. On the contrary, *P. odorata* is an herb, 40–90 cm tall and has small leaves. Publications with a clear taxonomic description of the plant, which permits to differentiate between the two species, are included in this study.

2 Common Names

In the Spanish speaking Caribbean, *P. carolinensis* is known as *salvia* (Dominican Republic, Liogier 1990, 1996, 2000; Cordero 1986; Mañon et al. 1992; Gupta 1995; Puerto Rico, Nuñez 1992; Alvarado-Guzmán et al. 2009; Gupta 1995; Cuba,

Escandón and Méndez 2006; Hernández and Volpato 2004; Beyra et al. 2004; Hammer et al. 1990; Garcia et al. 2010; Gupta 1995; Roig and Mesa 1928, Roig and Mesa 1965; Florida (USA), Hodges and Bennett 2006; Nicaragua, Gupta 1995; Barrett 1994; Panama, Gupta 1995; Venezuela, Gupta 1995). The popularity of this plant is indicated by the presence of quite some detailed names like: *Sauge rouge* (Haíti, Beauvoir et al. 2001 in Duke et al. 2009)*, la sauge* (Haíti, Liogier 1990, 1996, 2000)*, salvia*, *salvia blanca* (Dominican Republic, Liogier 1990, 1996, 2000; Cordero 1986)*, salvia cimarron* (Cuba, Pino et al. 2009; Gupta 1995; Roig and Mesa 1928, 1965)*, salvia de las Antillas* (Dominican Republic, Cordero 1986)*, salvia de playa* Cuba, Barreto et al. 2002, 2007; Fuentes et al. 1989; Godínez and Volpato 2008; Pino et al. 2005, 2009; Milanés et al. 1999; Pérez et al. 2007; Rosales et al. 1999; Gupta 1995; Roig and Mesa 1928, 1965)*, salvia del pais* (Cuba, Pino et al. 2009; Milanés et al. 1999; Fernández and Torres 2006; Gupta 1995; Roig and Mesa 1928, 1965)*, salvia olorosa* Puerto Rico, Nuñez 1992)*, salvia real* (Middle America, Morton 1981) and *salvia santa* (Middle America, Morton 1981). The name *salvia* may refer to the European species *Salvia officinalis* L. The leaves are much alike, upper surface green and lower surface grayish and hairy. Both have a bitter taste. The common names, *cure for all* (Florida, USA, Woodmansee and Green 2006; Wilder and Roche 2009; Barbados, Honychurch 1986; Peter n.d.)*, cureforal* (Panama, Gupta 1995)*, guerit-tout (*French Guiana, DeFilipps et al. 2008), *Guétit-tout* (Haíti, Beauvoir et al. 2001 in Duke et al. 2009) and *geritout* (Trinidad and Tobago, Seaforth et al. 1983) are an allusion of the wide spectrum of medicinal application. Several common names refer to the medicinal uses of *P. carolinensis. Cough bush* (Bahamas, Austin 2004) indicate its use as an expectorant. In the former Aztec region of Mexico and Central America the common names *siguapote* (El Salvador, Honduras, and Guatemala, Gupta 1995)*, siguapate* (Honduras, Ticktin and Dalle 2005; House et al. 1990)*, ciguapate* (Guatemala, Kufer et al. 2005; Nicaragua, Gupta 1995) and *seguapeti* (El Salvador, Gupta 1995) signify in the Nahuath-lenguage, women medicine. *Chal-che* (Belize, Acevedo-Rodriguez 1996; Mexico, Steggerda 1943), *Chalche'* (Mexico, Ankli et al. 1999; Gupta 1995) means 'wash-quickly', referring to its use before, during and after childbirth (Austin 2004). *Sour bush* (Republic of Kiribati, Space and Imada 2004; Hawaii, Wood and LeGrande 2006; Starr et al. 2006; Englund et al. 2002; Bahamas, Eldridge 1975; U.S.Virgin Islands, Acevedo-Rodriguez 1996) refers to the bitter taste of its leaves as does *bitter tabacco* (Jamaica, Austin 2004). Smoking the leaves, like tabacco, may have given rise to the variety of common names with tabacco or tabac as its noun: *Bitter tabacco* (Jamaica, Austin 2004), *Indian tabacco* (Turks and Caicos Islands, Morton 1977)*, tabac a Jacot* (Haíti, Beauvoir et al. 2001 in Duke et al. 2009)*, tabac a jacquot* (Martinique, Slama et al. 2003)*, tabac diable* (Martinique, Honychurch 1986)*, tabac du diable* (French Guiana, DeFilipps et al. 2008)*, tabac marron* (Middle America, Morton 1981)*, tabac sauvage* (Haíti, Liogier 1990, 1996, 2000)*, tabac zombie* Dominica, Honychurch 1986; Quinlan and Quinlan 2007) and *tabacco cimarron* (Panama, Austin 2004). The meaning of the common name *tabac a jacot* is explained by Austin (2004). In a game called 'Simon says' one imitates as a parrot (jacquot) imitates people. Thus *P. carolinensis* imitates the real tabacco.

Sweet scent (Puerto Rico, Nuñez 1992; U.S.Virgin Islands, Acevedo-Rodriguez 1996), *sweet scanted fleabane* (Middle America, Morton 1981), *coniza olorosa* (*odorous coniza*, Dominican Republic, Cordero 1986) and *conyse odorante (odorous conyse* Haíti, Liogier 1996, 2000), point out the odour of the plant. Formerly many species of *Pluchea* were placed in the genus *Conyse* (Greek, a flea) and hence the name *coniza* and *conyse*. In the Asteracea family several plants are called 'fleabanes' and are considered to repel fleas. Here we have several names who might refer to this action: *Bushy fleabane* (Middle America, Morton 1981), *hairy fleabane* (leaves are hairy dorsal, Middle America, Morton 1981)), *shrubby fleabane* (Middle America, Morton 1981) and *sweet scanted fleabane* (Middle America, Morton 1981; Austin 2004).

3 Crude Drug Used

All parts of *P. carolinensis* are used as a drug. However, the leaves are the part of the plant that is mostly used in the traditional applications. These can be fresh as well as dried. The most common application form is as a tea (infusion or decoction). Externally the leaves (fresh, boiled or warmed) can be placed on the affected area. They are collected generally in the wild. In some countries, Cuba, Venezuela and Panama, the plant is cultivated in home gardens (Morton 1981). In so-called Botánicas, herb stores in Latin America, the plant (generally dried leaves) is rather popular (Hodges and Bennett 2006).

4 Major Chemical Constituents and Bioactive Compounds

Sesquiterpenes of the type eudesmane and cuauthemone are widespread in the genus *Pluchea* (Ahmed et al. 1996, 1998; Jakupovic et al. 1985). The name cuauthemone has been derived from the Mexican medicinal plant named Cuauhtematl (*P. odorata*) (Nakanishi et al. 1974). This compound demonstrates growth inhibition against bean and corn seeds (pers. com. Dr. M.R. Grarciduenar in Nakanishi et al. 1974). This bicyclic eudesmene – type sesquiterpene was first synthesized by Goldsmith and Sakano (1976). The absolute configuration was elucidated by Torres-Valencia et al. (2003).

The essential oils of the leaves and flowers were separately investigated by Pino et al. (2005, 2009). There exists quite a difference between the main constituents of these two essential oils. The essential oil of the leaves contains principally: juniper camphor (37.6%), 3-thujopsanone (8.1%), ß-caryophyllene (7.6%), spathulenol (7.4%) and ß-chamigrene (5.9%), whereas the essential oil of the flowers is characterized by: selin-11-en-4α-ol (kongol) (43.4%), 2,5-dimethoxy-p-cymene (12.5%), caryophylle oxide (6.8%), nerylisovalerate (6.4%) and ß-chamigrene. This oil also contains different aldehydes and esters to give the floral odour. The main component of the leaf oil, the sesquiterpene juniper camphor, is known as one of the principal ingredients of "Juniper Berry oil". This oil is widely used as a diuretic, stomachic, carminative in indigestion, kidney and bladder disorders, flatulence and rheumatism (British Herbal Medical Ass. 1983, Grieve n.d.). Recently isolated compounds, caffeic acid, chlorogenic acid, ferulic acid, quercetin and rosmarinic acid (Perera 2012) showed activity against *Leishmania amazonensis* (Montrieux et al. 2014). Several other biologically active compounds are present in *P. carolinensis* with the following pharmacologically activities: Taraxasteryl acetate, analgesic activity, (Palacios et al. 2008; Bahadir et al. 2010) and preventive effect on experimental hepatitis (Iijima et al. 1995); Isorhamnetin, anti-cancer activity (Lee et al. 2008; Ma et al. 2007; Teng et al. 2006); Kaempferol, bioactive dietary constituent (Calderón et al. 2011); Tannins, astringent (Haslam 1996).

In Table 1 the chemical constituents of *P. carolinensis* are mentioned.

Table 1 Chemical constituents of *Pluchea carolinensis*

Constituent	Plant part	References
Flavonols		
Isorhamnetin	Leaves	Perera et al. (2006a, b) Perera (2012), and Scholz et al. (1994)
Eupalitin	Leaves	Perera et al. (2006a)
Isorhamnetin-3-*O*-sulfate	Leaves	Perera et al. (2007)
3′,4′,5,6,7-pentahydroxy-3-methoxyflavone	Leaves	Perera et al. (2007)
Quercetin	Leaves	Perera et al. (2010), Perera (2012), and Scholz et al. (1994)
Quercitrin	Leaves	Perera (2012)
Quercetagetin	Leaves	Perera (2012)
Kaempferol	Leaves	Perera et al. (2010) and Perera (2012)
Myricetin	Leaves	Perera et al. (2010), Perera (2012)
Luteolin	Leaves	Perera (2012)
Herbacetin	Leaves	Perera (2012)
Sterols		
4,22-stigmastadien-3-one	Roots and stems	Lin (2009)
Terpenes		
3ß-Acetoxyurs-13 (18)-ene	Roots and stems	Lin (2009)
3ß-Angeloyl cuauhtemone	Aerial parts	Jakupovic et al. (1985)
3ß-Angeloyloxy-4-hydroxy-11-hydroperoxide-6,7-dehydroeudesman-8-one	Aerial parts	Jakupovic et al. (1985)
3-Thujopsanone	Essential oil[a] (leaf)	Pino et al. (2005)

(continued)

Table 1 (continued)

Constituent	Plant part	References
3α-(2′,3′-dihydroxy-2′-methylbutanoyl) 4,11-dihydroxy-6,7-dehydroeudesman-8-one	Aerial parts	Ahmed et al. (1998)
3α-(2′,3′-epoxy-2′-methylbutanoyl) 4α,11-dihydroxy-6,7-dehydroeudesman-8-one	Aerial parts	Ahmed et al. (1998)
3α-(2′,3′-epoxy-2′-methylbutanoyl) cuauhtemone	Aerial parts	Ahmed et al. (1998) and Jakupovic et al. (1985)
3α-(3′-chloro-2′-hydroxy-2′-methylbutanoyl) cuauhtemone	Aerial parts	Ahmed et al. (1998)
3α-(3′-chloro-2′-hydroxy-2′-methylbutanoyl)-4α,11-dihydroxy-6,7-dehydroeudesman-8-one	Aerial parts	Ahmed et al. (1998)
3α-Angeloyl cuauhtemone	Aerial parts	Ahmed et al. (1998)
3α-Angeloyloxy-4,11-dihydroxy-6,7-dehydroeudesman-8-one	Aerial parts	Jakupovic et al. (1985)
3α-Angeloyloxy-4-hydroxy-11-hydroperoxide-6,7-dehydroeudesman-8-one	Aerial parts	Jakupovic et al. (1985)
4α-Acetoxy-3α-(2′,3′-epoxy-2′-methylbutanoyl) cuauhtemone	Aerial parts	Ahmed et al. (1998) and Jakupovic et al. (1985)
4α-Acetoxy-3α-(2′,3′-epoxy-2′-methylbutanoyl)-11-hydroperoxide-6,7-dehydroeudesman-8-one	Aerial parts	Jakupovic et al. (1985)
4α-Acetoxy-3α-(2′,3′-epoxy-2′-methylbutanoyl)-11-hydroxy-6,7-dehydroeudesman-8-one	Aerial parts	Ahmed et al. (1998)
4α-Acetoxy-3α-(3′-chloro-2′-hydroxy-2′-methylbutanoyl)-11-hydroxy-6,7-dehydroeudesman-8-one	Aerial parts	Ahmed et al. (1998)
4α-Acetoxy-3α-angeloyloxy-11-hydroperoxide-6,7-dehydroeudesman-8-one	Aerial parts	Jakupovic et al. (1985)
5-Angeloyloxycarvotagetone	Aerial parts	Jakupovic et al. (1985)
5-*O*-Acetylcuauhtemonyl 6-*O*-2′,3′-epoxy-2′-methylbutyrate	Aerial parts	Ahmed et al. (1996)
α-Atlantone	Essential oil (leaf)	Pino et al. (2005)
Bicyclogermacrene	Essential oil (leaf)	Pino et al. (2005)
δ-Cadinene	Essential oil (flower)	Pino et al. (2009)
ß-Caryophellene	Essential oil (flower, leaf)	Pino et al. (2005, 2009)
Caryophyllene oxide	Essential oil (flower, leaf)	Pino et al. (2005, 2009)
ß-Chamigrene	Essential oil (flower, leaf)	Pino et al. (2005, 2009)
Cuauhtemone	Aerial parts	Ahmed et al. (1998)
Cubebol	Essential oil (flower, leaf)	Pino et al. (2005, 2009)
α-Gurjunene	Leaf	Sardans et al. (2010)

(continued)

Table 1 (continued)

Constituent	Plant part	References
γ-Gurjunene	Leaf	Sardans et al. (2010)
Juniper camphor (selin-7(11)-en-4α-ol)	Essential oil (leaf)	Pino et al. (2005)
Linalool	Essential oil (leaf)	Pino et al. (2005)
ß-Maaliene	Essential oil (flower)	Pino et al. (2009)
α-Pinene	Essential oil (leaf)	Pino et al. (2005)
α-Pinene	Leaf	Sardans et al. (2010)
Selin-11-en-4α-ol	Essential oil (flower)	Pino et al. (2009)
Selina-4,7-diene	Essential oil (flower, leaf)	Pino et al. (2005, 2009)
α-Selinene	Leaf	Sardans et al. (2010)
Spathulenol	Essential oil (leaf)	Pino et al. (2005)
Taraxasteryl acetate	Aerial parts	Jakupovic et al. (1985)
2-(hex-5-en-1-3-diynyl)-5-(prop-1-ynyl) thiophene	Aerial parts	Jakupovic et al. (1985)
2-(but-3-en-1-ynyl)-5-(penta-1-3-diynyl) thiophene	Aerial parts	Jakupovic et al. (1985)
Thymohydroquinone dimethyl ether	Aerial parts	Jakupovic et al. (1985)
Valencene	Essential oil (flower, leaf)	Pino et al. (2005, 2009)
Others		
3,4-O-dicaffeoylquinic acid	Aerial parts	Scholz et al. (1994)
	Leaves	Perera (2012)
4,5-O-dicaffeoylquinic acid	Aerial parts	Scholz et al. (1994)
	Leaves	Perera (2012)
3,5-O-dicaffeoylquinic acid	Aerial parts	Scholz et al. (1994)
	Leaves	Perera (2012)
3,4,5-O-tricaffeoylquinic acid	Aerial parts	Scholz et al. (1994)
	Leaves	Perera (2012)
1,3,4,5-O-tetracaffeoylquinic acid	Aerial parts	Scholz et al. (1994)
	Leaves	Perera (2012)
1,3-Di-O-[3,4-bis-(3,4-dihydroxyphenyl)-cyclobutane-1,2-dicarbonyl]-4,5-di-O-caffeoylquinic acid	Aerial parts	Scholz et al. (1994)
Caffeic acid	Roots and stems	Lin (2009)
	Leaves, flowers and stem	Perera (2012)
Caffeic acid methyl ester	Roots and stems	Lin (2009)

(continued)

Table 1 (continued)

Constituent	Plant part	References
2,6-Dimethoxy-1,4-benzoquinone	Roots and stems	Lin (2009)
2,5-Dimethoxy-p-cymene	Essential oil (flower, leaf)	Pino et al. (2005, 2009)
Neryl isobutyrate	Essential oil (flower)	Pino et al. (2009)
Tridecanal	Essential oil (flower)	Pino et al. (2009)
Neryl isovalerate	Essential oil (flower)	Pino et al. (2009)
Tannins	Leaves	Seaforth et al. (1983)
Chlorogenic acid	Leaves	Perera (2012)
Ferulic acid	Leaves	Perera (2012)
Rosmarinic acid	Leaves	Perera (2012)

[a]Only the main constituents of the essential oil are mentioned (>1.0%)

5 Morphological Description

"Erect shrub 1–2.5 m tall, much branched, branches densely tomentose. Leaf oblong-ovate to elliptic, 6–15 cm long, 2–6 cm wide, thinly tomentose and glandular on both surfaces, upper surface green, lower surface grayish, apex mucronulate-obtuse, margins entire or nearly so, base attenuate, petioles 1–2.5 cm long. Capitula 5–7 mm (when fresh) or ca. 10 mm (in dried specimens) in diameter, 6 mm long, peduncles 3–8 mm long, densely congested into terminal and axillary corymbs. Involucres ovate to campanulate, bracts greenish-purplish, 4-5-seriate; the outer very widely elliptic to very widely obovate, rounded at apex, 2–4 mm long, 1.5–2 mm wide, tomentose abaxially, ciliate at margins; the inner lanceolate to linear-lanceolate, acute at apex, 4–5 mm long, 0.5–1 mm wide, sparingly pubescent to glabrous. Receptacles flat, glabrous. Outer florets numerous, corolla filiform, pale greenish white, pinkish toward the summit, 3.5–4 mm long, tip 3-lobed; pappus white, slightly shorter than corolla; mature achenes not available for examination. Central florets ca. 20–25; corolla whitish, pinkish toward the summit, 4–5 mm long, sparingly glandular hairy at base; anthers obtuse at apex, shortly tailed at base; anthers and style exserted; achenes vestigial as a small, cartilaginous ring" (Peng et al. 1998).

6 Geographical Distribution

This tropical plant, *P. carolinensis*, is widely distributed in Central America, the Caribbean, the North of South America, and naturalized in Florida, Hawaii, Islands of the Pacific, and Taiwan (Villaseñor and Villareal 2006; van Belle n.d.; Dillon 2006; Peng et al. 1998; Anonymous 2010; Starr et al. 2006; Fosberg and Sachet

1987). Recently it has been described for the first time in the north of Peru (Dillon 2006).

7 Ecological Requirements

P. carolinensis is a shrub that is adapted to a wide variety of soils. It grows on wet and dry soils. However, it does not like shade. It is common in disturbed areas (US Forest Service n.d.). The plant is cultivated in gardens in Venezuela, Panama and Cuba and is sold on markets (Morton 1981). In Cuba, a phytosanitary study was performed to determine the illnesses, insects and present overgrowths in the nursery (Escandón and Méndez 2006). To-date, there is no further data on cultivation present in the literature.

8 Collection Practice

In Venezuela, Panama and Cuba the plant is cultivated in gardens and patios (Morton 1981). The leaves are generally collected in the wild.

9 Traditional Use (Part(s) Used) and Common Knowledge

In the literature, we have found a total of 186 recipes describing the traditional uses of *P. carolinensis* and these are distributed over 21 countries in North America, Central America, the Caribbean and South America. The plant is also found in the Islands of the Pacific and Taiwan, but no medicinal uses are reported. In North America, we have found information on its medicinal uses in Florida, the city of New York (USA) and Mexico. In Central America six countries, Guatemala, Belize, Honduras, Nicaragua, Costa Rica and Panama present data. Medicinal uses are recorded for Cuba, Dominican Republic, Trinidad and Tobago, Turks and Caicos, Bahamas, Puerto Rico, Dominica, Jamaica, Martinique and Haiti. *P. carolinensis* has its habitat in countries in the northern part of South America like Ecuador, Colombia and Venezuela. Recently it was described in the north of Peru (Dillon 2006). Remarkably, only from French Guiana have we found documents on the medicinal uses of this plant. Apparently, the medicinal value of this plant has not found its way in the traditional health systems in these countries. This is also the case in the islands of the Pacific and Taiwan. Here the plant is more considered as a rather aggressive, invasive weed (Global Invasive Species Database 2008). By far, most recipes refer to ailments, illnesses of the Respiratory tract (27%), followed by Pains (15%), Women Diseases (14%), the Digestive tract (12%), Fever (8%), Rheumatism (7%), Wounds (6%), Winds (3%), Liver (2%) and Sundries (6%).

In conditions of the Respiratory tract, the leaves are mostly used for *Throat* (Dominican Republic, Cordero 1986; Liogier 1990; Roersch, Unpublished results (Unp. Res.); Cuba, Godínez and Volpato 2008; Florida (USA), Hodges and Bennett 2006; Haíti, Liogier 2000; Belize, Arvigo and Balick 1993; Dominica, Quinlan and Quinlan 2007; Bahamas, Eldridge 1975; French Guiana, DeFilipps et al. 2008), *Cough* (Haíti, Liogier 2000; Belize, Arvigo and Balick 1993; Dominica, Quinlan and Quinlan 2007; Bahamas, Eldridge 1975), *Expectorant* (Haíti, Liogier 2000), *Hoarseness* (Cuba, Beyra et al. 2004; Godínez and Volpato 2008; Mexico, Morton 1981; Dominican Republic, Roersch, Unp.res.), *Flu* (Dominican Republic, Mañon et al. 1992, Roersch, Unp. Res.; Belize, Arvigo and Balick 1993; Martinique, Longuefosse and Nossin 1996; Guatemala, Gupta 1995; Honduras, Gupta 1995; French Guiana, DeFilipps et al. 2008), *Colds* (Belize, Arvigo and Balick 1993; Caribbean, Honychurch 1986; Guatemala, Kufer et al. 2005; Cuba, Beyra et al. 2004; Bahamas, Eldridge 1975; Morton 1981, French Guiana, DeFilipps et al. 2008; Trinidad and Tobago, Seaforth et al. 1983), *Chest colds with wheezing* (Turks and Caicos Islands, Morton 1977), *Pneumonia* (Cuba, Hernández and Volpato 2004), *Bronchopneumonia* (Dominican Republic, Roersch, Unp. Res.), *Bronchitic rattle* (Martinique, Longuefosse and Nossin 1996), *Catarrh* (Cuba, Volpato et al. 2009; Hernández and Volpato 2004; Beyra et al. 2004; Godínez and Volpato 2008; Florida (USA), Hodges and Bennett 2006; Mexico, Morton 1981), *Asthma* (Belize, Arvigo and Balick 1993), *Sinusitis* (Panama, Gupta 1995; Honduras, House et al. 1990). The second category is Pains. Generally the leaves are used as *Analgesic* (Puerto Rico, Nuñez 1992; Costa Rica, Gupta 1995), *pain* (Bahamas, Eldridge 1975; Nicaragua, Gupta 1995), *Ear pain* (Mexico, Heinrich et al. 1992), *Toothache* (Cuba, Beyra et al. 2004; Bahamas, Eldridge 1975; Morton 1981; Nicaragua, Gupta 1995; Florida (USA), Hodges and Bennett 2006; Dominican Republic, Roersch, Unp. Res.), *Thoracic pain* (Martinique, Longuefosse and Nossin 1996), *Chest pain* (Mexico, Steggerda 1943), *Headache* (Dominican Republic, Liogier 2000; Roersch, Unp. Res.; Guatemala, Kufer et al. 2005; Cuba, Godínez and Volpato 2008; Roig and Mesa 1928; Nicaragua, Gupta 1995; Panama, Gupta 1995; Trinidad and Tobago, Seaforth et al. 1983; Florida (USA), Hodges and Bennett 2006), *Migraine* (Cuba, Beyra et al. 2004; Puerto Rico, Alvarado-Guzmán et al. 2009), *Muscular pain* (Honduras, House et al. 1990), *Abdominal pain* (Honduras, House et al. 1990), *Azahar* (Guatemala, Kufer et al. 2005), *Whole body pain* (Nicaragua, Gupta 1995).

In the group of traditional ailments, i.e. Women Diseases, we have found 26 recipes of which 21 come from Mexico and Central America, where the local names *siguapate* (women medicine) and *Chalche* (wash-quickly) dominate. Mainly leaves are used for: *Pregnancy (to alleviate abdominal pain)* (Honduras, Ticktin and Dalle 2005), *Women in labor* (Mexico, Steggerda 1943), *After childbirth* (Belize, Arvigo and Balick 1993; Mexico, Steggerda 1943), *Expulsion of the placenta* (Mexico, Gupta 1995; Guatemala, Kufer et al. 2005; Bahamas, Eldridge 1975), *Desire of having a child* (Mexico, Ankli et al. 1999), *Fertility treatment or contraception* (Guatemala, Kufer et al. 2005), *Abortion* (Mexico, Ankli et al. 1999; Gupta 1995), *Miscarriage* (Honduras, Ticktin and Dalle 2005), *To induce menstruation* (Mexico, Gupta 1995), *Amenorrhea* (Mexico, Steggerda 1943), *Childbirth* (Haíti, Beauvoir et al. 2001 in Duke 2009), *menstruation (pain)* (Mexico, Ankli et al. 1999; Guatemala, Kufer et al. 2005), *menstrual problems* (Mexico, Bork et al. 1997; Heinrich et al. 1992; Cuba, Godínez and Volpato 2008; Honduras, House et al. 1990), *Menstruation* (Dominican Republic, Roersch, Unp. Res.; Mexico, Steggerda 1943), *To regulate menstruation* (Honduras, House et al. 1990) *Matrix prolapse* (Martinique, Longuefosse and Nossin 1996), *Uterine fibroids* (USA (New York), Balick et al. 2000), *Galactagogue* (Mexico, Gupta 1995).

P. carolinensis is used to cure ailments of the Digestive tract. We have found the following traditional applications: *Stomach disorders* (Jamaica, Liogier 1990; Guatemala, Gupta 1995; Dominican Republic, Roersch, Unp. Res.), *Digestive* (Cuba, Godínez and Volpato 2008), *Stomachache* (Mexico, Bork et al. 1997; Heinrich 1989 in Scholz et al. 1994; Heinrich et al. 1992; Nicaragua, Gupta 1995; Honduras, House et al. 1990), *Dyspepsia* (Haíti, Beauvoir et al. 2001 in Duke 2009), *Intestinal pain* (Dominican Republic, Roersch, Unp. Res.; Mexico, Steggerda 1943), *Gastrointestinal parasites* (Heinrich 1989 in Scholz et al. 1994; Heinrich et al. 1992), *Diarrhoea* (Bork et al. 1997; Heinrich 1989 in Scholz et al. 1994; Heinrich et al. 1992; Nicaragua, Gupta 1995), *Gastrointestinal disorders* (Mexico, Frei et al. 1998; Guatemala, Gupta 1995), *Carminative* (Puerto Rico, Nuñez 1992), *Colic* (Nicaragua, Gupta 1995; Honduras, House et al. 1990), *Spasm* (Nicaragua, Gupta 1995), *Stomach ailments* (Florida (USA), Hodges and Bennett 2006), *Flatulence* (Florida (USA), Hodges and Bennett 2006), and *Constipation* (Honduras, House et al. 1990).

The next category, Fever, contains recipes from Spanish, French and English speaking nations. *P. carolinensis* is used for: *Fever* (Haíti, Liogier 2000; Beauvoir et al. 2001 in Duke 2009; Guatemala, Kufer et al. 2005; Cuba, Beyra et al. 2004; Volpato et al. 2009; Godínez and Volpato 2008; French Guiana, DeFilipps et al. 2008; Trinidad and Tobago, Seaforth et al. 1983; Turks and Caicos islands, Morton 1977; Mexico, Steggerda 1943; Honduras, House et al. 1990), *Fever and Chills* (Martinique, Longuefosse and Nossin 1996), *Diaphoretic* (Puerto Rico, Nuñez 1992), *To bring out the heat* (Nicaragua, Barrett 1994), *To cool the heat of the blood* (Nicaragua, Barrett 1994).

Regarding Rheumatism the following cases are mentioned: *Rheumatic pains* (Belize, Arvigo and Balick 1993; Venezuela, Morton 1981), *Rheumatism* (Guatemala, Kufer et al. 2005; Cuba, Hernández and Volpato 2004; Roig and Mesa 1928; Haíti, Beauvoir et al. 2001 in Duke 2009; Bahamas, Eldridge 1975; Martinique, Longuefosse and Nossin 1996; Nicaragua, Gupta 1995; Mexico, Steggerda 1943; Dominican Republic, Roersch, Unp. Res.; Honduras, House et al. 1990), *Arthritic joints* (Belize, Arvigo and Balick 1993).

For curing Wounds and swellings, only the leaves are used: *Wounds, purulent* (Dominican Republic, Cordero 1986), *Wounds* (Jamaica, Liogier 2000), *Pyoderma* (Cuba, Beyra et al. 2004), *Ulcers* (Jamaica, Liogier 2000), *Skin infections* (Mexico, Bork et al. 1997), *Tumors* (Belize, Arvigo and Balick 1993), *Bruises* (Belize, Arvigo and Balick 1993), *Swellings* (Turks and Caicos Islands, Morton 1977; Belize, Arvigo and Balick 1993), *Rash* (Florida (USA), Hodges and Bennett 2006), and *Antiseptic* (Florida (USA), Hodges and Bennett 2006).

The following category is Winds. This culturally bound syndrome is called in Spanish *Aires* or *Viento*, which is exclusively mentioned in Spanish speaking countries: Nicaragua (Gupta 1995), Panama (Gupta 1995), Cuba (Roig and Mesa 1928), Dominican Republic (Roersch, Unp. Res.) and Honduras (House et al. 1990).

The smallest category is Liver conditions, where the leaves and flowers of *P. carolinensis* are used for: *Liver* (Dominican Republic, Cordero 1986), *Hepatic complaints* (Mexico, Frei et al. 1998) and *Gallbladder* (Dominican Republic, Cordero 1986).

Finally, there is a wide range of other traditional diseases, for which *P. carolinensis* is used. To name a few: *Malaria* (Honduras, Liogier 1990), *Sore muscles* (Belize, Arvigo and Balick 1993), *Twitching muscles* (Mexico, Steggerda 1943), *Strains or dislocations* (Caribbean, Honychurch 1986), *Rubefacient* (Puerto Rico, Nuñez 1992), *Head cold* (Dominica, Quinlan and Quinlan 2007) and *Ear infection* (Mexico, Gupta 1995).

In the Turks and Caicos Islands people smoke the dried leaves like tobacco (Morton 1977). In Miami, Florida, *P. carolinensis* is a very popular plant which is sold in special stores called Botánicas (Hodges and Bennett 2006). Botanicas are health stores mostly frequented by Latinos (Latin-Americans) who look for remedies to alleviate not only their health problems but also their love problems. Also all kinds of religious objects are offered from amulets to pictures of saints (Gómez-Beloz and Chávez 2001). *P. carolinensis*, Salvia, is used for Mal de ojo (evil eye), Mala suerte (Bad luck), Limpiezas (ritualistic cleansings), mental problems, and as a spiritual panacea (Hodges and Bennett 2006). In Santo Domingo, the capital of the Dominican Republic, Salvia is also present in the Botánicas as a remedy for the throat, hoarseness and in baths to bring good luck (Roersch, unpublished results).

10 Modern Medicine Based on Its Traditional Medicine Uses

Dried ethanol extract (98%) of the leaves dissolved in H_2O showed inhibitory activity against *Enterobacter faecalis* (MIC 100 mg/ml), *Staphylococcus aureus* (MIC 100 mg/ml), *Mycobacterium* sp. (MIC 100 mg/ml), *Mycobacterium fortuitum* (MIC 100 mg/ml), *Mycobacterium* sp. (MIC 10 mg/ml), *Pseudomonas* sp. (MIC 100 mg/ml), *Escherichia coli* (MIC 100 mg/ml), *Klebsiella* sp. (MIC 1,0 mg/ml) and *Klebsiella* sp. (MIC 0,1 mg/ml) (Pérez et al. 2007). Antimicrobial activity against *Staphylococcus aureus* and *Bacillus subtilis* by the EtOAc (MIC = 1,0 mg/ml) and *n*-BuOH (MIC = 1,0 mg/ml) crude extracts of the leaves was studied by Perera et al. (2006a, b). The $CHCl_3$ extract showed activity against *Bacillus subtilis* (MIC = 1,0 mg/ml). The isolated compounds of the EtOAc extract, isorhamnetin and eupalitin, didn't demonstrate activity (Perera et al. 2006a, b). The aqueous infusion of the aerial parts and the isolated compounds 3,4,5-O-tricaffeoylquinic acid and 1,3,4,5- O- tetracaffeoylquinic acid demonstrated inhibition against *Bacillus subtilis* (MIC = 360 μg/ml, 110 μg/ml, 80 μg/ml respectively), *Escherichia coli* (MIC = 3300 μg/ml, 330 μg/ml, 330 μg/ml respectively) and *Micrococcus luteus* (MIC = 3300 μg/ml, 660 μg/ml, 330 μg/ml, respectively) (Scholz et al. 1994).

Different extract of the aerial parts did not show antifungal activity in vitro against *Cladosporium cucumerinum* and *Penicillum oxalicum* (Scholz et al. 1994). The $CHCl_3$ extract of the aerial parts gave in vitro nematocidal activity against *Caenorhabditis elegans* (ED_{50} = 250–500 μg/ml). In vivo activity against *Trichostrongylus colubriformis* in jirds (*Meriones unguiculatus*) given orally and given subcutaneously at a single dose of 200 mg/ml reduced the worm burden by 30% and 40% respectively. 1,3,4,5-O-tetracaffeoylquinic acid had also in vitro nematocidal activity against *Caenorhabditis elegans* (ED_{50} = 125–250 μg/ml). However, in vivo (200 mg/ml) it didn't show activity against *Trichostrongylus colubriformis* in jirds (*Meriones unguiculatus*) given orally and given subcutaneously it had a reduced effect (worm burden reduced by 15%) (Scholz et al. 1994). The ethanolic and hydro-ethanolic leave extracts showed antifungal activity against *Candida* and *Trichophyton* spp. (200≤MIC≤400 μg/ml) (Biabiany et al. 2013).

$CHCl_3$ and EtOAc extract of the aerial parts and 1,3,4,5-O-tetracaffeoylquinic acid showed in vitro low antiamoebic activity against *Entamoeba histolytica* (IC_{50} = 250–500 μg/ml, 250–500 μg/ml and 125–250 μg/ml respectively) (Scholz et al. 1994).

Antioxidant activity was investigated using the L-epinephrine oxidation by hydroxyl radical generated in the Fenton reaction. This was inhibited by the crude phenol (conc. 40, 60, 80 and 100 μg/ml) and 50% ethanol extract (conc. 100, 200, 300 and 400 μg/ml) from the leaves of *P. carolinensis* in a dose-dependent manner (Fernández and Torres 2006). In vitro antioxidant activity, using the DPPH (2,2-diphenyl-1-picrylhydrazyl) and ABTS (2,2′-azino-bis(3-ethylbenzthiazoline-6-sulfonic acid) methods, were highest for the EtOAc and *n*-BuOH extracts of the leaves (Perera et al. 2010).

Antileishmanial activity was investigated with a Cuban species of *P. carolinensis*. However, the results are somewhat confusing. In their first publication, Garcia et al. (2010) inform that the ethanolic extract (80%) of the leaves hardly inhibit the growth of promastigotes of *L. amazonensis* at concentrations of 50 μg/ml (inhibition 13.5%) and 100 μg/ml (inhibition 12.7%), whereas in their second publication, Garcia et al. (2011) mention that *P. carolinensis* inhibits 50% of promastigote growth at a concentration of 30 μg/ml, without referring to their first article. The pure compounds from *P. carolinensis*, caffeic acid, chlorogenic acid, ferulic acid, quercetin and rosmarinic acid, showed inhibitory activity against promastigotes (IC_{50} = 0.2–0.9 μg/ml) and intracellular amastigotes (IC_{50} = 1.3–2.9 μg/ml). In BALB/c infected mice, caffeic acid, ferulic acid, and rosmarinic acid controlled lesion size development and parasite burden in footpads (Montrieux et al. 2014).

Tincture (30%) of the plant (part not specified) has a significant in vivo anti-inflammatory activity in acute and chronic processes by using carrageenan-induced rat paw edema (doses 80 mg/ml of tincture, orally) and the cotton-induced granuloma model (doses 80 mg/ml of tincture, orally) (Rosales et al. 1999). The ethanolic extract (96%) of the aerial parts did not show activation (conc. 100 μg/ml) of the transcription factor NF-κB in HeLa cell culture (Bork et al. 1997).

Weak antispasmogenic effect (conc. 0.1 ml) was observed using the Guinea-pig ileum with the aqueous extract of the leaves and succulent stems. From this extract, the high molecular weight material was precipitated with ethanol and the resulting extract also showed weak antispasmogenic effect (conc. 0.5 g/ml) using the Guinea-pig ileum. In addition vasodilator activity in the rat hind limb (conc. 0.01 g/ml) was demonstrated (Feng et al. 1962).

Aqueous infusion of the aerial parts had a weak antisecretory effect (conc. 250 μg/ml) on the isolated rabbit colon (inhibition of prostaglandin E_2 stimulated Cl^- secretion) and the EtOAc extract of the aerial parts did not have any antisecretory effect (conc. 750 μg/ml) on the isolated rabbit colon (inhibition of prostaglandin E_2 stimulated Cl^- secretion) (Scholz et al. 1994).

11 Toxicology

The toxicity of the ethanolic extract (70%) of the leaves of *P. carolinensis* was evaluated using the Toxicity Class Method. The only applied doses of 2000 mg/kg did not produce any deaths among the test animals (rats) during the observation period of 14 days. Afterwards, an anatomo-pathological assessment was performed and no macroscopic alterations were observed in the external surface and in the cavities, organs and tissues (Arteaga et al. 2008). Feng et al. (1962) found that mice were killed applying 0.5 ml (intraperitoneally) of a water extract (conc. 5 g/ml) or 1.0 ml water/ethanol extract (conc.1 g/ml) of the leaves.

12 Conclusions

The large number of common names given to *P. carolinensis* indicates the popularity of this medicinal plant. Even after migration, Latin-Americans (Hispano-Americans) visit their herb stores (Boticas) to purchase this plant. Also, the conformity in traditional uses between Spanish, French and English speaking cultures is remarkable. However, until now very few biological, pharmacological experiment are carried out to corroborate the traditional uses. Clinical experiments are completely absent. The principal traditional uses, illnesses of the Respiratory tract, are hardly confirmed with laboratory data. The second group of traditional ailments, Pains, has some confirmation. The principle use in Central America and Mexico, in Women diseases, have not received any attention so far. Generally, the biological/pharmacological part is poor. The experiments related to the anti-Leishmania activity of the extracts and pure compounds are promising.

To resume a very interesting medicinal plant with very little attention from the scientific world.

References

Acevedo-Rodriguez P, collaborators (1996) Flora of St. John, U.S. Virgin Islands. Mem N Y Bot Gard 78:1–581

Ahmed AA, Ali BA, Krawiec M, Watson WH (1996) 5-O-Acetylcuauhtemonyl 6-O-2′,3′-epoxy-2′-methylbutyrate. Acta Crystallogr C52:235–237

Ahmed AA, El-Seed HR, Mahmoud AA, El-Douski AEA, Zeid IF, Bohlin L (1998) Eudesmane derivates from Laggera crispata and Pluchea carolonesis. Phytochemistry 49(8):2421–2424

Alvarado-Guzmán JA, Gavillán-Suárez J, Germosén-Robineau L (2009) TRAMIL Ethnopharmacological survey: knowledge distribution of medicinal plant use in the southeast region of Puerto Rico. P R Health Sci J 28(4):329–339

Ankli A, Sticher O, Heinrich M (1999) Medical ethnobotany of the Yucatec Maya: healers' consensus as a quantitative criterion. Econ Bot 53(2):144–160

Anonymous (2010) US Forest Service, Pacific Island Ecosystems at Risk (PIER). Online resource at http://www.hear.org/pier/. Accessed: 15/02/11

Arteaga ME, Payo AL, González C, Bada AM, González BO, Curbelo A (2008) Evaluación del extracto fluido de Pluchea carolinensis (Jacq.) G.Don. por el método de clases de toxicidad en ratas. Revista Cubana de Plantas Medicinales (on line) 13(4)

Arvigo R, Balick M (1993) Rainforest remedies. One hundred healing herbs of Belize. Lotus Press, Twin Lakes

Austin DF (2004) Florida ethnobotany. CRC Press, Boca Raton

Bahadir O, Çitoğlu GS, Šmejkal K, Dall'Acqua S, Özbek H, Cvacka J, Zemlicka M (2010) Analgesic compounds from Scorzonera latifolia (Fisch. And Mey.) DC. J Ethnopharmacol 131:83–87

Balick MJ, Kronenberg F, Ososki AL, Reiff M, Fugh-Berman A, O'Conner B, Roble M, Lohr P, Atha D (2000) Medicinal plants used by Latino healers for women's health conditions in New York City. Econ Bot 54(3):344–357

Barreto A, Pérez E, Reyes G, Enríquez N, Primelles J, Sedeño E (2002) Aportes al conocimiento de la riqueza florística para la gestión ambiental de la Sierra de Najasa. Camagüey Cuba Rodriguésia 53(82):131–145

Barreto A, Godínez D, Enríquez N, Reyes G (2007) Riqueza florística del complejo orográfico Sierra de Najasa, provincia Camagüey, Cuba. Rodriguésia 58(1):59–71

Barrett B (1994) Medicinal plants of Nicaragua's Atlantic coast. Econ Bot 48(1):8–20

Beauvoir MG, Defillips RA, Wolpert BJ, Crepin J (2001) Selected medicinal plants of Haitian Vodou. Typescript. Smithsonian Institution, Washington, DC

Beyra A, León M, Iglesias E, Ferrándiz D, Herrera R, Volpato G, Godínez D, Guimarais M, Álvarez R (2004) Estudios etnobotánicos sobre plantas medicinales en la provincia de Camagüey (Cuba). An Jard Bot Madr 61(2):185–203

Biabiany M, Roumy V, Hennebelle T, François N, Sendid B, Pottier M, Aliouat E, Rouaud I, Lohézic-Le Dévéhat F, Joseph H, Bourgeois P, Sahpaz S, Bailleul F (2013) Antifungal activity of 10 Guadeloupian plants. Phytother Res 27:1640–1645

Bork PM, Schmitz ML, Kuhnt M, Escher C, Heinrich M (1997) Sesquiterpene lactone containing Mexican Indian medicinal plants and pure sesquiterpene lactones as potent inhibitors of transcription factor NF- κB. FEBS Lett 402:85–90

British Herbal Medicine Association (1983) British herbal pharmacopoeia. Bournemouth, England

Calderón JM, Burgos E, Pérez G, López M (2011) A review on the dietary flavonoid Kaempferol. Mini-Rev Med Chem 11:298–344
Cordero AB (1986) Manual de Medicina Domestica; Plantas Medicinales Dominicanas. Colección Ciencia y Tecnología no. 7, vol CCLII, 2nd edn. Publicaciones de la Universidad Autónoma de Santo Domingo, Santo Domingo
Defilipps RA, Maina SL, Crepin J (2008) Medicinal plants of the Guianas (Guyana, Surinam, French Guiana). Smithsonian National Museum of Natural History, Smithsonian Inst, Washington, DC
Dillon MO (2006) Nuevos registros de Asteraceae para la flora de Peru. New records in the Asteracea flora of Peru. Amaldoa 13(2):314–317
Duke JA, Bogenschutz-Godwin MJ, Ottensen AR (2009) Duke's handbook of medicinal plants of Latin America. CRC Press, Taylor & Francis Group, Boca Raton
Eldridge J (1975) Bush medicine in the Exumas and Long Osland, Bahamas. A field study. Econ Bot 29:307–332
Englund RA, Imada C, Preston DJ (2002) Stream and botanical survey of an unnamed tributary flowing into Pu'u Ka 'Ele reservoir and Pila'a stream, Pila'a, Kilauea, kaua'i, Final Report. Contribution No. 2002–001 to the Hawaii Biological Survey. Bishop Museum, Honolulu, Hawaii, USA
Escandón M, Méndez M (2006) Enfermedades, insectos, y malezas observadas en Pluchea carolinensis (Jacq.) G. Don (Asteraceae) cultivada en el Instituto de Ecología y Sistemática. www.fao.org/docs/eims/upload/cuba/5374/PLUCHEA_2006.pdf. Accessed on 10 Feb 2011
Feng PC, Hayes LJ, Magnus KE, Plimmer JR, Sherratt HSA (1962) Pharmacological screening of some West Indian medicinal plants. J Pharm Pharmacol 14:556–561
Fernández F, Torres M (2006) Evaluation of Pluchea carolinensis extracts as antioxidants by the epinephrine oxidation method. Fitoterapia 77:221–226
Fosberg FR, Sachet MH (1987) Flora of the Gilbert islands, Kiribati, checklist. The Smithsonian Institution, Washington, DC
Frei B, Baltisberger M, Sticher O, Heinrich M (1998) Medical ethnobotany of the Zapotecs of the isthmus-sierra (Oaxaca, Mexico): documentation and assessment of indigenous uses. J Etnopharmacol 62:149–165
Fuentes VF, Granda MM, Armas I, Izquierdo M, Martínez M, Rodríguez CA (1989) Estudios sobre la medicina tradicional en Cuba. IV Rev Cubana Farm 23(1–2):99–115
Garcia M, Monzote L, Montalvo AM, Scull R (2010) Screening of medicinal plants against Leishmania amazonensis. Pharm Biol 48(9):1053–1058
Garcia M, Perera WH, Scull R, Monzote L (2011) Antileishmanial assessment of leaf extracts from Pluchea carolinensis, Pluchea odorata and Pluchea rosea. Asian Pac J Trop Med:1836–1840
Gillis WT (1977) Pluchea revisited. Taxon 26:587–591
Global Invasive Species Database (2008) Pluchea carolinensis. Available from: http://www.issg.org/database/species/ecology.asp?si=1338&fr=1&sts=&lang=EN. Accessed 9 Feb 2011
Godfrey RK (1952) Pluchea, section Stylimnus, in North America. J Elisha Mitchell Sci Soc 68:238–279
Godínez D, Volpato G (2008) Plantas medicinales que se venden en el mercado El Río, Camagüey, Cuba. Rev Mex Biodiversidad 79(1):243–259
Goldsmith DJ, Sakano I (1976) Synthesis of Cuauhtemone. J Org Chem 41(12):2095–2098
Gómez-Beloz AG, Chávez N (2001) The Botánica as a culturally appropriate health care option for Latinos. J Altern Complement Med 7(5):537–546
Grieve M (n.d.) A modern herbal. www.botanical.com/botanical/mgmh/j/junipe11.html. Accessed 14 Mar 2012
Gupta M (1995) 270 Plantas medicinales Iberoamericanas. CYTED-SECAB, Bogota
Hammer K, Esquivel M, Fuentes V, Lima H (1990) Additional notes to the checklist of Cuban cultivated plants (1). Genet Resour Crop Ev 38(3):325–343
Haslam E (1996) Natural polyphenols (vegetable tannins) as drugs: possible modes of action. J Nat Prod 59:205–215

Heinrich M (1989) Ethnobotanik der Tieflandmixe (Oaxaca, Mexico) und phytochemische Untersuchung von Capraria biflora L. (Scrophulariaceae). Dissertationes Botanicae No. 144. J.Cramer, Berlin and Stuttgart, Germany

Heinrich M, Rimpler H, Antonio N (1992) Indigenous phytotherapy of gastrointestinal disorders in a lowland Mixe community (Oaxaca, Mexico): ethnopharmacological evaluation. J Ethnopharmacol 36:63–80

Hernández J, Volpato G (2004) Herbal mixtures in the traditional medicine of eastern Cuba. J Ethnopharmacol 90:293–316

Hodges S, Bennett BC (2006) The ethnobotany of Pluchea carolinensis (Jacq.)G.Don (Asteraceae) in the Botánicas of Miami, Florida. Econ Bot 60(1):75–84

Honychurch PN (1986) Caribbean wild plants and their uses: an illustrated guide to some medicinal and wild ornamental plants of the West Indies. Macmillan Publishers, Ltd., London

House PL, Witte S, Torres C (1990) Manual popular de 50 plantas medicinales de Honduras. CONS-H, CIIR, UNAH, Honduras

Iijima K, Kiyohara H, Tanaka M, Matsumoto T, Cyong J, Yamada H (1995) Preventive effect of Taraxasteryl acetate from Inula britannica subsp. japonica on experimental hepatitis in vivo. Planta Med 61(1):50–53

Jakupovic J, Misra LN, Chau Thi TV, Bohlmann F, Castro V (1985) Cuauthemone derivates from Tessaria integrifolia and Pluchea symphytifolia. Phytochemistry 24(12):3053–3055

Khan R, Jarvis CE (1989) The correct name for the plant known as Pluchea symphythifolia (Miller) Gillis (Asteracea). Taxon 38:659–662

Kufer J, Förther H, Pöll E, Heinrich M (2005) Historical and modern medicinal plant uses – the example of the Ch'orti' Maya and Ladinos in Eastern Guatemala. J Pharm Pharmacol 57:1127–1152

Lee HJ, Lee HJ, Ko SG, Bae HS, Kim CH, Ahn KS, Lu J, Kim SH (2008) Mitochondria-cytochrome C-caspase-9 cascade mediates isorhamnetin induced apoptosis. Cancer Lett 270(2):342–353

Lin LS (2009) Studies on the constituents of the roots and stems of Pluchea carolinensis (Jacq.) G.Don. www.ethesys.lib.pu.edu.tw/ETD-db/ETD-search/view_etd?URN=etd-0616103-213917. Accessed on 10 Feb 2011

Liogier AH (1990) Plantas medicinales de Puerto Rico y del Caribe. Iberoamericana Ediciones, Hato Rey, Puerto Rico

Liogier AH (1996) La Flora de la Española, Vol. LXXII, serie Científica 29, vol VIII. Universidad Central del Este, San Pedro de Macorís, Dominican Republic

Liogier AH (2000) Diccionario Botánico de Nombres Vulgares de la Española, 2nd edn. Jardín Botánico "Dr. Rafael M. Moscoso" y Universidad Nacional Pedro Henriquez Ureña, Santo Domingo, Dominican Republic

Longuefosse J-L, Nossin E (1996) Medical ethnobotany survey in Martinique. J Ethnopharmacol 53:117–142

Ma G, Yang C, Qu Y, Wei H, Zhang T, Zhang N (2007) The flavonoid component isorhamnetin in vitro inhibits proliferation and induces apoptosis in Eca-109 cells. Chem Biol Interact 167(2):153–160

Mañon DI, Haché Ll I, García SE (1992) Medicina Tradicional Dominicana (una contribución a su estudio). Universidad Nacional Pedro Henriquez Ureña, Academia Dominicana de Medicina, Museo del Hombre Dominicana, Instituto Dominicano de Ambiente y Sociedad, Santo Domingo, Dominican Republic

Milanés R, Alonso D, González G, Espín G (1999) Farmacognosia de la droga <Flores de Majagua> (*Hibiscus elatus* Sw., familia Malvaceae). I: farmacogeografia, farmacoetimología, farmacoergasia y farmacoetnología. Rev Cuba Plantas Medicinales 3(3):98–101

Montrieux E, Perera WH, García M, Maes L, Cos P, Monzote L (2014) In vitro and in vivo activity of major constituents from Pluchea carolinensis against Leishmania amazonensis. Parasitol Res 113:2925–2932

Morton JF (1981) Atlas of medicinal plants of middle America, Bahamas to Yucatan. Charles C. Thomas, Publisher, Springfield

Morton JF (1981) Atlas of medicinal plants of middle America, Bahamas to Yucatan. Charles C. Thomas, Publisher, Springfield, USA

Nakanishi K, Crouch R, Miura I, Dominguez X, Zamudio A, Villareal R (1974) Structure of a sesquiterpene, Cuauhtemone, and its derivative. Application of partially relaxed Fourier transform 13C nuclear magnetic resonance. J Am Chem Soc 96(2):609–611

Nuñez E (1992) Plantas Medicinales de Puerto Rico, Reimpresión. Editorial de la Universidad de Puerto Rico, Puerto Rico

Palacios E, Déciga M, Mata R (2008) Antinociceptive, hypoglycemic and spasmolytic effects of *Brickellia veronicifolia*. J Ethnopharmacol 118:448–454

Peng C-I, Chen C-H, Leu W-P, Yen H-F (1998) Pluchea Cass. (Asteraceae: Inuleae) in Taiwan. Bot Bull Acad Sinica 39:287–297

Perera WH (2012) Isolation and identification of antioxidant phytochemicals from Cuban species of the genera Erythroxylum P. Browne and Pluchea Cass. PhD Thesis, Laboratory of Molecular Biology and Plant Biotechnology, University of Liege, Belgium

Perera WH, González L, Payo, AL (2006a) Metabolitos secundarios y actividad antimicrobiana de Pluchea carolinensis. Revista Cubana de Farmacia; 40(2). http://scielo.sld.cu/scielo.php?script=sci_arttext&pid=S0034-75152006000200007&lng=es&nrm=iso. ISSN 0034-7515

Perera WH, González L, Payo AL, Nogueiras C, Delgado G, Oquendo M, Sarduy R (2006b) Antimicrobial activity of crude extracts and flavonoids from leaves of Pluchea carolinensis (Jacq.) G.Don. Pharmacol Ther 3:757–761

Perera WH, Nogueiras C, Payo A, Queiroz B, Sarduy R, Oquendo M (2007) Flavonols from the leaves of Pluchea carolinensis (Jacq.) G.Don (Asteraceae). Rev Latinoamer Quim 35(3):68–73

Perera WH, Tabart J, Gómez A, Sipel A, Payo AL, Kevers C, Dommes J (2010) Antioxidant capacity of three Cuban species of the genus Pluchea Cass. (Asteraceae). J Food Chem 34:249–261

Pérez C, Balcinde Y, Suárez C, Hernández V, Falero A, Hung BR (2007) Ensayo de la actividad antimicrobiana de Pluchea carolinensis (salvia de playa). Rev CENIC Cienc Biológicas 38(2):150–154

Peter SR (n.d.) Conservation of the biodiversity of medicinal plants in Barbados http://www.eclipsbiz.com/JIEP/S_Peter.pdf. Accessed on 03 Apr 2008

Pino JA, Marbot R, Payo A, Chao D, Herrera P, Martí MP (2005) Leaf oils of two Cuban Asteraceae species: Pluchea carolinensis Jacq. and *Ambrosia hispida* Pursh. J Essent Oil Res 17:318–320

Pino JA, Perera WH, Sarduy R, Oviedo R, Quijano CE (2009) Essential oil from flowers of Pluchea carolinensis (Jacq.) G.Don. J Essent Oil Res 21:45–47

Quinlan MB, Quinlan RJ (2007) Modernization and medicinal plant knowledge in a Caribbean horticultural village. Med Anthropol Q 21(2):169–192

Roig, Mesa JT (1928) Diccionario Botánico de Nombres Vulgares Cubanos. Imprenta y papeleria Ramela Bauzá y Cia, La Habana (On line by Ann Arbor Michigan, University of Michigan Library, 2005)

Roig y Mesa JT (1965) Diccionario Botánico de Nombres Vulgares Cubanos. 3, ampliada y corregida. Editora del Consejo Nacional de Universidades, La Habana, Cuba

Rosales VP, Gross MC, Rosales RA, García RC, León JE (1999) Evaluación farmacológica de Pluchea carolinensis Jacq. (Salvia de playa) en animales de experimentación. Rev Cuba Plantas Medicinales 3(2):65–67

Sardans J, Llusià J, Niinemets U, Owen S, Peñuelas J (2010) Foliar mono- and Sesquiterpene contents in relation to leaf economic spectrum in native and alien species in Oahu (Hawai'i). J Chem Ecol 36:210–226

Scholz E, Heinrich M, Hunkler D (1994) Caffeoylquinic acids and some biological activities of Pluchea symphytifolia. Planta Med 60(3):360–364

Seaforth CE, Adams CD, Sylvester YA (1983) Guide to the medicinal plants of Trinidad & Tobago. Commonwealth Secretariat, London

Sharma SK, Goyal N (2011) Biological studies of the plants from the genus Pluchea. Ann Biol Res 2(3):25–34

Slama R, Gillet C, William R, Longuefosse JL, Brunod R (2003) Histoire de la folie à la Martinique. Inf Psychiatr 79(6):493–499

Space JC, Imada CT (2004) Report to the Republic of Kiribati on invasive plant species on the islands of Tarawa, Abemama, Butaritari and Maiana. U.S.D.A. Forest Service, Pacific Southwest Research Station, Institute of Pacific Islands Forestry, and Bishop Museum, Pacific Biological Survey, Department of Natural Sciences, Honolulu, Hawaii, USA

Starr F, Starr K, Wood K. (2006) Lanai offshore islets botanical survey. Department of Land and Natural Resources, Division of Forestry and Wildlife and Offshore Islet Restoration Committee, Honolulu, Hawaii, USA

Steggerda M (1943) Some ethnological data concerning one hundred Yucatan plants. Anthropological Papers no. 29, Smithsonian Institution, Bureau of American Ethnology Bulletin 136:189–226

Teng BS, Lu YH, Wang ZT, Tao XY, Wei DZ (2006) In vitro anti-tumor activity of isorhamnetin isolated from *Hippophae rhamnoides* L. against BEL-7402 cells. Pharmacol Res 54(3):186–194

Ticktin T, Dalle SP (2005) Medicinal plant use in the practice of midwifery in rural Honduras. J Ethnopharmacol 96:233–248

Torres-Valencia JM, Quintero-Mogica DL, Léon GI, Suárez-Castillo OR, Villagómez-Ibarra JB, Maldonado E, Cerda-García-Rojas CM, Joseph-Nathan P (2003) The absolute configuration of cuauhtemone and related compounds. Tetrahedron Asymmetry 14:543–548

US Forest Service, Pacific Island Ecosystems at Risk (PIER) (n.d.) Online resource at http://www.hear.org/pier. Accessed 09-02-2011

van Belle J (n.d.) Inventarisatie Terrestrisch Milieu van het eiland Aruba. RuG, the Netherlands. (without date)

Villaseñor JL, Villareal JA (2006) El género Pluchea (familia Asteraceae, tribu Plucheea) in Mexico. Rev Mex Biodiversidad 77:59–65

Volpato G, Godínez D, Beyra A, Barreto A (2009) Uses of medicinal plants by Haitian immigrants and their descendants in the Province of Camagüey, Cuba. J Ethnobiol Ethnomed 5:16

Wilder GJ, Roche BJ (2009) A floristic inventory of Marco Island (Collier County), Florida. J Bot Res Inst Tex 3(2):873–899

Wood KR, LeGrande M (2006) An annotated checklist and new island records of flowering plants from Lehua Islet, Ni'ihau, Hawai'i. Records of the Hawaii Biological Survey for 2004–2005. In: Evenhuis NL, Eldredge LG (eds) Bishop Museum Occasional Papers, vol 87, pp 19–29

Woodmansee SW, Green SE (2006) A floristic survey and rare plant assessment of Caloosahatchee Creek Preserve, Lee County, Florida, Final Report (abridged online version). The Institute for Regional Conservation, Miami, Florida

Polygonum punctatum Elliott

Maria Izabela Ferreira, Gabriela Granghelli Gonçalves, and Lin Chau Ming

Polygonum punctatum Elliot
Photo: O.M. Montiel
Available in: http://www.tropicos.org/Image/100160110

M. I. Ferreira · G. G. Gonçalves · L. C. Ming (✉)
Horticulture Department, School of Agronomic Sciences,, Universidade Estadual Paulista (UNESP), Botucatu, São Paulo, Brazil
e-mail: linming@fca.unesp.br

U. P. Albuquerque et al. (eds.), *Medicinal and Aromatic Plants of South America*, Medicinal and Aromatic Plants of the World 5,
https://doi.org/10.1007/978-94-024-1552-0_35

Abstract *Polygonum punctatum* Elliot is found all over on the American continent in areas of flooded, sandy or fertile land. It belongs to Polygonaceae family and it is popularly known in Brazil as erva-de-bicho, cataia, persicária do Brasil, pimenteirad'água. In Spain is known as ajicillo, erva do bicho, caa-tai and in the United States of America dotted smartweed, water smartweed and water pepper. It is widely used in folk medicine and the uses of *P. punctatum* are referred in literature to treat hemorrhoids and rheumatism, besides presenting diuretic, abortive and emmenagogue action. There is a range of secondary metabolites groups in aerial parts, like tannins, free anthraquinones, saponins, pelargonidin, flavonoids and acids, polyphenols, coumarins, glycosides, terpenoids, sesquiterpenes and the major components, the sesquiterpenes polygodial and isotadeonal are the main active compounds. Pharmacological pre-clinical studies of the hydroalcoholic extract showed antihistaminic activity, anti-inflammatory, antipyretic and hypotensive activities emphasizing the popular indication for the treatment of intestinal pains and as a disinfectant in the treatment of skin infections. So, this species has potential to develop into an herbal medicine. Presently, however, there are just a few studies aimed at growing and improving its chemical quality.

Keywords *Polygonum punctatum* · Polygonaceae · *Erva-de-bicho* · Antihemorrhoidal drugs · Dotted smartweed

1 Taxonomic Characteristics

Polygonaceae is a cosmopolitan plant family, with most genera and species occurring in northern temperate regions and are herbs, shrubs, or rarely trees. The family consists of 31 genera and about 750 species. In the western hemisphere, 16 of these genera are restricted to western North America, with three disjuncts to Chile and Argentina (Melo 1999; Melo and França 2009). The genus *Polygonum* comprises about 300 species (Wang et al. 2005). Brazil is represented by 16 species including *Polygonum punctatum* Elliott (Melo and França 2009).

P. punctatum was described by Elliott and published in A Sketch of the Botany of South Carolina and Georgia 1 (5): 455–456, 1821 [1817]. It belongs to Equisetopsida class, Caryophyllales order, Polygonaceae family and *Polygonum* genus. It has an homonym, *Polygonum punctatum* Buch.-Ham. ex D. Don, published in Prodromus Florae Nepalensis 72. (1825) and two basionym *Discolenta punctate* (Elliott) Raf. and *Persicaria punctate* (Elliott) Samll. The species also has 35 synonyms, five from *Persicaria* genus and thirty from *Polygonum* genus (Tropicos 2015).

Synonyms *Persicaria punctata* (Elliott) Small, *Persicaria punctata* var. eciliata Small, *Persicaria punctata* var. robustior (Small) Small, *Persicaria punctata* var. tacubayana Nieuwl., *Persicaria robustior* (Small) E.P. Bicknell, *Polygonum acre* Kunth, Polygonum acre Lam., Polygonum acre var. aquatile Meisner in Martius,

Polygonum acre var. brachystachyum Meisn., Polygonum acre var. confertiflorum Meisn., Polygonum acre var. leptostachyum Meisn., Polygonum acre var. majus Meisn., Polygonum acre var. riparium Meisn., Polygonum antihaemorrhoidale fo. aquatile Mart, Polygonum antihaemorrhoidale fo. riparium Mart., *Polygonum var. aquatile* Mart., *Polygonum antihaemorrhoidale* var. riparium Mart., *Polygonum epilobioides* Wedd., *Polygonum hydropiperoides* Michx., *Polygonum punctatum* fo. longicollum Fassett, *Polygonum punctatum* fo. stipitatum Fassett, *Polygonum punctatum* var. aquatile (Mart.) Fassett, *Polygonum punctatum* var. confertiflorum (Meisn.) Fassett, *Polygonum punctatum* var. eciliatum Small, *Polygonum punctatum* var. ellipticum Fassett, *Polygonum punctatum* var. littorale Fassett, *Polygonum punctatum* var. majus (Meisn.) Fassett, *Polygonum punctatum* var. mexicanum Fassett, *Polygonum punctatum* var. parviflorum Fassett, *Polygonum punctatum* var. parvum Vict. & J. Rousseau, *Polygonum punctatum* var. riparium (Meisn.) Fassett, *Polygonum punctatum* var. robustius Small, *Polygonum punctatum* var. tacubayanum (Nieuwl.) Fassett, *Polygonum punctatum* var. typicum Fassett, Polygonum robustius (Small) Fernald.

2 Crude Drug Used

The infusion of the dried aerial parts is indicated as antihemorrhoidal. It must be prepared with 3 g of aerial parts in 150 mL of water, used externally, in a sitz bath, three times a day. It should not be used by pregnant and lactating women (Brasil 2011).

3 Major Chemical Constituents and Bioactive Compounds

Essential oils, flavonoids, triterpenoids, anthraquinones, coumarins, phenylpropanoids, tannins, and drimanes are secondary metabolites that are characteristic of the genus *Polygonum* (Fukuyama et al. 1980; Gilabert et al. 2014; López et al. 2006; Wang et al. 2005).

There is a range of secondary metabolites in the aerial parts of *P. punctatum*. Tannins, free anthraquinones, saponins, pelargonidin, flavonoids: quercetin, kaempferol, luteolin and acids: formic, acetic, valproic, lactic and malic (Teske and Trentini 1994). Polyphenols, coumarins, glycosides, were observed by Jácome et al. (2004) and volatile terpenoids such as sesquiterpenes: α-bisabolol (3.4%), various methylated phenol like α-tocopherol or vitamin E (3.6%), phytosterols: stigmasterol (2.1%) and β-sitosterol (29.9%) and the majors components, polygodial and isotadeonal (34.0%) were identified by Gilabert et al. (2014) showing that this species can be a promising source of drimane sesquiterpenes and phytoestrogens with important bioactivities.

The sesquiterpene polygodial is the active compound of *P. punctatum* and is responsible for most biological activities, especially the fungicidal activity of this species (Alves and Ribeiro 2001).

4 Morphological Description

This plant is 50–60 cm tall, branching occasionally and rather erect in habit. The alternate leaves are lanceolate-ovate or narrowly ovate, usually hairless, tapering to short petioles. At the base of each leaf, there is a sheath (ocrea) that wraps around the stem, which drops from the stem with age. The upper stems terminate in more or less erect spike-like racemes with small flowers that are sparsely distributed along its length. Each flower is about 3 mm long, white or greenish white, and its sepals have glandular dots that are either pale or dark-colored. The five sepals of the flower are more or less tightly folded against one other, while the short style is divided at its base into two or three segments. It has no noticeable floral scent. Each flower is replaced by an achene that is shiny, dark-brown to black, three-angled, and rather oblong (Hilty 2013; Lorenzi and Matos 2002; Melo 1999).

5 Geographical Distribution

P. punctatum is found throughout in the temperate, subtropical and tropical America, from Canada to Argentina (Pott and Pott 2000). In the USA it occurs in the south of California, Texas and Florida. In Canada, from Quebec to British Columbia. It also occurs in Mexico, Central America and West Indies (Mohlenbrock and Thomson 2009). In Brazil, it occurs in the North (Acre, Amazonas, Pará, Roraima), Northeast (Alagoas, Bahia, Ceará, Maranhão, Paraíba, Pernambuco, Piauí, Sergipe), Midwest (Distrito Federal, Goiás, Mato Grosso do Sul, Mato Grosso), Southeast (Espírito Santo, Minas Gerais, Rio de Janeiro) and in the South (Paraná, Rio Grande do Sul, Santa Catarina) (Melo 2014). It occurs in areas with climatic and environmental characteristics that are very different, such as the Amazon, Caatinga, Pantanal, Cerrado and Atlantic Forest, in the mixed ombrophilous forest (Melo 2000).

6 Ecological Requirements

Although widely distributed, this species occurs in humid environments. As an herbaceous species, emergent or amphibious, it is abundant in flooded fields, edge ponds, lowlands, wetlands, floodplain, clay or silty-organic soils and fertile sandy soils (Melo 2014). The plants often form colonies, of varying size, and require full or partial sun, moist to wet conditions, in mucky soil that is high in organic matter. This plant tolerates shallow standing water (Hilty 2013).

7 Collection Practice

The way of obtaining the plant material is still by collection in the natural populations. According to Ming et al. (2012) there is no commercial cultivation in Brazil, then harvesting is usually performed in moist or swampy areas.

Plant material recommended to use is aerial parts (leaves and stems), so it is important not to collect the flower.

In the USA the blooming period occurs from mid-summer to early fall, and lasts about 1–2 months (Hilty 2013).

The dried plant material should be stored away from light and heat, in tightly closed containers.

8 Traditional Use and Common Knowledge

P. punctatum is popularly known as *erva de bicho, cataia, persicária do Brasil, pimenteirad'água, barbasco*; in Spain is known as, *ajicillo, erva do bicho* and *caa-tai* and in the United States of America dotted *smartweed, water smartweed* and *water pepper* (Lorenzi and Matos 2002; Martínez-Crovetto 1981). It is used in folk medicine as an astringent, stimulant, diuretic, vermicide, antigonorrheic and antihemorrhoidal also being used locally against skin ulcers, erysipelas and arthritis (Lorenzi and Matos 2002; Mors et al. 2000). In traditional medicine from Toba Indians of the northeastern region in Argentina, *P. punctatum* is used as a disinfectant and also commonly used as a spice in Japanese cuisine (Martínez-Crovetto 1981). At traditional medicine, in rural areas of Colombia, a decoction of the aerial plant is used externally in the treatment of skin infections (Lopez et al. 2001).

9 Modern Medicine Based on Its Traditional Medicine Uses

The uses of *P. punctatum* are referred in literature to treat hemorrhoids and rheumatism, besides presenting diuretic, abortive and emmenagogue action (Lorenzi and Matos 2002). Aqueous extracts of *P. punctatum* have shown *in vitro* activity against infectious diseases. In an ethnopharmacological screening of medicinal plants used in Argentina, aqueous extracts of *P. punctatum* showed in vitro activity against Herpes Simplex Virus type 1 (HSV-1) and antiviral activity against respiratory syncytial virus (RSV) (Kott et al. 1999).

It has also been observed stronger antiviral and antimicrobial activities in the methanolic extract. According to Lopez et al. (2001) a complete virus inactivation was detected in Herpes Simplex Virus type 1 (HSV-1) in a minor dose described by Kott et al. (1999). In addition, a potent antimicrobial activity against *Streptococcus faecalis*, *Mycobacterium phlei*, *Bacillus subtilis* and *Staphylococcus aureus* was reported by Lopez et al. (2001), emphasizing the popular indication as a disinfectant and in the treatment of skin infections and the importance of further pharmacological studies.

Gilabert et al. (2014) provide evidence that support the antimicrobial use of *P. punctatum* against *Staphylococcus aureus* and *Pseudomonas aeruginosa*, as well as, demonstrates that the isotadeonal has been a bioactive compound able to control biofilm formation and bacterial growth of both human pathogens. Furthermore, the aqueous extract of the leaves has potential antidiarrhoeic effect by increasing the intestinal absorption of water (Almeida et al. 1995).

Toxicity assays of the methanolic and aqueous extracts, in a rat model, indicate low toxicity and relative safety of use, shown by a LC50>1 g/kg (Bhakuni et al. 1969).

The in vivo pharmacological studies with rats highlighted the bioactivity of *P. puctatum* extracts. According to Oliveira-Simões et al. (1989) the ethanol/water extract of the entire plant disclosed antihistaminic, anti-inflammatory, antipyretic and hypotensive activities. Alves and Ribeiro, (2001) reported anti-inflammatory activities of the decoction and the presence of polygodial, a sesquiterpene with a strong antibiotic compound (Kott et al. 1999; Lopez et al. 2001; Penna et al. 2001). It also displays anti-hyperalgesic properties in models of inflammatory and neurogenic pain (Mendes et al. 1998). All these reports support the ethnomedical use of this plant for the treatment of intestinal pains and infections.

10 Conclusions

P. punctatum is widely used in folk medicine. Preclinical studies validate the popular indication in the treatment of intestinal pains and as a disinfectant in the treatment of skin infections. The species seems a promising source of important bioactive compounds, such as drimane sesquiterpenes and phytoestrogens for the production of herbal medicines. Farther studies aimed at domestication and improving chemical quality are needed. With the growing market demand, its availability can be threatened dramatically, since these studies don't seem to take into account such important factors, such as plant regeneration, frequency and intensity of collection.

References

Almeida C, Karnikowski G, Foleto R, Baldisserotto B (1995) Antidiarrhoeic effect of plants used in popular medicine. Rev Saúde Publica 29(6):428–433

Alves T, Ribeiro F (2001) Polygodial, the fungitoxic component from the Brazilian medicinal plant Polygonum punctatum. Mem Inst Oswaldo Cruz 96(6):831–833

Bhakuni O, Dhar M, Dhar M, Dhawan B, Mehrotra B (1969) Screening of Indian plants for biological activity: part II. Indian J Exp Biol 7(1):250–262

Brasil (2011) In: Anvisa (ed). Agencia Nac. Vigil. SanitáriaFormulário de Fitoterápicos da Farmacopéia Brasileira. Agencia Nacional de Vigilancia Sanitária (ANVISA), Brasilia

Fukuyama Y, Sato T, Asakawa Y, Takemoto T (1980) A potent cytotoxic warburganal and related drimane-type sesquiterpenoids from *Polygonum hydropiper*. Phytochemistry 21(12):2895–2898

Gilabert M, Cartagenab E, Gilabert M, Cartagena E, Escobar G, Bardón A et al (2014) Volatile terpenoids from water pepper (polygonum punctatum) against pseudomonas aeruginosa and staphylococcus aureus virulence strategies. Glob J Agric Innov Res Dev 1:3–10

Hilty J (2013) Wetland wildflowers of Illinois [Internet]. World Wide Web Electron. Publ. Available from: http://www.illinoiswildflowers.info/wetland/plants/water_smartweed.htm

Jácome RLRP, Lopes DES, Recio RA, Macedo JF, Oliveira AB (2004) Caracterização farmacognóstica de Polygonum hydropiperoides Michaux e P. spectabile (Mart.) (Polygonaceae). Rev Bras Farmacogn Sociedade Brasileira Farmacogn 14(1):21–27

Kott V, Barbini L, Cruañes M, Muñoz JD, Vivot E, Cruañes J et al (1999) Antiviral activity in Argentine medicinal plants. J Ethnopharmacol 64:79–84

Lopez A, Hudson J, Towers GH (2001) Antiviral and antimicrobial activities of Colombian medicinal plants. J Ethnopharmacol 77:189–196

López SN, Sierra MG, Gattuso SJ, Furlán RL, Zacchino SA (2006) An unusual homoisoflavanone and a structurally-related dihydrochalcone from Polygonum ferrugineum (Polygonaceae). Phytochemistry 67(19):2152–2158

Lorenzi H, de Matos FA (2002) Plantas medicinais no Brasil: nativas e exóticas, 2nd edn. Nova Odessa, Plantarum

Martínez-Crovetto R (1981) Las plantas utilizadas en medicina popular en el noroeste de Corrientes (República Argentina). Miscelanea Educacion, Ministerio de Cultura y Educacion, San Miguel de Tucuman

Melo E (1999) Levantamento da família Polygonaceae no estado da Bahia, Brasil: espécies do semi-árido. Rodiguésia 50(76/77):29–47

Melo E (2000) Polygonaceae da cadeia do espinhaço, Brasil. Acta Bot Bras 14(3):273–300

Melo E (2014) Polygonaceae [Internet]. List. espécies da flora do Bras. Jard. Botânico do Rio Janeiro. Available from: http://floradobrasil.jbrj.gov.br/jabot/floradobrasil/FB13728

Melo E, França F (2009) Neotropical Polygonaceae – Neotropikey from Kew [Internet]. Milliken, W., Klitgård, B. Baracat, A. Interact. key Inf. Resour. Flower. plants Neotrop [cited 2015 Feb 14]. Available from: http://www.kew.org/science/tropamerica/neotropikey/families/Polygonaceae.htm

Mendes GL, Santos ARS, Campos MM, Tratsk KS, Yunes RA, Filho VC et al (1998) Anti-hyperalgesic properties of the extract and of the main sesquiterpene polygodial isolated from the barks of Drymis winteri (Winteraceae). Life Sci 63(5):369–381

Ming LC, Ferreira MI, Gonçalves GG (2012) Pesquisas agronômicas das plantas medicinais da Mata Atlântica regulamentadas pela ANVISA. Rev Bras Plantas Med Sociedade Brasileira Plantas Medicinais 14(spe):131–137

Mohlenbrock RH, Thomson PM (2009) Flowering plants: smartweeds to hazelnuts- illustrated Flora of Illinois. Book 13, 2nd edn. Southern Illinois University Press, Carbondale

Mors WB, Rizzini CT, Pereira NA, DeFilipps RA (2000) Medicinal plants of Brazil. Reference Publications, Michigan

Oliveira-Simões C, Ribeiro-do-Vale R, Poli A, Nicolau M, Zanin M (1989) Pharmacological investigation on Polygonum punctatum Elliott (P. acre H.B.K.) extracts. Part I. Tests in vivo. J Pharm Belg 44:275–284

Penna C, Marino S, Vivot E, Cruañes MC, de Muñoz DJ, Cruañes J et al (2001) Antimicrobial activity of Argentine plants used in the treatment of infectious diseases. Isolation of active compounds from Sebastiania brasiliensis. J Ethnopharmacol 77(1):37–40

Pott V, Pott A (2000) Plantas aquáticas do Pantanal. Brasilia, Embrapa Comunicacao Para Transferencia de Tecnologia

Teske M, Trentini AMM (1994) Compêndio de fitoterapia, 3rd edn. Herbarium Lab. Bot. Ltda, Curitiba

Tropicos (2015) Polygonum punctatum [Internet]. Missouri Bot. Gard. Available from: http://www.tropicos.org/Name/26000886

Wang K-J, Zhang Y-J, Yang C-R (2005) Antioxidant phenolic compounds from rhizomes of Polygonum paleaceum. J Ethnopharmacol 96(3):483–487

Ptychopetalum olacoides Benth.

Leonardo Frasson dos Reis and Fúlvio Rieli Mendes

Abstract The *Ptychopetalum olacoides* Benth. (Olacaceae) is an Amazonian tree popularly known as muirapuama or marapuama, among other names, which is used for several central nervous system related problems. The roots and occasionally the bark roots are the main medicinal parts employed and are prepared as an alcoholic infusion, tinctures, and tea. Phytochemical studies revealed that the roots contain tannins, flavonoids, and several terpenoids, while the presence of alkaloids is not clear. Most studies used ethanolic or hydroalcoholic extracts prepared with the roots of the plant. These studies indicate that the species has promising potential for treating central nervous system disorders, acting as an antidepressant, an anti-stress, a neuroprotective agent, and improving cognition. Although some herbal products contain *P. olacoides* in their composition, clinical studies are still needed to confirm the effects observed in pre-clinical studies.

Keywords *Ptychopetalum olacoides* · Olacaceae · Muirapuama · Neuroprotective · Neurotonic · Catuama

1 Taxonomic Characteristics

The *Ptychopetalum olacoides* Benth. is an Equisetopsida, subclass Magnoliidae, order Santalales, from the Olacaceae family (subfamily Olacoideae Sond, tribe 5) occurring exclusively in the north region of South America (Malécot and Nickrent 2008; Tropicos 2015). No botanical synonyms are accepted for the species, but the first botanical reports on the plant have erroneously referred to the species as *Acanthes viriles* and *Liriosma ovata* (Silva 1925).

L. F. dos Reis (✉)
Centro de Matemática, Computação e Cognição, Universidade Federal do ABC, São Bernardo do Campo, SP, Brazil

F. R. Mendes
Centro de Ciências Naturais e Humanas, Universidade Federal do ABC, São Bernardo do Campo, SP, Brazil

U. P. Albuquerque et al. (eds.), *Medicinal and Aromatic Plants of South America*, Medicinal and Aromatic Plants of the World 5,
https://doi.org/10.1007/978-94-024-1552-0_36

The *P. olacoides* presents great morphological similarity to the species *Ptychopetalum uncinatum* Anselmino, endemic to Brazil (Malécot et al. 2004). Fossil records dated to the Campanian period (Upper Cretaceous, about 83 million years ago) indicate the family as being evolutionarily close to the Anacolosidites family. The fact that the Olacaceae family presents great heterogeneity in its anatomical and pollen morphology, as well as in the form of nutrition, has led many researchers to consider this family as a polyphyletic group (Malécot and Nickrent 2008).

The species is popularly known as muirapuama, marapuama, muira puama, marapama, muiratam, muiratã, mirantã, pau-homem, potency wood, the tree of virility, and potenzholz (Silva 1926; Steinmetz 1962; Bonnard 1999; Lorenzi and Matos 2002; Siqueira et al. 2003). The names muirapuama and marapuama are also used to refer to other species: *Ptychopetalum uncinatum, Croton moritibensis, Croton echioides, and Liriosma ovata*, for which similar medicinal properties are attributed (Youngken 1921; Braz et al. 2012; Novello et al. 2012).

2 Crude Drug Used

The botanical drug generally is made up of root powder (Fig. 1) or the powdered bark of the root, but there are also reports of the use of bark and leaves of muirapuama (*P. olacoides*). Amazonian communities use preparations called "garrafadas" in

Fig. 1 Flask with powder of *P. olacoides* roots. (Photo by By Maša Sinreih in Valentina Vivod (Own work) via Wikimedia Commons, available at: http://commons.wikimedia.org/wiki/File%3APtychopetali2.JPG)

which the root of the plant is macerated in an alcoholic spirit or wine and consumed daily before meals (Siqueira et al. 1998; Piato et al. 2010). The plant drug is part of many commercially available herbal products, either as a single constituent (Testor-Plus®) or in combination with other plants (Catuama®, Herbal vX®, Masculex®) (Vaz et al. 1997; Bonnard 1999; Waynberg and Brewer 2000; Da Silva et al. 2002). The plant drug also was part of a nervine tonic called Esthenol (Silva 1925), no longer found in the market.

Muirapuama powder or its extract can be easily obtained via the Internet and by vendors of botanical material, and it is known that the commerce of adulterated product is common. In the Brazilian market the bark of *Croton echioides*, known as northeastern marapuama, is eventually sold as the bark or roots of *P. olacoides* (Novello et al. 2012). Rolim et al. (2006) developed an analytical method for quantificating total flavonoids in an emulsion containing *Trichillia catigua* and *P. olacoides*.

3 Morphological Description

P. olacoides is a small deciduous tree, ranging from 5 to 15 m tall with stems up to 25 cm in diameter. The leaves are alternate, oblong-elliptics, present a leathery appearance, are soft and bluish green in color, and when dried have a dark green to black color on the upper face and dark gray on the lower face (Lorenzi and Matos 2002). The inflorescences are racemes, with one or two axles, with strong perfume. The flowers are approximately 2 cm long, with a narrow calyx and five petals. The corolla is white, oblong, measuring 1.3–2 mm. The ovary's shape gradually widens in the end portion. The fruit is long, elliptical, initially green, changes color to pinkish lilac, and finally is blackish when ripe. The pericarp is thin and the endocarp is hard (Gruenwald et al. 2000). Figure 2 shows a classical botanical illustration of a flowering branch of *P. olacoides*.

4 Geographical Distribution and Collection Practice

The *P. olacoides* is endemic to the Amazon Rainforest (ombrophyllous forest) in the geographical area comprising the northern region of Brazil, Suriname, Guyana, and French Guiana, occurring in (Rossi 2015). The *P. olacoides* grows in poor, slightly acidic sandy soil, as is characteristic of the Amazon region.

The botanical material of *P. olacoides* is collected by pruning the branches and by removing the bark and roots. Although manufactured products containing *P. olacoides* are found in the market, only a few farmers are involved in cultivating the species. The material can be collected 3 years after planting; however, the method of production is still predominantly extractive, i.e., the plants are harvested directly from forest areas (Shanley et al. 2001).

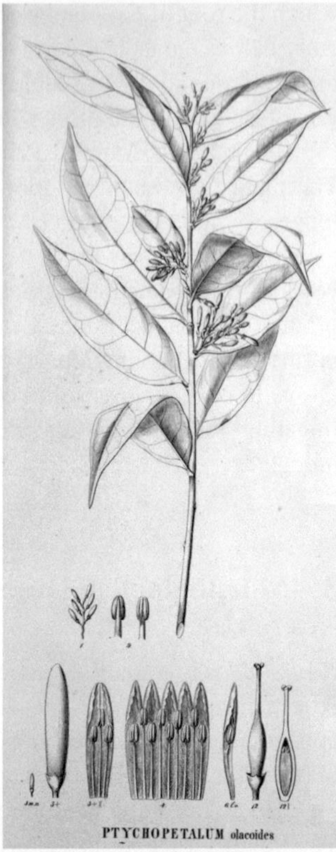

Fig. 2 Illustration of a flowered branch of *P. olacoides*. (Picture from *Flora Brasiliensis on-line*, available at: http://florabrasiliensis.cria.org.br/)

5 Major Chemical Constituents and Bioactive Compounds

Qualitative phytochemical studies indicated the presence of flavonoids, triterpenes, and saponins on the hydroalcoholic extract of muirapuama's bark (Paiva et al. 1998). Alkaloids, terpenic compounds, tannins, saponins, and flavonoids/compound phenolics were revealed in muirapuama's root extract in a preliminary phytochemical analysis carried out by Siqueira et al. (1998). The species is rich in terpenoids, in particular α-pinene (Bucek et al. 1987), a terpenoid present in other plant species, especially in conifers and in rosemary (*Rosmarinus officinalis*) (Chalchat et al. 1993). Among the volatile oils also reported are α-humulene, β-pinene, β-caryophyllene, camphene, and camphor, and in lower concentrations elixene, α-copaene, Δ-3-carene, linalool, and α-muurolene (Bucek et al. 1987). The presence of sterols such as β-sitosterol (Auterhoff and Pankow 1968; Gruenwald et al. 2000), a boldenone phytosterol precursor (Gallina et al. 2007), campesterol, stigmasterol, and lupeol was also described. Clerodane-type diterpenoids as ptychonolide and ptychonal, among others, were isolated from the methanolic extract of *P. olacoides* barks (Tang et al. 2008, 2009, 2011).

Although some studies have cited the presence of alkaloids (Peckolt 1901; Silva 1925; Siqueira et al. 1998), this class of substances has not been properly characterized (Tang et al. 2009; Piato et al. 2010). The belief that alkaloids are present in the species may have its origin in a misinterpretation of an old study that led to the purification of crystals called "muyrapuamina" which probably correspond to β-sitosterol or other sterols already identified in the species (Steinmetz 1962; Siqueira et al. 2003).

6 Traditional Use (Part(s) Used) and Common Knowledge

The medicinal use of *P. olacoides* has been described since the early twentieth century (Peckolt 1901; Youngken 1921; Silva 1925) and the plant properties were included in the first edition of the *Brazilian Pharmacopoeia* (Silva 1926). Other international publications have cited the medical use of the species (Anselmino 1932, 1933; Steinmetz 1962; Toyota et al. 1979), contributing to the species becoming internationally known. There is a preference for root or root bark (Siqueira et al. 2002), but other parts are utilized. The preparations are varied, the most common being the tea prepared from the bark of the roots, the intake of dried and ground plant parts, and alcoholic preparations, including tinctures. In Brazil alcoholic infusions are prepared with the roots, root bark, or bark and are used as an aphrodisiac, tonic, stimulant, and antitremor, while in Guyana people employ only the root as an aphrodisiac (Siqueira et al. 1998).

The roots and occasionally the muirapuama barks are traditionally used by the Amazon community as a tonic for treating a wide range of symptoms and diseases, including counteracting impotence, debility, asthenia, and neurasthenia (Mendes and Carlini 2007). Alcoholic infusion of the roots is cited for treating central nervous system (CNS) -related ailments and during highly stressful periods (Elisabetsky 1987). It is used by people recovering from CNS damages such as stroke, to treat nerve weakness in the elderly, for improving cognitive function and sexual performance, and as a remedy against fatigue and tremors (Siqueira et al. 2004). The decoction of the root is used in baths and massages to treat paralysis and beriberi. The root and bark tea is used to improve sexual function; for rheumatism, influenza, and for cardiac and gastrointestinal problems.

European explorers brought the plant drug to Europe, spreading its use in herbal medicine, especially in England. The species is described in *British Herbal Pharmacopoeia* as useful in treating dysentery and erectile dysfunction. In Germany *P. olacoides* is mentioned in the *German Pharmacopoeia* as a CNS tonic, to treat worms (*Ancylostoma duodenale*), menstrual problems, and rheumatism (Steinmetz 1962). In the United States it has gained great attention among herbalists and is used to treat erectile dysfunction, depression, menstrual cramps, neuralgia, and CNS disorders. A review describing the effect of herbals on human exercise performance cites the muirapuama for effects similar to testosterone, this effect being attributed to the presence of β- sitosterol (Bucci 2000).

7 Modern Medicine Based on Its Traditional Medicine Uses

Due to the widespread use of *P. olacoides* in traditional medicine as a neurotonic, several studies were conducted to evaluate its effects on the CNS (Duke 2000; Piato et al. 2010; Figueiró et al. 2010; Mendes 2011; Mendes et al. 2012; Howes and Houghton 2012). The psychopharmacological profile of the hydroalcoholic extract of the roots or rootbark indicated that the plant could interact with cholinergic, dopaminergic, and serotonergic systems (Siqueira et al. 1998). Da Silva et al. (2002) observed a moderate anxiogenic effect for doses from 30 to 300 mg/kg (ip) of muirapuama root ethanolic extract in the hole board test, without observing the decreased motor activity or motor incoordination in the rota-rod. The authors suggested that moderate anxiogenic action can be associated with the stimulating action of *P. olacoides*, contributing to an increased alertness as well as to physical and psychological resistance. In this regard, it was demonstrated that the ethanolic extract decreased anxiety and hyperglycemia induced by unpredictable chronic stress and increased the time to hypoxia-induced convulsion (Piato et al. 2010). These results suggest that the extract increases resistance to stress and has a normalizing function in the body, similar to adaptogenic plants.

The muirapuama is traditionally used by Amazonian communities to treat lassitude and lack of motivation, common symptoms of depression (Piato et al. 2008, 2009). The 70% hydroalcoholic extract of *P. olacoides* bark administered orally (100 mg/kg) decreased the immobility time in the forced swimming test, which was attributed to a possible antidepressant and anti-stress effect (Paiva et al. 1998). This effect was blocked by yohimbine, an α_2 adrenergic antagonist, suggesting that the mechanism of action involves these receptors. Oral treatment with an ethanolic extract of *P. olacoides* roots prevented the decrease of grooming and increase of serum corticosterone (both induced by unpredictable chronic mild stress, a model of depression in rats) similar to imipramine, suggesting an antidepressant-like effect (Piato et al. 2008). The extract also decreased the immobility time of mice in the forced swimming test and in the tail suspension test (Piato et al. 2009), two animal models of depression, confirming previous data. The pre-treatment with different drugs suggests that the antidepressant effect is possibly mediated by β-adrenergic and D_1 dopaminergic receptors (Piato et al. 2009).

Different studies have demonstrated the benefit of treatment with muirapuama on learning and memory in rodents. Acute administration of *P. olacoides* (50–100 mg/kg, ip or 800–1000 mg/kg, oral) improved the memory retrieval of young and old (14 months) mice in step-down avoidance inhibition test 24 h after training, without interfering with acquisition and consolidation (Da Silva et al. 2004). Further study showed a similar effect for short-term memory, evaluated 3 h after training (Da Silva et al. 2007). In this study, the authors also used a non-aversive paradigm (novel object recognition test) and showed that the extract increased the novel object recognition index 24 h after the training phase. The extract also reversed the amnesic effect of scopolamine (a cholinergic muscarinic antagonist) on short- and long-term memory on inhibitory step-down avoidance test and reversed the effect of

MK801 (a glutamatergic NMDA antagonist) on memory consolidation (Da Silva et al. 2009). The same authors also showed a synergistic effect between the extract of muirapuama and spiridone, a 5-HT_{2A} receptor antagonist, indicating that the promnesic effect should occur by multiple mechanisms (Da Silva et al. 2008).

Data suggest that the pro-cholinergic activity of the extract is also important for its promnesic function. The ethanolic extract of *P. olacoides* roots inhibited the acetylcholinesterase activity in the frontal cortex, hippocampus, and striatum both in *in vitro* study (incubation of the tissue with the extract) and in *ex vivo* (when animals were treated with the plant and the brain removed after 2 h to evaluate enzyme activity) (Siqueira et al. 2003). Oral treatment of mice with an ethanolic extract at a dose of 300 mg/kg inhibited the acetylcholinesterase activity in the hippocampus (CA_1 and CA_3 areas) and striatum, without altering the enzyme levels, indicating that the extract does not interfere with the enzyme synthesis (Figueiró et al. 2010).

The antioxidant activity of the extract also seems to contribute to its neuroprotective and pro-cognitive functions. The antioxidant potential of the ethanolic extract from root barks of *P. olacoides* was demonstrated against different challenges such as nitric oxide, superoxide, and peroxyl radicals (Siqueira et al. 2002). Siqueira et al. (2007) have shown that acute administration of the extract (100 mg/kg, ip) to 14-month-old mice decreased free radical production and lead to a decrease in lipid peroxidation in important cerebral areas. Moreover, the extract increased the activity of glutathione peroxidase and catalase in the hippocampus, while the catalase activity was also increased in the cortex, striatum, and cerebellum.

The neuroprotective action of the ethanolic extract of the rootbarks was demonstrated in hippocampal slices deprived of oxygen and glucose for 45 min, followed by reoxygenation. The incubation of the slices at a concentration of 0.6 mg/mL of extract increased cell viability by 65% (as assessed by MTT assay) and decreased by 30% the levels of free radicals formed (Siqueira et al. 2004). Furthermore, incubating hippocampal slices with the extract led to an increase of mitochondrial activity by approximately 40%, without affecting the levels of free radicals when the tissue was not deprived of oxygen and glucose (Siqueira et al. 2004). Clerodane diterpenoids isolated from the methanolic extract of muirapuama's bark exhibited neurite-outgrowth-promoting activities on NGF-mediated PC_{12} cells (Tang et al. 2008, 2009). Mice that had previously received an iv injection of β-amyloid ($A\beta_{1-42}$), when treated orally with muirapuama extract (800 mg/kg, 14 days) showed decreased Aß deposits and did not present cognitive impairment, evaluated in the step-down avoidance test (Figueiró et al. 2011). Moreover, the treatment reduced the astrogliosis and CA_1 hippocampus loss, although it did not affect the hippocampal BDNF levels. Considering the multifactorial nature of neurodegeneration, the several effects observed in *P. olacoides* make it a promising candidate for treating neurodegenerative diseases (Mendes et al. 2012; Howes and Houghton 2012).

Most pharmacological studies were conducted by the Elisabetsky group and employed an ethanolic extract from the roots of *P. olacoides*, whose preparation method was the object of a patent (Elisabetsky et al. 2005). A traditional product called Viriliflora® was composed by the tinctures of *P. olacoides, Tynanthus fasciculatus,* and *Anemopaegma mirandum*, Brazilian plants popularly used as aphrodisiacs

(Mendes 2011). A preclinical toxicology study of another herbal medicine containing *P. olacoides*, *Anemopaegma arvense*, *Paullinia cupana*, *Cola nitida*, *Passiflora alata*, and thiamin evaluated the effect of its oral administration in rabbits (Mello et al. 2010). Treatment for 30 days in a dose ten times that prescribed for human was considered innocuous (Mello et al. 2010).

Catuama®, an herbal medicine composed by the hydroalcoholic extract of *Ptychopelatum olacoides* leaves, *Trichilia catigua* bark, *Paullinia cupana* seeds, and *Zinziber officinalis* rhizomes, is indicated for managing several disorders, including mental and physical fatigue, stress, and muscular asthenia (Campos et al. 2004), but due to its composition the herbal formulation is also reputed to be an aphrodisiac.

Catuama® induced vasorelaxant action in a study with rodents and the effect was both endothelium-dependent and independent, depending on the tissue tested (Calixto and Cabrini 1997). One of the most reported traditional uses – also responsible for some of the popular names of muirapuama (Potency wood, Pau-homen) – is treating erectile dysfunction. A short-lived and dose-dependent relaxant effect on rabbit corpus cavernosum induced by Catuama® extract was shown by Antunes et al. (2001). Catuama® administered orally showed an antinociceptive effect in models of chemical and thermal nociception via interaction with the nitric oxide pathway and the opioid system (Vaz et al. 1997). Campos et al. (2004) observed an antidepressant-like effect in two animal models following acute and chronic administration of Catuama®. Using rat brain preparations, these authors showed that Catuama® increased the release of dopamine and serotonin and inhibited in a concentration-dependent manner the in vitro synaptosomal uptake of noradrenaline, dopamine, and serotonin, while the treatment of rats for 42 days also decreased the uptake of serotonin and dopamine (Campos et al. 2004).

A Phase I clinical study was performed with Catuama® in which 25 mL of the herbal was administered twice a day for 28 days. The haematological and biochemical analysis did not show alterations compared to the normal range and only minor symptoms and signals (such as insomnia and gastrointestinal issues) were related to use of the drug (Oliveira et al. 2005).

8 Conclusion

P. olacoides appears as a species with promising potential in treating central nervous system disorders. Most studies used ethanolic or hydroalcoholic extracts prepared with the roots of the plant, but the chemical composition of these extracts and the active ingredients responsible for their biological effects are not well understood. Although several studies have evaluated the biological properties of *P. olacoides* and the plant is present in various herbal products, more clinical studies are needed to confirm the effects observed in pre-clinical studies.

Acknowledgements The authors thank Prof. Wayne Losano for the grammar review.

References

Anselmino E (1932) Die Stammpflanzen der droge Muira-puama. Notizblatt Königl Bot Gartens Mus Berl 11(107):623–629 German

Anselmino E (1933) Die Stammpflanzen von Muira-puama. Arch Pharm (Weinheim) 271(5):296–314 German

Antunes E, Gordo WM, de Oliveira JF, Teixeira CE, Hyslop S, De Nucci G (2001) The relaxation of isolated rabbit corpus cavernosum by the herbal medicine Catuama and its constituents. Phytother Res 15(5):416–421

Auterhoff H, Pankow E (1968) Inhaltsstoffe von Muira puama. Arch Pharm (Weinheim) 301(7):481–489 German

Bonnard M (1999) The Viagra alternative: the complete guide to overcoming erectile dysfunction naturally. Healing Arts Press, Rochester, p 226

Braz R, Wolf LG, Lopes GC, Mello JCP (2012) Quality control and TLC profile data on selected plant species commonly found in the Brazilian market. Rev Bras Farmacog 22(5):1111–1117

Bucci LR (2000 Aug) Selected herbals and human exercise performance. Am J Clin Nutr 72(2 Suppl):624S–636S

Bucek EU, Fournier G, Dadoun H (1987) Volatile constituents of *Ptychopetalum olacoides* root oil. Planta Med 53(2):231

Calixto JB, Cabrini DA (1997) Herbal medicine Catuama induces endothelium-dependent and independent vasorelaxant action on isolated vessels from rats, Guinea-pigs and rabbits. Phytother Res 11(1):32–38

Campos MM, Fernandes ES, Ferreira J, Bortolanza LB, Santos ARS, Calixto JB (2004) Pharmacological and neurochemical evidence for antidepressant-like effects of the herbal product Catuama. Pharmacol Biochem Behav 78(4):757–764

Chalchat J-C, Garry R-P, Michet A, Benjilali B, Chabart JL (1993) Essential oils of Rosemary (*Rosmarinus officinalis* L.). The chemical composition of oils of various origins (Morocco, Spain, France). J Essent Oil Res 5(6):613–618

Da Silva AL, Bardini S, Nunes DS, Elisabetsky E (2002) Anxiogenic properties of *Ptychopetalum olacoides* Benth. (Marapuama). Phytother Res 16(3):223–226

Da Silva AL, Piato ALS, Bardini S, Netto CA, Nunes DS, Elisabetsky E (2004) Memory retrieval improvement by *Ptychopetalum olacoides* in young and aging mice. J Ethnopharmacol 95(2–3):199–203

Da Silva AL, Piato AL, Ferreira JG, Martins BS, Nunes DS, Elisabetsky E (2007) Promnesic effects of *Ptychopetalum olacoides* in aversive and non-aversive learning paradigms. J Ethnopharmacol 109(3):449–457

Da Silva AL, Ferreira JG, da Silva Martins B, Oliveira S, Mai N, Nunes DS et al (2008) Serotonin receptors contribute to the promnesic effects of *P. olacoides* (Marapuama). Physiol Behav 95(1–2):88–92

Da Silva AL, da Silva Martins B, Linck Vde M, Herrmann AP, Mai N, Nunes DS et al (2009) MK801- and scopolamine-induced amnesias are reversed by an Amazonian herbal locally used as a "brain tonic". Psychopharmacology 202(1–3):165–172

Duke JA (2000) Handbook of phytochemical constituents of GRAS herbs and other economic plants. Herbal Reference Library. CRC Press, Boca Raton, p 497

Elisabetsky E (1987) From indigenous disease concepts to laboratory working hypothesis: the case of "nerve tonics" from the Brazilian Amazon, Provisional Report Series, vol 19. International Foundation for Science, Stockholm, p S-11438

Elisabetsky E, Netto CA, da Silva AL, Siqueira IS, Nunes DS (2005, December 20) BR Patent No PI 0307637-4 A

Figueiró M, Ilha J, Pochmann D, Porciúncula LO, Xavier LL, Achaval M et al (2010) Acetylcholinesterase inhibition in cognition-relevant brain areas of mice treated with a nootropic Amazonian herbal (Marapuama). Phytomedicine 17(12):956–962

Figueiró M, Ilha J, Linck VM, Herrmann AP, Nardin P, Menezes CB et al (2011) The Amazonian herbal Marapuama attenuates cognitive impairment and neuroglial degeneration in a mouse Alzheimer model. Phytomedicine 18(4):327–333

Gallina G, Ferretti G, Merlanti R, Civitareale C, Capolongo F, Draisci R et al (2007) Boldenone, boldione, and milk replacers in the diet of veal calves: the effects of phytosterol content on the urinary excretion of boldenone metabolites. J Agric Food Chem 55(20):8275–8283

Gruenwald J, Brendler T, Jaenicke C (2000) PDR for herbal medicines. Thomson Medical Economics Co, Montvale, p 990

Howes MJR, Houghton PJ (2012) Ethnobotanical treatment strategies against Alzheimer's disease. Curr Alzheimer Res 9(1):67–85

Lorenzi H, Matos FJA (2002) Plantas medicinais no Brasil: nativas e exóticas, 2nd edn. Instituto Plantarum de Estudos da Flora, Nova Odessa, p 512

Malécot V, Nickrent DL (2008) Molecular phylogenetic relationships of Olacaceae and related Santalales. Syst Bot: Am Soc Plant Taxonomists 33(1):97–106

Malécot V, Nickrent DL, Baas P, van den Oever L, Lobreau-Callen DA (2004) morphological cladistic analysis of Olacaceae. Syst Bot Am Soc Plant Taxonomists 29(3):569–586

Mello JRB, Mello FB, Langeloh A (2010) Toxicity study of a phytotherapic with *Anemopaegma mirandum*, *Cola nitida*, *Passiflora alata*, *Paullinia cupana*, *Ptychopetalum olacoides* and thiamin in rabbits. Lat Am J Pharm 29(8):1431–1435

Mendes FR (2011) Tonic, fortifier and aphrodisiac: adaptogens in the Brazilian folk medicine. Rev Bras Farmacog 21(4):754–763

Mendes FR, Carlini EA (2007) Brazilian plants as possible adaptogens: an ethnopharmacological survey of books edited in Brazil. J Ethnopharmacol 109(3):493–500

Mendes FR, Negri G, Duarte-Almeida JM, Tabach R, Carlini EA (2012) The action of plants and their constituents on the central nervous system. In: Cechinel Filho V. Planta bioactives and drug discovery: principles, practice, and perspectives, 4th edn, pp 161–204

Novello C, Marques LC, Miyazaki CR, Milaneze-Gutierre MA, Carneiro-Torres DS, Sarragiotto MH, Mello JCP (2012) Morphoanatomy and pharmacognostic study of the wood of Croton echioides, the northeastern marapuama. Rev Bras Pharmacog 22(5):946–956

Oliveira CH, Moraes MEA, Moraes MO, Bezerra FAF, Abib E, De Nucci G (2005) Clinical toxicology study of an herbal medicinal extract of Paullinia cupana, Trichilia catigua, Ptychopetalum olacoides and Zingiber officinale (Catuama) in healthy volunteers. Phytother Res 19(1):54–57

Paiva LAF, Rao VSN, Silveira ER (1998) Effects of Ptychopetalum olacoides extract on mouse behaviour in forced swimming and open field tests. Phytother Res 12(4):294–296

Peckolt T (1901) Heil und Nutzpflanzen Brasiliens. Ber Dtsch Pharm Ges 11:40

Piato AL, Detanico BC, Jesus JF, Lhullier FLR, Nunes DS, Elisabetsky E (2008) Effects of Marapuama in the chronic mild stress model: further indication of antidepressant properties. J Ethnopharmacol 118(2):300–304

Piato AL, Rizon LP, Martins BS, Nunes DS, Elisabetsky E (2009) Antidepressant profile of *Ptychopetalum olacoides* Bentham (Marapuama) in mice. Phytother Res 23(4):519–524

Piato AL, Detanico BC, Linck VM, Herrmann AP, Nunes DS, Elisabetsky E (2010) Anti-stress effects of the "tonic" *Ptychopetalum olacoides* (Marapuama) in mice. Phytomedicine 17(3–4):248–253

Rolim A, Oishi T, Maciel CPM, Zague V, Pinto CASO, Kaneko TM et al (2006) Total flavonoids quantification from O/W emulsion with extract of Brazilian plants. Int J Pharm 308(1–2):107–114

Rossi AMA (2015) [Internet]. Olacaceae in Lista de Espécies da Flora do Brasil. Jardim Botânico do Rio de Janeiro. Access 2015. Available from: http://floradobrasil.jbrj.gov.br/

Shanley P, Luz L, Swingland IR (2001) The faint promise of a distant market: a survey of Belém's trade in non-timber forest products. Biodivers Conserv 11(4):615–636

Silva RAD (1925) Plantas medicinaes brasileiras. Estudo botanico e pharmacognostico. Muirapuama. Rev Bras Med Pharm 1(1):37–41 Portuguese

Silva RAD (1926) Pharmacopeia dos Estados Unidos do Brasil. Companhia Editora Nacional, Rio de Janeiro

Siqueira IR, Lara DR, Silva D, Gaieski FS, Nunes DS, Elisabetsky E (1998) Psychopharmacological Properties of *Ptychopetalum olacoides* bentham (Olacaceae). Pharm Biol 36(5):327–334

Siqueira IR, Cordova CAS, Creczynski-Pasa TB, Elisabetsky E, Nunes DS, Netto CA (2002) Antioxidant action of an ethanol extract of *Ptychopetalum olacoides*. Pharm Biol 40(5):374–379

Siqueira IR, Fochesatto C, da Silva AL, Nunes DS, Battastini AM, Netto CA et al (2003) *Ptychopetalum olacoides*, a traditional Amazonian "nerve tonic", possesses anticholinesterase activity. Pharmacol Biochem Behav 75(3):645–650

Siqueira IR, Cimarosti H, Fochesatto C, Nunes DS, Salbego C, Elisabetsky E et al (2004) Neuroprotective effects of *Ptychopetalum olacoides* Bentham (Olacaceae) on oxygen and glucose deprivation induced damage in rat hippocampal slices. Life Sci 75(15):1897–1906

Siqueira IR, Fochesatto C, Torres ILS, da Silva AL, Nunes DS, Elisabetsky E et al (2007) Antioxidant activities of Ptychopetalum olacoides ("muirapuama") in mice brain. Phytomedicine 14(11):763–769

Steinmetz EF (1962) Muira Puama ("Potency wood"). Pharm Biol 2:229–232

Tang W, Hioki H, Harada K, Kubo M, Fukuyama Y (2008) Clerodane diterpenoids with NGF-potentiating activity from *Ptychopetalum olacoides*. J Nat Prod 71:1760–1763

Tang W, Kubo M, Harada K, Hioki H, Fukuyama Y (2009) Novel NGF-potentiating diterpenoids from a Brazilian medicinal plant, *Ptychopetalum olacoides*. Bioorg Med Chem Lett 19(3):882–886

Tang W, Harada K, Kubo M, Hioki H, Fukuyama Y (2011) Eight new clerodane diterpenoids from the bark of *Ptychopetalum olacoides*. Nat Prod Commun 6(3):327–332

Toyota A, Ninomiya R, Kobayashi H, Kawanish K, Uhara Y, Kato A et al (1979) Studies of Brazilian crude drugs 1 muirapurama. Shoyakugaku Zasshi 33:57–64

Tropicos.org (2015) [Internet]. Missouri Botanical Garden. Access 2015 Apr 16. Available from: http://www.tropicos.org

Vaz ZR, Mata LV, Calixto JB (1997) Analgesic effect of the herbal medicine Catuama in thermal and chemical models of nociception in mice. Phyther Res 11(2):101–106

Waynberg J, Brewer S (2000) Effects of Herbal vX on libido and sexual activity in premenopausal and postmenopausal women. Adv Ther 17(5):255–262

Youngken HW (1921) Observations of Muira-Puama. J Am Pharm Assoc 10(9):690–692

Punica granatum L.

André dos Santos Souza, José Ribamar de Souza Jr., Daniel Carvalho Pires Sousa, and Ulysses Paulino Albuquerque

Punica granatum L.
Photo: Gerrit Davidse
Available in: http://www.tropicos.org/Image/54964

A. S. Souza (✉) · J. R. d. Souza Jr. · D. C. P. Sousa
Laboratório de Ecologia e Evolução de Sistemas Socioecológicos, Departmento de Botânica, Universidade Federal de Pernambuco, Recife, PE, Brazil

U. P. Albuquerque
Departamento de Botânica, Centro de Biociências, Universidade Federal de Pernambuco, Recife, Brazil

U. P. Albuquerque et al. (eds.), *Medicinal and Aromatic Plants of South America*, Medicinal and Aromatic Plants of the World 5,
https://doi.org/10.1007/978-94-024-1552-0_37

Abstract *Punica granatum* L., popularly known as "romã," "romanzeira," "mangrano" and "granado" in Latin America and as "pomegranate" in English, recently has been reported to have a high medicinal value. This plant is native to Southwest Asia, more specifically the Middle East, and grows mainly in arid and dry regions with direct solar incidence. The main plant organs used for medicinal purposes are the fruits, particularly the fruit peel, which is usually used to treat infections and inflammation, and the spongy membrane (rag) surrounding the seeds, which is used to produce juice. The main compounds responsible for the biological activity of fruits are polyphenols, ellagic acids and tannins. The wide range and versatility of the medicinal uses of *P. granatum* have made it the focus of several studies, specifically for its medicinal potential against inflammation and bacterial and fungal infections.

Keywords Pomegranate · Medicinal use · Inflamation · Phytochemical compounds

1 Taxonomic Characteristics

Punica granatum L. is popularly known as "romã," "romazeira," "mangrano" and "granado" in Latin American countries. It belongs to the family Lythraceae, which includes more than 30 genera and approximately 600 species.

Its botanical Synonyms are *Punica malus* L., *Punica nana* L., *Punica spinosa* Lam., *Punica florida* Salisb. and *Punica grandiflora* hort. ex Steud.

2 Crude Drug Used

The main organ of *P. granatum* that is used for medicinal purposes is the fruit. It is composed of approximately 50% peel, 40% pulp (edible part) and 10% seeds (Sadeghipour et al. 2014; Gavanji et al. 2014). The fruit peel has been reportedly used to treat infections, the fruit pulp has been used to treat intestinal problems (consumed unprocessed or in juices), and the flowers have been used to prevent diabetes and treat wounds when applied directly to the skin (Dipak et al. 2012).

3 Major Chemical Constituents and Bioactive Compounds

One hundred grams of *P. granatum* L. fruit contains approximately 78% water, 10 mg calcium, 1.6% protein, 70 mg phosphorus, 0.1% lipids, 0.3 mg iron, 0.7% minerals, 16 mg ascorbic acid, 14.5% carbohydrates, a negligible amount of B complex vitamins, 5.1% fiber, and 65 Kcal of energy (Bhowmik et al. 2013). Ellagic acid, a strong polyphenol, is one of the main bioactive constituents of *P. granatum* fruits and is responsible for the biological activity of these fruits (Dipak et al. 2012). Namely, the presence of this hydrosoluble tannin with proven biological activity may decrease the effects of intestinal disturbances (Qnais et al. 2007). Other important components that also exhibit biological activity are linolenic acids (alpha, oleic, palmitic, punicic, and stearic acid); eicosanoic, citric, malic, gallic, protocatechuic, chlorogenic, caffeic and ferulic acids; catechin; phloridzin and quercetin (Bhowmik et al. 2013; Quattrucci et al. 2013). The fruit and flowers contain considerable amounts of ellagitannins, punicic acid, flavonoids, anthocyanins and anthocyanidins (Quattrucci et al. 2013).

The amount of bioactive compounds that are present in pomegranates may vary with soil fertility, irrigation, and several agroclimatic factors. For example, pomegranate trees treated with potassium nitrate were observed to contain higher vitamin C concentrations than control trees without potassium nitrate treatment (Khayyat et al. 2012).

Because pomegranates are non-climacteric fruits, they may exhibit several changes after harvest that lead to physiological and biochemical changes. These changes include weight loss, peel darkening and aril discoloration (Mphahlele et al. 2014; Gil et al. 1996; Ghafir et al. 2010; Lee and Kader 2000). Thus, several postharvest treatments are necessary to maintain the nutritional quality of the fruits, including temperature control, atmospheric control, polypropylene packaging, acetylsalicylic acid application, and fruit dehydration (Mphahlele et al. 2014; Artes et al. 2000; Sayyari et al. 2010). Temperature and relative humidity were shown to directly affect the quantity of vitamin C and anthocyanins in pomegranates. In general, anthocyanins are labile compounds that are easily degraded in response to the environmental conditions. An optimal temperature, storage period and processing time are required to prevent the instability of the anthocyanins and other compounds that are present in pomegranates (see Pilano et al. 1985; Markakis 1982; García-Viguera et al. 1999; Martí et al. 2001; Mphahlele et al. 2014). The loss of anthocyanins has also been attributed to other factors, such as pH, acidity, sugar degradation products, oxygen and ascorbic acid (Withy et al. 1993).

4 Morphological Description

Pomegranate plants are ramified, woody shrubs that can reach 1.5–5 m in height. The leaves are small (3–7 cm in length and 2 cm in width), dark green, spear-shaped, tough, shiny, and membranaceous (Holland et al. 2009; Levin 2006). The flowers are red-orange, with five to eight petals, are approximately 3 cm in diameter and are arranged on the ends of branches (Fahan 1976). The fruit is a spherical (6–12 cm), globose berry with a sweet and slightly acid flavor, a pleasant odor, many angulose seeds arranged in layers and surrounded by pulpous arils, and a yellow or reddish coriaceous peel that is generally stained dark (Dipak et al. 2012; Catão 2006).

5 Geographical Distribution

P. granatum L. is native to Southwest Asia, more specifically to the Middle East. Spain is the main global producer of pomegranate, followed by Iran (Parvizi et al. 2014; Nuncio-Jauregui et al. 2014). Because pomegranate is an excellent antioxidant and a source of tannins, flavonoids, anthocyanidins and minerals that is well adapted to many different climate conditions, it has been grown in several countries, including India, Egypt, Lebanon, China, France, the United States, Oman, Syria, Italy, Greece, Cyprus, Israel, Chile, Portugal, Morocco, Russia, Japan, Brazil, and, more recently, South Africa (see Nuncio-Jauregui et al. 2014; Mphahlele et al. 2014; Dipak et al. 2012; Lorenzi and Souza 2001; Mmarm 2009).

6 Ecological Requirements

P. granatum primarily grows in arid and dry geographical areas (Gomes 2007; Nuncio-Jauregui et al. 2014). It grows better with direct sunlight and in slightly alkaline (pH <7.5) and clayey soils. However, it is quite adaptable and may also grow in temperate to subtropical climates with hot summers and cold winters, enabling it to grow from North to South America (Pedriali et al. 2010; Dipak et al. 2012) and even in desert regions (Aseri et al. 2008). Some studies have developed soil preparation strategies for its large-scale production in arid regions, which, in some cases, include the use of biofertilizers, namely nitrogen-fixing bacteria, that contribute to the growth and biomass production of *P. granatum* by producing vigorous plants that are capable of surviving in soils with extreme conditions (Aseri et al. 2008). Commercial orchards of *P. granatum* can be found in the Thar desert in India, despite its nutrient-deficient sandy soils; high wind speeds evaporation rates and solar radiation; and irregular rainfall distribution (Panwar and Tarafdar 2006).

7 Collection Practice

The fruit is the main plant organ that is harvested for use in both traditional (Quattrucci et al. 2013; Gavanji et al. 2014) and modern medicine (Legua et al. 2012; Khan and Hanee 2011; Qnais et al. 2007; Mansourian et al. 2014). The leaves are reported to be effective against obesity (Sadeguipour et al. 2014). A few studies have reported collection practices that lead to tree destruction, and currently, several different *P. granatum* cultivars are produced worldwide.

8 Traditional Use (Part(s) Used) and Common Knowledge

P. granatum is described as a holy fruit in the Old Testament of the Bible, the Koran, the Torah, and the Babylonian Talmud, to which the powers of fertility, abundance and good luck are attributed (Miguel et al. 2010). It can be consumed unprocessed or processed in juices, canned beverages, alcoholic beverages, jellies, or aromatized beverages (Legua et al. 2012).

Its use for the treatment of heart disease, heartburn, diarrhea, thrush, cancer, bone disease, diabetes, anemia, skin infections, wounds, bronchitis, and hair loss, in addition to as an aphrodisiac and blood tonic, has been reported. It has also been reported to have astringent, homeostatic, antibacterial, antimicrobial, antiviral and antiparasitic activities (see Bhowmik et al. 2013; Dipak et al. 2012; Sadeguipour et al. 2014; Mphahlele et al. 2014; Gavanji et al. 2014; Al-Olayan et al. 2014; Nuncio-Jauregui et al. 2014; Fawole and Opara 2013; Mansourian et al. 2014).

9 Modern Medicine Based on Its Traditional Medicine Uses

The *P. granatum* fruits are used as food that also possesses medicinal properties. The fruit peel is used by the pharmaceutical industry for the production of antibacterial drugs, the pulp juice and flower extracts are used to obtain antioxidant compounds, and the seeds are considered blood tonics (Bhowmik et al. 2013; Gavanji et al. 2014). The maximum oxidative potential and high polyphenol concentrations are observed with approximately 120–150 g of fruit (Nuncio-Jauregui et al. 2014).

The increasing interest in the healthy way of nutrition, over the last decades, has led to an increased production of *P. granatum* (Mphahlele et al. 2014), and its importance as a traditional medicinal plant has made it the focus of laboratory studies (Manera et al. 2013). Its main medicinal compounds are ellagitannins, which are antioxidant polyphenols with antidiarrheal, antiseptic, antimicrobial and homeostatic effects; punic acid, which showed anticancer activity in vitro; and flavonoids, which have anti-inflammatory, neuroprotective and antihyperglycemic properties (Dipak et al. 2012; Al-Olayan et al. 2014). Mansourian et al. (2014) observed that *P. granatum* extracts (100 mg/mL) are effective against *Candida albicans,* the main cause of thrush in patients with low immunological resistance. Khan and Hanee (2011) reported the presence of phenols (flavonoids and tannins) in the pericarp, leaves and flowers and complex polysaccharides in the fruit peel; all of these compounds were effective against *Escherichia coli, Pseudomonas aeruginosa* and *Staphylococcus aureus.* Sadeguipour et al. (2014) reported that the plant extracts exhibited significant antilipidemic activity and may be used to reduce the patients' lipid levels. The juice has high concentrations of vitamin C, A and E (Bhowmik et al. 2013) and was found to have a significant effect against diarrhea, confirming several traditional reports (Qnais et al. 2007). Pomegranate juice may also prevent the formation of and treat malignant cells by preventing their growth, increasing apoptosis, decreasing inflammation, decreasing metastasis, and decreasing their resistance to the drugs used to treat cancer (Lansky and Newman 2007). Julie (2008) also reported that pomegranate juice can be used to treat prostate cancer and arteriosclerosis (by inhibiting the lipid peroxidation of plasma lipoproteins) and to promote platelet aggregation, hyperlipidemia (via decreasing cholesterol and promoting its absorption), and fecal excretion. Pomegranate juice was also an effective treatment for hypertension by decreasing the activity of angiotensin converting enzyme (ACE); it was also used to treat myocardial ischemia.

Several new pomegranate-based cosmetic products are being commercialized by traditional producers. Some studies have shown its efficacy in cosmetic treatments. For example, the seed oil and aqueous extract of the fruit peel stimulate the production of keratinocytes, fibroblasts and collagen, which are necessary for the reconstruction of cutaneous tissue, particularly in diabetic patients (Aslam et al. 2006).

Conclusions The information presented here clearly indicates that pomegranate is a species of great importance for not only the food, pharmaceutical and cosmetic industries but also the traditional medicine, that specifically uses it to treat infections, inflammations and fungal diseases.

Acknowledgments We are especially grateful to the National Institute of Science and Technology in Ethnobiology, Bioprospecting and Nature Conservation, certified by CNPq, with financial support from FACEPE (Foundation for the Support of Science and Technology of the State of Pernambuco).

References

Al-Olayan EM, El-Khadragy MF, Metwally DM, Moneim AEA (2014) Protective effects of pomegranate (*Punica granatum*) juice on testes against carbon tetrachloride intoxication in rats. BMC Complement Altern Med 14:164

Artes F, Villaescusa R, Tudela JA (2000) Modified atmosphere-packaging of pomegranate. J Food Sci 65:1112–1116

Aseri GK, Jain N, Panwar J, Rao AV, Meghwal PR (2008) Biofertilizers improve plant growt, fruit yield, nutrition, metabolism and rhizosphere enzymae activities of Pomegranate (*Punica granatum* L.) in Indian Thar Desert. Sci Hortic 117:130–135

Aslam MN, Lansky EP, Varani J (2006) Pomeranate as a cosmeceutical source: Pomegranate fractions promote proliferation and procollagem synthesis and inhibit matrix metalloproteinase-1 production in human skin cells. J Ethnopharmacol 103:311–318

Bhowmik D, Gopinath H, Kumar BP, Kumar K (2013) Medicinal uses of Punica granatum and its health benefits. J Pharmacogn Phytochem 1(5):28–35

Catão RMR (2006) (org.). Atividade antimicrobiana "in vitro" do extrato etanolíco de *Punica granatum Linn.* (Romã) sobre isolados dos ambulatoriais de *Staphylococcus aureus*. Revista Brasileira de Análises Clínicas 38(2):111–114

Dipak G, Axay P, Manodeep C, Jagadish VK (2012) Phytochemical and Pharmacological profile of *Punica granatum*: an overview. Int Res J Pharm 3(2):65–68

Fahan A (1976) The flower. In: Plant anatomy. Hakkibutz Hameuhad Publi, Jerusalem, pp 321–394

Fawole OA, Opara UL (2013) Changes in physical properties, chemical and elemental composition and antioxidant capacity of pomegranate (cv. Ruby) fruit at five maturity stages. Sci Hortic 150:37–46

García-Viguera C, Zafrilla P, Romero F, Abellán P, Artés F, Tomás-Barberán FA (1999) Color stability of strawberry jam as affected by cultivar and storage temperature. J Food Sci 64:243–247

Gavanji S, Larki B, Bakhtari A (2014) The effect of extract of *Punica granatum* var. *pleniflora* for treatment of minor recurrent aphthous stomatitis. Integr Med Res 3:83–90

Ghafir SAM, Ibrahim IZ, Zaied SA (2010) Response of local variety 'Shlefy' pomegranate fruits to packaging and cold storage. 6th International Postharvest Symposium. Acta Hortic 877:427–432

Gil M, Tomas-Barberan FA, Hess-Pierce B, Holcroft DM, Kader AA (1996) Antioxidant activity of pomegranate juice and its relationship with phenolic composition and processing. J Agric Food Chem 48:4581–4589

Gomes P (2007) Fruticultura Brasileira. Nobel, 446p

Holland D, Hatib K, Bar-Ya'akov I (2009) Pomegranate: Botany, horticulture, breeding. Hortic Rev 35:127–191

Julie JMT (2008) Therapeutic applications of pomegranate (*Punica granatum* L.): a review. Altern Med Rev 13:123–144

Khan JA, Hanee S (2011) Antibacterial properties of *Punica granatum* peels. Int J Appl Biol Pharm Technol 2:23–27

Khayyat M, Tehranifar A, Zaree M, Karimian Z, Aminifard MH, Vazifeshenas MR, Amini S, Noori Y, Shakeri M (2012) Effects of potassium nitrate spraying on fruit characteristics of 'Malas Yazdi' pomegranate. J Plant Nutr 35:1387–1393

Lansky EP, Newman RA (2007) *Punica granatum* (pomegranate) and its potential for prevention and treatment of inflammation and cancer. J Ethnopharmacol 109:177–206

Lee SK, Kader AA (2000) Preharvest and postharvest factors influencing vitamin C content of horticultural crops. Postharvest Biol Technol 20:207–220

Legua P, Melgarejo P, Abdelmajid H, Martínez JJ, Martínez R, Ilham H, Hafida H, Hernández F (2012) Total phenols and antioxidant capacity in 10 moroccan pomegranate varieties. J Food Sci 71:115–120

Levin GM (2006) Pomegranate roads: a Soviet botanist's exile from Eden. Floreant Press, Forestville, p 183p

Lorenzi H, Souza HM (2001) Plantas ornamentais no Brasil – arbustivas, herbáceas e trepadeiras, 3rd edn. Plantarum, Nova Odessa, p 1088

Markakis P (1982) Stability of anthocyanins in foods. In: Markakis P (ed) Antho-cyanins as food colors. Academic, New York, pp 163–180

Martí N, Pérez-Vicente A, García-Viguera C (2001) Influence of storage temper-ature and ascorbic acid addition on pomegranate juice. J Sci Food Agric 82:217–221

Manera FJ, Legua P, Melgarejo P, Brotons JM, Hernández FCA, Martínez JJ (2013) Determination of a colour index for fruit of pomegranate varietal group "Mollar de Elche". Sci Hortic 150:360–364

Mansourian A, Boojarpour N, Ashnagar S, Beitollahi JM, Shamshiri AR (2014) The comparative study of antifungal activity of *Syzygium aromaticum, Punica granatum* and nystatin on *Candida albicans*: an in vitro study. J de Mycol Médicale 24:e163–e168

Miguel MG, Neves MA, Antunes MD (2010) Pomegranate (*Punica granatum* L.): a medicinal plant with myriad biological properties – a short review. J Med Plant Res 4(25):2836–2847

Mmarm. Anuario de estadística agroalimentaria. Madrid Indian Council of Agricultural Research (2009) Pawar inaugurates new national research centre on pomegranate. In: www.icar.org.in/prI25092005.htm

Mphahlele RR, Fawole AO, Stander MA (2014) Preharvest and postharvest factors influencing bioactive compounds in pomegranate (*Punica granatum* L.) – a review. Scientia Hoticulturae 178:114–123

Nuncio-Jáuregui N, Calín-Sánchez A, Carbonell-Barrachina A, Hernández F (2014) Changes in quality parameters, proline, antioxidant activity and color of pomegranate (Punica granatum L.) as affected by fruit position within tree, cultivar and ripening stage. Sci Hortic 165:181–189

Panwar J, Tarafdar JC (2006) Distribution of three endangered medicinal plant species and their colonization with arbuscular mycorrhizal fungi. J Arid Environ 65:337–350

Parvizi H, Sepaskhah AR, Ahmadi SH (2014) Effect of drip irrigation and fertilizer regimes on fruit yields and water productivity of a pomegranate (*Punica granatum* (L.) cv. Rabab) orchard. Agric Water Manag 146:45–56

Pedriali CA, Fernandes AU, Santos PA, Silva MM, Severino D, Silva MB (2010) Antioxidant activity, cito- and phototoxicity of pomegranate (*Punica granatum* L.) seed pulp extract. Cienc Tecnol Aliment 30(4):1017–1021

Pilano LS, Wrolstad RE, Heatherbell DA (1985) Influence of fruit composition, maturity and mold concentration on the color and appearance of strawberrywine. J Food Sci 50:1121–1125

Qnais EY, Elokda AS, Abu Ghalyun YY, Abdulla FA (2007) Antidiarrheal Activity of the Aqueous Extract of *Punica granatum*. (Pomegranate) Peels. Pharm Biol 45(9):715–720

Quattrucci A, Ovidi E, Tiezzi A, Vinciguerra V, Balestra GM (2013) Biological control of tomato bacterial speck using *Punica granatum* fruit peel extract. Crop Prot 45:18–22

Sadeghipour A, Eidi M, Ilchizadeh Kavgani A, Ghahramani R, Shahabzadeh S, Anissian A (2014) Lipid lowering effect of *Punica granatum* L. peel in high lipid diet fed male rats. Evid Based Complement Alternat Med 2014:432650

Sayyari M, Valero D, Babalar M, Kalantari S, Zapata PJ, Serrano M (2010) Prestorage oxalic acid treatment maintained visual quality, bioactive com-pounds, and antioxidant potential of pomegranate after long-term storage at 2°C. J Agric Food Chem 58:6804–6808

Withy LM, Nguyen TT, Wrolstad RE, Heatherbell DA (1993) Storage changes inanthocyanin content of red raspberry juice concentrate. J Food Sci 1993(58):190–192

Schinopsis brasiliensis Engl.

Ana Cláudia Dantas Medeiros, Laianne Carla Batista Alencar, and Délcio de Castro Felismino

Schinopsis brasiliensis Engl.
Photo available in:
http://www.arvoresdobiomacerrado.com.br/site/2017/03/30/schinopsis-brasiliensis-engl/

A. C. D. Medeiros (✉) · L. C. B. Alencar · D. de Castro Felismino
Laboratory of Development and Assays of Drugs, State University of Paraíba, Campina Grande, Paraíba, Brazil
e-mail: anaclaudia@uepb.edu.br; dcfelismino@ccbs.uepb.edu.br

U. P. Albuquerque et al. (eds.), *Medicinal and Aromatic Plants of South America*, Medicinal and Aromatic Plants of the World 5,
https://doi.org/10.1007/978-94-024-1552-0_38

Abstract *Schinopsis brasiliensis* Engl. (Anacardiaceae) is the main species representative of the *Schinopsis* genre, which is native to Brazil. Not an endemic tree, it is popularly known as braúna and baraúna, distributed in the Northeast, Midwest, and Southeast of the country. It has the plant characteristic of the Caatinga and great economic value for the northeastern region. It is widely used in traditional medicine for anti-inflammatory, analgesic, hemostatic, antiseptic, and antimicrobial purposes. From a scientific perspective few works that confirm its pharmacological activity were found in the scientific literature. Phytochemical studies showed the presence of polyphenols, flavonoids, and tannins. Two compounds were isolated and gallic acid was determinate as the chemical marker of *S. brasiliensis*.

Keywords *Schinopsis brasiliensis* Engler · Anacardiaceae · Traditional use · Gallic acid

1 Taxonomic Characteristics

According to the classification system based on The Angiosperm Phylogeny Group (APG) II (Chase 2003), the taxonomic position of *Schinopsis brasiliensis* Engl according to the following hierarchy: Family Anacardiaceae, order Sapindales, Malvids clade (in rosids, in core eudicots); Genre: *Schinopsis*; Specie: *S. brasiliensis* (Carvalho 2009).

Synonyms *Schinopsis brasiliensis* var. *glabra* Engl., *Schinopsis brasiliensis* Engl. var. *brasiliensis*, *Schinopsis glabra* (Engl.) F. A. Barkley & T. Mey. It is commonly known as braúna and baraúna (Ceará, Paraíba, Pernambuco, Sergipe and Bahia); Chamacoco and chamucoco (Mato Grosso do Sul); blackwood (Brazil); soto (Bolivia) and barauva (Paraguay) (Braga 1978; Dantas 2007).

2 Crude Drug Used

Fernandes et al. (2013) characterized the dried extract of the bark of *S. brasiliensis* by analytical methods. Tests conducted with thermal analysis showed an endothermic process at 80.99 °C, probably related to the loss of volatile constituents of the sample and the beginning of the process of decomposition, which occurs at a temperature of 126.14 °C. The X-ray diffraction shows a high degree of amorphization, particularly at angles between 10 and 30°. Furthermore, the infrared spectrum showed absorption bands indicative of the presence of several secondary metabolites in the extract, such as tannins, polyphenols, flavonoids, etc.

3 Major Chemical Constituents and Bioactive Compounds

Some chemical compounds have been isolated from *S. brasiliensis*. Among them highlight the alkyl, phenol, methyl 6-hydroxy-2-eicosanyl-4-methoxybenzoate and the unusual steroid 5α, 8α -epidioxyergosta-6,22-dien-3-β-ol (Cardoso et al. 2005). The phytochemical profile realized with the bark, leaves, flowers, fruits and roots of *S. brasiliensis* showed the presence of polyphenols (gallic acid and ellagic), flavonoids (aglycones), steroids, terpenoids, lignans, triterpenoids, cinnamic derivatives, condensed proanthocyanidins, and leucoanthocyanidins (Saraiva 2007; Saraiva et al. 2011; Cardoso 2001; Cardoso et al. 2003, 2004). Fernandes et al. (2015) developed and validated an analytical method for the identification of gallic acid as the chemical marker of *S. brasiliensis*.

The essential oil extracted from the leaves of *S. brasiliensis* had a good amount of myrcene and low amounts of other compounds such as β-caryophyllene, eucalyptol, and guaiol (Donati et al. 2014).

4 Morphological Description

The *S. brasiliensis* is a plant that is xerophytic and helióﬁtic, fully deciduous during the dry season. It is a tree rounded with a dense canopy, and a height of 15 m and 60 cm DAP (diameter at breast height, measured 1.30 m from the ground), in adulthood. It is one of the largest trees in the Caatinga (Carvalho 2009), providing branches with thorns. The trunk is straight and shaped, more or less cylindrical with a short shaft (Saraiva 2007; Dantas et al. 2008). The branching is dichotomous. The rind has a thickness up to 30 mm. It is externally dark gray, almost black, rough, and gives off in portions irregularly quadrangular. Pinnate leaves are composed with 7–17 leaflets subcoriaceous consistency, oblong, measuring 3–4 cm long and 2 cm wide, obtuse at the apex, dark green on the upper face, and the lower face is pale. When steeped, it has low odor resin (Carvalho 2009).

It presents inflorescence in panicles. The flowers are monoecious, small, measuring 3–4 mm in diameter, white, glabrous and gently fragrant. Flowering occurs in July, in Mato Grosso do Sul, from November to December in Ceará, and from November to February, in Pernambuco (Carvalho 2009). The fruit pods are of a woody nature, thick, and sickle-shaped, rounded, covered by the fine hair measuring 3–3.5 cm long [14], the type samara with the pericarp layers markedly differentiated membranous epicarp, mesocarp spongy, and waterproof cored water (Oliveira and Oliveira 2008). Its fruiting occurs between August and September.

Its seeds are obovóides tending to be kidney-shaped, light-yellow in color with a dull rough surface, and surrounded by a tough woody seed coat to be broken (Carvalho 2009). Dantas et al. (2008) obtained a curve of *S. brasiliensis* seed soaking making it possible to observe a three-phase model, where the phase FI was completed in 48 h and FIII started after 152 h of soaking, with root protrusion. The levels of soluble sugars amount and reducing sugars in seeds increase during imbibition, while the starch content decreased after the FII. Albumine, globulins and prolamins were constant during FI and FII and decreased after root protrusion and glutelin contents were practically null during seed germination.

5 Geographical Distribution

S. brasiliensis is native to Brazil (Saraiva 2007) and is not an endemic tree. The type of vegetation is Cerrado (lato sensu), Semideciduous forest (Silva-Luz and Pirani 2015), and Caatinga (stricto sensu) (Silva-Luz and Pirani 2015; Rodal and Nascimento 2006; Andrade et al. 2009), being of great economic value for the Northeast region (Saraiva 2007). In Brazil it is distributed in the Northeast (Lima and Lima 1998; Nascimento et al. 2003; Silva et al. 2004; Trovão et al. 2004; Lacerda et al. 2007; Oliveira et al. 2009; Ramalho et al. 2009; Santos and Melo 2010; Calixto and Drumond 2011; Barbosa et al. 2014) Midwest (Federal District (Silva and Scariot 2004), Tocantins, Mato Grosso do Sul, Goiás (Lima et al. 2008) and Southeast (Espírito Santo, Minas Gerais) (Santos et al. 2007; Santos et al. 2008). It also grows in Bolivia and Paraguay (Williams et al. 2001).

6 Ecological Requirements

It is a characteristic species of the wetlands of semiarid regions (Tigre 1970). It is more common in calcareous soils and can occur in rocky outcrops, which usually grows slowly (Maia 2004). It is rarely found in deep soils and low-lying arenaceous areas (Carvalho 2009; Maia 2004). The species can be found from 18 m sea level to about 1.000 m altitude and latitude 5° S, Rio Grande do Norte, 19° S, in Mato Grosso do Sul (Killeen et al. 1993; Carvalho 2009).

The hydric behavior at the end of the rainy season reveals that this plant it is in water-savings scheme is higher than its consumption for its metabolic needs. In the middle of the dry season of the year, *S. brasiliensis* has little restriction on its sweating in the most critical hours of the day. Considered dominant in the Caatinga, it has a low rate of association, demonstrating growth with virtually no affinity with each other. The genetic variability of this species is not evenly dispersed throughout the Brazilian semiarid regions, but in the ecoregions (Tigre 1970; Killeen et al. 1993; Maia 2004).

7 Collection Practice

This species is distinguished by its high commercial and medicinal value. Due to its high degree of resistance, this plant is widely used by the timber industry, in construction, in furniture and sleepers production and in the production of fuels (Braga 1978; Gonzaga et al. 2003; Albuquerque 2006; Ferraz et al. 2006; Albuquerque et al. 2007; Alves et al. 2007; Saraiva 2007; Lucena et al. 2008; Albuquerque et al. 2009). In the medical field, it has secondary metabolites ensuring anti-inflammatory activity, anti-hemorrhagic, antimicrobial and other uses (Cardoso et al. 2003, 2006; Saraiva 2007; Lima et al. 2008; Silva et al. 2012). Due to the systematic and irrational exploitation for these and other purposes, *S. brasiliensis* was included in the official list of species threatened with extinction flora (MMA 2008).

8 Traditional Use (Part(s) Used) and Common Knowledge

Several pieces of *S. brasiliensis* (leaf, bark and fruit) are used in traditional medicine as anti-inflammatory, analgesic, for healing fractures, and for flu, cough and fever (Almeida et al. 2005; Albuquerque 2006; Albuquerque et al. 2007; Agra et al. 2007; Gomes et al. 2012; Pereira Júnior et 2014). It has anti-hysterical and neurasthenic properties and is also used to treat diarrhea and uterine bleeding in combating (Gonzaga and Bandeira 2003; Dantas 2007; Agra et al. 2007; Farias et al. 2013) sexual impotence (Almeida et al. 2005; Albuquerque et al. 2007; Saraiva et al. 2011) injuries, fungal infections of the skin, antiseptic (Dantas 2007; Saraiva 2007) for prostate and as an anticoagulant (Gomes and Bandeira 2012), and for gastric disorders and liver problems (Ribeiro et al. 2014). The tea of crushed bark is used for pain of the teeth and head (Albuquerque et al. 2012).

9 Modern Medicine Based on Its Traditional Medicine Uses

Few studies confirming the pharmacological activity of *S. brasiliensis* were carried out to date, with the existing majority related to its antimicrobial activity. Studies with hydroalcoholic extracts made from bark and leaves of *S. brasiliensis* showed activity against the *Pseudomonas aeruginosa*, *Staphylococcus aureus*, *Streptococcus oralis*, *Streptococcus mutans*, *Streptococcus parasanguinis*, *Enterococcus faecalis*, *Klebsiella pneumoniae*, *Candida albicans*, *C. tropicalis*, *C. guilliermondii* and *C. krusei* (Silva et al. 2012; Guimarães 2010; Chaves et al. 2011; Santos 2013). While studies of ethanol extract also produced with bark and leaves and their fractions, hexane, methanol, dichloromethane and ethyl acetate showed activity against *Staphylococcus aureus*, *Enterococcus faecalis*, *Pseudomonas aeruginosa*, *Escherichia coli*, *Klebsiella pneumoniae*, *Salmonella typhimurium*, *Staphylococcus* sp., *Candida albicans* and *Candida krusei* clinical isolates and strains of Standard American Type Culture Collection (ATCC). The best results were obtained with the methanol fraction, which showed activity for all strains tested (Saraiva et al. 2011; Machado 2012). The essential oil produced from the leaves of *S. brasiliensis* showed weak activity only against *Staphylococcus aureus* (Donati et al. 2014). The antioxidant activity performed with methanol extract and essential oil produced from the leaves of *S. brasiliensis* showed a high antioxidant power (Saraiva et al. 2011; Donati et al. 2014).

Oliveira (2011) evaluated the antimalarial activity of the ethanol extract obtained from the bark of *S. brasiliensis*. The assay was performed in vivo using mice infected with *Plasmodium falciparum* and in vitro, using the same parasite. The in vivo study showed that the species studied reduced parasitemia by 86 and 95% at doses of 250 and 500 mg.Kg^{-1}, respectively. Meanwhile, in vitro performed with chloroform and hexane fractions were considered partially active, because it only inhibited the growth from 50 to 79% of the parasites.

The toxicity of bark extracts and leaves of *S. brasiliensis* were evaluated by in vivo and in vitro methods. In vivo assay showed that animals used present no behavioral changes after oral administration of the extract at a dose of 2000 mg. Kg^{-1}. However, during the observation period analgesia at 2 and 4 h after administration of the extract was observed (Silva 2011; Santos 2013).

Bioassays using extracts and fractions produced with the bark of this species showed that the dry extract and chloroform fraction showed toxicity against brine shrimp (LC_{50} 428 and 313 μg.mL^{-1}); that only the chloroform, hexane and ethyl acetate showed larvicidal potential against *Aedes aegypti* (LC_{50} 345.527 and 583 μg. mL^{-1}, respectively); while chloroform and ethyl acetate fractions were highly toxic to *Biomphalaria glabrata* (LC_{90} 68 and 73 μg.mL^{-1}, respectively) (Silva 2011; Santos et al. 2014). Meanwhile, the extract produced with the leaves showed moderate toxicity (LC_{50} 511.90 μg.mL^{-1}) (Santos 2013). It was also observed that the activity of the seeds of this species was larvicide, pulpicida, and the reduction in egg production by females of *Aedes aegypti*. Also observed was its toxicity in the microcrustacean *Ceriodaphnia* as dubious and their cytotoxicity in mice of the 3T3 fibroblast cells and in HeLa cells (Oliveira 2011; Barbosa et al. 2014; Santos et al. 2014).

10 Conclusions

The *S. brasiliensis* is a native species, found in semiarid regions and highly used by the traditional medicine of the Brazilian Northeast. Few studies have been published that confirm its pharmacological activity, as well as the isolation of new compounds of this species. Its dried extract was characterized by analytical methods and the gallic acid was identified as the chemical marker. Some studies of in vivo toxicity showed that the species did not show signs of toxicity at the dose tested.

References

Agra MF, Baracho GS, Nurit K, Basílio IJLD, Coelho VPM (2007) Medicinal and poisonous diversity of the flora of "Cariri Paraibano". Brazil J Ethnopharmacol 114(1):325–354

Albuquerque HN, Figueredo DJC, Cerqueira JS (2012) Os vegetais com potencial fitoterápico do complexo Aluízio Campos, Campina Grande – PB. Rev Bras Inform Cient 3(2):17–26

Albuquerque UP, Araújo TAS, Ramos MA, Nascimento VT, Lucena RFP, Monteiro JM, Alencar NL, Araujo EL (2009) How ethnobotany can aid biodiversity conservation: reflections on investigations in the semi-arid region of NE Brazil. Biodivers Conserv 18:127–150

Albuquerque UP, Medeiros PM, de ALS A, Monteiro JM, Lins EMFN, Melo JG, dos Santos JP (2007) Medicinal plants of the caatinga (semi-arid) vegetation of NE Brazil: a quantitative approach. J Ethnopharmacol 114(1):325–354

Albuquerque UP (2006) Re-examining hypotheses concerning the use and knowledge of medicinal plants: a study in the Caatinga vegetation of NE Brazil. J Ethnobiol Ethnom 30(2):1–10

Almeida CFCBR, Silva TCL, Amorim ELC, Maia MBS, Albuquerque UP (2005) Life strategy and chemical composition as predictors of the selection of medicinal plants from the Caatinga (Northeast Brazil). J Arid Environ 62(1):127–142

Alves AF, Alves AF, Guerra MEC, Medeiros SF (2007) Superação de dormência de sementes de braúna (*Schinopsis brasiliense* Engl.). Rev Ciência Agron 38(1):74–77

Andrade WM, Lima EA, Rodal MJN, Encarnação CRF, Pimentel RMM (2009) Influência da precipitação na abundância de populações de plantas da Caatinga. Rev Geogr 22(3):182–190

Barbosa PBBM, Oliveira JM, Chagas JM, Rabelo LMA, Medeiros GF, Giodani RB et al (2014) Evaluation of seed extracts from plants found in the Caatinga biome for the control of *Aedes aegypti*. Parasitol Res 113(1):3565–3580

Braga R (1978) Plantas do nordeste especialmente do Ceará, 3th edn. Mossoroense, Fortaleza

Calixto Júnior JT, Drumond MA (2011) Estrutura fitossociológica de um fragmento de Caatinga sensu stricto 30 anos após corte raso, Petrolina-PE, Brasil. Rev Caatinga 24(2):67–74

Cardoso MP, David JM, David JP (2005) A new alkyl phenol from *Schinopsis brasiliensis*. Nat Prod Res 19(5):431–433

Cardoso MP, David JM, David JP (2004) Estudo Fitoquímico do caule de *Schinopsis brasiliensis*. Anais da 27° Reunião Anual da Sociedade Brasileira de Química. Salvador, Brasil

Cardoso MP, David JM, David JP (2006). n- alquil e n-alquenil fenóis identificados de *Schinopsis brasiliensis* (Anacardiaceae). Anais da 29° Reunião Anual da Sociedade Brasileira de Química May 19–22, Águas de Lindóia, Brasil

Cardoso MP (2001) Contribuição ao estudo fitoquímico de *Schinopsis brasiliensis* (Anacardiaceae). Dissertation. Chemical Institute:UFBA

Cardoso MP, Farias MT, David JM, David JP (2003) Estudo Fitoquímico do extrato clorofórmico do caule de *Schinopsis brasiliensis*. Anais da 26° Reunião Anual da Sociedade Brasileira de Química May 26–29, Poços de Caldas, Brasil

Carvalho PER (2009) Braúna-do-Sertão: *Schinopsis brasiliensis*, available online at http://www.cnpf.embrapa.br/publica/comuntec/edicoes/CT222.pdf.Brazil:Embrapa

Chase M (2003) An update of the Angiosperm Phylogeny Group classification for the orders and families of flowering plants: APG II. Bot J Linn Soc 141:399–336

Chaves TP, Dantas IC, Felismino DC, Vieira KVM, Clementino ELC, Costa LS (2011) Atividade antimicrobiana das folhas de *Schinopsis brasiliensis* Engler. Biofar 5(2):11–17

Dantas BF, Soares FSJ, Lúcio AA, Aragão CA (2008) Alterações bioquímicas durante a embebição de sementes de baraúna (*Schinopsis brasiliensis* Engl.). Rev Bras Sem 30(2):214–219

Dantas IC (2007) O Raizeiro. EDUEPB, Campina Grande

Donati M, Mondin A, Chen Z, Miranda FM, Nascimento Junior BB, Schirato G, Pastore P, Froldi G (2014) Radical scavenging and antimicrobial activities of *Croton zehntneri*, *Pterodon emarginatus* and *Schinopsis brasiliensis* essential oils and their major constituents: estragole, trans-anethole, b-caryophyllene and myrcene. Nat Prod Res:2–8

Farias DF, Souza TM, Viana MP, Soares BM, Cunha AP, Vasconcelos IM, Ricardo NMPS, Ferreira PMP, Melo VMM, Carvalho AFU (2013) Antibacterial, antioxidant and anticholinesterase activities of plant seed extracts from Brazilian semiarid region. Biomed Res Int 2013:1–9

Fernandes FHA, Santana CP, Santos RL, Correia LP, Conceição MM, Macêdo RO et al (2013) Thermal characterization of dried extract of medicinal plant by DSC and analytical techniques. J Therm Anal Calorim 113(2):443–447

Fernandes FHA, Batista RSA, Medeiros FD, Santos FS, Medeiros ACD (2015) Development of a rapid and simple HPLC-UV method for determination of gallic acid in *Schinopsis brasiliensis*. Rev Bras Farmacognosia 25:208–211

Ferraz ISF, Albuquerque UP, Meunier IMJ (2006) Valor de uso e estrutura da vegetação lenhosa às margens do riacho do Navio, Floresta, PE. Brasil Acta Bot Bras 76(20):125–134

Gomes TB, Bandeira FPSF (2012) Uso e diversidade de plantas medicinais em uma comunidade quilombola no Raso da Catarina, Bahia. Acta Bot Bras 26(4):796–709

Gonzaga TWC, Mata MERMC, Silva H, Duarte MEM (2003 Out) Crioconservação de sementes de aroeira (*Astronium urundeuva* Engl.), e baraúna (*Schinopsis brasiliensis* Engl.). Rev Bras Prod Agroind 5(2):145–154

Guimarães GP (2010) Atividade antifúngica de plantas medicinais frente a espécies de Candidas de interesse médico. Monograph. UEPB

MMA (2008) Instrução Normativa N° 6, de 23 de Setembro de 2008. Ministro de Estado do Meio Ambiente. No uso das suas atribuições legais, resolve reconhecer como espécies da flora brasileira ameaçadas de extinção aquelas constantes do Anexo I a esta Instrução Normativa. MMA, Brasília

Killeen TJ, Garcia EE, Beck SG (eds) (1993) Guía de arboles de Bolivia. La Paz: Herbario Nacional de Bolívia. MBG, St. Louis

Lacerda AV, Barbosa FM, Barbosa MRV (2007) Estudo do componente arbustivo-arbóreo de matas ciliares na bacia do Rio Taperoá, semi-árido paraibano: uma perspectiva para a sustentabilidade dos recursos naturais. Oecol Bras 11(3):331–340

Lima PCF, Lima JLS (1998) Composição florística e fitossociologia de uma área de Caatinga em contendas do Sincorá, Bahia, microrregião homogênea da chapada Diamantina. Acta Bot Bras 12(3):441–450

Lima VVF, Vieira DLM, Sevilha AC, Salomão AN (2008) Germinação de espécies arbóreas de floresta estacional decidual do vale do rio Paranã em Goiás após três tipos de armazenamento por até 15 meses. Biota Neotropical 8(3):89–97

Lucena RFP, Nascimento VT, Araújo EL, Albuquerque UP (2008) Local uses of native plants in an area of Caatinga vegetation (Pernambuco, NE Brazil). Ethnobot Res Appl 6:3–13

Machado SMF (2012) Avaliação da atividade antimicrobiana dos extratos fracionados de cascas e folhas da *Schinopsis brasiliensis* Engler através de análise comparativa entre os métodos de difusão em disco e cavidade em placa. Monograph. UEPB

Maia GN (2004) Caatinga: árvores e arbustos e suas utilidades. Leitura & Arte, São Paulo

Nascimento CES, Rodal MJN, Cavalcanti AC (2003) Phytosociology of the remaining xerophytic woodland associated to an environmental gradient at the banks of the São Francisco river – Petrolina, Pernambuco, Brazil. Braz J Bot 26(3):271–287

Oliveira AMGC (2011) Avaliação da atividade antimalárica e citotóxica de plantas medicinais dos biomas Caatinga e Amazônico. Dissertation. UFRN

Oliveira MCP, Oliveira GJ (2008) Superação da dormência de sementes de *Schinopsis brasiliensis*. Ciênc Rural 38(1):251–254

Oliveira PTB, Trovão DMBM, Carvalho ECD, Souza BC, LMR F (2009) Floristística e fitossociologia de quatro remanescentes vegetacionais em áreas de serra no Cariri paraibano. Rev Caatinga 22(4):169–178

Pereira Júnior LR, Andrade AP, Araújo KD, Barbosa AS, Barbosa FM (2014) Espécies da Caatinga como alternativa para o desenvolvimento de novos fitofármacos. Flora 21(4):509–520

Ramalho CIR, Andrade AP, Félix LP, Lacerda AV, Maracajá PB (2009) Flora arbóreo-arbustiva em áreas de caatinga no semiárido Baiano, Brasil. Rev Caatinga 22(3):182–190

Ribeiro DA, Macêdo DG, Oliveira LGS, Saraiva ME, Oliveira SF, Souza MMA, Menezes IRA (2014) Potencial terapêutico e uso de plantas medicinais em uma área de Caatinga no estado do Ceará, nordeste do Brasil. Rev Bras Plantas Med 16(4):912–930

Rodal MJN, Nascimento LM (2006) The Arboreal component of a dry forestin northeastern Brazil. Braz J Biol 66(2A):479–491

Santos CCS, Araújo SS, Santos ALLM, Almeida ECV, Dias AS, Damascena NP, Santos DM, Santos MIS, Junior KALR, Pereira CKB, Lima ACB, Shan AYKV, Sant'ana AEG, Estevan CS, Araujo BS (2014) Evaluation of the toxicity and molluscicidal and larvicidal activities of *Schinopsis brasiliensis* stem bark extract and its fractions. Rev Bras Farmacog 24(3):298–203

Santos PM, Melo JIM (2010) Flora vascular de uma área de caatinga no estado da Paraíba – nordeste do Brasil. Rev Caatinga 23(2):32–40

Santos PM, Vieira FA, Gusmão E, YRF N (2007) Florística e estrutura de uma floresta estacional decidual, no pareque municipal da Sapucaia, Montes Claros (MG). Rev Cerne 13(3):248–256

Santos RL (2013) Desenvolvimento de um dentifrício a partir de extrato nebulizado de *Schinopsis brasiliensis* Engler. Dissertation. UEPB

Santos RM, Vieira FA, Santos PF, Medeiros MA (2008) Estrutura e florística de um remanescente florestal na Fazenda Ribeirão, município de Juvenília, MG, Brasil. Rev Caatinga 21(4):154–162

Saraiva AM, Castro RHÁ, Cordeiro RP, Peixoto TJS, Castro VTNA, Amorim ELC et al (2011) In vitro evaluation of antioxidant, antimicrobial and toxicity properties of extracts of *Schinopsis brasiliensis* Engl. (Anacardiaceae). Afr J Pharm Pharmacol 5(14):1724–1731

Saraiva AM (2007) Estudo Farmacognóstico e Determinação da Atividade Biológica de *Caesalpinia pyramidalis* Tull. E *Schinopsis brasiliensis* Engl. frente a cepas de *Staphylococcus aureus* MRSA Multirresistentes. Recife 184 p. Dissertation (PPGCF-UFPE)

Silva EC, Nogueira RJMC, Azevedo ADN, Brito JZ, Cabral EL (2004) Aspectos ecofisiológicos de dez espécies em uma área de caatinga no município de Cabaceiras, Paraíba, Brasil. Sér Bot 59(2):201–205

Silva LA, Scariot A (2004) Comunidade arbórea de uma floresta estacional decídua sobre afloramento calcário na bacia do Rio Paraná. Rev Árvore 18(1):61–67

Silva MSP, Brandão DO, Chaves TP, Formiga ALNF, Costa EMMB, Santos VL, Medeiros ACD (2012) Study bioprospecting of medicinal plant extracts of the semiarid northeast: contribution to the control of oral microorganisms. Evid Based Complement Alternat Med 2012(2012):1–6

Silva MSP (2011) Ensaios pré-clínicos com extratos de plantas medicinais do semiárido nordestino: contribuição para o tratamento de infecções da cavidade bucal. Dissertation. UEPB

Silva-Luz CL, Pirani JR (2015) *Anacardiaceae* in list of plant species in Brazil : *Schinopsis brasiliensis,* available online at http://www.floradobrasil.jbrj.gov.br/jabot/floradobrasil/FB4395

Tigre CB (1970) Silvicultura para as matas xerófilas. DNOCS, Fortaleza

Trovão DMBM, Silva SC, Silva AB, Vieira RLJ (2004) Estudo comparativo entre três fisionomias de Caatinga no estado da Paraíba e análise do uso das espécies vegetais pelo homem na área de estudo. Rev Biol Ciências Terra 4(2):1–5

Williams RS, Miller R, Gangstad J (2001 Out) Characteristics of ten tropical hardwoods from certified forests in Bolivia part I weathering characteristics and dimensional change. Wood Fiber Sci 33(4):618–626

Stryphnodendron adstringens (Mart.) Coville

Letícia Mendes Ricardo and Maria G. L. Brandão

Stryphnodendron adstringens (Mart.) Coville
Photo source: data bank from Laboratório de Ecologia e Evolução de sistemas socioecológicos

L. M. Ricardo
CEPLAMT, Museu de História Natural e Jardim Botânico & Faculdade de Farmácia, Universidade Federal de Minas Gerais, Belo Horizonte, Minas Gerais, Brazil

Departamento de Assistência Farmacêutica e Insumos Estratégicos, Secretaria de Ciência, Tecnologia e Insumos Estratégicos, Ministério da Saúde, Brazil

M. G. L. Brandão (✉)
CEPLAMT, Museu de História Natural e Jardim Botânico & Faculdade de Farmácia, Universidade Federal de Minas Gerais, Belo Horizonte, Minas Gerais, Brazil
e-mail: mbrandao@ufmg.br

U. P. Albuquerque et al. (eds.), *Medicinal and Aromatic Plants of South America*, Medicinal and Aromatic Plants of the World 5,
https://doi.org/10.1007/978-94-024-1552-0_39

Abstract Barbatimão is the common name of the Brazilian plant *Stryphnodendron adstringens* (Mart.) Coville. The barks of this plant have been used for centuries in the traditional Brazilian medicine as astringent. This activity is attributed to the presence of high concentration of tannins. Numerous studies have consolidated the biological activities as a fungicide, anti-inflammatory and as wound healing. Topical applications of barbatimão ointment stimulates the proliferation of keratinocytes. A pharmaceutical formulation containing *S. adstringens* has been developed to heal skin wounds.

Keywords Barbatimão · *Stryphnodendron adstringens* · Tannins

1 Taxonomic Characteristics

Stryphnodendron adstringens (Mart.) Coville belongs to the Mimosoideae, subfamily of the family Leguminosae, that includes mostly trees of tropical and subtropical South America (Lorenzi 1998).

Synonyms *Acacia adstringens* Mart.; *Mimosa barbadetimam* Vell.; *Mimosa virginalis* Arruda; *Stryphnodendron barbatimam* Mart. e *S. barbatimam* (Vell.) Mart

2 Crude Drug Used

The crude drug is consisted of the dried barks with a minimum of 8% of tannins, as described in the Brazilian Official Pharmacopoeia (Brasil 2010).

3 Major Chemical Constituents and Bioactive Compounds

The tannins from the barks of *S. adstringens* are considered its bioactive compounds. They are constituted by pirogalol ($C_6H_6O_3$; 126,11), from which a minimum of 0.2 mg/g correspond to gallic acid ($C_7H_6O_5$; 170,1) and 0.3 mg/g to galocatequine ($C_{15}H_{14}O_7$; 306,27) (Lopes et al. 2009; Audi et al. 2004; Santos et al. 2002). Besides the tannins, the barks have mucilage, flavonoids and saponins (Glehn and Rodrigues 2012; Bardal 2011). The hydroalcoholic leaf extract of *S. adstringens* contains tannins, steroids, simple phenols, flavonoids, flavanones, flavonols and saponins (Pinho et al. 2012).

4 Morphological Description

S. adstringens is a regular shrub or small tree with crooked branches, coveredby little foliage; rough bark; bipinnate leaves, oval leaflets, small, sometimes nude, red or almost white flowers arranged in cylindrical spikes, axillary. Fruit sessile, thick and fleshy, linear, oblong, 10 cm long. The bark is presented in arched fragments with dimensions and varied formats. In cross-section, on average, 0.6 mm thick when dried, 10 mm and 12 mm thick when hydrated. The inner phloem region is of lighter brown color as compared to the suber region that has intense reddish-brown color (Brasil 2010; Sanches et al. 2007).

5 Geographical Distribution

The species *S. adstringens* (Mart.) Coville is found in all regions of Brazil, especially in areas of caatinga and cerrado (savanna) (Flora do Brasil 2013; Correia et al. 2012).

6 Collection Practice

It is noticed that extraction of the bark of trees as a practice that partially removes the bark disrupting wood vessels and causing premature death of the trees (Correia et al. 2012). There is a need for conservation of *S. adstringens* since is listed as endangered due to its commercial value as a tanning source and timber.

7 Traditional Use (Part(s) Used) and Common Knowledge

Barks of the plant have been used in Brazil for centuries, as cicatrizing, astringent, anti-diarrheic, to treat leucorrhoea and as anti-hemorrhagic (Brandão et al. 2008, 2009, 2012; Albuquerque et al. 2007; Rodrigues and Carvalho 2001). More recent ethnobotanical studies have revealed the current use of the plant as wound healing agent (Ferrão et al. 2014; Lima et al. 2012; Oliveira and Menini Neto 2012; Sousa et al. 2011; Freitas and Fernandes 2006; Tresvenzol et al. 2006; Maciel and Neto 2006). These effects are directly correlated to the presence of high concentrations of tannins in the barks. Other uses described in the bibliography are: antidiabetic (David and Pasa 2015), antioxidant (Sousa et al. 2011), for the treatment of amoeba (Freitas and Fernandes 2006), malaria and as a febrifuge (Vila Verde et al. 2003).

8 Modern Medicine Based on Its Traditional Medicine Uses

In vitro studies with extracts from barks of *S. adstringens* present a potential antimicrobial effect against *Staphylococcus aureus* (Pinho et al. 2012; Souza et al. 2007a, b), *S. epidermitis* and *Escherichia coli* (Souza et al. 2007a, b). Another study showed positive results against *Prevotella nigrescens, Actinomyces naeslundii, Porphyromonas gingivalis, Enterococcus faecalis and Haemophilus actinomycetemcomitans*, microorganisms present in endodontic infections (Miranda 2010). Extracts from barks and stem barks of the plant also show antifungal (Glehn and Rodrigues 2012; Bardal 2011; Oliveira 2011; Melo-Silva et al. 2009; Ishida et al. 2006), anti-viral (Felipe et al. 2006), antiprotozoal (Herzog-Soares et al. 2006; Herzog-Soares et al. 2002; Luize et al. 2005; Holetz et al. 2002) and larvicidal activities in vitro (Vinaud et al. 2005). Ex vivo studies using isolated rat liver perfused with the extract from barks confirm that barbatimão impairs hepatic energy metabolism by different mechanisms (Rebecca et al. 2003).

In vivo assays with extracts and fractions from barks show activities in wound healing. The ethanolic extract promoted the epithelialization after 14 days of treatment (Coelho et al. 2010). In another study, a product prepared with 1% of extract promoted the epithelialization in 4, 7 and 10 days (Hernandes et al. 2010). Anti-inflammatory activity was observed in models of oedema in Wistar, but not equivalent to indometacina and dexametasona (Coutinho et al. 2004; Santos et al. 2002; Lima et al. 1998). Gastroprotective effects in models of gastric lesions induced by stress were observed for extracts prepared with ethyl acetate and n-butanol (200 mg/Kg). The effects were similar as observed for the controls cimetidina (32 mg/kg) (Audi et al. 1999). Other studies show the activity against gastric hypersecreting (Martins et al. 2002) and as antinociceptive (Melo et al. 2007).

One of the clinical trials was aimed at studying the effects of topical administration of a product containing 3% of extract from *S. adstringens* in cicatrization of decubitus ulcers. After 6 months, it was observed the cicatrization of 100% of the lesions, being 70% of the patients cured after 2 months (Minatel et al. 2010). In another clinical double-blind, randomized and placebo-controlled study that was performed with a cream containing the *S. adstringens* bark extract, the terminal hair growth suppressing activity established (Vicente et al. 2009).

Currently, in Brazil, there is only one phytomedicine registered by Brazilian Health Regulatory Agency (ANVISA) containing *S. adstringens*. This product is an ointment for topical use and is indicated for wound healing in instances of several types of lesions. It contains 60 mg of dry extract/g of ointment, corresponding to 27 mg of total tannins.

9 Conclusions

This review shows that *Strypnodendron adstringens* is an important medicinal plant from Brazil. The traditional use of its bark has been confirmed by numerous studies. These activities are due to the presence of high concentration of tannins. The collection practice that partially removes the bark is leading to premature death of the trees. Methods for the sustainable production and utilization of the species should be elaborated in order to avoid this.

References

Albuquerque UP, Monteiro JM, Ramos MA, Amorim ELC (2007) Medicinal and magic plants from a public market in northeastern Brazil. J Ethnopharmacol 110(1):76–91

Audi EA, Toledo DP, Peres PG, Kimura E, Pereira WKV, Mello JCP et al (1999) Gastric antiulcerogenic effects of *Stryphnodendron adstringens* in rats. Phytother Res 13(3):264–266

Audi EA, Toledo CEM, Santos FS, Bellanda PR, Alves-do-Prado W, Ueda-Nakamura T et al (2004) Biological activity and quality control of extract and stem bark from *Stryphnodendron adstringens*. Acta Farm Bonaer 23(3):328–333

Bardal D (2011) Atividade antimicrobiana de barbatimão *Stryphnodendron adstringens* (Martius) Coville em agentes causadores da mastite. Universidade Federal de Minas Gerais, Montes Claros

Brandão MGL, Zanetti NNS, Oliveira P, Grael CFF, Santos ACP, Monte-Mór RLM (2008) Brazilian medicinal plants described by 19th century European naturalists and in the Official Pharmacopoeia. J Ethnopharmacol 120(2):141–148

Brandão MGL, Cosenza GP, Grael CFF, Netto Junior NL, Monte-Mór RLM (2009) Traditional uses of American plant species from the 1st edition of Brazilian Official Pharmacopoeia. Rev Bras Farmacogn 19(2):478–487

Brandão MGL, Pignal M, Romaniuc S, Grael GFF, Fagg CW (2012) Useful Brazilian plants listed in the field books of the French naturalist Auguste de Saint-Hilaire (1779–1853). J Ethnopharmacol 143(2):488–500

Brasil (2010) Farmacopeia Brasileira, vol 2. Agência Nacional de Vigilância Sanitária, Brasília 904 p

Coelho JM, Antoniolli AB, e Silva D N, TMMB C, ERJC P, Odashiro AN (2010) O efeito da sulfadiazina de prata, extrato de ipê-roxo e extrato de barbatimão na cicatrização de feridas cutâneas em ratos. Rev Col Bras Cirurgiões 37(1):45–51

Correia VS, Cerdeira AL, Fachin AL, Bertoni BW, Pereira PS, França SC, Momm HG, Moraes RM, Pereira MAS (2012) Geographical variation and quality assessment of *Stryphnodendron adstringens* (Mart.) Coville within Brazil. Genet Resour Crop Evol 59:1349–1356

Coutinho H, Pinto DS, Ribeiro JEG, Friedman H (2004) Ação antiedematosa do *Stryphnodendron barbadetiman* (Barbatimão) a 1 por cento em comparação com a clorexidina a 0,12 por cento. Rev Odonto Cienc 19(45):201–206

David M, Pasa MC (2015) As plantas medicinais e a etnobotânica em Várzea Grande, MT, Brasil. Dermatol Int 16(1):97–108

Felipe AMM, Rincão VP, Benati FJ, Linhares REC, Galina KJ, Toledo CEM (2006) Antiviral effect of *Guazuma ulmifolia* and *Stryphnodendron adstringens* on poliovirus and bovine herpesvirus. Biol Pharm Bull 29(6):1092–1095

Ferrão BH, Oliveira HB, Molinari RF, Teixeira MB, Fontes GG, Amaros MOF, Rosa MB, Carvalho CA (2014) Importância do conhecimento tradicional no uso de plantas medicinais em Buritis, MG, Brasil. Ciênc e Nat 36:321–334

Flora do Brasil (2013) *Stryphnodendron* in Lista de Espécies da Flora do Brasil, available online at http://floradobrasil.jbrj.gov.br/jabot/floradobrasil/FB19133/. Rio de Janeiro: Jardim Botânico

Freitas JC, Fernandes MEB (2006) Uso de plantas medicinais pela comunidade de Enfarrusca, Bragança, Pará. Bol Mus Para Emílio Goeldi 1(3):11–26

Glehn EAV, Rodrigues GPS (2012) Antifungigrama para comprovar o potencial de ação dos extratos vegetais hidroglicólicos sobre *Candida* sp. (Berkhout). Rev Bras Plantas Medicinais 14(3):435–438

Hernandes L, Pereira LMS, Palazzo F, Mello JCP (2010) Wound-healing evaluation of ointment from *Stryphnodendron adstringens* (barbatimao) in rat skin. Braz J Pharm Sci 46:431–436

Herzog-Soares JD, Alves RK, Isac E, Bezerra JCB, Gomes MH, Santos SC et al (2002) Atividade tripanocida in vivo de *Stryphnodendron adstringens* (barbatimão verdadeiro) e *Caryocar brasiliensis* (pequi). Rev Bras Farmacogn 12(1–2):1–2

Herzog-Soares JD, Isac E, Castro AM, Bezerra JCB (2006) Bioatividade de *Stryphnodendron adstringens, S.* polyphyllum, *Caryocar brasiliense*, plantas do cerrado brasileiro, sobre Trypanosoma cruzi in vivo. Biosci J 22(3):113–118

Holetz FB, Nakamura TU, Dias Filho BP, Cortez DAG, Mello JCP, Nakamura CV (2002) Effect of plant extracts used in folk medicine on cell growth and differentiation of *Herpetomonas samuelpessoai* (Kinetoplastida, Trypanosomatidae) cultivated in defined médium. Acta Sci 24(3):657–662

Ishida K, Mello JCP, Garcia Cortez DA, Dias Filho BP, Ueda-Nakamura T, Nakamura CV (2006) Influence of tannins from *Stryphnodendron adstringens* on growth and virulence factors of Candida albicans. J Antimicrob Chemother 58(5):942–949

Lima JCS, Martins DTO, Souza PT Jr (1998) Experimental evaluation of stem bark of *Stryphnodendron adstringens* (Mart.) Coville for antiinflammatory activity. Phytother Res 12(3):218–220

Lima SCS, Arruda GO, Renovato RD, Alvarenga MRM (2012) Representations and uses of medicinal plants in elderly men. Rev Lat Am Enfermagem 20(4):778–786

Lopes GC, Sanches ACC, Toledo CEM, Isler AC, Mello JCP (2009) Determinação quantitativa de taninos em três espécies de *Stryphnodendron* por cromatografia líquida de alta eficiência. Braz J Pharm Sci 45(1):135–143

Lorenzi H (1998) Árvores brasileiras. Manual de identificação e cultivo de plantas arbóreas nativas do Brasil. Plantarum, Nova Odessa, p 189

Luize PS, Tiuman PS, Morello LG, Maza PK, Ueda-Nakamura T, Dias Filho BP et al (2005) Efeito de extratos de plantas medicinais no crescimento de *Leishmania* (L.) *amazonensis* e *Trypanosoma cruzi*. Rev Bras Cien Farm 41(1):85–94

Maciel MRA, Neto GG (2006) Um olhar sobre as benzedeiras de Juruena (Mato Grosso, Brasil) e as plantas usadas para benzer e curar. Bol Mus Para Emílio Goeldi 1(3):61–77

Martins DTO, Lima JCS, Rao VSN (2002) The acetone soluble fraction from bark extract of *Stryphnodendron adstringens* (Mart.) Coville inhibits gastric acid secretion and experimental gastric ulceration in rats. Phytother Res 16(5):427–431

Melo JO, Endo TH, Bersani-Amado LE, Svidzinski AE, Baroni S, Mello JCP (2007) Efeito da casca de *Stryphnodendron adstringens* (barbatimão) em modelos de nocicepção animais. Braz J Pharm Sci 43(3):465–469

Melo-Silva F, Paula JE, Espindola LS (2009) Evaluation of the antifungal potential of Brazilian Cerrado medicinal plants. Mycoses 52(6):511–517

Minatel DG, Pereira AMS, Chiaratti TM, Pasqualin L, Oliveira JCN, Couto LB, Lia RCC, Cintra JM, Bezzon MFA, Franca SC (2010) Clinical study for the validation of the efficacy of ointment containing barbatimao (*Stryphnodendron adstringens* (Mart.) Coville) on healing of decubitus ulcers. Rev Bras Med 67(7):250–256

Miranda MA (2010) Atividade antimicrobiana das soluções de Barbatimão, Mamona e Clorexidina utilizadas na Endodontia. Avaliação comparativa in vitro. Universidade de São Paulo, Ribeirão Preto

Oliveira JR (2011) Ensaio de citotoxicidade de extratos naturais após determinação da concentração microbicida mínima para *Staphylococcus* spp., *Streptococcus mutans* e *Candida* spp. Universidade Estadual Paulista "Julio de Mesquita Filho", São José dos Campos

Oliveira ER, Menini Neto L (2012) Levantamento etnobotânico de plantas medicinais utilizadas pelos moradores do povoado de Manejo, Lima Duarte–MG. Rev Bras Plantas Med 14(2):311–320

Pinho L, Souza PNS, Sobrinho EM, Almeida AC, Martins ER (2012) Atividade antimicrobiana de extratos hidroalcoolicos das folhas de alecrim-pimenta, aroeira, barbatimão, erva-baleeira e do farelo da casca de pequi. Cienc Rural 42(2):326–331

Rebecca MA, Ishii-Iwamoto EL, Kelmer-Bracht AM, Caparroz-Assef SM, Cuman RKN, Pagadigorria CLS (2003) Effect of *Stryphnodendron adstringens* (barbatimão) on energy metabolism in the rat liver. Toxicol Lett 143(1):55–63

Rodrigues V, Carvalho D (2001) Levantamento etnobotânico de plantas medicinais no domínio dos cerrados na região do Alto Rio Grande–Minas Gerais. Ci Agrotecnologia 25(1):102–123

Sanches ACC, Lopes GC, Toledo CEM, Sacramento LVS, Sakuragui CM, Mello JC (2007) Estudo morfológico comparativo das cascas e folhas de *Stryphnodendron adstringens*, *S.* polyphyllum e S. obovatum – Leguminosae. Lat Am J Pharm 26(3):22–35

Santos SC, Costa WF, Ribeiro JP, Guimarães DO, Ferri PH, Ferreira HD et al (2002) Tannin composition of barbatimão species. Fitoterapia 73(4):292–299

Sousa FC, Oliveira ENA, Santos DC, Oliveira FAA, Mori E (2011) Uso de plantas medicinais (fitoterápicos) por mulheres da cidade de Icó-CE. Bio Far 5(1):161–170

Souza TM, Moreira RRD, Pietro RCLR, Isaac VLB (2007a) Avaliação da atividade anti-séptica de extrato seco de Stryphnodendron adstringens (Mart.) Coville e de preparação cosmética contendo esse extrato. Rev Bras Farmacogn 17(1):71–75

Souza TM, Severi JA, Silva VYA, Santos E, Pietro RCLR (2007b) Bioprospecção de atividade antioxidante e antimicrobiana da casca de *Stryphnodendron adstringens* (Mart.) Coville (Leguminosae-Mimosoidae). Rev Cienc Farm Básica Apl 28(2):221–226

Tresvenzol LM, Paula JR, Ricardo AF, Ferreira HD, Zatta DT (2006) Estudo sobre o comércio informal de plantas medicinais em Goiânia e cidades vizinhas. Rev Elet Farm 3(1):23–28

Vicente RA, Silva VR L e, Baby AR, Velasco MV, Bedin V (2009) Double-blind, randomized, placebo-controlled trial of a cream containing *the Stryphnodendron adstringens* (Martius) Coville bark extract for suppressing terminal hair growth. J Europ Acad Dermatol Venereol 23(4):410–414

Vila Verde GM, Paula JR, Caneiro DM (2003) Levantamento etnobotânico das plantas medicinais do cerrado utilizadas pela população de Mossâmedes (GO). Rev Bras Farmacogn. 13(suppl):66–68

Vinaud MC, Santos SC, Ferri PH, Lino Junior RS, Bezerra JCB (2005) Avaliação da atividade larvicida de plantas fitoterápicas do cerrado do gênero *Stryphnodendron* spp. sobre miracídios e cercarias de *Schistosoma mansoni*. Rev Patol Trop 34(2):137–143

Tabebuia avellanedae Lorentz ex Griseb.

Rainer W. Bussmann

Tabebuia avellanedae Lorentz ex Grieseb.
Photo: Indiana Coronado
Available in: http://www.tropicos.org/Image/100134182

R. W. Bussmann (✉)
Ilia State University, Institute of Botany and Bakuriani Alpine Botanical Garden, Department of Ethnobotany, Tbilisi, Georgia
e-mail: rainer.bussmann@iliauni.edu.ge

U. P. Albuquerque et al. (eds.), *Medicinal and Aromatic Plants of South America*, Medicinal and Aromatic Plants of the World 5,
https://doi.org/10.1007/978-94-024-1552-0_40

Abstract *Tabebuia avellanedae* Lorentz ex Grieseb. (Lapacho, Pau'd'Arco), has long been reported as used in traditional medicine in Central and Latin America for disorders as varied as leishmaniasis, bacterial infections, fever, malaria and syphilis. In the early 1960 reports of cancer being cured with Lapacho extract appeared in Brazil. The taxonomy of the genus *Tabebuia* is however complicated, and various species are used interchangeably in traditional medicine. At least *Tabebuia serratifolia* (Vahl) Nichols has to be seen as bioequivalent to *T. avellanedae*. Lapacho bark is the crude drug, in most cases prepared as infusion or tea. Lapachol and ß-Lapachol are recognized as the main bioactive compounds, and a large number of studies have focused on the anti-tumor, anti-bacterial and anti-inflammatory activity. However, so far little conclusive evidence for efficacy could be provided. The main problem of many studies had been the lack of exact taxonomic identification of the source material, the use of the wrong plant parts, and a focus of very few compounds, rather than traditional preparations. Much more research is needed to assess the actual efficacy of *Tabebuia* preparations.

Keywords Lapacho · Pau'd'Arco · *Tabebuia avellanedae* · *Tabebuia serratifolia* · Bignoniaceae

1 Taxonomic Characteristics

Tabebuia avellanedae Lorentz ex Griseb. has long been reported as "Lapacho" and "Pau d'Arco" from Latin America. The taxonomy of the species, and the genus *Tabebuia* in general is however difficult. *Tabebuia* is often linked to the genus *Tecoma* or separated into *Tabebuia* and *Handroanthus*, and most species have been described under a plethora of synonyms.

Recent taxonomic studies suggest that yellow-flowered, lapachol containing species are best recognized in their own genus, *Handroanthus* (Grose and Olmstead 2007). This, however, does not include pink-flowered lapachol containing species like *T. avellanedae*, although the species sometimes is included in *Handronanthus* nevertheless.

Tabebuia avellanedae is by far the most commonly used scientific name for the species in all but the most recent literature, and thus the older, broad concept of the genus is followed here.

Synonyms *Gelseminum avellanedae* (Lorentz ex Griseb.) Kuntze; *Handroanthus avellanedae* (Lorentz ex Griseb.) Mattos; *Tabebuia avellanedae* Lorentz ex Griseb.; *Tabebuia dugandii* Standl.; *Tabebuia impetiginosa* (Mart. ex DC.) Standl.; *Tabebuia ipe* var. *integra* (Sprague) Sandwith; *Tabebuia nicaraguensis* S.F.Blake; *Tabebuia palmeri* Rose; *Tabebuia schunkevigoi* D.R.Simpson; *Tecoma adenophylla* Bureau & K.Schum.; *Tecoma avellanedae* (Lorentz ex Griseb.) Speg.; *Tecoma avellanedae* var. *alba* Lillo; *Tecoma impetiginosa* Mart. ex DC.; *Tecoma integra* (Sprague) Hassl.; *Tecoma ipe* var. *integra* Sprague; *Tecoma ipe* var. *integrifolia* Hassl.; *Tecoma ipe* f. *leucotricha* Hassl.

2 Crude Drug Used

The United States Food and Drug Administration (FDA) recognizes Lapacho Tea as a dietary supplement, Generally Regarded as Safe (GRAS) (FDA 1999). The pharmaceutical definition of the crude drug is Tabebuiae cortex. Lapacho bark is normally prepared as tea, although the material does need to be steeped for at least 8–10 min. Since the main compounds are not readily water-soluble (Taylor 2005). Traditionally the inner bark of the tree is used.

3 Major Chemical Constituents and Bioactive Compounds

Given the long use of the species, and the large commercial interest in particular of its use as a nutritional supplement, a large number if studies focused on elucidating the compounds of *Tabebuia avellaedae* and other species. Lapachol and ß-lapachone are regarded as the most common, and earliest isolated quinones in *Tabebuia* (Thomson 1971; de Oliveira et al. 1993).

The current list for compounds found in *Tabebuia* bark also includes acetaldehydes, alpha-lapachone, ajugols, anisic acid, anthraquinones, benzoic acids, benzenes, carboxaldehydes, chromium, chrysanthemin, dehydro-alpha-lapachone, dehydroisolapachone, deoxylapachol, flavonoids, furanonaphthoquinones, hydrochlorolapachol, 2-hydroxy-3-methyl-quinone, 6-hydroxy-mellein, iso-8-hydroxy-lariciresinol, kigelinone, lapachenol, lapachenole, various lapachones, menaquinones, 4-methoxyphenol, naphthoquinones, paeonidin-3-cinnamyl-sophoroside, phthiolol, quercetin, tabebuin, tectoquinone, vanillic acid, vanillin, veratric acid, veratric aldehyde, and xyloidone (Koyama et al. 2000a, b; Kreher et al. 1988; Lemos et al. 2007; Pertino et al. 2011; Suo et al. 2013; Steinert et al. 1995, 1996; Wagner et al. 1989; Warashina et al. 2004, 2005, 2006; Yamashita et al. 2009).

The characteristic compounds of the inner bark and the wood are naphthochochinones, mainly lapachol (3.6%), ß-lapachone, its cyclisation product and in lower concentrations (<0.01%) cumarins and saponines (10). Lapachol and lapachone are the biologically most active substances. For a complete list of compounds see Table 1.

Table 1 Characteristic compunds of Lapacho

(+)-2-(1′-Hydroxy-ethyl)-naphtho-(2,3,B)-furan-4,9-Dione	Bark
(−)-5-Hydroxy-2-(1′-Hydroxy-ethyl)-naphtho-(2,3,B)-furan-4,9-Dione	Bark
(−)-6-Hydroxy-mellein	Bark
1-(1-Hydroxy-ethyl)-Furonaphthoquinone	Bark
2-(1-Hydroxy-ethyl)-naphtho-(2–3-B)-furan-4-9-Dione	Bark
2-Acethyl-5-Hydroxy-naphtho-(2–3-B)-furan-4-9-Dione	Bark
2-Acethyl-8-Hydroxy-naphtho-(2–3-B)-furan-4-9-Dione	Bark
2-Acethyl-naphtho-(2–3-B)-furan-4-9-Dione	Bark
2-Dehydro-alpha-lapachone	Bark
2-ethyl-naphtho(2,3-B)-furan-4-9-Dione	Bark
3,4,5-Trimethoybenzoic-acid	Bark
5-Hydroxy-2-(1-Hydroxy-ethyl)-naphtho(2,3-B)-furan-4-9-Dione	Bark
6-O-(3–4-Dimethoxy-benzoyl)-ajugol	Bark
6-O-(P-Hydroxy-benzoyl)-ajugol	Bark
8-Hydroxyisolariciresinol	Bark
Anisaldehyde	Bark
Anisic-acid	Bark
Benzo[B]furan-6-Carboxaldehyde	Bark
Dehydro-alpha-isolapachone	Bark
Kigelinone	Bark
RS-8-Hydroxy-2-(1′-hydroxy-ethyl)-naphtho-(2,3,B)-furan-4,9-Dione	Bark
Vanillic-acid	Bark
Vanillin	Bark
Veratric-acid	Bark
Veratric-aldehyde	Bark
Xyloidone	Bark
Alpha-Lapachone	Bark, Wood
Beta-lapachone	Bark, Wood
Dehydro-alpha-lapachone	Bark, Wood
4-Hydroxy-benzoic-acid	Bark, Wood
Lapachenole	Bark, Wood
Lapachol	Bark, Wood
Anthraquinone-2-Aldehyde	Wood
Anthraquinone-2-Carboxylic-acid	Wood
1-Hydroxyanthraquinone	Wood
1-Methoxy-anthraquinone	Wood
2,3-Dimethyl-1,4-Naphthoquinone	Wood
2-Acetoxy-methyl-anthraquinone	Wood
2-Hydroxy-3-Methyl-anthraquinone	Wood
2-Hydroxy-methyl-anthraquinone	Wood

(continued)

Table 1 (continued)

Deoxylapachol	Wood
Lapachol-methyl-ether	Wood
Menaquinone-1	Wood
O-hydroxybenzoic-acid	Wood
Phthiolol	Wood
Quercetin	Wood
Tabebuin	Wood
Tectoquinone	Wood
P-hydroxy-benzoic-acid	Plant
chrysanthemin	Flower
Cyanidin-3-O-beta-d-rutinoside	Flower
Peonidin-3-Cinnamyl-sophoroside	Flower

4 Morphological Description

The genus *Tabebuia* includes about 100 species of large, flowering trees that are common to South American. *T. avellanedae* grows to 50 m high and the base of the tree can be 2–3 m in diameter. It is deciduous and shed its opposite leaves in the dry season. The red flowers are 3–11 cm wide, in dense clusters. The calyx is campanulate to tubular, mostly five-lobed, and trumpet-like. The corolla is pink or red. The outside of the flower tube is either glabrous or pubescent. The fruit is a dehiscent pod, 10–50 cm long with numerous seeds and often persists on the tree through the dry season to shed seeds just at the start if the rains. The wood is very hard, and denser than water.

5 Geographical Distribution

The genus *Tabebuia* belongs to the Bighnoniaceae and contains around 100 species, six of which are common in Central America, 75 in the Caribbean and 25 in South America (10). *Tabebuia* species are widely used as ornamentals in tropical landscaping.

T. avellanedae has a particularly wide distribution that ranges from Northern Mexico to northern Argentina.

6 Ecological Requirements

Many *Tabebuia* species can be classified as late succession pioneer trees, and *T. avellanedae* is no exception. The species requires full light, but has been collected from secondary humid rainforest to semi-humid forests, and is known to survive well in pastures, where it is planted for reforestation. Specimens are known from sea-level up to about 3000 m altitude.

7 Collection Practice

For the harvest of Lapacho bark *T. avellanedae* trees sometimes felled and debarked, or simply debarked and the resulting material rasped once-twice per year (Schultes and Raffauf 1990; García-Barriga 1992). Although Lapacho collection and commerce have increased rapidly over the last decades, due to the hype about the species anti-tumor properties (Gómez Castellanos et al. 2009), *T. avellanedae* can be regarded as not threatened, its range, in fact, has expanded due to its use for reforestation.

8 Traditional Use (Part(s) Used) and Common Knowledge

T. avellanedae has been reported as being by several groups of Central and Latin American indigenous peoples to treat a wide variety of conditions, ranging from malaria, leishmaniasis, fevers, fungal and bacterial infections, to syphilis (Schultes and Raffauf 1990; Duke 1985; Duke and Vasquez 1994). The species is now mainly known by several common names in Portuguese and Spanish, and common names used in English are borrowed from South American common names. Popular common names for both species include pau d'arco (or palo de arco) (Grenand et al. 2004; Rodrigues 2006), lapacho, tahuarí (Duke and Vasquez 1994), tajibo (taheebo), and ipé (Grose and Olmstead 2007). These common names are best understood as folk genera; many species of *Tabebuia* are indicated by these names, which are not specific to *T. avellanedae,* but applied to a large variety of species with both pink and yellow flowers. Folk species may be distinguished by applying a modifying adjective to the common name. For example, Duke and Vasquez (1994) list six species of *Tabebuia,* three with an unmodified "tahuarí" as the common name, and the remaining three with an adjective in addition to "tahuarí". Although for example *Tabebuia serratifolia* (Vahl) Nichols. may be known as ipé-amarelo (Grenand et al. 2004) or pau d'arco amarelo (Jones 1995) in Portuguese, even these more specific common names may be applied to any of the 30 or more species of *Tabebuia* with yellow (amarelo) flowers. Ipé appears to be the most popular Portuguese name for *Tabebuia* spp. when they are being treated as a source of timber or as ornamentals. The name pau d'arco appears to be mostly used when medicinal uses of *Tabebuia* are considered. Boom (1990) reports the use of the bark among the Panare to treat stomachache. Muñoz et al. (2000) report its use as febrifuge by the Chacobo. The Palikur of French Guiana use the leaves to treat colds, coughs and flu, and the bark to treat leishmaniasis, dysentery, and (in a mixture with three other species) to treat diabetes. The Wayapi of French Guiana use the bark as a febrifuge (Grenand et al.

2004). Label data of an herbarium specimen at the Missouri Botanical Garden indicate that the Tacana of Bolivia use the bark to eliminate internal tumors (de Walt 1995). de Melo et al. (2011) document anti-cancer use in modern ethnomedicine.

Use is also reported among Mestizo/Creole populations. Rodrigues (2006) reports the use of *T. avellanae* bark among a Brazilian mestizo population for gastrointestinal disturbances, inflammation and tropical diseases. Grenand et al. (2004) reports French Guianese Creoles using flowers to treat colds, coughs and flu. The label data of a herbarium specimen collected by Schunke (1993), indicates the use of bark and wood in Peru to treat uterine cancer and liver cirrhosis. Another specimens, collected by Plowman (1967), reports that a bark decoction is used for "various maladies, especially cancer" in Colombia. Jones (1995) mentions use as an astringent and to treat cutaneous ulcers, and quotes a report by Wade Davis that the species is a popular cure for cancer.

There are numerous reports in the literature of the ethnomedicinal use of other species of *Tabebuia*. Given that common names such as pau d'arco represent folk generic concepts that refer to multiple scientifically recognized species, it is possible that *T. serratifolia* may be used interchangeably with other *Tabebuia* species. *Tabebuia* species are similar biochemically, so are likely to similarly efficacious (Gentry 1992). Lapachol is produced by all of the 30 species Grose and Olmstead (2007) segregates into the genus *Handroanthus*.

9 Modern Medicine Based on Its Traditional Medicine Uses

Beginning in the late 1960s, there were a number of news reports about the anticancer potential of lapachol containing species of *Tabebuia* (Jones 1995). Herbarium specimens collected by Schunke and Plowman, and Davis (as quoted in Jones 1995) all post-date the 1967 news-magazine article which Jones (1995) believes was responsible for increased interest in pau d'arco. All these species, however, belong to *Tabebuia serratifolia* (Vahl) Nichols. Gentry (1992) reported "indigenous uses of *Tabebuia* bark against cancer include that of [...] *Tabebuia* in Colombia." However, the source cited by Gentry discusses the use of lapachol containing species collectively under the heading of *Tabebuia serratifolia* (or under the common name palo de arco), and notes that use of palo de arco to treat cancer had only been occurring in Colombia for about 3 years (Garcia-Barriga 1975). There is no secure indication whatsoever that indeed *T. avellanedae* was the species that first entered into modern medicinal practice based on traditional use.

Modern research on the medicinal properties of *Tabebuia* goes back to the 1960s when the US National Cancer Institute started a large scale global plant screening program in order to isolate new anti-cancer compounds (Cragg and Newman 2005). One of the compounds of interest turned out to be lapachol, isolated from *T. avellanedae* (Cassady and Douros 1980). Gómez Castellanos et al. produced a review of earlier medicinal research on Lapacho (2009).

A variety of authors found scant anti-cancer efficacy of *Tabebuia* compounds. de Santana et al. (1968) were the first team to report anti-cancer activity. Choi et al. (2003) report on efficacy of ß-lapachone against prostate cancer by down-regulating pRB regulation and Cdk inhibitor p21 induction. de Sousa et al. (2009)and Costa et al. (2011) found lapachol and other compounds as tumor inhibitor in *Drosophila*, while Queiroz et al. (2008) and Higa et al. (2011) documented activity in mice, and Moon et al. (2010) and Inagaki et al. (2013) produces cytotoxicity against leukemia cells. Kim et al. (2007) found anti-invasive and anti-metastatic properties of ß-lapachone, while a general antitumor effect of the molecule was reported by a variety of teams (Lamberti et al. 2013, Lee et al. 2005, 2006, 2012, 2013). Mukherjee et al. (2009) produced growth inhibition of human estrogen receptors in breast cancer cells by applying *Tabebuia* extract. Finally, tumor apoptosis was shown by Woo and Choi (2005), Woo et al. (2006), and Yamashita et al. (2007).

Although most research focused on anti cancer properties, some teams found indication of activity in other areas such anti-oxidant activity (Awale et al. 2005; Moreira Vasconcelos et al. 2014; Park et al. 2003); immuno-stimmulation (Böhler et al. 2008); anti-inflammatory effects (Byeon et al. 2008; Lee et al. 2012); wound-healing (Coelho et al. 2010; Kung et al. 2008; Suo et al. 2012); anti-depressant (Freitas et al. 2010, 2013); anti-vascular (Garkavtsev et al. 2011); anti-leishmanial (González-Coloma et al. 2012; Menna-Barreto et al. 2005); anti-bacterial (Höfling et al. 2010; Macedo et al. 2013; Machado et al. 2003; Moreira Vasconcelos et al. 2014; Park et al. 2006; Pereira et al. 2006); anti-triglyceric (Kiage-Mokua et al. 2012); larvicidal (Kim et al. 2013); anti-fungal (Melo e Silva et al. 2009); anti-ulcer (Pereira et al. 2013; Twardowschy et al. 2008), molluscididal (Silva et al. 2007), reduction of autoimmune effects (Xu et al. 2013)

Toxic effects explaining anti-conceptive properties were found by de Cássia da Silveira and de Oliveira (2007), de Miranda et al. (2001), Lemos et al. (2012), and Moreira Vasconcelos et al. (2014).

10 Conclusions

Based on the above stipulations, it is evident that *T. avellanedae*, as well as other species like *Tabebuia serratifolia*, are known by several common names, all of which may also be applied to other species of *Tabebuia*. Referring to all these species as "pau d'arco" or "lapacho" reflects traditional folk taxonomy. There are a variety of reported ethnomedicinal uses for various species. The earliest reports of traditional medicinal use most likely refer to *Tabebuia serratifolia* (Vahl) Nichols. Many other species of *Tabebuia* are also used medicinally, and various scientifically recognized species with similar biochemistry may be used interchangeably under the folk concept of "pau d'arco" and "lapacho". From a scientific perspective the uses as well as vernacular names of *Tabebuia serratifolia* (Vahl) Nichols. are entirely interchangeable with the uses and traditional names of *T. avellanedae* Lorentz ex Griseb.

Although lapachol and ß-lapachol are recognized as the main bioactive compounds, and a large number of studies have focused on anti-tumor, anti-bacterial and anti-inflammatory activity, so far little conclusive evidence for efficacy could be provided. The main problem of many studies had been the lack of exact taxonomic identification of the source material. In addition, many studies focused on material consisting of any woody part of the tree, rather than the inner bark layer that is reported in traditional use. The focus on very few compounds regarded as bioactive, rather than traditional preparations, might also have had serious effects on efficacy. Much more research is needed to assess the actual efficacy of *Tabebuia* preparations.

References

Awale S, Kawakami T, Tezuka Y, Ueda JY, Tanaka K, Kadota S (2005) Nitric oxide (NO) production inhibitory constituents of *Tabebuia avellanedae* from Brazil. Chem Pharm Bull (Tokyo) 53(6):710–713

Böhler T, Nolting J, Gurragchaa P, Lupescu A, Neumayer HH, Budde K, Kamar N, Klupp J (2008) *Tabebuia avellanedae* extracts inhibit IL-2-independent T- lymphocyte activation and proliferation. Transpl Immunol 18(4):319–323

Boom B (1990) Useful plants of the Panare Indians of the Venezuelan Guyana. Adv Econ Bot 8:57–76

Byeon SE, Chung JY, Lee YG, Kim BH, Kim KH, Cho JY (2008) In vitro and in vivo anti-inflammatory effects of taheebo, a water extract from the inner bark of *Tabebuia avellanedae*. J Ethnopharmacol 119(1):145–152

Cassady JM, Douros JD (1980) Miscellaneous natural products with antitumor activity. Anticancer agents based on natural product models. Academic, New York

Choi YH, Kang HS, Yoo MA (2003) Suppression of human prostate cancer cell growth by beta-lapachone via down-regulation of pRB phosphorylation and induction of Cdk inhibitor p21(WAF1/CIP1). J Biochem Mol Biol 36(2):223–229

Coelho JM, Antoniolli AB, Nunes e Silva D, Carvalho TM, Pontes ER, Odashiro AN (2010) Effects of silver sulfadiazine, ipê roxo (*Tabebuia avellanedae*) extract and barbatimão (*Stryphnodendron adstringens*) extract on cutaneous wound healing in rats. Rev Col Bras Cir 37(1):45–51. Portuguese

Costa WF, Oliveira AB, Nepomuceno JC (2011) Lapachol as an epithelial tumor inhibitor agent in Drosophila melanogaster heterozygote for tumor suppressor gene wts. Genet Mol Res 10(4):3236–3245

Cragg GM, Newman DJ (2005) Plants as a source of anticancer agents. J Ethnopharmacol 100:72.79

de Cássia da Silveira ESR, de Oliveira GM (2007) Reproductive toxicity of lapachol in adult male Wistar rats submitted to short-term treatment. Phytother Res 21(7):658–662

de Melo JG, Santos AG, de Amorim EL, do Nascimento SC, de Albuquerque UP (2011) Medicinal plants used as antitumor agents in Brazil: an ethnobotanical approach. Evid Based Complement Alternat Med:365359. http://doi.org/10.1155/2011/365359

de Miranda FG, Vilar JC, Alves IA, Cavalcanti SC, Antoniolli AR (2001) Antinociceptive and antiedematogenic properties and acute toxicity of *Tabebuia avellanedae* Lor. ex Griseb. inner bark aqueous extract. BMC Pharmacol 1:6

de Oliveira AB, Raslan DS, de Oliveira GG, Maia JGS (1993) Lignans and naphthoquinones from *Tabebuia incana*. Phytochemistry 34:1409–1412 de Santana CF, de Lima OG, d' Albuquerque IL, Lacerda AL, Martins DG (1968) Antitumoral and toxicological properties of extracts of bark and various wood components of Pau d'arco (*Tabebuia avellanedae*). Rev Inst Antibiot (Recife) 8(1):89–94. Portuguese

de Sousa NC, de Rezende AA, da Silva RM, Guterres ZR, Graf U, Kerr WE, Spanó MA (2009) Modulatory effects of *Tabebuia impetiginosa* (Lamiales, Bignoniaceae) on doxorubicin-induced somatic mutation and recombination in Drosophila melanogaster. Genet Mol Biol 32(2):382–388

de Walt S (1995) Collection #160. Herbarium specimen label data, available online at http://www.tropicos.org/Specimen/1323429. MBG, St. Louis, 1995

Duke J (1985) CRC Handbook of medicinal herbs. CRC Press, Boca Ratón

Duke J, Vasquez R (1994) Amazonian ethnobotanical dictionary. CRC Press, Ann Arbor

FDA (1999) Econimuc characterization of the dietary supplement industry. Final Report. US FDA, Washington, DC 1999

Freitas AE, Budni J, Lobato KR, Binfaré RW, Machado DG, Jacinto J, Veronezi PO, Pizzolatti MG, Rodrigues AL (2010) Antidepressant-like action of the ethanolic extract from *Tabebuia avellanedae* in mice: evidence for the involvement of the monoaminergic system. Prog Neuro-Psychopharmacol Biol Psychiatry 34(2):335–343

Freitas AE, Machado DG, Budni J, Neis VB, Balen GO, Lopes MW, de Souza LF, Veronezi PO, Heller M, Micke GA, Pizzolatti MG, Dafre AL, Leal RB, Rodrigues AL (2013) Antidepressant-like action of the bark ethanolic extract from *Tabebuia avellanedae* in the olfactory bulbectomized mice. J Ethnopharmacol 145(3):737–745

Garcia-Barriga H (1975) Flora medicinal de Colombia Tomo 3. Instituto de Ciencias Naturales, Bogota, pp 142–148

Garkavtsev I, Chauhan VP, Wong HK, Mukhopadhyay A, Glicksman MA, Peterson RT, Jain RK (2011) Dehydro-alpha-lapachone, a plant product with antivascular activity. Proc Natl Acad Sci U S A 108(28):11596–11601

Gentry A (1992) A synopsis of Bignoniaceae ethnobotany and economic botany. Ann Miss Bot Gard 79(1):53–64

Gómez Castellanos JR, Prieto JM, Heinrich M (2009) Red Lapacho (*Tabebuia impetiginosa*) – a global ethnopharmacological commodity? J Ethnopharmacol 121(1):1–13

González-Coloma A, Reina M, Sáenz C, Lacret R, Ruiz-Mesia L, Arán VJ, Sanz J, Martínez-Díaz RA (2012) Antileishmanial, antitrypanosomal, and cytotoxic screening of ethnopharmacologically selected Peruvian plants. Parasitol Res 110(4):1381–1392

Grenand P, Moretti C, Jacquemin H, Prévost M (2004) Pharmacopées Traditionnelles en Guyane: Créoles, Wayãpi, Palikur. IRD Éditions, Paris

Grose SO, Olmstead RG (2007) Taxonomic revisions in the polyphyletic genus *Tabebuia* s.l. (Bignoniaceae). Syst Bot 32(3):660–670

Higa RA, Aydos RD, Silva IS, Ramalho RT, Souza AS (2011) Study of the antineoplastic action of *Tabebuia avellanedae* in carcinogenesis induced by azoxymethane in mice. Acta Cir Bras 26(2):125–128

Höfling JF, Anibal PC, Obando-Pereda GA, Peixoto IA, Furletti VF, Foglio MA, Gonçalves RB (2010) Antimicrobial potential of some plant extracts against CaIndida species. Braz J Biol 70(4):1065–1068

Inagaki R, Ninomiya M, Tanaka K, Watanabe K, Koketsu M (2013) Synthesis and cytotoxicity on human leukemia cells of furonaphthoquinones isolated from *Tabebuia* plants. Chem Pharm Bull (Tokyo) 61(6):670–673

Jones K (1995) Pau d'Arco: Immune power from the rain forest. Healing Arts Press, Rochester

Kiage-Mokua BN, Roos N, Schrezenmeir J (2012) Lapacho tea (*Tabebuia impetiginosa*) extract inhibits pancreatic lipase and delays postprandial triglyceride increase in rats. Phytother Res 26(12):1878–1883

Kim SO, Kwon JI, Jeong YK, Kim GY, Kim ND, Choi YH (2007) Induction of Egr-1 is associated with anti-metastatic and anti-invasive ability of beta-lapachone in human hepatocarcinoma cells. Biosci Biotechnol Biochem 71(9):2169–2176

Kim MG, Jeon JH, Lee HS (2013) Larvicidal activity of the active constituent isolated from Tabebuia *avellanedae* bark and structurally related derivatives against three mosquito species. J Agric Food Chem 61(45):10741–10745

Koyama J, Morita I, Kino A, Tagahara K (2000a) Micellar electrokinetic chromatography (MEKC) separation of furanonaphthoquinones from *Tabebuia impetiginosa*. Chem Pharm Bull (Tokyo) 48(6):873–875

Koyama J, Morita I, Tagahara K, Hirai K (2000b) Cyclopentene dialdehydes from *Tabebuia impetiginosa*. Phytochemistry 53(8):869–872

Kreher B, Lotter H, Cordell GA, Wagner H (1988) New Furanonaphthoquinones and other Constituents of *Tabebuia avellanedae* and their immunomodulating activities in vitro. Planta Med 54(6):562–563

Kung HN, Yang MJ, Chang CF, Chau YP, Lu KS (2008) In vitro and in vivo wound healing-promoting activities of beta-lapachone. Am J Phys Cell Phys 295(4):C931–C943

Lamberti MJ, Vittar NB, da Silva Fde C, Ferreira VF, Rivarola VA (2013) Synergistic enhancement of antitumor effect of β-Lapachone by photodynamic induction of quinone oxidoreductase (NQO1). Phytomedicine 20(11):1007–1012

Lee JH, Cheong J, Park YM, Choi YH (2005) Down-regulation of cyclooxygenase-2 an telomerase activity by beta-lapachone in human prostate carcinoma cells. Pharmacol Res 51(6):553–560

Lee JI, Choi DY, Chung HS, Seo HG, Woo HJ, Choi BT, Choi YH (2006) Beta- lapachone induces growth inhibition and apoptosis in bladder cancer cells by modulation of Bcl-2 family and activation of caspases. Exp Oncol 28(1):30–35

Lee MH, Choi HM, Hahm DH, Her E, Yang HI, Yoo MC, Kim KS (2012) Analgesic and anti-inflammatory effects in animal models of an ethanolic extract of Taheebo, the inner bark of *Tabebuia avellanedae*. Mol Med Rep 6(4):791–796

Lee S, Kim IS, Kwak TH, Yoo HH (2013) Comparative metabolism study of β- lapachone in mouse, rat, dog, monkey, and human liver microsomes using liquid chromatography-tandem mass spectrometry. J Pharm Biomed Anal 83:286–292

Lemos TL, Monte FJ, Santos AK, Fonseca AM, Santos HS, Oliveira MF, Costa SM, Pessoa OD, Braz-Filho R (2007) Quinones from plants of northeastern Brazil: structural diversity, chemical transformations, NMR data and biological activities. Nat Prod Res 21(6):529–550

Lemos OA, Sanches JC, Silva IE, Silva ML, Vinhólis AH, Felix MA, Santos RA, Cecchi AO (2012) Genotoxic effects of *Tabebuia impetiginosa* (Mart. Ex DC.) Standl. (Lamiales, Bignoniaceae) extract in Wistar rats. Genet Mol Biol 35(2):498–502

Macedo L, Fernandes T, Silveira L, Mesquita A, Franchitti AA, Ximenes EA (2013) β-Lapachone activity in synergy with conventional antimicrobials against methicillin resistant *Staphylococcus aureus* strains. Phytomedicine 21(1):25–29

Machado TB, Pinto AV, Pinto MC, Leal IC, Silva MG, Amaral AC, Kuster RM, Netto-dos Santos KR (2003) In vitro activity of Brazilian medicinal plants, naturally occurring naphthoquinones and their analogues, against methicillin- resistant *Staphylococcus aureus*. Int J Antimicrob Agents 21(3):279–284

Melo e Silva F, de Paula JE, Espindola LS (2009) Evaluation of the antifungal potential of Brazilian Cerrado medicinal plants. Mycoses 52(6):511–517

Menna-Barreto RF, Henriques-Pons A, Pinto AV, Morgado-Diaz JA, Soares MJ, De Castro SL (2005) Effect of a beta-lapachone-derived naphthoimidazole on *Trypanosoma cruzi*: identification of target organelles. J Antimicrob Chemother 56(6):1034–1041

Moon DO, Kang CH, Kim MO, Jeon YJ, Lee JD, Choi YH, Kim GY (2010) Beta-lapachone (LAPA) decreases cell viability and telomerase activity in leukemia cells: suppression of telomerase activity by LAPA. J Med Food 13(3):481–488

Moreira Vasconcelos C, Chaves Vasconcelos TL, Póvoas FTX, Evangelista Pires dos Santos RF, da Costa Maynart WH, Gomes de Almeida T, da Silva Oliveira JF, Dalboni França AD, Sales Santos Veríssimo RC, Lins TH, de Araújo-Júnior JX, de Assis Bastos ML (2014) Antimicrobial, antioxidant and cytotoxic activity of extracts of *Tabebuia impetiginosa* (Mart. ex DC.) Standl. J Chem Pharm Res 6(7):2673–2681

Mukherjee B, Telang N, Wong GY (2009) Growth inhibition of estrogen receptor positive human breast cancer cells by Taheebo from the inner bark of *Tabebuia avellandae* tree. Int J Mol Med 24(2):253–260

Muñoz V, Sauvain M, Bourdy G, Callapa J, Rojas I, Bergeron S, Rojas I, Bravo J, Balderrama L, Ortiz B, Gimenez A, Deharo E (2000) A search for natural bioactive compounds through a multidisciplinary approach in Bolivia. Part I. Evaluation of the antimalarial activity of some plants used by Chacobo Indians. J Ethnopharmacol 69:127–137

Park BS, Lee KG, Shibamoto T, Lee SE, Takeoka GR (2003) Antioxidant activity and characterization of volatile constituents of Taheebo (*Tabebuia impetiginosa* Martius ex DC). J Agric Food Chem 51(1):295–300

Park BS, Lee HK, Lee SE, Piao XL, Takeoka GR, Wong RY, Ahn YJ, Kim JH (2006) Antibacterial activity of *Tabebuia impetiginosa* Martius ex DC (Taheebo) against *Helicobacter pylori*. J Ethnopharmacol 105(1–2):255–262

Pereira EM, Machado Tde B, Leal IC, Jesus DM, Damaso CR, Pinto AV, Giambiagi-de Marval M, Kuster RM, Santos KR (2006) *Tabebuia avellanedae* naphthoquinones: activity against methicillin-resistant staphylococcal strains, cytotoxic activity and in vivo dermal irritability analysis. Ann Clin Microbiol Antimicrob 5:5

Pereira IT, Burci LM, da Silva LM, Baggio CH, Heller M, Micke GA, Pizzolatti MG, Marques MC, Werner MF (2013) Antiulcer effect of bark extract of *Tabebuia avellanedae*: activation of cell proliferation in gastric mucosa during the healing process. Phytother Res 27(7):1067–1073

Pertino MW, Theoduloz C, Palenzuela JA, Afonso Mdel M, Yesilada E, Monsalve F, González P, Droguett D, Schmeda-Hirschmann G (2011) Synthesis and pharmacological activity of diterpenylnaphthoquinone derivatives. Molecules 16(10):8614–8628

Plowman T (1967) Collection #126. Herbarium specimen label data, available online at http://www.tropicos.org/Specimen/1048249. MBG, St. Louis, 1967

Queiroz ML, Valadares MC, Torello CO, Ramos AL, Oliveira AB, Rocha FD, Arruda VA, Accorci WR (2008) Comparative studies of the effects of *Tabebuia avellanedae* bark extract and beta-lapachone on the hematopoietic response of tumour- bearing mice. J Ethnopharmacol 117(2):228–235

Rodrigues E (2006) Plants and animals utilized as medicines in the Jaú National Park (JNP), Brazilian Amazon. Phytother Res 20:378–391

Schultes RE, Raffauf RF (1990) The healing forest. Dioscorides Press, Portland, pp 107–109

Schunke J (1993) Collection #14378. Herbarium specimen label data, available online at http://www.tropicos.org/Specimen/2996041. MBG, St. Louis. 1993

Silva TM, Da Silva TG, Martins RM, Maia GL, Cabral AG, Camara CA, Agra MF, Barbosa-Filho JM (2007) Molluscicidal activities of six species of Bignoniaceae from north-eastern Brazil, as measured against Biomphalaria glabrata under laboratory conditions. Ann Trop Med Parasitol 101(4):359–365

Steinert J, Khalaf H, Rimpler M (1995) HPLC separation and determination of naphthol[2,3-b]furan-4,9-diones and related compounds in extracts of *Tabebuia avellanedae* (Bignoniaceae). J Chromatogr A693:281–287

Steinert J, Khalaf H, Rimpler M (1996) High-performance liquid chromatographic separation of some naturally occurring naphthoquinones and anthraquinones. J Chromatogr A723:206–209

Suo M, Isao H, Kato H, Takano F, Ohta T (2012) Anti-inflammatory constituents from *Tabebuia avellanedae*. Fitoterapia 83(8):1484–1488

Suo M, Ohta T, Takano F, Jin S (2013) Bioactive phenylpropanoid glycosides from *Tabebuia avellanedae*. Molecules 18(7):7336–7345

Taylor L (2005) The healing power of rainforest herbs. Square One Publishers, Garden Park City

Thomson RH (1971) Naphthaquinones. Naturally occurring quiones, vol 203. Academic, London

Twardowschy A, Freitas CS, Baggio CH, Mayer B, dos Santos AC, Pizzolatti MG, Zacarias AA, dos Santos EP, Otuki MF, Marques MC. Antiulcerogenic activity of bark extract of *Tabebuia avellanedae*, Lorentz ex Griseb. J Ethnopharmacol. 2008;118(3):455–459

Warashina T, Nagatani Y, Noro T (2004) Constituents from the bark of *Tabebuia impetiginosa*. Phytochemistry 65(13):2003–2011

Warashina T, Nagatani Y, Noro T (2005) Further constituents from the bark of *Tabebuia impetiginosa*. Phytochemistry 66(5):589–597

Warashina T, Nagatani Y, Noro T (2006) Constituents from the bark of *Tabebuia impetiginosa*. Chem Pharm Bull (Tokyo) 54(1):14–20

Woo HJ, Choi YH (2005) Growth inhibition of A549 human lung carcinoma cells by beta-lapachone through induction of apoptosis and inhibition of telomerase activity. Int J Oncol 26(4):1017–1023

Woo HJ, Park KY, Rhu CH, Lee WH, Choi BT, Kim GY, Park YM, Choi YH (2006) Beta-lapachone, a quinone isolated from *Tabebuia avellanedae*, induces apoptosis in HepG2 hepatoma cell line through induction of Bax and activation of caspase. J Med Food 9(2):161–168

Xu J, Wagoner G, Douglas JC, Drew PD (2013) β-Lapachone ameliorization of experimental autoimmune encephalomyelitis. J Neuroimmunol 254(1–2):46–54

Yamashita M, Kaneko M, Iida A, Tokuda H, Nishimura K (2007) Stereoselective synthesis and cytotoxicity of a cancer chemopreventive naphthoquinone from *Tabebuia avellanedae*. Bioorg Med Chem Lett 17(23):6417–6420

Yamashita M, Kaneko M, Tokuda H, Nishimura K, Kumeda Y, Iida A (2009) Synthesis and evaluation of bioactive naphthoquinones from the Brazilian medicinal plant, *Tabebuia avellanedae*. Bioorg Med Chem 17(17):6286–6291

Uncaria tomentosa (Willd. ex Schult.) DC. and Uncaria guianensis (Aubl.) J.F. Gmell

Izaskun Urdanibia and Peter Taylor

Uncaria tomentosa (Willd. ex Schult.) DC.
Photo: T. Croat
Available in: http://www.tropicos.org/Image/18420

I. Urdanibia (✉) · P. Taylor
Centre for Experimental Medicine, Venezuelan Institute for Scientific Research (IVIC), Caracas, Venezuela
e-mail: iurdanib@ivic.gob.ve; ptaylor@ivic.gob.ve

U. P. Albuquerque et al. (eds.), *Medicinal and Aromatic Plants of South America*, Medicinal and Aromatic Plants of the World 5,
https://doi.org/10.1007/978-94-024-1552-0_41

Abstract The jungles of Central and South America contain two predominant species of cat's claw (Uña de Gato), *Uncaria tomentosa* (Willd. ex Schult.) DC. and *Uncaria guianensis* (Aubl.) J.F. Gmell, which are used in traditional medicine mainly for their anti-inflammatory properties. However, a wealth of compounds have been isolated from these two vines of the Rubiaceae family, including alkaloids, flavonoids and terpenoids, showing a wide range of activities: anti-inflammatory, anti-oxidative, hypotensor, antiviral, smooth muscle relaxant, antispasmodic, gastrointestinal mucosa protector, antiarrhythmic, anticonvulsant, analgesic, anti-leishmaniosis, cytostatic, cytotoxic, hypoglycaemizing, anticholestatic, antihistaminic, hepatoprotective, diuretic, antiulcer, immunostimulating and sedative effects. Some of these activities have been confirmed in both in vitro and in vivo models.

Keywords *Uncaria tomentosa* · *Uncaria guianensis* · Cat's claw · Medicinal plant · Peru · Inflammation · Cancer

Abbreviations

POA	Pentacyclic Oxindole Alkaloids
TOA	Tetracyclic Oxindole Alkaloids
NF-κB	Nuclear Transcription Factor
TNF-α	Tumour Necrosis Factor alpha
IL-1	Interleukin-1
PGE_2	Prostaglandin E_2
NO	Nitric Oxide
COX-1	Cyclooxygenase-1
iNOS	inducible Nitric Oxygen Synthase
MAPK	Mitogen-Activated Protein Kinase
MMP	Matrix Metalloproteinases
VEGF	Vascular Endothelial Growth Factor

1 Taxonomic Characteristics

Uncaria tomentosa (Willd. ex Schult.) DC. and *Uncaria guianensis* (Aubl.) J.F. Gmel. are the best known South American species of the Uncaria genus, which total about 40 worldwide. They belong to the Cinchonoideae subfamily of the Rubiaceae family.

Both species are known in folk and complementary medicine under various traditional names: vilcacora, uña de gato, cat's claw, cat's crew, saventaro, hawk's claw, samento, unganangi, garabato amarillo, rangaya, bejuco de agua, tuajuncara and Katzenkralle (Falkiewicz and Łukasiak 2001; Heitzman et al. 2005; Quintela and Lock de Ugaz 2003). The names uña de gato and cat's claw are shared with other unrelated plants.

2 Crude Drug Used

U. tomentosa and *U. guianensis* are most frequently used and prepared in traditional medicine as an aqueous extraction in hot water of the inner bark or the root bark, or macerated in an alcoholic beverage (Gattuso et al. 2004; Sandoval et al. 2002). Powdered bark is also available commercially in capsules.

3 Major Chemical Constituents and Bioactive Compounds

U. tomentosa and *U. guianensis* contain a mixture of indole and oxindole alkaloids, glycosides, terpenoids and tannins. The chemical composition of the plant may vary depending on the collection site and the period of the year in which it was collected (Heitzman et al. 2005). For this reason, the diverse pharmacological properties reported for *U. tomentosa* and *U. guianensis* in the literature may be attributed to quantitative and qualitative differences in the composition of different collections.

Two chemotypes have been reported for U. tomentosa, which are botanically indistinguishable, but which show different profiles of chemical constituents (Guthrie et al. 2011; Reinhard 1999). One contains principally tetracyclic oxindole alkaloids (rhynchophylline, isorhynchophylline, corynoxeine, isocorynoxeine, rotundifoline, isorotundifoline), and the other, pentacyclic oxindole alkaloids (pteropodine, isopteropodine, mitraphylline, isomitraphylline, speciophylline, uncarine F) (Falkiewicz and Łukasiak 2001; Keplinger et al. 1999; Laus 2004; Reinhard 1999). Awareness of the existence of these two chemotypes comes from the traditional medicine of the Asháninka Indians of Peru, who distinguished savéntaro (saveshi: plant, antearo: potent), which contains more pentacyclic oxindole alkaloids (POA), from a less potent plant. It has been proposed that in extracts of the less potent chemotype, tetracyclic alkaloids (TOA) counteract the immune-modulating activity of the pentacyclic compounds (Reinhard 1999).

Alkaloids represent the principle group of compounds isolated from the *Uncaria* genus although the alkaloid content of leaf, bark and roots is variable. Rhynchophylline, isorhynchophylline and mitraphylline are the major alkaloids, while rotundifoline, isorotundifoline, corynoxeine and isocorynoxeine are present in lesser quantities. The stereoisomeric alkaloids, pteropodine (uncarine C), isopteropodine (uncarine E), speciophylline (uncarine D), uncarine F and isomitraphylline have been reported, as well as gluco-indole alkaloids (3, 4-dehydro-5-carboxystrictosidine, 5α-carboxystrictosidine and lyaloside). Isomitraphylline, dihydrocorynantheine, hirsutine and hirsuteine, have also been identified together with their N-oxides (Falkiewicz and Łukasiak 2001; Heitzman et al. 2005; Keplinger et al. 1999; Laus 2004; Quintela and Lock de Ugaz 2003). Pteropodine (uncarine C), isopteropodine (uncaring E), mitraphylline, isomitraphylline, speciophylline (uncarine D) and uncarine F show anti-proliferative and cytotoxic effects on several tumour cell lines. The most potent activity of this kind has been demonstrated by uncarine F, with a 50% inhibitory concentration (IC_{50}) of 1.7–29 μmol/l. The mechanisms of action proposed are: (a) inhibition of Cγ1 phospholipase (Gattuso et al. 2004; Lee et al. 2000), (b) translocation of Bcl-2 and Bax family proteins to mitochondria, resulting in the release of the *c* cytochrome, leading to caspase-9 and -3 activation, and (c) activation of caspase-8 and -3 via the Fas signalling cascade (Cheng et al. 2007). On the other hand, it has also been reported that alkaloids, such as the POAs, uncarine C and isomitraphylline, are able to condense and to contract chromosomes, inhibiting mitosis in onion root cells (Kuraś et al. 2009).The immunomodulatory POAs have been reported to increase the number of immune cells such as B, T and NK cells, granulocytes and memory lymphocytes, and increase phagocytosis by granulocytes and macrophages, possibly due to the ability of *U. tomentosa* compounds to inhibit Nuclear Transcription Factor κB (NF-κB) activation and oxidative stress (Åkesson et al. 2003; Bacher et al. 2006; García Prado et al. 2007; Kaiser et al. 2013; Keplinger et al. 1999; Pilarski R et al. 2007). In the central nervous system, isorhynchophylline depressed locomotor activity by antagonizing central dopaminergic receptors (Sakakibara et al. 1999). Pteropodine, isopteropodine and mitraphylline affect cognitive processes in rats by positively modulating 5-HT2 and muscarinic M1 receptors. The interruption of memory caused by cholinergic agents is also improved by these alkaloids (Abdel-Fattah et al. 2000; Kang et al. 2002). Rhynchophylline, isorhynchophylline, hirsuteine, corynantheine and dihydrocorynantheine show hypotensive effects. The mechanism of action, proposed for rhynchophylline and isorhynchophylline, is ascribed to voltage-dependent calcium channel blocking (Falkiewicz and Łukasiak 2001; Heitzman et al. 2005; Laus 2004).

Terpenoids a variety of this family of compounds have been isolated from different parts of *U. tomentosa* and *guianensis*: polyhydroxylated triterpenes (uncaric acid, floridic acid, and 3β,6β,19α-trihydroxy-23-oxo-urs-12-en-28-oic acid), triterpenes (3β,19α-dihydroxy-6,23-dioxo-urs-12-en-28-oic acid and 3β,19α,23-trihydroxy-6-oxo-urs-12-en-28-oic acid), three polyoxygenated triterpenes, quinovic acid glycosides, ursolic acid and oleanolic acid, nor-triterpene glycosides derived from pyroquinovic acid (tomentosides A and B) and 5α-carboxystrictosidine (Falkiewicz and Łukasiak 2001; Keplinger et al. 1999; Quintela and Lock de Ugaz 2003). A quinovic acid glycoside from *U. tomentosa* was reported to show activity against rhinovirus type 1B infection and vesicular stomatitis virus (Heitzman et al. 2005; Laus 2004).

Flavonoids the procyanidins A1, B2, B3 and B4, kaempherol, dihydrokaempherol, quercetin, epicatechin and cinchonain Ia and Ib have been isolated from different parts of *U. tomentosa* and *guianensis* (Falkiewicz and Łukasiak 2001; Laus

2004; Quintela and Lock de Ugaz 2003). Procyanidins from the bark and root show anti-oxidant properties, quench free radicals, scavenge the peroxynitrite radical and inhibit oxidative DNA sugar damage, suggesting hydroxyl radical scavenging activity.

The compounds mentioned above show a wide range of biological activities, with anti-inflammatory properties, in both in vitro and in vivo models, being the most widely reported, followed by reports of cytotoxic activity. Extracts and compounds inhibit the production of pro-inflammatory mediators such as Tumour Necrosis Factor alpha (TNF-α), Interleukin-1 and -6 (IL-1, IL-6), prostaglandin E_2 (PGE_2), nitric oxide (NO), the activation of cyclooxygenase-1 and -2 (COX-1 and -2), and the expression of the inducible nitric oxygen synthase (iNOS). Inhibition of NF-κB and the inhibition of mitogen-activated protein kinase (MAPKs) phosphorylation have been proposed as possible mechanisms of action. The principle compounds reported to have anti-inflammatory activity are mitraphylline, rhynchophylline, quinolic, ursolic and oleanolic acids, cinchonains and procyanidins (Aguilar et al. 2002; Åkesson et al. 2003; Cao et al. 2012; Carvalho et al. 2006; Dreifuss et al. 2010; Fazio et al. 2008; Heitzman et al. 2005; Rojas-Duran et al. 2012; Sandoval-Chacón et al. 1998; Sandoval et al. 2002; Song et al. 2012; Urdanibia et al. 2013; Yuan et al. 2009).

Although anticancer activities have been reported for this plant (Heitzman et al. 2005), the reported in vitro activities against tumour cells are for the most part observed at relatively high concentrations, (Bacher et al. 2006; De Martino et al. 2006), and may not be sufficiently powerful to fully explain its traditional use against tumours. We offer here an alternative explanation. The pro-tumoural effect of chronic inflammation has been extensively studied (Coussens and Werb 2002). NF-κB, which is inhibited by Uncaria compounds as described above, represents an important link between chronic inflammation and cancer (Li et al. 2005) and has been suggested as a possible target for the therapy of both (Bremner and Heinrich 2002). Thus it is possible that *U. tomentosa* and *guianensis* may diminish tumour growth and metastasis via a reduction in pro-tumoural inflammatory processes in the tumour microenvironment (Caballero et al. 2005; Dreifuss et al. 2010, 2013;Fazio et al. 2008; Pilarski et al. 2010; Urdanibia et al. 2013). In our laboratory, we demonstrated that *U. guianensis* decreased the number of infiltrating macrophages and neutrophils in mouse tumours, cells which favour all stages of carcinogenesis, through the production of inflammatory mediators such as TNF-α, NO, IL-6, IL-10, PGE_2, IL-8, matrix metalloproteinases (MMP) and Vascular Endothelial Growth Factor (VEGF). These mediators contribute to an increase in vascular permeability, adhesion molecule expression on the endothelial cell, recruitment of more immune cells to the tumour, production of cytokines, tumour cell proliferation, angiogenesis, and extracellular matrix degradation (Balkwill and Mantovani 2001; Condeelis and Pollard 2006; Coussens and Werb 2002; Philip et al. 2004). In our study, a decrease in tumour-infiltrating immune cells was concomitant with a reduction in COX-2, iNOS, TNF-α, IL-6, and NF-κB, suggesting that these anti-inflammatory activities of *U. guianensis* are possibly responsible for the observed inhibition of tumour growth and metastasis (Urdanibia et al. 2013). However, it is important not to attribute the activities of Uncaria spp. solely for isolated compounds. Very little solid evidence is available but it is possible that the biomedical value of Uncaria preparations may come from the combined effect of two or more compound working synergistically (Heitzman et al. 2005; Pilarski R et al. 2007; Sandoval et al. 2002).

4 Morphological Description

U. tomentosa and *guianensis* are woody vines which may grow up to 30 m long, with a main stem of up to 25 cm in diameter. The name, cat's claw (uña de gato) comes from the thorns in the shape of curved claws which characterize this genus and which help the plant to climb through the vegetation. Those of *U. tomentosa* are straight or slightly curved, up to 10 mm in length, but those of *U. guianensis* are more claw-like and may reach 25 mm. Both species show a longitudinally striated outer bark, cinnamon in colour with a fibrous inner bark. The leaves are simple, opposite and distinct ovate to elliptic. The Latin name *tomentosa* describes the small hairs that cover the leaves and stipules of that species in contrast to the glabrous leaves of *U guianensis*. The lateral branches of the inflorescence are ramified in *U. tomentosa*, but simple in *U. guianensis.* The whole corolla of *U. tomentosa* is densely covered with short hairs on the outer side, whereas in *U. guianensis*, the long narrow corolla tube is largely glabrous on the outer side, only the uppermost part, together with the conic part and the lobes, being bearded with whitish hairs. The flowers and fruits of *U. tomentosa* are nearly sessile, the hairs on the fruits are evenly dense and persisting, while the hairs, and outermost layers of the ripe fruits and their stalks, of *U. guianensis* are shed off successively. The flowers of *U. tomentosa* are small and yellow-white while those of *U. guianensis* are orange-red. The fruits are dry and dehiscent, elliptical capsules, with numerous oblong seeds in both species (Gattuso et al. 2004; Keplinger et al. 1999).

5 Geographical Distribution

U. tomentosa is widely distributed in the Amazon and in Central America (Belize, Bolivia, Brazil, Colombia, Costa Rica, Equator, Guatemala, Guiana, French Guiana, Honduras, Nicaragua, Panama, Peru and Venezuela), at 5–750 m above sea level, latitude 15°30′00″N–13°36′00″S. *U. guianensis* does not grow so far north, being more restricted to the Amazon region (Bolivia, Brazil, Colombia, Equator, Guiana, French Guiana, Peru, Suriname and Venezuela), at 7–1010 m above sea level, latitude 08°04′00″N–17°32′00″S (Gattuso et al. 2004; Zevallos Pollito and Tomazello 2010).

6 Ecological Requirements

U. tomentosa prefers humid conditions, soils rich in nutrients near streams and in the glades of primary forests, old secondary forests, roads and closed trails. *U. guianensis* may be found at lower altitudes in poorer soils, resisting a wider range of dry to humid conditions, in more open vegetation and in both primary and secondary forests (Gattuso et al. 2004; Zevallos Pollito and Tomazello 2010).

7 Collection Practice

Production Practices although cat's claw may be propagated asexually by cuttings, it is generally collected in the wild (Hughes and Worth 1999).

Harvesting today, the root is not normally harvested because of the destructiveness of this method of harvest. The primary product in trade comes from the stem bark. Although there are different chemotypes found in the field, there are no known morphological differences to distinguish them. Generally, it is recommended that the vine is cut at 15–100 cm above ground and left to regenerate. Vines are only harvested at 8 or more years old, otherwise, the diameter of the vine is not sufficient for bark removal. As a regular practice, the cut vine is stripped of its bark in the field to avoid the weight of the whole vine, and the inner stem is discarded (Hughes and Worth 1999).

Processing the Association for the Conservation of the Patrimony of Cutivireni (ACPC) recommends the following processing procedure for a quality product. The damaged (infected or punctured) inner bark is discarded, and drying is conducted on clean raised surfaces to avoid mould growth. It may be dried in both sun or shade, and packaged in waterproof sacks for shipping (Hughes and Worth 1999).

8 Traditional Use (Parts Used) and Common Knowledge

The therapeutic uses of *U. tomentosa* and *U. guianensis* come from the aqueous extract of the bark or root bark, and include a wide range of treatments. It is reported that Amazonian tribes such as Asháninka, Aguaruna, Cashibo and Shipibo use as a remedy for abscesses, allergies, arthritis, asthma, diabetes, cancer, chemotherapy side effects, contraception, disease prevention, fevers, gastric ulcers, haemorrhages, inflammations, menstrual irregularity, recovery from child birth, rheumatism, skin impurities, urinary tract inflammation, chemotherapy side-effects, viral infections, weakness, wounds, and others (Åkesson et al. 2003; Allen-Hall et al. 2007; Heitzman et al. 2005; Keplinger et al. 1999; Pilarski et al. 2009).

Cat's claw is generally prepared in traditional medicine as an infusion, for example, the liquid obtained from boiling 10 g of the leaf with 200 ml of water, is ingested three times a day. A tincture, prepared with 10% bark w/w in 70° alcohol, is often mixed with other medicinal plants (Sánchez Schwartz 1995). Currently, cat's claw is available in many types of presentation, dried powders or cuts of the root and stem, encapsulated powdered material or lyophilized aqueous extracts, tinctures, tablets, ointments and gels (Reinhard 1999).

9 Modern Medicine Based on Its Traditional Medicine Uses

There are only a few formal studies the curative properties of cat's claw in humans. When a water-soluble *U. tomentosa* extract was given daily (5 mg/kg for six consecutive weeks) to four healthy adult males, no toxicity was observed and white blood cell numbers were significantly elevated. A significant increase in DNA repair was also found in one human volunteer study (Heitzman et al. 2005: Laus 2004; Sheng et al. 2001). In patients with rheumatoid arthritis, the incidence of painful joints was reduced 24 weeks of treatment with an *U. tomentosa* extract (Mur et al. 2002). In another study of the possible anti-inflammatory properties of Uncaria, an aqueous extract of *U. guianensis* relieved pain in patients with osteoarthritis of the knee (Piscoya et al. 2001). Patients with invasive ductal carcinoma stage II, treated with a standard FAC regimen (Fluorouracil, Doxorubicin and Cyclophosphamide) were also treated simultaneously with dry *U. tomentosa* extract resulting in a reduction of the neutropenia caused by chemotherapy. Cellular DNA damage was also restored in these patients, concluding that *U. tomentosa* is an effective adjuvant treatment (Santos Araujo et al. 2012).

10 Conclusions

U. tomentosa and *U. guianensis* are used in traditional medicine for their healing properties; they are very similar plants, but with notable differences in terms of geographical distribution, growth requirements, morphology and chemical constituents. Although different activities have been reported for the two species, the traditional use to treat inflammation predominates. However more clinical studies are required to place the traditional use of cat's claw on a sound scientific basis.

Acknowledgments We would like to thank Ilsa Coronel for her help in the preparation of this manuscript.

References

Abdel-Fattah MAF, Matsumoto K, Tabata K, Takayama H, Kitajima M, Aimi N, Watanabe H (2000) Effects of Uncaria tomentosa total alkaloid and its components on experimental amnesia in mice: elucidation using the passive avoidance test. J Pharm Pharmacol 52(12):1553–1561

Aguilar JL, Rojas P, Marcelo A, Plaza A, Bauer R, Reininger E, Klaas CA, Merfort I (2002) Anti-inflammatory activity of two different extracts of Uncaria tomentosa (Rubiaceae). J Ethnopharmacol 81(2):271–276

Åkesson C, Lindgren H, Pero RW, Leanderson T, Ivars F (2003) An extract of Uncaria tomentosa inhibiting cell division and NF-κB activity without inducing cell death. Int Immunopharmacol 3(13–14):1889–1900

Allen-Hall L, Cano P, Arnason JT, Rojas R, Lock O, Lafrenie RM (2007) Treatment of THP-1 cells with Uncaria tomentosa extracts differentially regulates the expression if IL-1α and TNF-α. J Ethnopharmacol 109(2):312–317

Bacher N, Tiefenthaler M, Sturm S, Stuppner H, Ausserlechner MJ, Kofler R, Konwalinka G (2006) Oxindole alkaloids from Uncaria tomentosa induce apoptosis in proliferating, G0/G1-arrested and bcl-2-expressing acute lymphoblastic leukaemia cells. Br J Haematol 132(5):615–622

Balkwill F, Mantovani A (2001) Inflammation and cancer: back to Virchow? Lancet 357(9255):539–545

Bremner P, Heinrich M (2002) Natural products as targeted modulators of the nuclear factor-kappaB pathway. J Pharm Pharmacol 54(4):453–472

Caballero M, Arsenak M, Abad MJ, Cesari IM, Taylor P (2005) Effect of plant extracts on B16-BL6 melanoma cell growth and metastasis in C57BL/6 mice. Acta Cient Venez 55(7–8):21–27

Cao W, Wang Y, Lv X, Yu X, Li X, Li H, Wang Y, Lu D, Qi R, Wang H (2012) Rhynchophylline prevents cardiac dysfunction and improves survival in lipopolysaccharide-challenged mice via suppressing macrophage I-κBα phosphorylation. Int Immunopharmacol 14(3):243–251

Carvalho MV, Penido C, Siani AC, Valente LMM, Henriques MGMO (2006) Investigations on the anti-inflammatory and anti-allergic activities of the leaves of Uncaria guianensis (Aublet) J. F Gmelin Inflammopharmacol 14(1–2):48

Cheng AC, Jian CB, Huang YT, Lai CS, Hsu PC, Pan MH (2007) Induction of apoptosis by Uncaria tomentosa through reactive oxygen species production, cytochrome c release, and caspases activation in human leukemia cells. Food Chem Toxicol 45(11):2206–2218

Condeelis J, Pollard JW (2006) Macrophages: obligate partners for tumor cell migration, invasion, and metastasis. Cell 124(2):263–266

Coussens LM, Werb Z (2002) Inflammation and cancer. Nature 420(6917):860–867

De Martino L, Martinot JLS, Franceschelli S, Leone A, Pizza C, De Feo V (2006) Proapoptotic effect of Uncaria tomentosa extracts. J Ethnopharmacol 107(1):91–94

Dreifuss AA, Bastos-Pereira AL, Ávila TV, BdS S, Rivero AJ, Aguilar JL, Acco A (2010) Antitumoral and antioxidant effects of a hydroalcoholic extract of cat's claw (*Uncaria tomentosa*) (Willd. Ex Roem. & Schult) in an in vivo carcinosarcoma model. J Ethnopharmacol 130(1):127–133

Dreifuss AA, Bastos-Pereira AL, Fabossi IA, dos Reis Lívero FA, Stolf AM, de Souza CEA, de Oliveira Gomes L, Constantin RP, Furman AEF, Strapasson RLB (2013) Uncaria tomentosa exerts extensive anti-neoplastic effects against the Walker-256 tumour by modulating oxidative stress and not by alkaloid activity. PLoS One 8(2):e54618

Falkiewicz B, Łukasiak J (2001) Vilcacora Uncaria tomentosa (Willd.) DC. and Uncaria guianensis (Aublet) Gmell. – a review of published scientific literature. Case Rep Clin Pract Rev 2(4):305–316

Fazio AL, Ballén D, Cesari IM, Abad MJ, Arsenak M, Taylor P (2008) An ethanolic extract of Uncaria tomentosa reduces inflammation and B16-BL6 melanoma growth in C57BL/6 mice. Bol Latinoam Caribe Plantas Med Aromáticas 7(4):217–224

García Prado E, García Gimenez MD, De la Puerta Vázquez R, Espartero Sánchez JL, Sáenz Rodríguez MT (2007) Antiproliferative effects of mitraphylline, a pentacyclic oxindole alkaloid of Uncaria tomentosa on human glioma and neuroblastoma cell lines. Phytomedicine 14(4):280–284

Gattuso M, Di Sapio O, Gattuso S, Li Pereyra E (2004) Morphoanatomical studies of Uncaria tomentosa and Uncaria guianiensis bark and leaves. Phytomedicine 11(2–3):213–223

Guthrie OW, Gearhart CA, Fulton S, Fechter LD (2011) Carboxy alkyl esters of Uncaria tomentosa augment recovery of sensorineural functions following noise injury. Brain Res 1407:97–106

Heitzman ME, Neto CC, Winiarz E, Vaisberg AJ, Hammond GB (2005) Ethnobotany, phytochemistry and pharmacology of Uncaria (Rubiaceae). Phytochemistry 66(1):5–29

Hughes K, Worth A (1999) Cat's claw [Internet]. Purdue University Center for New Crops and Plant Products [cited 2015 Jan 15]. Available from: http://www.hort.purdue.edu/newcrop/cropfactsheets/catsclaw.html#Crop

Kaiser S, Dietrich F, Resende PE, Verza SG, Moraes RC, Morrone FB, Batastini AMO, Ortega GG (2013) Cat's claw oxindole alkaloid isomerization induced by cell incubation and cytotoxic activity against T24 and RT4 human bladder cancer cell lines. Planta Med 79(15):1413–1420

Kang TH, Matsumoto K, Tohda M, Murakami Y, Takayama H, Kitajima M, Aimi N, Watanabe H (2002) Pteropodine and isopteropodine positively modulate the function of rat muscarinic M1 and 5-HT2 receptors expressed in xenopus oocyte. Eur J Pharmacol 444(1–2):39–45

Keplinger K, Laus G, Wurm M, Dierich MP, Teppner H (1999) Uncaria tomentosa (Willd.) DC. – Ethnomedicinal use and new pharmacological, toxicological and botanical results. J Ethnopharmacol 64(1):23–34

Kuraś M, Pilarski R, Nowakowska J, Zobel A, Brzost K, Antosiewicz J, Gulewicz K (2009) Effect of alkaloid-free and alkaloid-rich preparations from Uncaria tomentosa bark on mitotic activity and chromosome morphology evaluated by Allium test. J Ethnopharmacol 121(1):140–147

Laus G (2004) Advances in chemistry and bioactivity of the genus Uncaria. Phytother Res 18(4):259–274

Lee JS, Kim J, Kim BY, Lee HS, Ahn JS, Chang YS (2000) Inhibition of phospholipase Cγ1 and cancer cell proliferation by triterpene esters from Uncaria rhynchophylla. J Nat Prod 63(6):753–756

Li Q, Withoff S, Verma IM (2005) Inflammation-associated cancer: NF-KB is the lynchpin. Trends Immunol 26(6):318–325

Mur E, Hartig F, Eibl G, Schirmer M (2002) Randomized double blind trial of an extract from the pentacyclic alkaloid-chemotype of uncaria tomentosa for the treatment of rheumatoid arthritis. J Rheumatol 29(4):678–681

Philip M, Rowley DA, Schreiber H (2004) Inflammation as a tumor promoter in cancer induction. Semin Cancer Biol 14(6):433–439

Pilarski R, Poczekaj-Kostrzewska M, Ciesiołka D, Szyfter K, Gulewicz K (2007) Antiproliferative activity of various *Uncaria tomentosa* preparations on HL-60 promyelocytic leukemia cells. Pharmacol Rep 59(5):565–572

Pilarski R, Bednarczyk M, Gulewicz K (2009) Evaluation of biological activity of *uncaria tomentosa* (willd.) DC. Using the chicken embryo model. Folia Biol (Krakow) 57(3–4):207–212

Pilarski R, Filip B, Wietrzyk J, Kuraś M, Gulewicz K (2010) Anticancer activity of the *Uncaria tomentosa* (Willd.) DC. preparations with different oxindole alkaloid composition. Phytomedicine 17(14):1133–1139

Piscoya J, Rodriguez Z, Bustamante SA, Okuhama NN, Miller MJ, Sandoval M (2001) Efficacy and safety of freeze-dried cat's claw in osteoarthritis of the knee: mechanisms of action of the species Uncaria guianensis. Inflamm Res 50(9):442–448

Quintela JC, Lock de Ugaz O (2003) Uña de gato, Uncaria tomentosa (Wild.) DC. Rev Fitoter 3(1):5–16

Reinhard KH (1999) Uncaria tomentosa (Willd.) D.C.: cat's claw, una de gato, or saventaro. J Altern Complement Med 5(2):143–151

Rojas-Duran R, Gonzalez-Aspajo G, Ruiz-Martel C, Bourdy G, Doroteo-Ortega V, Alban-Castillo J, Robert G, Auberger P, Deharo E (2012) Anti-inflammatory activity of mitraphylline isolated from Uncaria tomentosa bark. J Ethnopharmacol 143(3):801–804

Sakakibara I, Terabayashi S, Kubo M, Higuchi M, Komatsu Y, Okada M, Taki K, Kamei J (1999) Effect on locomotion of indole alkaloids from the hooks of Uncaria plants. Phytomedicine 6(3):163–168

Sánchez Schwartz C (1995) Uña de Gato Uncaria tomentosa (Willd.)DC. Rev Peru Reumatol 1(2):000–000

Sandoval M, Okuhama NN, Zhang XJ, Condezo LA, Lao J, Angeles FM, Musah RA, Bobrowski P, Miller MJ (2002) Anti-inflammatory and antioxidant activities of cat's claw (*Uncaria tomentosa* and *Uncaria guianensis*) are independent of their alkaloid content. Phytomedicine 9(4):325–337

Sandoval-Chacón M, Thompson JH, Zhang XJ, Liu X, Mannick EE, Sadowska-Krowicka H, Charbonnet RM, Clark DA, Miller MJ (1998) Antiinflammatory actions of cat's claw: the role of NF-KB. Aliment Pharmacol Ther 12(12):1279–1289

Santos Araujo MDC, Farias IL, Gutierres J, Dalmora SL, Flores N, Farias J, Cruz I, Chiesa J, Morsch VM, Chitolina Schetinger MR (2012) Uncaria tomentosa-adjuvant treatment for breast cancer: clinical trial. Evid Based Complement Alternat Med 2012:a676984. https://doi.org/10.1155/2012/676984

Sheng Y, Li L, Holmgren K, Pero RW (2001) DNA repair enhancement of aqueous extracts of Uncaria tomentosa in a human volunteer study. Phytomedicine 8(4):275

Song Y, Qu R, Zhu S, Zhang R, Ma S (2012) Rhynchophylline attenuates LPS-induced pro-inflammatory responses through down-regulation of MAPK/NF-κB signaling pathways in primary microglia. Phytother Res 26(10):1528–1533

Urdanibia I, Michelangeli F, Ruiz MC, Milano B, Taylor P (2013) Anti-inflammatory and antitumoural effects of Uncaria guianensis bark. J Ethnopharmacol 150(3):1154–1162

Yuan D, Ma B, Yang JY, Xie YY, Wang L, Zhang LJ, Kano Y, Wu CF (2009) Anti-inflammatory effects of rhynchophylline and isorhynchophylline in mouse N9 microglial cells and the molecular mechanism. Int Immunopharmacol 9(13):1549–1554

Zevallos Pollito PA, Tomazello M (2010) Survey and characterization of two species of the genus Uncaria Schreb. (rubiaceae) occurring in the state of Acre, Brazil. Ecol Apl 9(1):19–30

Valeriana carnosa Sm.

Soledad Molares and Ana H. Ladio

Valeriana carnosa Sm.
Photo: Jean-Pierre Bérubé
Available in: https://www.flickr.com/photos/jpdu12/sets/72157647845316863

S. Molares
CIEMEP (Centro de Investigación Esquel de Montaña y Estepa Patagónica), Universidad Nacional de la Patagonia San Juan Bosco-CONICET, Esquel, Chubut, Argentina

A. H. Ladio (✉)
Laboratorio Ecotono, INIBIOMA (Instituto de Investigaciones en Biodiversidad y Medio Ambiente), Universidad Nacional del Comahue-CONICET, San Carlos de Bariloche, Rio Negro, Argentina
e-mail: ladioah@comahue-conicet.gob.ar

U. P. Albuquerque et al. (eds.), *Medicinal and Aromatic Plants of South America*, Medicinal and Aromatic Plants of the World 5,
https://doi.org/10.1007/978-94-024-1552-0_42

Abstract *Valeriana carnosa* Sm. stands out as one of the key elements of the indigenous pharmacopoeia used in the extreme south of the American Continent. Its rhizomes and roots have been used since ancestral times in hepatic, respiratory, circulatory, urinary, digestive and anti-inflammatory remedies. They have also been used as painkillers, sedatives and for the treatment of cultural syndromes particular to Latin-American medicine such as the "*susto*" and the "*evil eye*". The breadth of its reputed uses has led to its being known as "the plant that cures the seven illnesses".

The crude drug is prepared from the roots and rhizomes, principally as a decoction. Several studies indicate that the principal active ingredients are valepotriates, lignans, flavonoids, tannins, phenolic acids and essential oils. Research carried out on *V. carnosa* reveals the presence of active ingredients similar to those of *V. officinalis*, a species found in many pharmacopoeias which is used as a sedative and sleep inducer. However, little conclusive evidence of efficacy can be provided for the remaining local uses. The key problem of various studies has been their emphasis on very few compounds, rather than traditional preparations. Much more research is required to evaluate the actual efficacy of preparations.

Keywords *Valeriana carnosa* · Valerianaceae · Subterranean organs · Mapuche pharmacopoeia · Ñamkulawen

1 Introduction

The roots and other subterranean organs of numerous Patagonian species have long been recognized as being of great value to rural Creole, Mapuche and Tehuelche populations both in Argentina and Chile (Ladio and Lozada 2009; Molares and Ladio 2009a; Ochoa and Ladio 2011), and also constitute an important part of many regional rites and legends (Ochoa and Ladio 2014).

From the perspectives of economic botany and ethnopharmacology, the main value of these species is based on the fact that their subterranean organs often contain starch and other carbohydrates of importance to the human diet, and also therapeutic compounds derived from plant secondary metabolism (Gurib-Fakim 2006). Amongst these species, *Valeriana carnosa* Sm. stands out as one of the principal elements in the indigenous pharmacopoeias of the southern cone of America, and its roots and rhizomes have been known and used since ancient times (Molares and Ladio 2009b). The local perception of this plant is that it has wide-ranging curative powers: "*it's a cure-all*". This attribute confers on the species high cultural and symbolic value for the Mapuche people, and its reputation and use has spread throughout the formal and informal medicinal herb market of Patagonian cities (Ladio 2006).

2 Taxonomic Characteristics

V. carnosa (synonym: *Valeriana magellanica* Lam.) belongs to the Valerianaceae family, which consists of 400 species and 17 genera, mainly found in the Northern Hemisphere and along the Andes mountain range. Of the approximately 250 species of Valerianaceae found in South America, 40 taxa are restricted to the Andes of Argentina and Chile (Bell et al. 2012). It has been suggested that Holartic *Valeriana* genera have been present on the South American continent for some time (>13 MY), and have exploited new niche opportunities, migrating from a temperate to a more Mediterranean-style climate (Bell et al. 2012). Most of the species are herbaceous or small shrubs with foul-smelling roots. The name of the genus stems from the Latin *valere*, "to be healthy", a reference to the medicinal uses of its plants, particularly those associated with treating nervous conditions and hysteria (Borsini et al. in Correa 1999). Their epithet *carnosa* makes references to the consistency of the leaves (Ferreyra et al. 2006).

3 Crude Drug Used

The crude drug consists of dried pieces of the roots and rhizomes, which are sold in bulk or hand packed in paper or cellophane bags for sale in drugstores and herbalist's shops. The recommended method of use is decoction of a handful of the material, followed by ingestion of one cupful, orally, over a variable timeframe (Cuassolo 2009; Cuassolo et al. 2011). Kutschker et al. (2002) describe a dosage of a daily cupful drunk on an empty stomach for a week.

V. carnosa and other species of the Patagonian region, such as *Valeriana clarionifolia*, are known as "ñamkulawen" and are used in similar ways in traditional medicine. According to diagnostic anatomical data provided by Bach et al. (2014), *V. carnosa* showed a primary pentarch aktinostele root, pith in the secondary structure and a rhizome with anomalous structure. *V. clarionifolia,* in contrast, has also rhizome and showed a protostele as a primary root structure and a secondary structure without pith. During the maceration process, the *V. carnosa* rhizome presented cork with irregular polygonal cells with acute and obtuse angles, while in *V. clarionifolia* rectangular cork cells with right angles were observed. Starch grains are simple, spherical in *V. carnosa* and polyhedral in *V. clarionifolia*. In addition, Molares and Ladio (2012) studied cross sections of *V. carnosa* primary root and observed a well-developed periderm consisting of cells with thickened, birefringent walls, from irregular to polygonal; cells of this tissue and phloem parenchyma with essential oils in the form of droplets (Sudan IV+); cortex with large air spaces between oval cells with brown contents (Fig. 1a–c). These anatomical characteristics could be used to recognize the crude drug commercialized in the region. *V. carnosa* is not included in the Argentine Pharmacopoeia (http://www.anmat.gov.ar), nor does it appears on the list of toxic species not recommended for consumption.

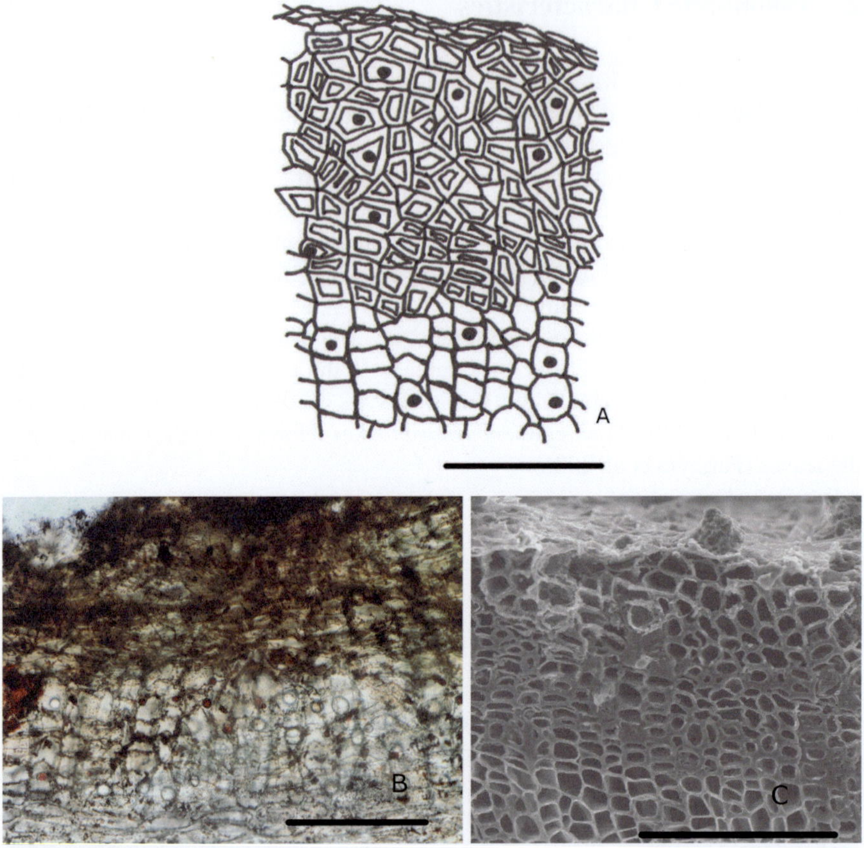

Fig. 1 Morpho-anatomy of transversal cut of primary root of *V. carnosa* Sm. (**a**) Diagram of a sector of the periderm and phloem parenchyma with drops of essential oils. (**b**) Positive reaction of Sudan IV on essential oil drops. (**c**) Inactive phloem and periderm viewed with an environmental scanning electron microscope. Scale in (**a**) 100 mm, in (**b**, **c**) 200 μm. (Taken from Molares and Ladio 2012)

4 Major Chemical Constituents and Bioactive Compounds

Several studies on the *Valeriana* genus indicate that the main active ingredients are the valepotriates, lignans, flavonoids, tannins, phenolic acids and essential oils (Kutschker et al. 2010). In particular, the essential oils have been researched; they primarily consist of elemol, bornyl-acetate, bornyl-isovalerate, isovalerate, and valerenone (Baby et al. 2005). Of all the Patagonian species belonging to this genus, the dry extract of the whole *V. carnosa* plant has been most studied (Cuadra and Fajardo 2002). It has been found that its valepotriate composition pattern, and especially its valtrates, is similar to *V. officinalis*, which is known for its tranquilizing and sleep inducing effect (Kutschker et al. 2010). However, according to Castillo and Martínez (2007), the chemical composition of *V. carnosa* varies according to the time of collection, preparation and packaging. In addition, Cuadra and Fajardo (2002) have isolated caffeoyl methyl ester and two pinoresinol-type lignans. Fajardo et al. (2010) have also suggested that in terms of its biological activity, it would present cytotoxic activity and negative toxicological activity.

5 Morphological Description

Evergreen herb of up to 80 cm in height, simple or branching from the base. Fleshy rhizome up to 50 cm long, with weak branches. Basal leaves 6–21 × 3–7 cm, obovate or elliptic, smooth-edged or coarsely toothed, glabrous and fleshy; 3–12 cm long petioles. Upper leaves are sessile or petiolate, 0.6–4.5 cm, obovate, oblong, triangular or lanceolate, smooth-edged or toothed. Axillary or terminal inflorescences, paniculiform, lax. Bracts are 3–9 mm in length, whole, oblong-lanceolate, ovate. Bracteoles are 2.5–4 mm in length, entire or auriculate, oblong-lanceolate, acute, glabrous or have long hairs on the edges, at the base. Hermaphrodite flower: 4 mm corolla, bell-shaped or funnel and bell-shaped, gibbous at the base; oblong lobes. Included stamens. Female flower: 2–3 mm corolla, bell-shaped, ovate lobes. Exerted styles, thickened at the tip. The fruit measures 5–7 × 2–3.5 mm, and is pyriform, with thick veins, glabrous; pappus formed by 14–15 feathery setae (Borsini et al. in Correa 1999). (Figs. 2 and 3).

6 Geographical Distribution

V. carnosa is an endemic species which is widely distributed and common to the whole of Patagonia (Borsini et al. in Correa 1999). In Chile it inhabits the southern mountain range, in the VI, VII, VIII, IX, X, XI and XII regions; in Argentina it inhabits the Mendoza, Neuquén, Río Negro, Chubut, Santa Cruz and Tierra del Fuego provinces. Its altitudinal range is from 0 to 2700 m.a.s.l. (Zuloaga et al. 2008). In phytogeographic terms, it is found in the Sub-Antarctic, Patagonian and High-Andean provinces (Borsini et al. in Correa 1999).

7 Ecological Requirements

The species flourishes in xeric, open, sunny environments in the rocky soils of the forest, steppe and the Patagonian Andean forest-steppe ecotone. It is also found in sandy sites, on low, sunny slopes of the Patagonian Andes. It flowers in spring-summer (Ferreyra et al. 2006).

Fig. 2 Diagram of the aerial parts of the plant (**a**), floral structures (**b**, **c**) and fruit (**d**) of *V. carnosa* Sm. (Taken from Borsini et al. in Correa 1999)

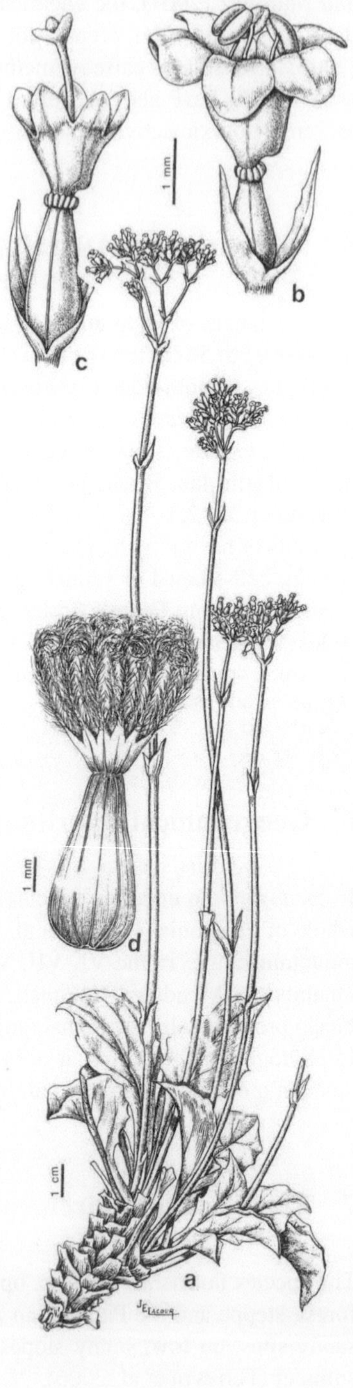

Fig. 3 General appearance of *V. carnosa* in a Patagonian forest-steppe ecotone habitat

8 Collection Practice

Gathering carried out by the settlers is characterized by the search for specimens in stony areas with a high level of light exposure, preferably at the highest altitude possible, with the help of simple tools like knives and spades. In the process of identification and selection of specimens, cultural practices of sensory perception come into play. These include the recognition of organoleptic qualities directly associated with this species, such as its bitter and unpleasant smell (*"like dirty feet"*) and its strong, bitter, repulsive flavor (*"füre"*), which is rather spicy (*"trapi"*) and astringent (*"seco"*) (Molares and Ladio 2009a). Various studies indicate that the collection of this species is associated with the care of livestock. The people take advantage of the time during which their animals are grazing to look for the plant in places far from their dwellings (Estomba et al. 2006; Richeri et al. 2013). With regard to the identification and collection of *V. carnosa* and *V. clarionifolia* by Patagonian inhabitants, studies reveal levels of organoleptic differentiation between the two species, which are of great cultural and ethnopharmacological value. For example, it was discovered that locals are capable of differentiating between *Valeriana* species, and that even though they recognize them as related (which can be deduced by the fact that both have the same common name), they can tell them apart by their smell and taste, which consequently determine their different uses and value (Molares and Ladio 2012). Unlike *V. carnosa, V. clarionifolia* is used for a limited number of ailments, mainly to relieve lower back pain and treat kidney and bladder disorders and cultural syndromes. In a curiously similar way, by means of laboratory tests with electronic noses, differences have been found between the aromatic profiles of *V. carnosa* and *V. clarionifolia,* which are determined by the chemical differences between the species (Baby et al. 2005).

The collected pieces of *V. carnosa* are usually taken to the dwellings where they are dried in the open air and in the shade, undercover, to be preserved later in mesh or paper bags. This practice ensures availability of the dried resource all year round, and is particularly useful in winter when the search for medicinal herbs in the mountains can become difficult due to the accumulation of snow (Molares and Ladio 2012).

Although *V. carnosa* gathering is very important and its commercialization has increased rapidly over the last decades (Cuassolo 2009), this species can be regarded as not threatened. However, settlers say that it is increasingly difficult to find plants, and that longer distances must be travelled in the search for them (Estomba et al. 2005, 2006). For this reason, the study of this plant's cultivation requirements must be encouraged (Cuassolo 2009).

9 Traditional Use (Part(s) Used) and Common Knowledge

V. carnosa has long been reported as "Ñamkulawen", in the Mapuzungun language, ("White hawk medicine" in English), probably in reference to the high sites where the species grows and where the ñamku can be seen in flight. This local name (Ñamkulawen) is shared with *V. clarionifolia* Phil. but this plant has different reputed attributes, as explained above. Another local name is "Valeriana", which is used by both Creole and rural settlers (in Spanish).

The root has been cited as a remedy used for hepatic, respiratory, circulatory, urinary and digestive disorders as well as having analgesic, anti-inflammatory, anti-tumoral, anti-depressive and wound-healing properties (e.g. Martinez Crovetto 1980; Estomba et al. 2005, 2006; Molares and Ladio 2009a, b, 2012; Richeri et al. 2013). It has also gained great prestige for its usefulness in treating cultural syndromes like the "*susto*", "*evil eye*" and "*frío*" (Molares 2010). *V. carnosa* is also used in mixtures with other species, like "nalka" (*Gunnera tinctorea* (Molina) Mirb.) to strengthen its medicinal attributes (Molares 2010), or with "carqueja" (*Baccharis sagittalis* (Less.) DC.) and "palo piche" (*Fabiana imbricata* Ruiz et Pav.) to make "body cleansers" (Toledo and Kutschker 2012), which are used in a process which is both symbolic and practical, where the wellbeing of the person is sought by eliminating all the elements (physical, social and spiritual) which may be causing harm (Molares 2010). All these properties, grouped in seven ethnocategories according to the particular precepts of the Mapuche culture, have led to the plant also being recognized as "the remedy that cures the seven diseases" (Molares and Ladio 2012).

The local indigenous communities use the plant through decoction. They boil a piece of root, approximately 3 cm in length per liter of water, and then drink a cup each day until the liter is finished. According to our sources, perception of the strong bitter taste of this decoction is an indicator of high therapeutic effectiveness, but also of potential danger, and because of this it is only consumed by adults and the dosages used are highly controlled and sporadic (Molares and Ladio 2009a). Traditionally, its use is not recommended for children or pregnant women (Kutschker et al. 2002). In addition, the dosage must be small because it causes sleepiness (Weigandt et al. 2004) and an excessive dosage can even be fatal (Molares and Ladio 2009a).

10 Modern Medicine Based on Its Traditional Medicinal Uses

Research carried out on *V. carnosa* and *V. clarionifolia* reveals the presence of active ingredients similar to those of *V. officinalis*, which is present in many pharmacopoeias for oral consumption as a sedative and sleep inducer for humans (Gratti et al. 2010). Kutschker et al. (2002) describe uses of the plant in modern medicine which are based on traditional methods, such as the preparation of tinctures using the steeped roots. The roots are placed in a jar with 300 ml of alcohol, left for 15 days and then filtered. The recommended dosage is 1–2 ml as a sedative.

11 Conclusions

V. carnosa is one of the most prominent medicinal plants in the Mapuche tradition, and from an ethnopharmacological viewpoint, one of the most versatile medicinal plants in Patagonia, when taking into account the wide range of therapeutic alternatives it can offer for the treatment of the different ailments of the region (Richeri et al. 2013).

The similarity between the active compounds found in *V. carnosa* and *V. clarionifolia* and those of *V. officinalis* is promising since this species is included worldwide in many pharmacopoeias and consumed orally as a sedative and sleep inducer in humans. However, little conclusive evidence for the efficacy of the other local uses can be provided. The key problem of various investigations has been an emphasis on very few compounds, rather than traditional preparations. Much more research is required to evaluate the actual efficacy of the preparations. The scientific research and cultural revalorization of the role played by *V. carnosa* in local herbal medicines is of considerable ethnopharmacological interest, and highly relevant to the medicinal security of Patagonian communities. However, there is evidence to indicate that the abundance of this species in natural environments is decreasing, mainly due to disturbance of the environments (Estomba et al. 2006; Ladio et al. 2007) and lack of regulation of its commercialization in Patagonian cities (Cuassolo 2009). Given that the roots are the organs of medical interest in this valuable species, the establishment of conservation strategies in situ and studies that provide guidelines for its cultivation and preservation ex-situ are of the utmost importance.

References

Baby RE, Cabezas M, Kutschker A, Messina, Walsöe de Reca NE (2005) Discrimination of different valerian types with an electronic nose. J Argentinean Chem Soc 93(1–3):43–50

Bach HG, Varela BG, Fortunato RH, Wagner ML (2014) Pharmacobotany of two *Valeriana* species (Valerianaceae) of Argentinian Patagonia known as "Ñancolahuen". Lat Am J Pharm 33(6):891–896

Bell CD, Kutschker A, Arroyo MTK (2012) Phylogeny and diversification of Valerianaceae (Dipsacales) in the southern Andes. Mol Phylogenet Evol 63(3):724–737. https://doi.org/10.1016/j.ympev.2012.02.015

Borsini OE, Rossow RA, Correa MN (1999) Valerianaceae. In: Correa MN (ed) Parte VI. Dicotyledones Gamopetalas. Flora Patagónica. INTA, Buenos Airess, pp 449–468

Castillo García E, Martínez Solís I (2007) Manual de Fitoterapia. Elsevier, España., 536 pp

Cuadra P, Fajardo V (2002) A new lignan from the Patagonian Valeriana carnosa Sm. Bol Soc Chil Quimica 47(4):361–366

Cuassolo F (2009) Estudio Etnobotánico de las plantas medicinales nativas y exóticas comercializadas en la Ciudad de Bariloche. Universidad Nacional del Comahue, Patagonia

Cuassolo F, Ladio AH, Ezcurra C (2011) Aspectos de la comercialización y control de calidad de las plantas medicinales más vendidas en una comunidad urbana del NO de la Patagonia Argentina Aspects. Bol Latinoam Caribe Plant Med Aromat 9(3):166–176

Estomba D, Ladio AH, Lozada M (2005) Plantas medicinales utilizadas por una comunidad Mapuche en las cercanías de Junín de los Andes, Neuquén. Bol Latinoam Plant Med Aromát 4(6):107–112

Estomba D, Ladio AH, Lozada M (2006) Medicinal wild plant knowledge and gathering patterns in a Mapuche community from North-western Patagonia. J Ethnopharmacol 103:109–119. https://doi.org/10.1016/j.jep.2005.07.015

Fajardo V, Gallardo A, Araya M, Joseph-Nathan P, Oyarzún A, Cuadra P, Sanhueza V, Manosalva L, Villarroel L, Darias J (2010) Químicas y algunos antecedentes y ensayos simples de la actividad biológica de plantas de la zona austral de Chile. Dominguezia 26(2.) – 2010):40–21

Ferreyra M, Ezcurra C, Clayton S (2006) Flores de alta montaña de los Andes patagónicos. Editorial L.O.L.A, Buenos Aires

Gratti A, Beeskow A, Fernández S (2010) El género *Valeriana* en la estepa patagónica argentina. Aportes al conocimiento fitoquímico. Dominguezia 26(2.) – 2010):55–56

Gurib-Fakim A (2006) Medicinal plants: traditions of yesterday and drugs of tomorrow. Mol Asp Med 27(1):1–93. https://doi.org/10.1016/j.mam.2005.07.008 http://www.anmat.gov.ar/webanmat/fna/fna.asp. 2013. Farmacopea Argentina. 7ed; 2014.

Kutschker A, Menoyo H, Hechem V (2002) Plantas medicinales de uso popular encomunidades del oeste del Chubut. Ed. Bavaria. INTA-UN de la Patagonia S.J.B.-GTZ, Bariloche

Kutschker A, Ezcurra C, Balzaretti V (2010) Valeriana (Valerianaceae) de los Andes australes: biodiversidad y compuestos químicos. In: Pochettino ML, Ladio AH, Arenas PM (eds) Tradiciones y Transformaciones en Etnobotánica. CYTED, La Plata, pp 219–224

Ladio AH (2006) Gathering of wild plant foods with medicinal use in a Mapuche community of Northwest Patagonia. In: Pieroni A, Price LL (eds) Eating and healing: traditional food. Harworth Press, Philadelphia, pp 297–321

Ladio AH, Lozada M (2009) Human ecology, ethnobotany and traditional practices in rural populations inhabiting the Monte region: resilience and ecological knowledge. J Arid Environ 73(2):222–227. https://doi.org/10.1016/j.jaridenv.2008.02.006

Ladio AH, Lozada M, Weigandt M (2007) Comparison of traditional wild plant knowledge between aboriginal communities inhabiting arid and forest environments in Patagonia, Argentina. J Arid Environ 69(4):695–715. https://doi.org/10.1016/j.jaridenv.2006.11.008

Martínez Crovetto R (1980) Apuntes sobre la vegetación de los alrededores del Lago Cholila. Publicación Técnica Fac Cien Agrarias 1:1–22

Molares S (2010) Flora medicinal aromática de la Patagonia: características anatómicas y propiedades organolépticas utilizadas en el reconocimiento por parte de la terapéutica popular. Tesis Doctoral. Universidad Nacional del Comahue. Bariloche

Molares S, Ladio AH (2009a) Chemosensory perception and medicinal plants for digestive ailments in a Mapuche community in NW Patagonia, Argentina. J Ethnopharmacol 123(3):397–406. https://doi.org/10.1016/j.jep.2009.03.033

Molares S, Ladio AH (2009b) Ethnobotanical review of the Mapuche medicinal flora: use patterns on a regional scale. J Ethnopharmacol 122(2):251–260. https://doi.org/10.1016/j.jep.2009.01.003

Molares S, Ladio AH (2012) Plantas aromáticas con órganos subterráneos de importancia cultural en la patagonia argentina: una aproximación a sus usos desde la etnobotánica, la percepción sensorial y la anatomía. Darwiniana 50(1):7–24

Ochoa JJ, Ladio AH (2011) Pasado y presente del uso de plantas silvestres con órganos de almacenamiento subterráneos comestibles en la Patagonia. Bonplandia 20(2):265–284

Ochoa JJ, Ladio AH (2014) Ethnoecology of Oxalis adenophylla Gillies ex Hook. & Arn. J Ethnopharmacol 155:533–542

Richeri M, Cardoso MB, Ladio AH (2013) Soluciones locales y flexibilidad en el conocimiento ecológico tradicional frente a procesos de cambio ambiental: estudios de caso en Patagonia. Ecol Austral 23:184–193

Toledo C, Kutschker A (2012) Plantas Medicinales en el Parque Nacional Los Alerces, Chubut, Patagonia Argentina. Bol Soc Argent Bot 47(3–4):461–470

Weigandt M, Ladio AH, Lozada M (2004) Plantas medicinales utilizadas en la comunidad Mapuche Curruhuinca. Ediciones Imaginaria, Bariloche. 75 pp

Zuloaga FO, Morrone O, Belgrano JM (2008) Catálogo de las plantas vasculares del Cono Sur. Monographs in systematic botany from the Missouri Botanical Garden. Ed. Missouri Botanical Garden Press. http://www2.darwin.edu.ar

Ximenia americana L.

Ana Cláudia D. Medeiros and Francinalva D. de Medeiros

Ximenia americana L.
Photo source: Data bank from Laboratório de Ecologia e Evolução de sistemas socioecológicos

A. C. D. Medeiros (✉) · F. D. de Medeiros
Laboratory of Development and Assays of Drugs, State University of Paraíba,
Campina Grande, Paraíba, Brazil
e-mail: anaclaudia@uepb.edu.br

U. P. Albuquerque et al. (eds.), *Medicinal and Aromatic Plants of South America*, Medicinal and Aromatic Plants of the World 5,
https://doi.org/10.1007/978-94-024-1552-0_43

Abstract *Ximenia americana* L. (Olacaceae) is widespread throughout the tropics, especially in Africa and Brazil. It is used as food or supplements and in the cosmetic industry. It is also used for traditional medicine as anti-inflammatory, analgesic, antipyretic, antimalarial, measles, mouth wounds, rheumatism, diarrhea, lung abscess, muscle cramps, and HIV. This species showed high sensitivity in tumor cell lines and the cell lines of MCF7 breast cancer, BV173 CML, and CC531 colon carcinoma. Santana et al. developed and validated an analytical method for the identification of gallic acid as a chemical marker of *X. americana.* It were also showed compounds such as sambunigrin, quercitrin, avicularin, and ximenynic acid. The fruit is a rich source of vitamin C and contains hydrocyanic acid riproximin.

Keywords *Ximenia americana* L. · Traditional use · Gallic acid · Ximenynic acid

1 Taxonomic Characteristics

Ximenia americana L. is commonly known as: “Wild Plum”, “Blue Sour Plum”, “Tallow Nut”, “Hog Plum”, “False Sandalwood”, “Seaside Plum”, “Small Sourplum”, “Sour Plum”, “Tallow Nut”, “Tallow Wood”, “Wild Lime”, “Wild Olivein” in English (Rossi 2015; Abdalla et al. 2013; Feyssa et al. 2012), and “Ameixa”, “Ameixa-da-Baía”, “Ameixa-da-Terra”, “Ameixa-de-espinho”, “Ameixa-do-Pará”, “Ameixeira-do-Brasil”, “Ameixeira-do-Pará”, “Ameixa-brava”, and “Muirapuama” in Brazil (Oliveira et al. 2010; Silva et al. 2008; Luna et al. 2005; Quintans-Júnior et al. 2002). “Tsada” and “Chabbuli” in west Africa (Maikai et al. 2008a) and “Ghène”, “N’ghani” and “Léaman” in Ivory Coast and “Kleinsuurpruim”, “Inkoy”, “Kol”, “Mulebe”, “Mungomba”, “Mulutulwa”, “Musongwasongwa”, “Mutente”, “Museka”, “Nogbé”, “Séno”, “Ntogé”, “Séné”, “Madarud”, “Madarau”, in other regions in Africa. “Cerise de Mer”, “Macaby”, “Citron de Mer”, “Croc”, “Prunier de Mer” in French. “Hicaco”, “Espino de Brujo”, “Ciruelillo”, “Caimito de Monte”, “Cagalero”, “Albaricoque”, “Albaria”, “Tigrito”, “Almendro de Costa” in Spanish (Orwa et al. 2015).

X. Americana belongs to family Olacaceae, a small plant family of the order *Santalales* (in core eudicots) (Bremer et al. 2009). The family consists of about 28 genera, with 200 species (Malécot et al. 2004). The genus *Ximenia* comprises about eight species: *Ximenia roiigi*, *Ximenia aegyptiaca*, *Ximenia parviflora*, *Ximenia coriaceae*, *Ximenia aculeata*, *Ximenia caffra*, *Ximenia americana* and *Ximenia aegyptica* (Brasileiro et al. 2008).

Synonyms *Amyris arborescens* P.Browne; *Heymassoli inermis* Aubl.; *Heymassoli spinosa* Aubl.; *Pimecaria odorata* Raf.; *Ximenia aculeata* Crantz; *Ximenia americana* var. oblonga DC.; *Ximenia americana* var. ovata DC.; *Ximenia arborescens* Tussac ex Walp.; *Ximenia elliptica* Spreng.; *Ximenia fluminensis* Roem.; *Ximenia inermis* L.; *Ximenia montana* Macfad.; *Ximenia multiflora* Jacq.; *Ximenia oblonga* Lam. ex Hemsl.; *Ximenia spinosa* Salisb.; *Ximenia verrucosa* Roem.

2 Crude Drug Used

Fernandes et al. (2013) characterized the dried extract of the bark of *X. americana* by analytical methods. Tests conducted with thermal analysis showed an endothermic process in 83.16 °C, likely related to the loss of volatile constituents of the sample and the beginning of the process of decomposition, which occurs at a temperature of 218.42 °C. The dried extract of *X. americana* L. showed high-intensity diffraction peaks and a slight increase of the peaks on 70 °C.

There are some products on the market, such as Xymelys 45® containing *X. americana* bark extract, as it is a cosmetic designed to protect ultrasensitive skin, oxidative stress, and free radicals. It also has strong astringent activity (Nachat-Kappes et al. 2014). The *X. americana* tea of the bark's vegetal drug powder is marketed in Brazil and used externally to heal wounds and ulcers, in addition to internally, to treat kidney and heart diseases (Lall and Kishore 2014).

The seed's yield is between 60% and 70% oil, which is edible and has anti-inflammatory properties. It is marketed for the treatment of dry skin conditions, in emollient, moisturising, and anti-aging skin care products, anti-acne products, products for dry, fragile and damaged hair, and for soaps, lipsticks, and lip balms (ITC 2012; Eromosele and Eromosele 2002). Studies on ximenynic acid (Ximenoil®) have revealed improvement in blood circulation (Indena 2015; Vermaak et al. 2011).

3 Major Chemical Constituents and Bioactive Compounds

The compounds found in *X. americana* included the following classes: saponins, glycosides, flavonoids, tannins, phenolic compounds, alkaloids, quinones, terpenoids, cardiac glycosides, phlobatannins and anthraquinones. Furthermore, this species is rich in fatty acids and glycerides (Abdalla et al. 2013; Cartaxo et al. 2010; Maikai et al. 2008b; Saçande and Vautier 2006; Ogunleye and Ibitoye 2003).

In *X. americana* it was studied if gallic acid could be the chemical marker (Brandão et al. 2014). The current list of compounds found by liquid-liquid extraction includes cyanogenic glycoside sambunigrin, gallic acid, gallotannins β-glucogalline, 1,6-digalloyl-β-glucopyranose. It furthermore includes the following flavonoids: quercetin, quercitrin (quercetin-3-O-α-rhamnopyranoside), avicularin (quercetin-3-O-α-arabinofuranoside), quercetin-3-O-β-xylopyranoside, quercetin-3-O-(6″-galloyl)-β-glucopyranoside and kaempferol-3-O-(6″-galloyl)-β-glucopyranoside and 3- methyl-1-oxo isochroman-8-Carboxylic acid (Abdalla et al. 2013; Le et al. 2011). In addition, in the leaves gathered from southern Niger, observed was high calcium content, iron, magnesium, manganese, and zinc, low protein content, the presence of linolenate, and high levels of palmitate (Mevy et al. 2006; Freiberger et al. 1998). Identified in the seed oil was the presence of oleic, hexacos-17-enoic (ximenic), linoleic, linolenic, stearic acids together with smaller quantities of triacont-21-enoic (lumequic), octadec-11-en-9-ynoic (ximenynic acid), arachidonic, erucic, and nervonic acids (Vermaak et al. 2011) and oleanene palmitates, β-sitosterol and C18 acetylenic fatty acids as yellow oils, octadeca-5-ynoic acid (tariric acid) and 10Z,14E,16E-octadeca – 10,14,16-triene-12-ynoic acid, a ene-ene-yneene acetylenic fatty acid (Indena 2015; Eromosele and Eromosele 2002).

The fruit is a rich source of vitamin C, of which the green fruit had higher content, 28, 74% compared to the more matured fruits (Vermaak et al. 2011; Silva et al. 2008). The fruits contain hydrocyanic acid (Arbonnier 2004). The seed's cyanide derivatives (Abdalla et al. 2013) and the fruit kernels exhibit high riproximin concentrations (Bayer et al. 2012).

4 Morphological Description

The Olacaceae family has great diversity in their morphology vegetation (mainly leaves) and reproduction, such as the welding of petals, ovary type, and relative number of stamens/petals (Cabral and Agra 1999).

X. americana is a shrub or small tree up to 6 m tall, commonly less than 4 m (Feyssa et al. 2012). Branches normally arch down and are often armed with straight spines. Its leaves are simple, alternate, or cluster on spur shoots (Kew 2015; Abdalla et al. 2013).

Leaf-lamina 2–8 × 1–4 cm, oblong-elliptic, obtuse to retuse at the apex, coriaceous, lateral nerves three to six pairs, sub inconspicuous on both surfaces, petiole 3–6 mm long, canaliculate, puberulous or pubescent above. Flowers are small, greenish-white, fragrant, 5–10 mm long, and branched inflorescences in pedunculate racemose or umbelliform cymes. Fruits, up to 3 cm long, are drupaceous, ellipsoid or subglobose, shiny, and edible. The fruits are green but turn golden yellow or red when ripe and when eaten is refreshing and has an almond, acid taste. It contains one large endospermic seed, which has up to 60% oil content (Kew 2015; Abdalla et al. 2013; Feyssa et al. 2012; Maikai et al. 2008b).

5 Geographical Distribution

X. americana is widespread throughout the tropics: Africa, India, and South East Asia, to Australia, New Zealand, Pacific Islands, West Indies, Central and South America (Feyssa et al. 2012; Mevy et al. 2006; Sacande and Vautier 2006), and is especially common in Africa and Brazil (Abdalla et al. 2013; Monte et al. 2012; Mora et al. 2009).

It is a plant of diverse habitats, mainly found in semi-arid bushland, in many types of dry woodland, sandy open woodland, dry hilly areas, coastal bushlands, and along watercourses and on stony slopes (Feyssa et al. 2012; Sacande and Vautier 2006).

6 Ecological Requirements

This is a mostly solitary tree dispersed in open country, savannah, gallery forest, along coastal areas, in the understory of dry forests, in dry woodlands, or on riverbanks. The species is drought resistant and the soil type requirements are often poor and dry, including clays, clay loam, loamy sands, sandy clay loam, and sands (Kew 2015; Orwa et al. 2015). It occurs at altitudes up to 2000 m and where rainfall exceeds 500 mm per year (Sacande and Vautier 2006; Feyssa et al. 2012) and grows on many soil types. It is able to absorb water and nutrients from other plants through the roots but does not depend on this for survival (Sacande and Vautier 2006).

The flowers and fruit of the *X. americana* ripen throughout the year; flowering and fruiting periods do not seem to be governed by climatic regimes, but flowering typically occurs in the dry season. In many places, it flowers and fruits throughout the year, and the trees may produce fruit after 3 years of growth. The fruits are dispersed by animals (Orwa et al. 2015). In spite of the multipurpose use of *Ximenia* and its large distribution, the species is under wide-scale threat in regions of Ethiopia (Feyssa et al. 2002).

7 Collection Practice

X. americana stands out for the use for the preparation of food compositions or food supplements. The ximenynic acid is widely used in the cosmetic industry and has been applied as an emollient, conditioner, skin softener, body and hair oil, as well as included as an ingredient in lipsticks and lubricants (Monte et al. 2012; Vermaak et al. 2011).

Its wood is compact, durable, lightweight, and very elastic, being quite used to manufacture tool handles and agricultural tools. The flowers are used by the perfume industry (Brasileiro 2008).

8 Traditional Use and Common Knowledge

X. americana L. is a medicinal plant used for a wide variety of diseases. Standing out among them are malaria (Gronhaug et al. 2008; Ogunleye and Ibitoye 2003), measles (Omer and Elnima 2003), mouth wounds, rheumatism, diarrhea (Koné et al. 2004), lung abscess, muscle cramp (Wondimu et al. 2007), antimicrobial (Maroyi 2013) and HIV/AIDS (Nagata et al. 2011). Magai et al. (2005) showed that the leaves and roots were used for Schistosomiasis and throat infection, but the healers reported that this plant presented toxic signs as salivation.

An Ethnobotanical survey carried with plants used in African medicine showed that the pulverized root of *X. americana* L. was used for leprosy and associated with

Guiera senegalensis, which is used against syphilis. The fruits, as well as the leaves, are consumed as anthelmintic, active against worms and diarrhea (Magassouba et al. 2007). The leaf, stem bark and root extracts were used against *Trypanosoma brucei brucei* and *T. congolens* (Ibrahim et al. 2014).

Others studies showed that *X. americana* was used for inflammations in general for healing, urinary tract infection, diarrhea, anti-parasitic, mental illness, leprotic ulcers, antiseptic, diuretic, ovarian and prostatic inflammations, pains, bloodshed, itching, burning, gastritis, fracture, inflammation, analgesic, anti-pyretic, cancer, hepatoprotective, ulcers, skin infections, purgative backache, hemorrhage, rash, toothache, and menstrual colic. The parts used are bark and leaves. And the forms used were infusion, decoction, tincture, syrup, and cataplasm (Chaves et al. 2014; Le et al. 2011; Oliveira et al. 2010; Cartaxo et al. 2010; Albuquerque et al. 2007).

9 Modern Medicine Based on Its Traditional Medicine Uses

The anticonvulsant activity of *X. americana* were investigated in mice using two methodologies: the pentylenotetrazole (PTZ) test and the maximal electroshock. This species caused a significant increase in latency for appearance seizures induced by PTZ, the same effect showed by drugs used in epilepsy (Júnior et al. 2011).

The development of new anti-cancer drugs is a public health problem and the traditional use of plants is a potentially rich source of information for detecting new molecules with antineoplastic activity (Adwan et al. 2014). This species was investigated in 17 tumor cell lines and three of these cell lines (MCF7 breast cancer, BV173 CML, and CC531 rat colon carcinoma) showed a particularly high sensitivity, with ratios lower than 0.1 of the average IC50. A physicochemical characterization showed that the active antineoplastic component of the plant material are proteins with galactose affinity (Puri et al. 2012; Voss et al. 2006).

Sawadogo et al. (2012) showed the traditional use of medicinal plants for cancer, report the use of *X. americana* L. against cervical cancer of the uterus. A novel cytotoxic type II ribosome-inactivating protein, riproximin, was recently detected with high selectivity for certain tumor cell lines. The compound was in isolation from parts of *X. Americana* (Adwan et al. 2014; Bayer et al. 2012; Ong et al. 2008).

The antioxidant activity of *X. americana* L. was analyzed for three different methods and the fruit of this species showed high total polyphenolic and antioxidant capacities. The correlations indicated that total phenolics and flavonoids are the major contributors to the antioxidant activity of these fruits (Maikai et al. 2010; Lamien-Meda et al. 2008).

The analgesic activity of *X. americana* L. was evaluated with the aqueous extract of the bark. The results showed that the extract possessed only a weak effect on the tail-flick response and on the early phase of the formalin test. The same researcher also evaluated the antipyretic activity of freeze-dried aqueous extract, by comparing the action with acetylsalicylic acid 100 mg. The results showed that the extract at a dose of 25 mg/kg after 2 h of administration had similar effects to acetylsalicylic acid. The extract administered at a dose of 100 mg has slightly larger antipyretic activity than the standard drug used in this study (Soro et al. 2009a, b).

The *X. americana* bark extract showed activity against *Enterococcus faecalis, Staphylococcus aureus and Steptococcus oralis*. The zones of inhibition observed at *E. faecalis* were statistically different compared with that of the chlorhexidine. The extract showed that it can be used as an alternative substance for endodontic treatments (Silva et al. 2012; Costa et al. 2010). The same extract produced significant blood glucose reduction in hyperglycaemic rats alloxan-induced after 6 h of administration (Ezuruike and Prieto 2014).

Toxic effects of *X. americana* extract was evaluated and it was observed that there were no deaths during the period of observation of the animals, which was 14 days, according to the Brazilian regulation. But, changes were observed in the behavior of animals after administration of 2000 $mg.kg^{-1}$ oral use, in the form of forced breathing and analgesia (Brandão 2014). Other studies evaluate hepatic and haematological effects of aqueous extracts of the root, stem, and leaves of this plant, observing increased serum transaminase and alanine transaminase aspartate, which suggests damage to liver cells (James et al. 2008; Wurochekke et al. 2008).

10 Conclusions

X. americana is known by several common names as Ameixa and Wild Plum. There are a variety of reported ethnomedicinal uses for this species mainly against inflammation, infections, and diarrhea. Its dry extract was characterized by analytical methods and gallic acid was identified as a chemical marker. The ximenynic acid was isolated and is widely used in the cosmetics industry. A physicochemical characterization showed that the active antineoplastic components of the plant material are its proteins with galactose affinity. Studies of in vivo toxicity showed no death cases recorded during the observation period of the animals, though changes were observed in their behavior and alterations in the hepatic and hematological parameters.

References

Abdalla AA, Shyaula SL, Ishak CY, Ayoub SMH (2013) Bioassay and phytochemical studies on *Ximenia americana* L. bark ethanolic extract. J For Prod Ind 2(3):63–68

Adwan H, Bayer H, Pervaiz A, Sagini M, Berger MM (2014) Riproximin is a recently discovered type II ribosome inactivating protein with potential for treating cancer. Biotechnol Adv 32(1):1077–1090

Albuquerque UP, Medeiros PM, de Almeida ALS, Monteiro JM, Lins Neto EMF, Melo JG, dos Santos JP (2007) Medicinal plants of the caatinga (semi-arid) vegetation of NE Brazil: a quantitative approach. J Ethnopharmacol 114(3):325–354

Arbonnier M (2004) Trees, shrubs and lianas of West African dry zones. Margraf Publishers, Alemanha

Bayer H, Ey N, Wattenberg A, Voss C, Berger MR (2012) Purification and characterization of riproximin from *Ximenia americana* fruit kernels. Protein Expr Purif 82(1):97–105

Brandão DO (2014) Desenvolvimento de uma formulação de uso intracanal com atividade antimicrobiana obtida a partir de uma planta do semiárido brasileiro. Universidade Estadual da Paraíba, Campina Grande

Brandão DO, Fernandes FHA, Ramos FJL, Silva PCD, Santana CP, Medeiros FD, Véras JG, Medeiros ACD (2014) Validation of UPLC method for determination of gallic acid from *Ximenia americana* L. Planta Med 80(16):1537–1537

Brasileiro MT (2008) Padronização, atividade biológica e desenvolvimento de formas farmacêuticas semi-sólida a base de *Ximenia americana* L. Universidade Federal de Pernambuco, Recife

Brasileiro MT, Egito MA, Lima JR, Randau KP, Pereira GC, Neto PJR (2008) *Ximenia americana* L: botânica, química e farmacologia no interesse da tecnologia farmacêutica. Rev Bras Farmacog 89(2):164–167

Bremer B, Bremer K, Chase MW, Fay MF, Reveal JL, Soltis DE, Soltis PS, Stevens PF (2009) An update of the Angiosperm Phylogeny Group classification for the orders and families of flowering plants: APG. Bot J Linn Soc 161:105–121

Cabral S, Agra MF (1999) Flora paraibana: Olacaceae Mirb. EX DC. Rev Nordest Biol 13(1/2):1–11

Cartaxo SL, Souza MMA, Albuquerque UP (2010) Medicinal plants with bioprospecting potential used in semi-arid northeastern Brazil. J Ethnopharmacol 131(2):326–342

Chaves EMF, Chaves EBF, Coelho-de-Souza, Figueiredo LS, Barros RFM, Kubo R (2014) Um olhar sobre *Ximenia americana* L. e suas potencialidades. Acta Tecn 9(1):70–77

Costa EMMB, Barbosa AS, Arruda TA, Oliveira PT, Dametto FR, Carvalho RA, Melo MD (2010) Estudo in vitro da ação antimicrobiana de extratos de plantas contra *Enterococcus faecalis*. J Bras Patol Med Lab 46(3):175–180

Eromosele CO, Eromosele IC (2002) Fatty acid compositions of seed oil of *Haematostaphis barteri* and *X. americana*. Bioresour Technol 82(3):303–304

Ezuruike UF, Prieto JM (2014) The use of plants in the traditional management of diabetes in Nigeria: pharmacological and toxicological considerations. J Ethnopharmacol 155(2):857–824

Fernandes FHA, Santana CP, Santos RL, Correia LP, Conceição MM, Macêdo RO, Medeiros ACD (2013) Thermal characterization of dried extract of medicinal plant by DSC and analytical techniques. J Therm Anal Calorim 113(2):443–447

Feyssa DH, Njoka JT, Asfaw Z, Nyangito MM (2012) Uses and management of *Ximenia Americana*, Olacaceae in semi-arid east Shewa, Ethiopia. Pak J Bot 44(4):1177–1184

Freiberger CE, Vanderjagt DJ, Pastuszyn A, Glew RS, Mounkaila G, Millson M, Glew RH (1998) Nutrient content of the edible leaves of seven wild plants from Niger. Plant Foods Hum Nutr 53(1):57–69

Gronhaug TE, Glæserud S, Skogsrud M, Ballo N, Bah D, Diallo D, Paulsen BS (2008) Ethnopharmacological survey of six medicinal plants from Mali, West-Africa. J Ethnobiol Ethnomed 4(26):1–11

Ibrahim MA, Mohammed A, Isah MB, Aliyu AB (2014) Anti-trypanosomal activity of African medicinal plants: a review update. J Ethnopharmacol 154(2014):26–54

Indena (2015) Ximilene® and Ximenoil® – microcirculation improvers. Available online at http://www.indena.com/pdf/ximilene_ximenoil.pdf

International Trade Centre (ITC) (2012) The North American market for natural products: prospects for Andean and African products. ITC, Geneva, p xiii, 99 pages

James DB, Owolabi AO, Ibiyeye OH, Magaji J, Ikugiyi YA (2008) Assessment of the hepatic effects, heamatological effect and some phytochemical constituents of *Ximenia americana* (leaves, stem and root) extracts. Afr J Biotechnol 7(23):4274–4278

Júnior WSF, Ladio AH, Albuquerque UP (2011) Resilience and adaptation in the use of medicinal plants with suspected anti-inflammatory activity in the Brazilian northeast. J Ethnopharmacol 138:238–252

Kew (2015) Royal botanic gardens, 2015. Plant Information Centre. Available online at http://epic.kew.org/epic/

Koné WM, Atindehou KK, Terreaux C, Hostettmann K, Traoré D, Dosso M (2004) Traditional medicine in North Côte-d'Ivoire: screening of 50 medicinal plants for antibacterial activity. J Ethnopharmacol 93(1):43–49

Lall N, Kishore N (2014) Review: are plants used for skin care in South Africa fully explored? J Ethnopharmacol 153:61–84

Lamien-Meda A, Lamien CE, Compaoré MMY, Meda RTT, Kiendrebeogo M, Zeba B, Millogo JF, Nacoulma OG (2008) Polyphenol content and antioxidant activity of fourteen wild edible fruits from Burkina Faso. Molecules 13(3):581–594

Le NH, Malterud KE, Diallo D, Paulsen BS, Nergard CS, Wangensteen H (2011) Bioactive polyphenols in *Ximenia americana* and the traditional use among Malian healers. J Ethnopharmacol 139(3):858–862

Luna JS, Santos AF, Lima MRF, Omena MC, Mendonça FAC, Bieber LW (2005) Sant'Ana AEG. A study of the larvicidal and molluscicidal activities of some medicinal plants from northeast Brazil. J Ethnopharmacol 97(2):199–106

Magai A, Diallo D, Fane S, Sanogo R, Paulsen BS, Cisse B (2005) A survey of toxic plants on the market in the district of Bamako, Mali: traditional knowledge compared with a literature search of modern pharmacology and toxicology. J Ethnopharmacol 96(1–2):183–193

Magassouba FB, Diallo A, Kouyaté M, Mara F, Mara O, Bangoura O et al (2007) Ethnobotanical survey and antibacterial activity of some plants used in Guinean traditional medicine. J Ethnopharmacol 14(1):44–53

Malécot V, Nickrent DL, Baas P, Oever LVD, Lobreau-Callen D (2004) A morphological cladistic analysis of Olacaceae. Syst Bot 29(3):569–586

Maikai VA, Kobo PI, Adaudi AO (2008a) Acute toxicity studies of aqueous stem bark extract of *Ximenia americana*. Afr J Biotechnol 7(10):1600–1603

Maikai VA, Nok J, Adaudi AO, Alawa CBI (2008b) In vitro antitrypanosomal activity of aqueous and methanolic crude extracts of stem bark of *Ximenia Americana* on *Trypanosoma congolense*. J Med Plants Res 2(3):55–58

Maikai VA, Kobo PI, Maikai BVO (2010) Antioxidant properties of *Ximenia americana*. Afr J Biotechnol 9(45):7744–7746

Maroyi A (2013) Traditional use of medicinal plants in south-central Zimbabwe: review and perspectives. J Ethnobiol Ethnomed 9(31):1–18

Mevy JP, Bessiere JM, Greff S, Zombre G, Viano J (2006) Composition of the volatile oil from the leaves of *Ximenia americana* L. Biochem Syst Ecol 34(7):549–553

Monte FJQ, Lemos TLG, Araújo MRS, Gomes ES. (2012) *Ximenia americana*: chemistry, pharmacology and biological properties, a review. Phytochemicals – a global perspective of their role in nutrition and health. Available online at http://www.intechopen.com/books/phytochemicalsa-global-perspective-of-their-role-in-nutrition-and-health

Mora VHF, Franco-Mora O, López-Sandoval JA, Pérez-López DJ, Balbuena-Melgarejo A (2009) Characterization of wild plum (*Ximenia americana* L. var. americana; Olacaceae) fruit growing at Tepexi of Rodríguez, Puebla, Mexico. Genet Resour Crop Evol 56(5):719–727

Nachat-Kappes R, Favre-Mercuret M, Cabannes M, Rios L, Ranouille E, Duclos V et al (2014) Xymelys 45 reduces inflammaging: a new way to prevent skin ageing. Cosmet Sci Technol Available online at http://www.cosmeticsciencetechnology.com/articles/samples/3191.pdf

Nagata JM, Jew AR, Kimeu JM, Salmen CR, Bukusi EA, Cohen CR (2011) Medical pluralism on Mfangano Island: use of medicinal plants among persons living with HIV/AIDS in Suba District, Kenya. J Ethnopharmacol 135(2):501–509

Ogunleye DS, Ibitoye SF (2003) Studies of antimicrobial activity and chemical constituents of *Ximenia americana*. Trop J Pharm Res 2(2):239–241

Oliveira FCS, Barros RFM, Moita Neto JM (2010) Plantas medicinais utilizadas em comunidades rurais de Oeiras, semiárido piauiense. Rev Bras Plant Med 12(3):282–201

Omer MEFA, Elnima EI (2003) Antimicrobial activity of *Ximenia americana*. Fitoterapia 74:122–126

Ong PL, Weng BC, Lu FJ, Lin ML, Hung RP (2008) The anticancer effect of protein-extract from *Bidens alba* in human colorectal carcinoma SW480 cells via the reactive oxidative species- and glutathione depletion-dependent apoptosis. Food Chem Toxicol 46(5):1535–1547

Orwa C, Mutua A, Kindt R, Jamnadass R Anthony S (2015) Agroforestree Database: a tree reference and selection guide version 4.0. Available online at http://floradobrasil.jbrj.gov.br/jabot/floradobrasil/FB10971

Puri M, Kaur L, Perugini MA, Gupta RC (2012) Ribosome-inactivating proteins: current status and biomedical applications. Drug Discov Today 17(13–14):774–783

Quintans-Júnior LJ, Almeida RN, Falcão ACGM, Agra MF, Sousa MFV, Barbosa-Filho JM (2002) Avaliação da atividade anticonvulsivante de plantas do nordeste brasileiro. Acta Farm Bonaer 21(3):179–184

Rossi L (2015) *Olacaceae* in Lista de Espécies da Flora do Brasil. Jardim Botânico do Rio de Janeiro. Available online at http://www.floradobrasil.jbrj.gov.br/jabot/floradobrasil/FB10971

Sacande M, Vautier H (2006) *X. americana* seed leaflet. Millennium Seed Bank Project, Kew. Forest & landscape, Denmark Let, 112, 2006. available online at http://curis.ku.dk/ws/files/20497181/ximenia_112.pdf

Sawadogo WR, Schumacher M, Teiten MH, Dicato M, Diederich M (2012) Traditional West African pharmacopeia, plants and derived compounds for cancer therapy. Biochem Pharmacol 84(10):1225–1240

Silva GG, Souza PA, Morais PLD, Santos EC, Moura RD, Menezes JB (2008) Caracterização do fruto de ameixa silvestre (*Ximenia americana* L). Rev Bras Frutic 30(2):311–314

Silva MSP, Brandão DO, Chaves TP, Formiga ALNF, Costa EMMB, Santos VL, Medeiros ACD (2012) Study bioprospecting of medicinal plant extracts of the semiarid northeast: contribution to the control of oral microorganisms. Evid Based Complement Alternat Med 12:1–6

Soro TY, Traoré F, Datte JY, Nene-Bi (2009a) Activité antipyrétique de l'extrait aqueux de *Ximenia americana*, Phytothérapie. Phytothérapie 7(6):297–203

Soro TY, Traorea F, Sakande J (2009b) Activité analgésique de l'extraitaqueux de *Ximenia americana* (Linné) (Olacaceae). C R Biol 332:371–377

Vermaak I, Kamatou GPP, Komane-Mofokeng B, Viljoen AM, Beckett K (2011) African seed oils of commercial importance – cosmetic applications. S Afr J Bot 77(4):920–933

Voss C, Eyol E, Berger MR (2006) Identification of potent anticancer activity in *Ximenia americana* aqueous extracts used by African traditional medicine. Toxicol Appl Pharmacol 211(13):177–187

Wondimu T, Asfaw Z, Kelbessa E (2007) Ethnobotanical study of medicinal plants around 'Dheeraa' town, Arsi Zone, Ethiopia. J Ethnopharmacol 112(1):152–161

Wurochekke AU, Anthony AE, Obidah W (2008) Biochemical effects on the liver and kidney of rats administered aqueous stem bark extract of *Ximenia americana*. Afr J Biotechnol 7(16):2777–2780

MIX
Papier aus verantwortungsvollen Quellen
Paper from responsible sources
FSC® C105338

If you have any concerns about our products,
you can contact us on
ProductSafety@springernature.com

In case Publisher is established outside the EU,
the EU authorized representative is:
Springer Nature Customer Service Center GmbH
Europaplatz 3, 69115 Heidelberg, Germany

Printed by Libri Plureos GmbH
in Hamburg, Germany